Optical MEMS, Nanophotonics, and Their Applications

SERIES IN OPTICS AND OPTOELECTRONICS

Series Editors: **Robert G W Brown**, University of California, Irvine, USA
E Roy Pike, Kings College, London, UK

Optical MEMS, Nanophotonics, and Their Applications

Edited by

Guangya Zhou
Department of Mechanical Engineering,
National University of Singapore, Singapore

and

Chengkuo Lee
Department of Electrical and Computer Engineering,
National University of Singapore, Singapore

CRC Press
Taylor & Francis Group
Boca Raton London New York

CRC Press is an imprint of the
Taylor & Francis Group, an **informa** business

CRC Press
Taylor & Francis Group
6000 Broken Sound Parkway NW, Suite 300
Boca Raton, FL 33487-2742

First issued in paperback 2020

© 2018 by Taylor & Francis Group, LLC
CRC Press is an imprint of Taylor & Francis Group, an Informa business

No claim to original U.S. Government works

ISBN-13: 978-1-4987-4133-0 (hbk)
ISBN-13: 978-0-367-78164-4 (pbk)

Visit the Taylor & Francis Web site at
http://www.taylorandfrancis.com

and the CRC Press Web site at
http://www.crcpress.com

Contents

Series preface

This international series covers all aspects of theoretical and applied optics and optoelectronics. Active since 1986, eminent authors have long been choosing to publish with this series, and it is now established as a premier forum for high-impact monographs and textbooks. The editors are proud of the breadth and depth showcased by published works, with levels ranging from advanced undergraduate and graduate student texts to professional references. Topics addressed are both cutting edge and fundamental, basic science and applications-oriented, on subject matter that includes: lasers, photonic devices, nonlinear optics, interferometry, waves, crystals, optical materials, biomedical optics, optical tweezers, optical metrology, solid-state lighting, nanophotonics, and silicon photonics. Readers of the series are students, scientists, and engineers working in optics, optoelectronics, and related fields in the industry.

Proposals for new volumes in the series may be directed to Lu Han, executive editor at CRC Press, Taylor & Francis Group (lu.han@taylorandfrancis.com).

Preface

Microelectromechanical systems (MEMS) technology, which features integration and miniaturization of mechanical and electrical elements at the microscale using IC-like microfabrication processes, is now well-known and has already had a profound impact on a range of engineering and scientific fields. Optical MEMS that involve the use of MEMS devices in optical and photonic systems to dynamically manipulate the light have been rapidly developed in the past decades. Optical MEMS technology not only provides the advantages of system miniaturization, performance enhancement, and lost cost production, but also enables new applications and innovations. Examples include micromirrors for portable displays, ultra-compact endoscopic probes, and micromirror arrays for ground- and space-based telescopes. Some of these optical MEMS-based devices and systems are already commercially available on the market.

More recently, as the nanofabrication technology has made significant progress, further scaling of MEMS down to nanoelectromechanical systems (NEMS) is now possible. At the same time, research in nanophotonics that focuses on the interaction of light with engineered nanostructures has gained significant momentum. In our view, the fusion of MEMS/NEMS with nanophotonic elements also opens up new possibilities and creates novel functional photonic devices and systems featuring dynamic tunability, enhanced performance, and a higher level of integration. Fascinating new devices, systems, and applications, for example, NEMS-driven variable silicon nanowire waveguide couplers and NEMS tunable photonic crystal (PhC) nanocavities, are emerging at a fast pace.

This edited book describes some of the most recent results in the field of optical MEMS and nanophotonics. It is challenging to provide a complete reference to address the rapid developments in this area, since new devices, systems, and applications are reported all the time. It is our intention to provide the engineers and researchers who are interested in micro- and nanofabrication, MEMS/NEMS, micro-optics, and nanophotonics the necessary fundamentals in this field as well as a source of information on recent photonic NEMS/MEMS devices and their uses.

This book aims to cover a broad spectrum and is divided into three sections. The first section addresses typical optical MEMS devices containing microsized features. It also features the use of this technology in imaging, communication, and sensing applications. This section presents a set of devices and systems, including PZT-driven micromirrors for ultra-portable laser scanning project displays, optical phased array technology using MEMS-driven high-contrast-grating (HCG) mirrors for fast beam steering, electrostatically driven programmable mirror arrays for space instruments, ultra-fast laser scanning using MEMS-driven vibratory gratings and their applications in hyperspectral imaging, MEMS-based tunable Fabry–Pérot filters, electrothermally driven micromirrors and their applications in endoscopes, and MEMS-driven tunable lenses and adjustable apertures for miniaturized cameras. The second section of the book covers the recent research in nanophotonics, again with an emphasis on its integration with NEMS/MEMS. This section reports on photonics crystal-based sensors, NEMS-driven silicon nanowire waveguide variable couplers, metasurfaces and ultrathin optical devices, sensors based on nanophotonic resonators, and MEMS tunable THz metameterials. The last section of the book presents particular types of micro- and nanophotonic devices that integrate with biological and/or microfluidic systems. This section reports optofluidic devices and their applications, implantable microphotonic devices, as well as portable microfluidic-based photocatalysis systems. These devices and systems are mostly targeted for biomedical and health care applications.

Last but not least, we wish to thank all the authors who contributed to this book. Their fascinating contributions, suggestions, and commitment made this book possible.

Guangya Zhou
Chengkuo Lee
Singapore

Editors

Guangya Zhou received his B.Eng. and PhD degrees in optical engineering from Zhejiang University, Hangzhou, China, in 1992 and 1997, respectively. His PhD thesis was on micro-optics and diffractive optics. He was a post-doctoral fellow at State Key Laboratory of thin film and microfabrication technology at Shanghai Jiao Tong University from 1997 to 1999, conducting research on optical MEMS for telecommunication systems. In 2000, he joined the School of Electrical and Electronic Engineering at Nanyang Technological University as a research fellow, where his research was on III-V semiconductor-based optical MEMS devices. He joined the Department of Mechanical Engineering, National University of Singapore in 2001 as a research fellow and in 2005 as an assistant professor. Since 2012, he has been an associate professor in the same department. His research covers optical MEMS scanners, MEMS spectrometers and hyperspectral imagers, optical MEMS-based ultra-compact endoscope probes, silicon nanophotonics, NEMS tunable photonic crystals, and nanoscale optomechanics. He has published over 100 research papers in peer-reviewed international journals in the fields of micro-optics, nanophotonics, micro-/nanofabrication, MEMS, and NEMS. He is the main inventor of the MEMS-driven vibratory grating scanner, MEMS-based miniature zoom lens system with autofocus function, and miniature MEMS-based adjustable aperture. The latter two were successfully licensed to a start-up company, where he currently also works as a technical advisor.

Chengkuo Lee received his MS degree in materials science and engineering from National Tsing Hua University, Hsinchu, Taiwan, in 1991, his M.S. degree in industrial and system engineering from Rutgers University, New Brunswick, NJ, in 1993, and his PhD in precision engineering from The University of Tokyo, Tokyo, Japan, in 1996. He was a Foreign Researcher with the Nanometer-Scale Manufacturing Science Laboratory of the Research Center for Advanced Science and Technology, The University of Tokyo, from 1993 to 1996. He was with the Mechanical Engineering Laboratory, AIST, MITI of Japan, as a JST Research Fellow, in 1996. Thereafter, he became a Senior Research Staff Member of the Microsystems Laboratory, Industrial Technology Research Institute, Hsinchu. In 1997, he joined Metrodyne Microsystem Corporation, Hsinchu, and established the MEMS Device Division and the first micromachining lab for commercial purposes in Taiwan. He was the Manager of the MEMS Device Division between 1997 and 2000. He was an Adjunct Assistant Professor with the Electro-Physics Department, National Chiao Tung University, Hsinchu, in 1998, and an Adjunct Assistant Professor with the Institute of Precision Engineering, National Chung Hsing University, Taichung, Taiwan, from 2001 to 2005. In 2001, he cofounded Asia Pacific Microsystems, Inc., Hsinchu, where he first became Vice President of Research and Development, before becoming Vice President of the Optical Communication Business Unit and Special Assistant to the Chief Executive Officer in charge of international business and technical marketing for the MEMS foundry service. He was a Senior Member of the Technical Staff at the Institute of Microelectronics, A*STAR, Singapore, from 2006 to 2009. Currently, he is an Associate Professor with the Department of Electrical and Computer Engineering, National University of Singapore, Singapore. He is the co-author of *Advanced MEMS Packaging* (McGraw-Hill, 2010). He has contributed to more than 230 international conference papers and extended abstracts and 160 peer-reviewed international journal articles in the fields of optical MEMS, NEMS, nanophotonics, and nanotechnology. He holds nine U.S. patents.

Contributors

Xianzhong Chen
Institute of Photonics and Quantum Sciences
School of Engineering and Physical Sciences
Heriot-Watt University
Edinburgh, Scotland, United Kingdom

Xudong Fan
Department of Biomedical Engineering
University of Michigan
Ann Arbor, Michigan

Kazuhiro Hane
Tohoku University
Sendai, Japan

Chong Pei Ho
Department of Electrical and Computer
    Engineering
National University of Singapore
Singapore

Bo Li
Department of Electrical and Computer
    Engineering
National University of Singapore
Singapore

Liying Liu
Department of Optical Science and
    Engineering
School of Information Science and
    Engineering
Fudan University
Shanghai, China

Jun Ohta
Graduate School of Materials Science
Nara Institute of Science and Technology
Ikoma, Japan

Sagnik Pal
Department of Electrical and Computer
    Engineering
University of Florida
Gainesville, Florida

Sung-Yong Park
Department of Mechanical Engineering
National University of Singapore
Singapore

Prakash Pitchappa
Department of Electrical and Computer
    Engineering
National University of Singapore
Singapore

Takashi Tokuda
Graduate School of Materials Science
Nara Institute of Science and Technology
Ikoma, Japan

Hiroshi Toshiyoshi
Research Center for Advanced Science and
    Technology
The University of Tokyo
Tokyo, Japan

Ning Wang
National Engineering Laboratory for Fiber
    Optic Sensing Technology
Wuhan University of Technology
Wuhan, People's Republic of China

Youmin Wang
Department of Electrical Engineering and
    Computer Sciences
University of California, Berkeley
Berkeley, California

Dandan Wen
Institute of Photonics and Quantum Sciences
School of Engineering and Physical Sciences
Heriot-Watt University
Edinburgh, Scotland, United Kingdom

Ming-Chiang Wu
Department of Electrical Engineering and
    Computer Sciences
University of California, Berkeley
Berkeley, California

Huikai Xie
Department of Electrical and Computer
    Engineering
University of Florida
Gainesville, Florida

Lei Xu
Department of Optical Science and
    Engineering
School of Information Science and Engineering
Fudan University
Shanghai, People's Republic of China

Fuyong Yue
Institute of Photonics and Quantum Sciences
School of Engineering and Physical Sciences
Heriot-Watt University
Edinburgh, Scotland, United Kingdom

Frederic Zamkotsian
Laboratoire d'Astrophysique de
    Marseille (LAM)
French National Research (*CNRS*) and
    Aix Marseille Univ
Marseille, France

Xiaoyang Zhang
Department of Electrical and
    Computer Engineering
University of Florida
Gainesville, Florida

Xingwang Zhang
Department of Mechanical Engineering
National University of Singapore
Singapore

Xuming Zhang
Department of Applied Physics
Hong Kong Polytechnic University
Hong Kong, People's Republic of China

Liang Zhou
Department of Electrical and Computer
    Engineering
University of Florida
Gainesville, Florida

Yongchao Zou
Department of Mechanical Engineering
National University of Singapore
Singapore

SECTION I

Optical MEMS for communication, imaging, and sensing applications

Optical MEMS: An introduction

Guangya Zhou and Chengkuo Lee

Contents

1.1 MEMS processes

Fabrication of devices in micro-/nanometer scales requires microelectromechanical systems (MEMS) technology which allows for static or moveable elements for sensing and actuation purposes. This chapter provides an overview of the basic MEMS fabrication methods. Developed from the modern complementary metal-oxide-semiconductor (CMOS) process, MEMS enables the integration of different functions with electronic circuits, thus is a key technology for the fabrication of future devices. A standard MEMS fabrication process includes lithography, thin-film formation, etching, and bonding and packaging. Various materials such as silicon, polymers, metals, and ceramics are used and must be processed in micro-/nanometer scale. The following section describes the fundamental MEMS process steps in detail.

1.1.1 Lithography

Photolithography is widely used for micro-/nanometer sized patterning. A mask, which contains the desired pattern layer on a glass plate, is required in the process. Typically, a glass mask is fabricated with an electron beam lithography technique where a computer-aided design (CAD) can be written on a glass plate with a metal layer (typically chromium) followed by an etching process to form the pattern layer [1]. Figure 1.1 illustrates the standard photolithographic process to transfer the desired pattern onto a thin film on a substrate. First, a photosensitive material (photoresist) is coated on the thin film. Next, the mask with the desired pattern is brought into contact or close to contact and aligned with the substrate using a mask aligner. This step ensures that the pattern is exposed at the correct location for the subsequent fabrication steps. The photoresist is then exposed to ultraviolet (UV) light through the glass mask layer. There are two types of photoresist, positive and negative. For a positive photoresist, the exposed part is more soluble in the developing solution hence it is removed in the development. On the other hand, a negative photoresist is polymerized through exposure and becomes difficult to dissolve, hence the exposed part remains on the substrate after development. After the development, the remaining photoresist acts as a protective layer in the pattern transfer process. Finally, the photoresist is stripped out leaving the thin film with the desired pattern on the substrate.

Recently, nanoimprint lithography (NIL) has been studied and introduced as a patterning method in MEMS technology. A mold is required in the patterning process, which acts as a stamp and contains the desired pattern. NIL can be categorized into thermal NIL, UV NIL, and Soft NIL. Figure 1.2 illustrates the standard process of thermal and UV NIL. First, a layer of polymer is applied on top of the substrate. The polymer is designed to cure through application of heat or exposure to UV light. Second, the mold is in contact and is pressed to the substrate which is coated with the curable polymer. Heat is provided for the curing of the polymer in the case of

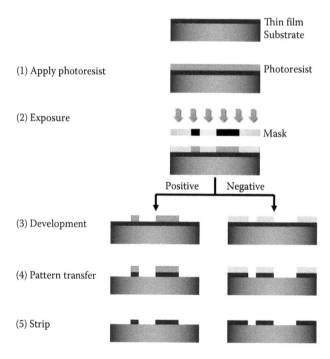

Figure 1.1 Schematic illustration of photolithographic process sequence for pattern transfer.

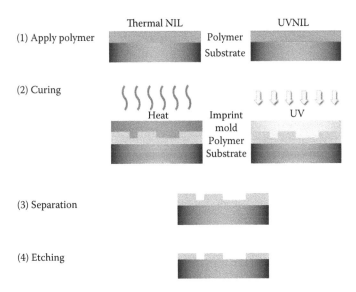

Figure 1.2 Schematic illustration of nanoimprint lithography (NIL) process sequence.

thermal NIL, while UV light is provided to cure the polymer for UV NIL. The substrate or the mold material needs to be transparent to UV light when adapting to the UV NIL process. After curing, the mold is separated from the substrate and the residual polymer at the bottom is etched away. The NIL process is especially unique in soft NIL where the mold is made from a soft material such as poly(dimethylsiloxane) (PDMS). This soft mold enables large area patterning using methods such as roll-to-sheet or roll-to-roll imprinting.

1.1.2 Thin film formation

The deposition method of the thin film in MEMS can be categorized as physical vapor deposition (PVD) or chemical vapor deposition (CVD). PVD use a physical method such as vaporizing or using ion bombardment of the targeted material and forcing it to turn into the gaseous phase. The gaseous phase source is then deposited as a thin-film on a substrate. On the other hand, CVD creates a thin film through a chemical reaction on the substrate surface. The most common PVD method is thermal evaporation as shown in Figure 1.3a. Heat is provided to melt the source material which turns into the liquid phase and eventually evaporates in the vacuum chamber. The vaporized source material then travels through the chamber to the substrate, thus forming a thin-film on the

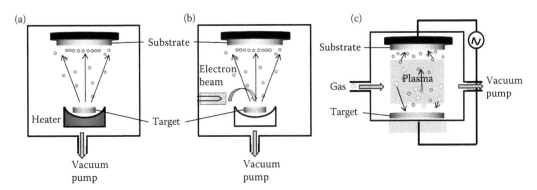

Figure 1.3 Schematic illustration of major physical vapor deposition (PVD): (a) thermal evaporation, (b) electron beam (E-beam) evaporation, and (c) sputtering.

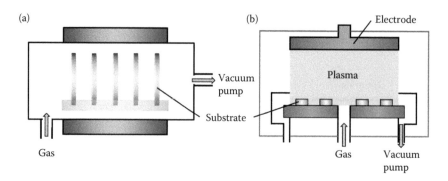

Figure 1.4 Schematic illustration of major physical vapor deposition (PVD). (a) Low pressure CVD (LPCVD), (b) plasma-enhanced CVD (PECVD).

surface. The heat can be provided by introducing current into the conductive supporting material which has a high melting point. Hence, by putting the target source material on top, thermal evaporation can be achieved. While these supporting materials allow evaporation, they may lead to contamination in the chamber. Electron beam (E-beam) evaporation is a method similar to thermal evaporation, but with a different melting method. As illustrated in Figure 1.3b, an electron gun is used to provide high energy for melting the target source material. This minimizes the possibility of contamination from supporting materials while also allowing finer control of the deposition rate. Plasma is also used as a method to directly turn the solid target into atoms, thus forming a thin film on the substrate. This process is called sputtering. Figure 1.3c shows the schematic of the AC sputtering process. A gas source is introduced into the vacuum chamber where AC voltage is applied between the top and bottom electrodes, which generates plasma. An ion bombardment process occurs, thus the target source material is physically etched away from the surface. These atoms are then deposited on the substrate to create a thin film.

The chemical vapor deposition (CVD) process also takes place in a chamber where the substrates are placed in a holder. Reactive gases are introduced into the chamber and form a thin film on the substrate surface. The produced by-product then exits the chamber. Due to this film-forming mechanism, both the reactive chemicals and the by-product in the CVD process need to be volatile while creating a non-volatile solid film. The energy for the reaction is supplied by thermal methods, photons, or electrons. The advantage of CVD over PVD is its high deposition rate and good step coverage. CVD can be conducted in atmospheric pressure and in reduced pressures. However, a lower chamber pressure allows better control in the deposition rate while also reducing the contamination. Figure 1.4a illustrates low-pressure CVD (LPCVD) which is operated in reduced pressure and allows for batch processing. The heater around the chamber provides the thermal energy for the chemical reaction. Plasma can also be an energy source for the chemical reaction. This system is called plasma-enhanced CVD (PECVD) as illustrated in Figure 1.4b. The gases in the chamber are highly reactive due to the plasma created by the applied bias. This allows for film formation on the substrates placed in the chamber. PECVD allows for film deposition with a high rate but at lower temperatures compared to LPCVD.

1.1.3 Etching

The deposited thin film can be partially removed in a desired pattern through the etching process. The etching process can be categorized into a wet process and a dry process. While the etchant of wet etching is in the liquid state, the etchant for dry etching is usually in a gas or plasma state. Wet etching can be conducted by immersing the substrate into the etchant liquid or spraying the etchant on the substrate. The etchant will then react with the substrate material creating a by-product to be washed away. Wet etching has the advantages of a high etching rate and high selectivity over the

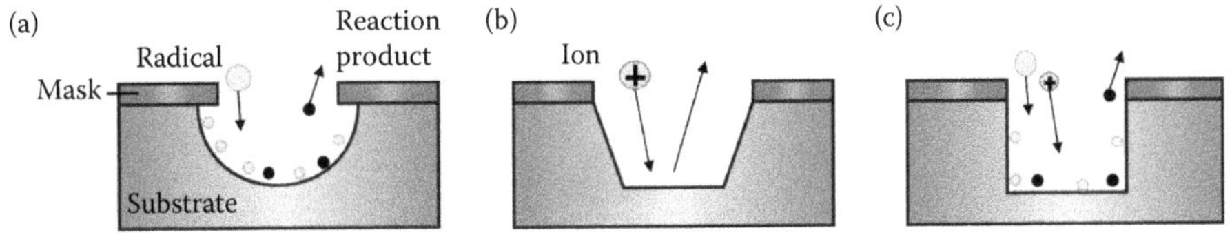

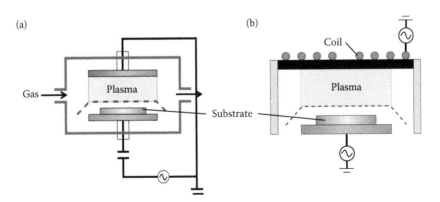

Figure 1.5 Schematic illustration of etching mechanisms: (a) chemical etching, (b) physical etching, (c) reactive ion etching (RIE).

selected mask layer, while the etching is not directional. Hence, it is mostly used for structure releasing in the MEMS process. On the other hand, dry etching is more commonly used in the etching of the actual structure in MEMS utilizing the etchant in the form of a plasma. In chemical dry etching, gaseous etchants encounter collisions with charged particles, resulting in a dissociation and excitation [2]. These reactive molecules then etch away the substrate materials. The physical dry etching process is similar to the sputtering process for thin-film deposition. The reactive ions bombard the substrates and etch away the material. A commonly used method in dry etching is reactive ion etching (RIE), where both chemical and physical etching take place as illustrated in Figure 1.5c.

There are two main types of process set ups for RIE, capacitive coupled plasma (CCP) RIE and inductive coupled plasma (ICP) RIE. As illustrated in Figure 1.6a, CCP-RIE contains two parallel electrodes in a side vacuum chamber. RF bias is through a blocking capacitor to the bottom electrode where the substrate to be etched is placed. Gas is fed into the vacuum chamber and plasma is generated. The plasma is then driven to the substrate surface and etches the material. The ICP system contains two bias generators, one of which controls the plasma density while the other controls the ion energy (Figure 1.6b). The induction coil at the top of the vacuum chamber generates plasma above the substrate and hence provides better control of the plasma density while also allowing higher plasma density for the CCP system. The other bias is provided to the bottom electrode which drives plasma to the substrate surface and hence controls the ion energy for the etching process.

Deep reactive ion etching (DRIE) is known as a process for Si etching with vertical side walls and a high aspect ratio. It was invented by Larmer and Urban at Robert Bosch Corporate Research [3], thus the process is often called the Bosch process. The Bosch process requires switching between several gases in the ICP-RIE set up, thus an ICP-RIE machine with the capability to program and switch gases is called a DRIE. Figure 1.7 depicts the process steps in the Bosch process. Starting from an Si with a masking layer and an exposed area to etch, sulfur hexafluoride (SF_6)-based etching gas is first introduced into the chamber. In the SF_6 plasma, ions contribute to directional physical etching while the fluorine radicals contribute to isotropic chemical etching. The gas in the chamber is then switched to octafluorocyclobutane (C_4F_8) which forms a passivation fluorocarbon layer on the

Figure 1.6 Schematic illustration of a major reactive ion etching (RIE) system: (a) capacitive coupled plasma (CCP) RIE, (b) inductive coupled plasma (ICP) RIE.

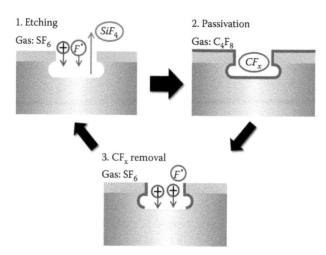

Figure 1.7 Schematic illustration of the etching steps of the Bosch process for high aspect ratio etching.

whole surface. After the passivation process, the gas is switched back to SF_6, but with bias control to enhance the ion bombardment effect of the SF_6 plasma. This allows for the removal of the passivation layer at the bottom of the etched silicon. The process then returns back to the first etching condition to conduct another Si etching. The remaining passivation layer on the side wall functions as a protective layer in the etching process, thus the side wall remains vertical. The process then passes to step three in Figure 1.7 and is repeated in cycles until the desired depth is achieved.

1.1.4 Bonding

A bonding technique is necessary for MEMS devices that require sealing or consist of multiple layers of wafer. This chapter covers direct wafer bonding and anodic bonding. Direct wafer bonding is also called silicon to silicon fusion bonding and applies a high temperature to both wafers [4]. The surfaces to be bonded need to be cleaned and smoothed enough to ensure firm contact between the two wafers. Surface activation is required in direct bonding where the wafer surface is chemically activated. This can be done through immersing the wafer into ammonium hydroxide (NH_4OH) or by exposure to oxygen (O_2) plasma [2]. The wafers are then brought into contact at room temperature. The temperature is then increased to (typically) 1100°C which results in strong siloxane Si–O–Si bonds between the two silicon wafers. Anodic bonding is a bonding method for glass and silicon bonding. The temperature for this process is lower than that of silicon to silicon fusion bonding, typically 300–500°C, where an electrical DC voltage is applied across both wafers as illustrated in Figure 1.8. Glass is mainly composed of silica, and many silica-based glasses

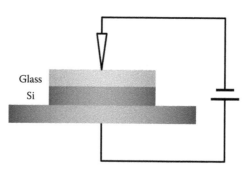

Figure 1.8 Schematic illustration of the anodic bonding.

contain ions such as sodium oxide (Na$_2$O) or calcium oxide (CaO). The DC voltage drives these ions to the cathode side while leaving negative ions in the glass–silicon boundary. A direct siloxane (Si–O–Si) bond is then formed between the glass and silicon.

1.2 Actuation schemes

Recent developments in the rapidly emerging discipline of MEMS have shown immense promise in actuators and micro-optical systems [1]. In conjunction with properly designed mirrors, lenses, and gratings, various micro-optical systems driven by microactuators can provide many unique functions in light manipulation such as reflection, beam steering, filtering, focusing, collimating, diffracting, etc. In the next few sections, the four major actuation schemes, that is, electrothermal, electrostatic, piezoelectric, and electromagnetic, for in-plane or out-of-plane movements, are introduced. Each actuation scheme has its inherent advantages and disadvantages, while its design feasibilities are often limited by the fabrication method used.

1.2.1 Electrothermal actuation

Electrothermal actuation makes use of the difference in the thermal expansions of materials to achieve mechanical actuation. The thermal expansion of a solid material is characterized by the coefficient of thermal expansion (CTE), αT, and has a unit of strain per change in temperature (K^{-1}). With a small temperature change of ΔT, the introduced mechanical strain ε is defined as the product,

$$\varepsilon = \alpha T \cdot \Delta T \tag{1.1}$$

One of the basic actuator structures for thermal actuation, as shown in Figure 1.9, is an electrothermal bimorph that consists of a cantilever with two different material layers [2–5]. The actuation relies on the difference in the linear expansion coefficients of the two materials, with one layer expanding by a larger amount than the other. This results in stress at the interface of these two layers, leading to bending of the cantilever. The elevated temperature can be created by heating up the cantilever when a bias current flows through an embedded resistor

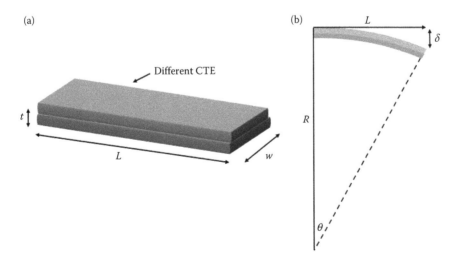

Figure 1.9 Schematic illustration of electrothermal bimorph: (a) material composition and (b) vertical displacement δ.

in the cantilever, which is the Joule heating effect. The radius of bending curve R and vertical displacement are

$$\delta = R - R\cos\theta \cong R - R\left(1 - \frac{1}{2}\theta^2\right) = \frac{L^2}{2R} \tag{1.2}$$

$$\frac{1}{R} = \frac{6w_1w_2E_1E_2t_1t_2(t_1+t_2)(\alpha_1-\alpha_2)\Delta T}{\left(w_1E_1t_1^2\right)^2 + \left(w_2E_2t_2^2\right)^2 + 2w_1w_2E_1E_2t_1t_2\left(2t_1^2 + 2t_2^2 + 3t_1t_2\right)} \tag{1.3}$$

where α_1, α_2, w_1, w_2, E_1, E_2, t_1, and t_2 are CTE, width, Young's modulus, and thickness of the two layers, respectively. Thus, the δ can be rewritten into another form [6],

$$\delta = \frac{3w_1w_2E_1E_2t_1t_2(t_1+t_2)(\alpha_1-\alpha_2)\Delta TL^2}{\left(w_1E_1t_1^2\right)^2 + \left(w_2E_2t_2^2\right)^2 + 2w_1w_2E_1E_2t_1t_2\left(2t_1^2 + 2t_2^2 + 3t_1t_2\right)} \tag{1.4}$$

L is the length of the bimorph. The equivalent stiffness EI for the bimorph is

$$EI = \frac{w_2^3 t_1 E_1 E_2}{12(E_1 t_1 + E_2 t_2)} K \tag{1.5}$$

$$K = \left[4 + 6\frac{t_1}{t_2} + 4\left(\frac{t_1}{t_2}\right)^2 + \frac{E_1}{E_2}\left(\frac{t_1}{t_2}\right)^3 + \frac{E_2 t_2}{E_1 t_1}\right] \tag{1.6}$$

The total force generated at the free end of the bimorph due to thermally induced deflection is

$$F = \frac{3EI}{L^3} \frac{3\Delta\alpha\Delta T\left(1 + \frac{t_1}{t_2}\right)L^2}{t_2\left\{\left(\frac{w_1}{w_2}\frac{E_1}{E_2}\right)\left(\frac{t_1}{t_2}\right)^3 + \left(\frac{w_2}{w_1}\frac{E_2}{E_1}\right)\left(\frac{t_2}{t_1}\right) + 2\left[2\left(\frac{t_1}{t_2}\right)^2 + 3\left(\frac{t_1}{t_2}\right) + 2\right]\right\}} \tag{1.7}$$

In addition to out-of-plane actuation, there are other applications demanding in-plane displacement, which will involve designs that are different from the above-mentioned bimorph actuator. For example, in-plane actuation is made possible by designing a U-shaped electrothermal actuator consisting of two arms of uneven widths [7] from a single material. As shown in Figure 1.10a, when an electrical current is applied from one anchor to the other, the arm with the larger electrical resistance heats up more. This results in a higher temperature and larger volume expansion in the thinner arm, that is, the so-called hot arm. The other thicker arm is relatively cold and is referred to as the cold arm. Eventually, the U-shaped thermal actuator will deflect laterally toward the cold arm side due to asymmetrical thermal expansion when the actuator is DC biased. Other variations of the classical single hot–cold arms design have also surfaced, with some research groups focusing on two hot arms and one cold arm design [8–10], and one group having integrated a piezoresistive lateral displacement sensor embedded into the actuator [11]. Other designs for in-plane electrothermal actuators, such as V-shaped chevron beam actuators illustrated in Figure 1.10b, have also been reported [12–14].

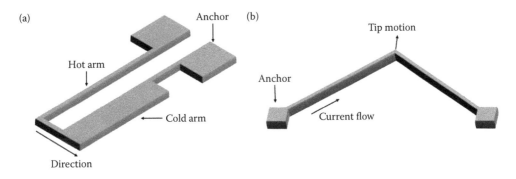

Figure 1.10 Schematic diagram of (a) in-plane U-shaped actuator, which deploys hot-cold arms of different width and (b) in-plane V-shaped chevron beam actuator that buckles along tip when current flows through.

Compared to other actuation schemes, an electrothermal actuator can achieve large forces (~100 µN) and static displacements (~100 µm) at relatively low voltages (~5 V) [1]. However, a large amount of thermal energy is required, therefore, it consumes substantial electrical energy (~1 W). In addition, it has a slower response and AC operation of the thermal actuator is generally limited to a frequency response of less than 1 kHz. This is due to the time constant associated with heat transfer. High temperature and complicated thermal management are further drawbacks of thermal actuation. For example, the upper practical limits for temperatures in polysilicon and single-crystal-silicon-based electrothermal actuator are approximately 600°C and 800°C, respectively, above which material property changes such as localized plastic yielding and material grain growth become an issue.

1.2.2 Electrostatic actuation

There are two major types of electrodes that are commonly used for electrostatic actuation: parallel plates [15,16] and interdigitated combs as illustrated in Figures 1.11 and 1.12. A typical configuration usually consists of a movable electrode connected to suspended mechanical springs while a fixed electrode is anchored onto the substrate. When a voltage is applied to the capacitive electrodes, the electrostatic attractive force actuates the movable electrode to the stationary electrode, causing the area of overlap and the capacitance between the two electrodes to increase. As a result, the spring suspending the movable electrode is deformed. Thus, the force balance between the spring restoring force and the electrostatic force determines the displacement of the movable

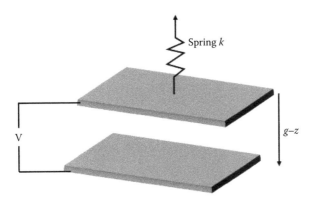

Figure 1.11 Parallel plates with applied voltage as electrostatic actuator; g is gap and z is displacement.

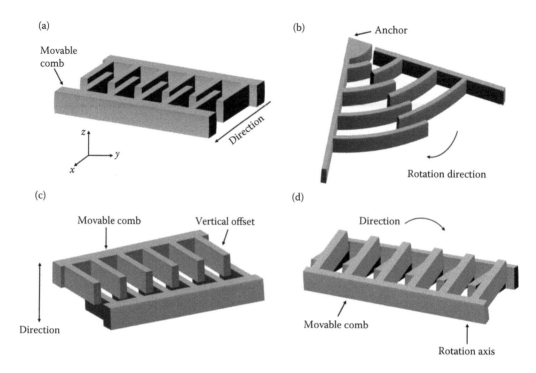

Figure 1.12 Schematic diagram illustrating the various types of electrostatic actuators commonly adopted in literature. They are (a) in-plane parallel lateral comb, (b) in-plane rotary combs, because of symmetrical structure only a part is drawn here, (c) out-of-plane staggered vertical combs, and (d) out-of-plane angular vertical combs.

electrode. Ignoring fringing effects, the force at a certain voltage V can be given by considering the energy stored inside the parallel plates W,

$$W = \frac{1}{2}\frac{\varepsilon A}{g-z}V^2 \tag{1.8}$$

$$\frac{\partial W}{\partial z} = F = \frac{1}{2}\frac{\varepsilon A}{(g-z)^2}V^2 \tag{1.9}$$

where g is the gap length, z the displacement, A the area of the plate, and ε the permittivity between plates. As the drive voltage reaches a threshold called the pull-in voltage, the displacement remains unchanged. If the movable electrode is attached to a spring, the pull-in occurs when the restoring force balances the electrostatic force,

$$V^2 = \frac{2Kz(g-z)^2}{\varepsilon A} \tag{1.10}$$

The derivative of voltage to displacement is zero now, which leads to the deflation at the pull-in voltage,

$$\frac{d}{z}(z(g-z)^2) = (g-z)(g-3z) = 0 \tag{1.11}$$

$$Z_{PI} = \frac{g}{3} \qquad (1.12)$$

Note that this deflation only depends on the gap. Substituting it into Equation 1.10 gives the pull-in voltage

$$V_{PI} = \sqrt{\frac{8}{27} \frac{Kg^3}{\varepsilon A}} \qquad (1.13)$$

Another type of electrostatic actuator is interdigitated combs. There are currently four categories of comb drive designs: lateral combs [17,18], rotary combs [19–21], staggered vertical combs (SVC) [22–27], and angular vertical combs (AVC) [28–33]. In the lateral combs, Figure 1.12a, the forces along the y direction are balanced and the combs move in the x direction. In an SVC actuator as shown in Figure 1.12c, a vertical offset between the moving combs and the fixed combs for out-of-plane rotation is required. In order to create the vertical offset between the two sets of combs, various fabrication techniques such as wafer bonding [22,25,27], integration of polysilicon and surface micromachining [23], and double-side alignment lithography on an SOI wafer [24,26] have been used. In the case of AVC illustrated in Figure 1.12d, the movable combs are often fabricated in the same layer as the fixed fingers and then tilted upward by various post-fabrication methods such as plastic deformation [30,31], residual stress [28], reflow of PR [29], and manual assembly [32,33]. In the lateral and vertical comb actuation setups, the force is independent of the displacement, unlike that of the parallel plate actuator setup. In addition, the force is inversely proportional to the gap distance making the force generated much smaller than that of the parallel plate actuator. This can be compensated by adding more fingers and applying a higher voltage.

In general, parallel plate and comb actuators are the available designs that may be used in bulk micromachined optical MEMS devices, while polysilicon-based comb actuators are often used in surface micromachined structures. Briefly speaking, parallel plate actuation can provide a large force ($\sim$50 μN) with a small displacement ($\sim$5 μm), but the force is highly nonlinear and unstable within the displacement range. On the other hand, interdigitated comb actuation provides a moderate level of force ($\sim$10 μN) with a reasonable displacement ($\sim$30 μm). Compared with other forms of actuation mechanisms, electrostatic actuation offers a fast response time ($\sim$1 ms) with negligible power consumption and can be easily integrated with an electronic control. However, it faces many challenging issues such as low mechanical stability due to pull-in, non-linearity, and a very high actuation voltage ($\sim$50 V).

1.2.3 Piezoelectric actuation

The piezoelectric effect is understood as the linear electromechanical interaction between the mechanical and the electrical states in a crystalline material. An applied DC voltage across the electrodes of a piezoelectric material will result in a net strain that is proportional to the magnitude of the electric field. A lack of a center of symmetry in a piezoelectric crystal means that a net movement of positive and negative ions with respect to each other as a result of stress will produce an electric dipole as shown in Figure 1.13. Adding up these individual dipoles over the entire crystal gives a net polarization and an effective field within the material. Conversely, a mechanical deformation of the crystal is produced when an electric field is applied, which makes this phenomenon extremely useful in driving optical MEMS devices [34,35].

In general, the piezoelectric effect is often described in terms of a piezoelectric charge coefficient, d_{ij}, which relates the static voltage or electric field in the i direction to the displacement of the applied force in the j direction. When a piezoelectric material is deposited on top of a microstructure, for example an Si cantilever, axes 1 and 3 are defined as the longitudinal and normal

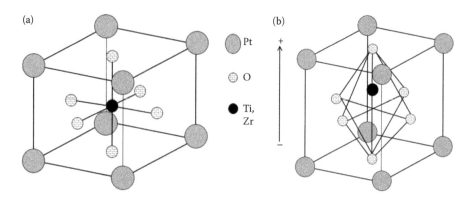

Figure 1.13 Schematic diagram illustrating the change in perovskite crystal structure (a) before and (b) after voltage is applied across it.

directions of the cantilever, respectively. The piezoelectric charge coefficients are given as d_{33} when both the voltage and force are along the vertical axis (axis 3), while d_{31} is used when the voltage is along the vertical axis, but the force generated is along the longitudinal axis (axis 1). The piezoelectric charge coefficient, which is the proportionality constant between strain and electric field, indicates that a higher value would be highly desirable for actuation purposes.

Considering a layer piezoelectric with two electrodes on both sides, the bending radius is

$$\frac{1}{R} = \frac{2d_{31}E_3(t_p + t_e)A_pE_pA_eE_e}{4(E_pI_p + E_eI_e)(A_pE_p + A_eE_e) + (A_pE_pA_eE_e)(t_p + t_e)^2} \tag{1.14}$$

Similar to a bimorph electrothermal actuator, the vertical displacement is

$$\delta = \frac{L^2d_{31}\dfrac{V}{t_p}A_eE_eA_pE_p(t_p + t_e)}{4(A_eE_e + A_pE_p)(E_pI_p + E_eI_e) + (t_e + t_p)^2A_eE_eA_pE_p} \tag{1.15}$$

where t_p, t_e, A_p, A_e, E_p, E_e, I_p, and I_e are the thicknesses, cross-section areas, Young's moduli, and moments of the piezoelectric (p) and elastic (e) layers. L is the length of the layer.

Most of the piezoelectric materials have a perovskite crystal structure and they include quartz (SiO$_2$), lithium niobate (LiNbO$_3$), aluminum nitride (AlN), zinc oxide (ZnO), and lead zirconate titanate (PZT), while the most well-known polymer-based piezoelectric material is polyvinylidene fluoride (PVDF). Among these materials, PZT has the largest piezoelectric charge coefficients (d_{31} and d_{33}) as shown in Table 1.1 [36]. Due to its excellent piezoelectric properties, PZT has often been used in numerous optical MEMS applications such as adaptive optics [37,38], optical

Table 1.1 Piezoelectric coefficient of selected piezoelectric materials

Material	Piezoelectric coefficients
Barium titanate	$d_{33} = 85.6$ pm/V; $d_{31} = 34.5$ pm/V
Aluminum nitride	$d_{33} = 4.5$ pm/V
Zinc oxide	$d_{33} = 12.4$ pm/V
Lead zirconate titanate	$d_{33} = 360$ pm/V; $d_{31} = 180$ pm/V
Polyvinylidene fluoride	$d_{33} = 20$ pm/V; $d_{31} = 30$ pm/V

Source: Adapted from Y. Eun et al. *Sensors and Actuators A: Physical*, vol. 165, pp. 94–100, 2011.

communication [39,40], and beam scanning [41–43]. However, unlike AlN, PZT is not CMOS compatible, hence making mass production by CMOS foundries impossible.

1.2.4 Electromagnetic actuation

Lorentz force is generated when a current-carrying element is placed within a magnetic field and it occurs in a direction equivalent to the cross product of the current and magnetic field. A simple example is the rotation of a soft magnetic cantilever, whose thickness t and width w are much shorter than its length L. If it is placed in a uniform magnetic H_0 field, the magnetic torque on the beam is [44,45]

$$\tau_{magnetic} = m \times B_0 = \mu m \times H_0 \tag{1.16}$$

where B_0 is the magnetic induction and m the magnetic moment, defined as

$$B_0 = \mu H_0 m = MV \tag{1.17}$$

where μ is the permeability constant, M the magnetization in the beam, and V the volume. When the magnetic force is large enough to deform the beam, the relationship between the corresponding torque and displacement $\Delta\delta$ at the free end is

$$\tau_{magnetic} = \frac{3EI\Delta\delta}{L^2} = \frac{Ewt^3\Delta\delta}{4L^2} \tag{1.18}$$

E is Young's modulus of the beam and I the moment of inertia. To enlarge the deformation, a larger magnetic field can be applied, thus producing stronger magnetization in the cantilever.

Although Lorentz force actuation may be applied to MEMS devices in a number of ways, the prevailing approach is to have metal coils integrated on a micromirror and actuated by an AC current at resonance when the mirror is placed near a permanent magnet [46–52]. Asada et al. proposed the two-dimensional scanner using electromagnetic coupling between a driving coil and a detection coil [46]. Figure 1.14a shows a three-axis actuated micromirror developed by Cho et al., where actuation coils made of gold are electroplated on the mirror plate and cantilever actuators

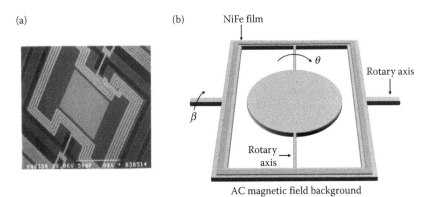

Figure 1.14 (a) A SEM photo showing the electroplated gold electromagnetic coils on the mirror plate and actuated by AC current at resonance in the presence of permanent magnet. (b) A schematic diagram illustrating a permanent magnetic film integrated on the mirror plate and actuated by the surrounding AC magnetic field. (From I.-J. Cho and E. Yoon. *Journal of Micromechanics and Microengineering*, vol. 19, p. 085007, 2009.)

[52]. Another approach, as shown in Fig. 1.14b, is to integrate a permanent magnet (hard ferromagnet) or a permalloy layer (soft ferromagnet) on a movable mirror while a Lorentz force is introduced through the interaction between the magnetic layer and surrounding AC magnetic field of an external solenoid [53–57]. The availability of permanent magnetic materials that are compatible with MEMS processing is limited and this produces a difficulty in the necessary process development. Thus, it is common for the magnetic field to be generated externally, while the discrete and movable electromagnetic actuators often comprise metal coils.

Similar to electrostatic actuation, electromagnetic actuation provides moderate switching speed (~10 ms) and low power consumption (~100 mW), but the assembly of permanent external magnets and coils make it extremely challenging. Fabricating ferroelectric materials can also be challenging, as these thin films may not be compatible with the standard CMOS processes.

1.3 Optical MEMS for imaging

We now discuss the applications of MEMS-driven mirrors in imaging systems. Such systems usually use single-pixel-based photodetectors (can be a single fiber coupled to a detector) and employ MEMS mirrors to rapidly steer their instantaneous field of view (IFOV) from point to point in a sequential way across the system field of view (FOV) to achieve image acquisition. There are several advantages for such systems. For example, using MEMS mirrors and single-pixel detectors, the imaging system can be constructed in a very compact package. This is attractive to biomedical imaging applications where the sizes of the imagers are often critical. In addition, it is easier for such systems to incorporate different types of imaging modalities, for instance, realization of 3D imaging deep into biological samples through laser confocal scanning [58] and the optical coherence tomography (OCT) approach [59].

Taking the scanning confocal imaging shown in Figure 1.15a as an example, a point light source (usually from a laser) is collimated with lens 1, reflected by a beamsplitter, and focused into the sample with lens 2. The scattered or emitted light by the sample is then collimated again with lens 2, passed through the beamsplitter, and focused by lens 3 to a confocal plane. A small pinhole is

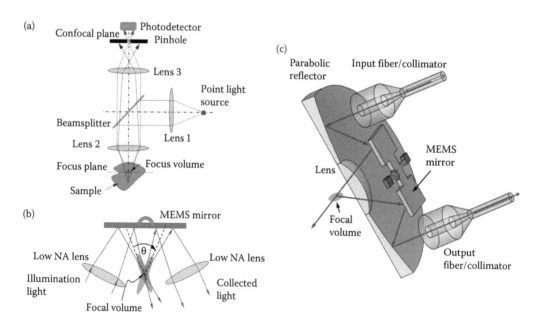

Figure 1.15 Schematic illustrations of (a) a scanning confocal microscope, (b) a dual-axis scanning configuration, and (c) a miniature dual-axis scanning confocal microscope.

then placed at the confocal plane, which forms a conjugated pair with the focus spot in the sample. Light passing through the pinhole is then received by a photodetector. In this configuration, only light scattered/emitted within a tiny volume (focus volume) inside the sample can pass through the pinhole and be detected, while light from outside the focus volume (above or below the focus plane) is blocked by the pinhole. To achieve imaging, an optical scanner (not shown in the figure for simplicity) is usually employed, which deflects the light beam and scans the focus spot across the two dimensions in the focus plane. In this way, a 2D image of the sample slicing at the focus plane is obtained. Light from the layers above or below the focus plane is rejected by the confocal pinhole mechanism. A 3D image of the sample can then be obtained through a layer-by-layer image acquisition by further moving the sample toward and/or away from the imaging optics. Benchtop confocal imaging microscopes are now widely available in the market.

With MEMS technology, a number of miniature confocal microscopes are reported [60–63]. They feature significantly reduced package sizes and can be handheld for clinical applications such as *in vivo* skin imaging [64]. Some of them are constructed with diameters small enough to be put into endoscopic platforms [65,66]. In such MEMS confocal microscopes, the beamsplitter in Figure 1.15a is replaced with a MEMS scanning mirror and an optical fiber is usually used as the point light source. The light is scanned by the MEMS mirror and focused onto the sample. The scattered or emitted light from the focus volume is collimated, de-scanned by the same MEMS mirror, and focused into the same input fiber. In this case, the fiber acts as both the point light source (due to its small core diameter) and the pinhole. This configuration is widely used and is also applicable to nonlinear optical microscopy such as coherent anti-Stokes Raman scattering (CARS) imaging [67], Raman spectroscopy [68], and two-photon imaging [69,70].

Recent developments in this field have also extended the design to a dual-axis scanning configuration [28,71–73], as shown in Figure 1.15b. It is well known there are certain limitations for the confocal imaging systems shown in Figure 1.15a. For example, lenses with large numerical apertures (NA) are needed for high resolution (especially for high axial resolution) imaging, which unfortunately leads to short working distances, low penetration depths, and reduced contrast. Dual-axis confocal imaging as shown in Figure 1.15b elegantly solves these problems by constructing the light illumination and collection in two different paths [30,74]. In this way, low NA lenses can be used leading to increased working distances. In addition, since the collection optics is tilted to an angle θ with respect to the illumination optics, the small focus volume is still obtained by the intersection of the illumination and collection paths despite the long depth of focus of the low NA lenses. This configuration also significantly rejects light from the illumination path outside of the focus volume thus leading to enhanced contrast. A schematic showing the construction of the dual-axis MEMS confocal microscope using two fibers, a parabolic reflector, a lens, and a MEMS-driven scanning mirror is illustrated in Figure 1.15c.

An endoscopic OCT system is another example of successful applications based on optical MEMS imaging technology. A standard OCT setup using a fiber-based Michelson interferometer is illustrated in Figure 1.16a [75]. The setup utilizes low coherence light to capture 3D images of biological tissues or samples, which are typically high scattering media. Compared with confocal imaging, OCT imaging usually results in a larger penetration depth into the tissue with a slightly lower resolution. OCT systems use low coherence or broadband light sources with typical coherent lengths of only a few micrometers. Light is then simultaneously sent into the reference and measurement arms with a fiber coupler (or a beamsplitter for a free-space Michelson interferometer). Light reflected from the reference and measurement arms is then recombined to form interference. It is well known that the interference pattern can only be formed if the optical path difference (OPD) between the reference and measurement paths is smaller than the coherent length of the light source, which is only a few microns for OCT systems. This defines the depth resolution and creates a "coherent gate" effect that isolates the light signals scattered within a thin slice at a specific depth under that sample surface and rejects other light signals from other locations. The 2D image of the slice is obtained again through point-by-point scanning using a mirror, and the full 3D image of the sample is obtained by an additional depth scanning by moving the reference

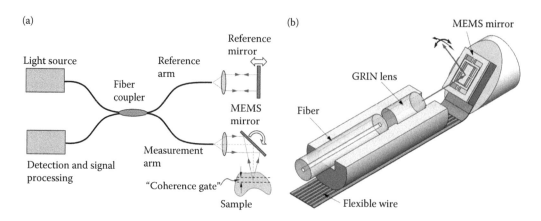

Figure 1.16 (a) Optical coherence tomography (OCT) in a basic configuration, and (b) a side-view MEMS-based endoscopic OCT probe.

mirror as shown in the figure. In addition to this standard configuration, there are many other OCT configurations such as Fourier domain OCT and swept-source OCT [76].

As shown in Figure 1.16a, MEMS-driven mirrors are usually employed in OCT systems to act as scanning mirrors in the measurement arm to rapidly steer the field of view. The application of optical MEMS scanning results in compact systems such as handheld OCT systems for *in vivo* tissue imaging (skin and eye imaging, for example) [77] and endoscopic OCT probes for imaging of internal organs [78]. A typical design of a side-view MEMS OCT probe is shown schematically in Figure 1.16b. Light is launched into the probe using an optical fiber. It is then focused by a gradient index (GRIN) rod lens, reflected by a MEMS mirror, and continues to focus into the tissue/sample under examination. The light reflected from the tissue/sample is de-scanned by the same mirror, collected by the same GRIN lens and fiber, and sent into the fiber Michelson interferometer for signal processing. Apart from side-view probes, full 360° circumferential scanning probes [79,80] and forward-view MEMS OCT probes are also demonstrated [81]. The use of MEMS scanning mirrors here is very attractive due to the following reasons. First, the ultra-small footprints of MEMS mirrors facilitate the miniaturization of OCT probes, which is critical for endoscopy applications. Second, the low inertia and fast scanning characteristics of these mirrors also enhance the image acquisition rate making real-time OCT imaging possible. Third, MEMS mirrors are low cost due to IC-like fabrication processes thus making such MEMS endoscopic OCT probes affordable. A wide variety of MEMS mirrors for OCT imaging have been reported, including those driven by electrostatic [82,83], electromagnetic [84], piezoelectric [85], and electrothermal actuators [86–88]. Each of those driving mechanisms has their own advantages and disadvantages. Overall, however, the electrothermal MEMS mirrors seem to be more suitable for MEMS endoscopic OCT probes due to their very low driving voltage (important for safe operation inside the human body) and large optical scan angle despite their relatively slow response time.

1.4 Optical MEMS for displays

Optical MEMS also plays an important role in the display industry [89]. Display devices can be categorized as projection displays, transmissive displays, and reflective displays according to their operation modes. Optical MEMS technology has been successfully applied to all of these displays. In addition, based on their working principles, optical MEMS displays can be divided into scanning-based and non-scanning-based displays. The former can be further divided into point source 2D scanning and linear array 1D scanning-based displays. In this section, we will briefly discuss some of the most successful devices and systems.

1.4.1 Scanning-based MEMS displays: Point source 2D scanning

There are generally two configurations for MEMS displays using the point source 2D scanning approach, with one having two MEMS mirrors each scanning a single direction and the other having a single MEMS mirror scanning in both directions. This type of scanning projection display has very small form factors and thus can be incorporated into wearable and mobile projection systems [90]. The performances of these 2D scanning-based displays, including frame rate, resolution, and brightness, are generally determined by those of the optical MEMS scanners and lasers employed.

As shown in Figure 1.17a, using two 1D optical scanners with one scanning the incident optical beam in the horizontal direction at a faster rate and the other scanning the beam in the vertical direction at a slower rate, one can project a raster scan pattern on a project screen. By modulating the light intensity according to its position on the screen, images and videos can be displayed. Three lasers (red, green, and blue) can be used and projected simultaneously with the same mirrors for color displays. This configuration is among the early types of the point source 2D scanning MEMS displays demonstrated [91]. The light beam wandering on the second mirror and the raster scan pattern distortion caused by the inherent spatial separation between the two scanning mirrors are usually among the major limitations of this kind of configuration.

A more compact solution has been developed that uses a single MEMS mirror scanning in both directions as shown in Figure 1.17b. Examples include the mobile projection systems developed by Microvision Inc. [46] and Fraunhofer Institute of Photonic Microsystems [92]. These MEMS mirrors typically have gimballed suspensions that allow them to simultaneously rotate about a fast axis and a slow axis. The MEMS driving mechanisms include electrostatic [93], electromagnetic [49], and piezoelectric [94] actuations. An example of electromagnetic actuation is illustrated schematically in Figure 1.17b. As shown, an electric coil is integrated in a movable frame of the scanner and the driving current is modulated. The scanner is placed in a magnetic field produced by an external permanent magnet. The modulated current generates oscillating torques that drive the movable frame (together with the mirror) into a slower sawtooth scanning and the inner mirror into a faster oscillation about its axis at its resonant frequency.

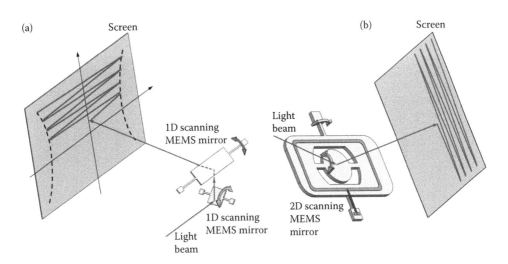

Figure 1.17 Schematics of the point source 2D scanning displays, (a) with two 1D scanning MEMS mirrors, and (b) with a single MEMS dual-axis mirror scanning in two directions.

1.4.2 Scanning-based MEMS displays: Linear array 1D scanning

In point source 2D scanning MEMS displays, one direction is usually scanned at a very fast rate (about tens of kHz) to meet the frame rate requirement of the display. This creates significant acceleration and deceleration forces on the MEMS mirror plate, thus the mirror is subjected to dynamic deformation which can potentially blur the light spot size and reduce the display resolution. Hence, considerations have to be taken in the mirror design to keep this dynamic deformation small enough (i.e., smaller than $\lambda/8$) to meet the Rayleigh wavefront criterion for a good imaging system. On the other hand, one can also construct a display system using a linear array of light sources, instead of a point light source, to display a line image on the screen and then scan this line in its perpendicular direction across the screen. This configuration only requires a slow 1D scanning mirror.

A typical linear array device is the grating light valve (GLV) [95]. Figure 1.18a shows one of its pixels, which usually consists of six micron-sized coated high reflectivity ribbons. At the initial state, all ribbons are in the same plane and the pixel essentially behaves like a mirror that reflects the incident light back creating a "dark" state. On the other hand, when the alternating ribbons are actuated, pulling down toward the substrate with a precise displacement of a quarter wavelength ($\lambda/4$), the pixel then behaves like a diffraction grating. Furthermore, the lights reflecting off the top and bottom ribbons are in destructive interference, thus distributing almost all light energy into the $\pm$ first diffraction order and creating a "bright" state for the pixel. This state switching is very fast and can happen at the order of tens of nanoseconds owing to the extremely small inertia of the ribbons. Since the GLV can only have "dark" and "bright" states, a pulse width modulation (PWM) scheme [96] is used to generate grayscale images with various levels of light intensity at each pixel. In other words, when a GLV pixel is switched to the "bright" state longer in a video frame time, it appears brighter. Figure 1.18b then shows schematically how the pixels are arrayed into a linear array, and a projection system employing a GLV device is illustrated in Figure 1.18c. Since GLV utilizes the first diffraction order as the light source for projection, its zeroth diffraction order has to be blocked. Usually, this can be achieved through employment of a 4-f optical filtering system in the projection optics as shown in Figure 1.18d [97].

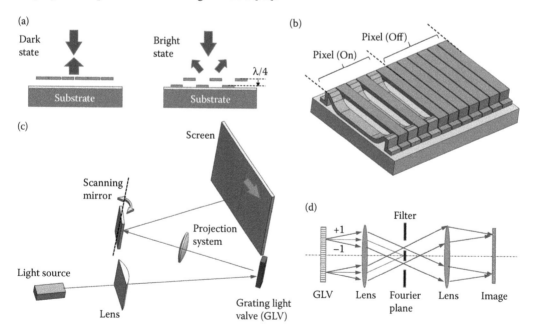

Figure 1.18 (a) Operation principle of the grating light valve (GLV), (b) schematic showing two GLV pixels, (c) schematic of the linear array 1D scanning display using GLV, and (d) optical filtering system for the GLV-based projection display.

1.4.3 Non-scanning-based MEMS displays

Non-scanning-based MEMS displays do not have any light scanning mechanisms and usually have a 2D array of pixels. The most well-known example of this kind of MEMS display is the digital light processing (DLP™) projection displays from Texas Instruments (TI) [98]. At the core of this projection display is the digital micromirror device (DMD), which contains a large number of bistable micromirrors (can be over a million depending on the display format). As shown in Figure 1.19a, each square micromirror has a width of about 15 μm and represents one pixel. It can be driven by its integrated COMS circuitry to rotate +10 degrees about an axis defined by a pair of torsional springs hidden underneath the mirror representing a "bright" or "ON" state or −10 degrees representing a "dark" or "OFF" state. The switching time is very fast at the microsecond level due to the low mirror inertia. Grayscale images are again achieved through a PWM scheme by controlling each mirror's "ON" state time during each frame. A schematic showing the use of a DMD in a projection display system is illustrated in Figure 1.19b. Absorbers can be used to absorb the light reflected from those micromirrors in the "dark" state. To display color images or videos, two approaches can be used [99]. One uses a rotating color wheel filter to separate each video frame into its constituent R, G, and B fields, and sequentially illuminate a single DMD with the R, G, and B light beams. The viewer's eye integrates sequential images to obtain a full color image. The other approach uses three DMDs instead of one. White light is first divided into RGB light beams using filters, which are then directed to their own DMD chips. Upon reflection from those chips, they are recombined and projected onto the screen. TI's DLP technology is currently applied to a wide range of applications including TVs, cinema projection systems, and mobile projectors.

Digital micro-shutters (DMS) developed by Pixtronix is another MEMS-based display technology in this category [100]. As shown in Figure 1.20a, DMS technology is built upon standard thin film transistor (TFT) liquid crystal display (LCD) manufacturing technology and incorporates MEMS. As shown in the figure, the working principle of a DMS device mainly consists of the following key features; an optical architecture with a light-recycling LED backlight, an optical aperture layer, and a MEMS layer containing an array of micro-shutters. The MEMS micro-shutter is actuated by the electrostatic force and can move laterally to open or block the aperture, thus creating "bright" and "dark" states for each pixel. Field sequential color with high color changing

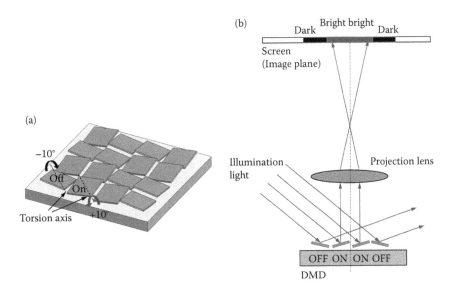

Figure 1.19 (a) Schematic of a DMD chip and (b) schematic of a projection display using TI's DMD.

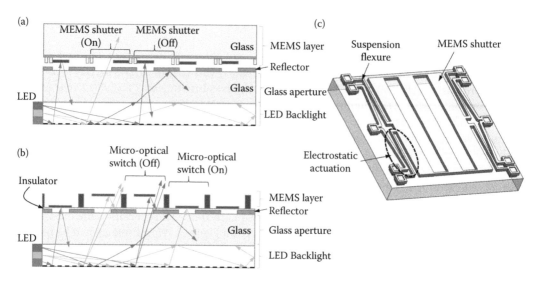

Figure 1.20 Schematics of (a) a transmissive display using digital micro-shutters (DMS), (b) a transmissive display using micro-optical switches, and (c) an electrostatically-driven MEMS micro-shutter.

frequencies is employ to create color images, thus no additional optical filters and polarizers are needed as compared with traditional LCD displays. The use of reflective surfaces of the apertures and micro-shutters to recycle the light energy also makes the display devices energy efficient. A typical MEMS micro-shutter design using four suspension flexures is illustrated in Figure 1.20c. Another similar concept of such a MEMS displays is also proposed, a schematic of which is illustrated in Figure 1.20b [101]. As shown, instead of using lateral MEMS shutters, the "bright" and "dark" states of the pixels are created by moving micromirrors up or down perpendicular to the aperture surface for opening or closing apertures. In the "bright" state, the light is coming out of the pixel through multiple reflections or direct leakage.

As discussed above, TI's DLP™ is mainly used for projection displays, while Pixtronix's DMS is mainly used for transmissive displays. There is yet another MEMS non-scanning display technology termed a interferometric modulator display (IMOD) developed by Qualcomm, which is mainly used for reflective displays [102,103]. IMOD operates and generates colors using the light interference effect, that is, in the same way as butterfly wings and bird feathers create colors using microscopically structured surfaces. As shown in Figure 1.21a, each pixel of the IMOD display is similar to a Fabry–Pérot interferometer, which consists of a thin film stack and a reflective mirror separated by an air gap. The light reflected off of the thin film and the mirror interfere constructively at certain wavelengths and destructively at others, thereby producing a colored effect on the pixel. If the air gap is reduced, the constructive wavelength blue-shifts thus changing the pixel color. The early versions of the IMOD displays are operated in digital mode, where each pixel is further divided into R, G, and B subpixels, each with a designated air gap, as shown in Figure 1.21b. If a driving voltage is applied across the flexible membrane-like mirror and the film stack, the electrostatic force drives the mirror to move up toward the thin film stack causing the air gap to collapse. This shifts the constructive interference wavelength to ultra-violet (UV) wavelengths, and the subpixel will display a "dark" state. When the voltage is released, the mirror returns to its original position, and the subpixel returns to its designated color. The switching time between the "color" and "dark" states is fast, in the range of 10 microseconds. A new generation of IMOD displays has also been developed by Qualcomm that uses a single pixel to produce multiple colors [104]. Each pixel is identical, and through precisely controlling the air gap, it can continuously tune its reflective color across the visible spectrum. An IMOD display can be viewed using only ambient light and thus has low power consumption and is energy efficient.

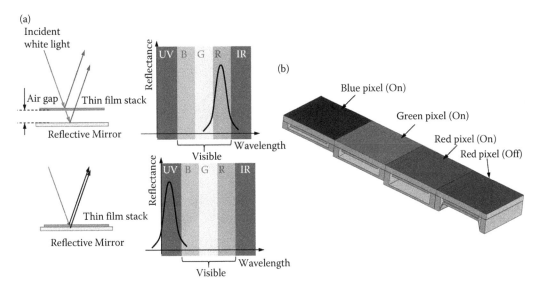

Figure 1.21 (a) Operation principle of the interferometric modulator display (IMOD) and (b) a schematic showing the structures of the red, green, and blue subpixels.

1.5 Optical MEMS for sensing

Optical MEMS technology is also extensively used in sensors. Based on their sensing mechanisms, these devices can be categorized into two groups, coherent and noncoherent optical sensors. Coherent-based sensors rely on various light interference effects for sensing, hence coherent light sources such as lasers are typically used. On the other hand, noncoherent-based sensors do not depend on light interference, and they typically detect variations of light intensities as functions of measurands.

1.5.1 Noncoherent optical MEMS sensors

Figure 1.22 shows typical examples of noncoherent optical MEMS sensors. One sensing principle is based on the well-known "light lever" effect, in which a measurand changes the orientation of a mirror thereby altering the direction of a light beam reflected off of it. The change of the beam direction is then converted into beam spot movement on a position-sensitive photodetector (PSD). A four-quadrant photodiode, which consists of four identical light-sensitive surfaces separated by small gaps, is usually employed for this type of position sensing. The system is initially set up so that the light energy is equally distributed to the four quadrants and the outputs of the diodes are equal. When the beam spot moves, the diode outputs change. In this case, the difference between the left and right quadrants indicates horizontal movement and the difference between top and bottom quadrants indicates vertical movement. This sensing configuration can be very sensitive owing to the fact that the movement of the light spot is significantly amplified if the mirror-to-detector distance is large. A large number of sensors are demonstrated with MEMS fabricated cantilevers using this sensing mechanism, as shown in Figure 1.22a. These include atomic force microscopes (AFM) and chemical/biological sensors including those used for the detection of antigen and antibody bonding events [105]. While out-of-plane motion can easily cause light beam deviation with the help of a cantilever or a mirror, in-plane motion is slightly more complicated and such a light beam deviation scheme can be achieved with the help of a diffraction grating as shown in Figure 1.22b. The in-plane rotation of the grating changes the orientation of the grating lines thereby inducing a change of direction for a selected non-zeroth-order diffracted beam. Such a grating-based sensing

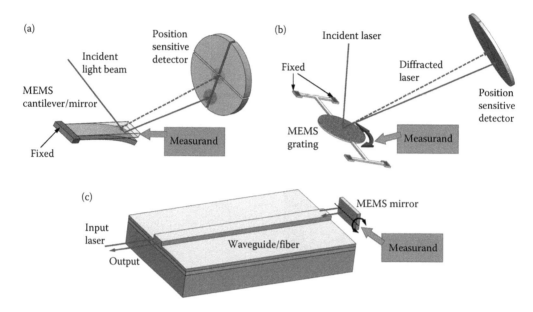

Figure 1.22 Schematics of the noncoherent optical MEMS sensors. (a) Sensing mechanism using "light lever" effect with a cantilever, (b) sensing mechanism using "light lever" effect with a MEMS grating, and (c) sensing mechanism based on light coupling to waveguides/fibers.

mechanism has been utilized to demonstrate a sensitive integrated MEMS tribometer [106]. These optical MEMS sensors based on the "light lever" effect are very sensitive. However, they have large footprints because the mirror-to-detector distances required are usually large.

There is another type of noncoherent optical MEMS sensor that uses light waveguiding and a coupling effect as shown schematically in Figure 1.22c. As shown, light is launched into an input waveguide (or fiber) and directed toward a mirror. Upon reflection from the mirror, the light is coupled back into an output waveguide (or fiber). The output waveguide can be a new waveguide or the same waveguide as the input waveguide. In this configuration, it is well known that the light coupling efficiency is very sensitive to the motion of the mirror. Hence, translational motion (varying the gap spacing between the mirror and the waveguide) or rotation (tipping or tilting) of the mirror caused by the measurand can drastically change the light intensity in the output waveguide. Several sensors such as magnetic field sensors and accelerometers are reported based on such a sensing mechanism [107]. It is noted that this type of optical coupling-based sensors, unlike those based on the "light lever" effect, can be miniaturized and integrated at a chip scale.

1.5.2 Coherent optical MEMS sensors

A majority of optical MEMS sensors are coherent-based sensors utilizing lasers and various light interference effects. They are essentially miniaturized optical interferometers developed using MEMS technology. In the following sections, we mainly focus on those that incorporate MEMS mirrors. For each interferometer type, we discuss two types of construction approaches, namely the out-of-plane (utilizing mirror surfaces parallel to the substrate) and in-plane (utilizing mirror surfaces perpendicular to the substrate) construction approaches. For the latter, silicon-on-insulator (SOI) wafers are typically used and mirrors are formed using deep-etched side walls. It should be noted that there are many other types of optical MEMS coherent-based sensors built on waveguide technology, such as a chemical/biological sensor using an integrated Mach–Zender interferometer as shown in Figure 1.23 [108]. The change of the refractive index in the measurement arm in close proximity to the waveguide surface varies the effective optical path length due

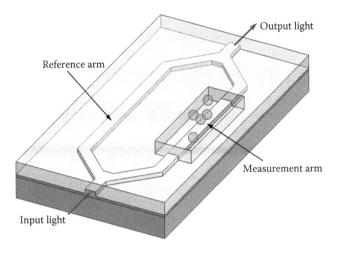

Figure 1.23 A waveguide-based Mach–Zehnder interferometer for biochemical sensing.

to the presence of a given chemical/biological analyte, thus changing the light interference and the interferometer's output.

1.5.2.1 Optical MEMS sensors based on Michelson interferometers

The Michelson interferometer is perhaps one of the most commonly used interferometers in optics. As shown in Figure 1.24a, the amplitude of the collimated incident light radiation is equally divided in a beamsplitter into two waves. They are reflected by two mirrors and recombined to form interference at the detection plane where photodetectors are located. For optical MEMS sensing applications, one mirror is usually fixed and the other is a movable MEMS mirror. The intensity of the interference signal detected varies as a function of the optical path difference (OPD) δ between the reference and measurement arms. The relationship is as follows:

$$I(\delta) = 0.5I_\sigma \left(1 + \cos 2\pi\sigma\delta\right), \qquad (1.19)$$

where σ and I_σ are, respectively, the wavenumber ($\sigma = 1/\lambda$, λ being the wavelength) and intensity of the light source. Clearly, if a measurand (force, acceleration, etc.) causes the MEMS movable mirror

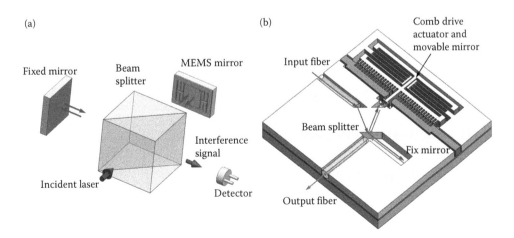

Figure 1.24 Optical MEMS-based Michelson interferomters in (a) out-of-plane construction and (b) in-plane construction using silicon-on-insulator (SOI) micromachining.

to move inducing a variation in the OPD δ, the output intensity from the interferometer varies, thus the measurand is detected. From the above equation, it is clear that the sensors built on such a light interference effect are inherently sensitive owing to the fact that the wavelengths of the light sources (lasers) used are usually at sub-micrometer levels.

Michelson interferometers are more frequently used for construction of Fourier transform infrared (FTIR) spectrometers, which are powerful tools for chemical analysis. In an FTIR system, the movable mirror is scanned over a range and the interferometer output light intensity variation $I(\delta)$ is recorded as a function of the OPD δ, which is usually called the interferogram. It can be shown that the light source power spectrum I_σ and the recorded interferogram are related by a Fourier transform [109]:

$$I_\sigma \propto \int_{-\infty}^{\infty} [I(\delta) - 0.5I(0)]\exp(-i2\pi\sigma\delta)d\delta, \qquad (1.20)$$

where $I(0)$ is the intensity of the interferometer output at zero OPD. It can also be shown that the spectral resolution of the FTIR spectrometer is inversely proportional to the maximum OPD scan range provided by the movable mirror [109].

Conventional FTIR spectrometers are bulky, expensive, and limited to lab use. In recent years, the market has seen a high demand for low-cost miniaturized field-applicable FTIR spectrometers with relatively lower solution, but capable of delivering cost effective real-time information for sensing applications in a wide variety of disciplines and industries, including environmental monitoring and food/beverage safety screening. In response to this, a number of miniaturized versions of FTIR spectrometers implemented with optical MEMS technology have been reported [110]. Similar to other optical MEMS interferometers, these miniature FTIRs are constructed in two ways. As shown in Figure 1.24a, for out-of-plane construction, the beam splitter, fixed mirror, and the MEMS movable mirror are separate components. They are aligned and assembled to form a Michelson interferometer. The MEMS mirror is typically operated in an out-of-plane mode moving translationally perpendicular to the device surface. A large stable travel range is also required in order to obtain sufficient OPD scanning for a high enough spectral resolution. The reported MEMS mirrors for FTIR spectrometers are typically driven by electrostatic, electrothermal [111,112], and electromagnetic actuation methods. Millimeter scale travel ranges have been demonstrated for some of those MEMS mirrors. Although this type of MEMS FTIR spectrometer has a relatively large footprint, they can potentially achieve good performances due to their high-quality mirror surfaces and large OPD scan ranges.

Another way of constructing optical MEMS miniature FTIR spectrometers is to build Michelson interferometers in a silicon device layer on an SOI wafer [110,113,114]. As shown in Figure 1.24b, the input and output fibers are, respectively, used to guide the light to the interferometer and collect the interference signal and guide it to the detector. Light propagates in the silicon device layer beneath the layer surface; hence this layer has to be thick enough, which is usually about a hundred micrometers. Deep reactive ion-etched silicon side-wall surfaces are utilized as the mirrors and beam splitters. Thermal oxidation and HF treatment can be used to reduce the side-wall surface roughness and improve the optical quality. Coatings can also be used to enhance the mirror's reflectivity. These types of optical MEMS FTIR spectrometers have compact sizes, but their spectral resolutions are usually limited.

1.5.2.2 Optical MEMS sensors based on lamellar grating interferometers

As shown in Figure 1.25a, a lamellar grating consists of an array of identical micromirrors, where alternating mirrors are grouped together. One group of the mirrors is usually fixed, while the other is movable. Driven by a measurand, the movable mirrors move with respect to the fixed mirrors, thus introducing an OPD between the light beams reflected off the surfaces of the two groups of

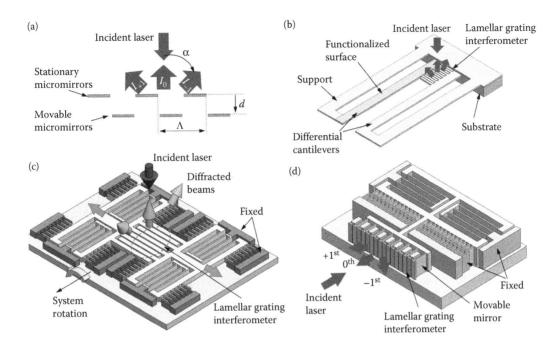

Figure 1.25 (a) Operation principle of the lamellar grating interferometer, (b) schematic illustration of a lamellar grating-based biochemical sensor, (c) a lamellar grating interferometer-based MEMS gyroscope, and (d) a lamellar grating-based FTIR spectrometer.

mirrors. The lamellar grating essentially behaves like a diffraction grating with varying grating groove depths. As a result, the intensities of the diffracted beams are dependent on the mirror separation gap d [115]. Assuming normal light incidence, the relationship is as follows:

$$I \propto \left(\frac{\sin K}{K}\right)\left(\frac{\sin 2NK}{2K}\right)(1 + \cos 2\pi\sigma\delta), \qquad (1.21)$$

where $K = (\pi\sigma\Lambda)\sin(\alpha)/2$, σ is the incident wavenumber $\sigma = 1/\lambda$, λ is the wavelength, and Λ and N are the grating period and the number of periods illuminated, respectively. The diffraction angle α and the OPD δ are given as follows:

$$\sin\alpha = \frac{m\lambda}{\Lambda}, \qquad (1.22)$$

$$\delta = d\left(1 + \cos\alpha + \frac{\Lambda}{2d}\sin\alpha\right), \qquad (1.23)$$

where m is the diffraction order.

It is clear that all diffracted beam intensities vary as the OPD δ changes, hence any order can be used for sensing applications to detect the movement d. However, the zeroth-order is usually preferred. The lamellar grating interferometer is a very attractive choice for sensing applications due to the fact that it is a common path interferometer, that is, the reference arm (for the light reflected off the fixed mirrors) and the measurement arm (for the light reflected off the movable mirrors) almost share the same optical path except for the small mirror gap d. Hence, external disturbances are identical in both arms, which results in cancellation and stable interference signals. Sensors constructed using common path interferometers usually tend to have higher signal-to-noise ratios (SNR) compared with those based on non-common path interferometers such as Michelson interferometers.

Lamellar grating interferometers can also be used in FTIR spectrometers [116]. In this case, the zeroth-order diffracted beam from the grating is recorded as the interferogram $I(\delta)$ while the movable mirrors move a sufficient amount to scan the OPD. The spectrum of the light source is again obtained through a Fourier transformation of the interferogram using Equation 1.20. In fact, a lamellar grating interferometer has a number of outstanding advantages for FTIR, including the absence of beam splitters, extended wavelength range, robustness (common-path interferometer), and high efficiency.

Again, lamellar grating interferometers based on optical MEMS can be built in two ways, namely out-of-plane and in-plane configurations. Figure 1.25b shows a micromachined biosensor using a lamellar grating interferometer in an out-of-plane configuration [117,118]. As shown, the sensor consists of a pair of suspended silicon nitride micro cantilevers, with one having a functionalized surface with receptors that are specific to the ligand under detection and the other without functionalization serving as a reference. At the tips of the cantilevers, interdigitated fingers are attached to act as the movable and fixed mirrors thus forming the lamellar grating interferometer. The interferometer is illuminated with a laser beam and the first-order diffraction is monitored. The presence of the ligand under detection bends the sensing cantilever thus creating a gap between the interdigitated fingers, inducing an OPD, and varying the intensity of the selected diffraction order. Figure 1.25c shows an optical MEMS gyroscope based on an out-of-plane lamellar grating interferometer [119]. As shown, the two proof masses are driven into in-plane anti-phase oscillations along a direction parallel to the device substrate with the same frequency. Interdigitated finger-like mirrors are also integrated on the two masses forming a lamellar grating interferometer. When the device experiences a rotation about an axis as indicated in the figure, the Coriolis forces on the two proof masses are also anti-phase, that is, pulling down one and pushing up the other. This creates a separation gap between the two sets of mirrors that is proportional to the rate of rotation. This dynamically changing gap is then detected by illuminating the interferometer with a laser beam and monitoring its diffraction orders. There are a number of sensors built on the basis of such out-of-plane configured lamellar grating interferometers, which include nano-g resolution accelerometers [120,121], electromagnetically-driven lamellar grating FTIR spectrometers [122], AFM cantilevers [123], and many others.

Figure 1.25d illustrates an example of a lamellar grating interferometer in an in-plane configuration [115]. The interferometer is developed using SOI microfabrication technology and designed for use as an FTIR spectrometer. As shown, deep-etched silicon side walls in the device layer of an SOI wafer are used as mirrors with one set anchored to the substrate and the other movable set attached to a MEMS electrostatic comb drive. The spectrometer can scan an OPD of about 145 μm resulting in a spectral resolution of a few nanometres in the visible spectrum range.

1.5.2.3 Optical MEMS sensors based on Fabry–Pérot interferometers

A Fabry–Pérot interferometer or cavity is another well-known interferometer commonly used in optical MEMS sensors. As shown in Figure 1.26a, the interferometer consists of two parallel mirrors separated with a gap d. The mirrors are identical and lossless with reflectance R. The overall coefficients of transmission (T_{FP}) and reflection (R_{FP}) of the system can be expressed as follows [124]:

$$T_{FP} = \frac{(1-R)^2}{(1-R)^2 + 4R\sin^2\left(\dfrac{\delta}{2}\right)},$$

(1.24)

$$R_{FP} = \frac{4R\sin^2\delta}{(1-R)^2 + 4R\sin^2\left(\dfrac{\delta}{2}\right)},$$

(1.25)

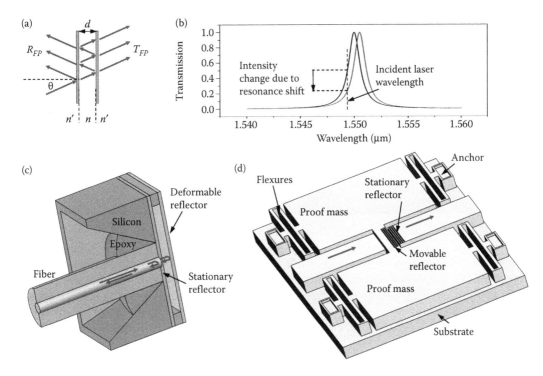

Figure 1.26 (a) Operation principle of a Fabry–Pérot interferometer, (b) typical optical readout mechanisms for sensors based on Fabry–Pérot interferometers, (c) a fiber-end-based Fabry–Pérot interferometric sensor, and (d) an in-plane constructed Fabry–Pérot interferometer for acceleration sensing.

where $\delta = (4\pi nd/\lambda) \cos(\theta)$ is the round-trip phase difference. When the resonant conditions, that is, $\delta = 2k\pi$ (k is an integer), are met, the transmission and reflection spectra show, respectively, a strong peak and a dip. The higher the reflectance of the mirror R, the sharper the transmission peak and reflection dip.

Both the refection and transmission from a Fabry–Pérot interferometer can be used for sensing, and normal incidence ($\theta = 0°$) is usually utilized. As an example, Figure 1.26b shows the transmission coefficient of the interferometer as a function of wavelength λ. Here, we assume normal incidence, the mirror reflectance $R = 0.98$, and the medium between the two mirrors is air ($n = 1$). The black and red curves show, respectively, the transmission spectra when the mirror gap d is at 6.2 and 6.21 μm, that is, when there is an increase of about 10 nm. It is quite clear that the sensors built on the Fabry–Pérot interferometers can be very sensitive owing to the fact that even a minute gap change can induce a drastic shift in the resonance transmission peak. Two detection methods are commonly used to obtain the minute gap variation from the interference signal. The first one is to directly track the wavelength shift of the resonant peak with a spectrometer [125]. The second method is to lock the incident laser beam to the half-maximum of the initial resonance peak (or a quadrature point) and monitor the output laser intensity from the interferometer [126]. As shown in the figure, the resonance shift due to the cavity gap change induces a large laser intensity variation. The latter has a number of advantages including high sensitivity and fast signal response, however, its dynamic sensing range is limited.

A majority of the sensors based on Fabry–Pérot interferometers are constructed in an out-of-plane configuration due to the high-quality mirror surfaces. A large number of sensors are demonstrated including pressure [127], acoustic and ultrasound [128–130], temperature [131], and chemical detectors [126]. Some are the stand-alone types with integrated photodetectors [126], and some are integrated with optical fibers for light input and output as shown in Figure 1.26c. For the fiber-based type, the movable reflector can be micromachined membranes coated with high-reflective coatings, high contrast gratings [132], or photonic crystal mirrors [133]. The stationary mirror can be fabricated on the substrate or directly coated on the fiber end face. The fiber and machined chip are usually aligned and fixed with epoxy. The measurand to be detected changes the gap between two mirrors thereby affecting the sensor output. There are also optical MEMS sensors reported based on Fabry–Pérot interferometers in an in-plane configuration including chemical sensors [134] and accelerometers [135]. The latter is illustrated schematically in Figure 1.26d. It is fabricated in a thick device layer on an SOI wafer. As shown, the two distributed Bragg reflectors (DBR) are formed by etching air slots in silicon structures. The accelerometer is constructed by fixing one DBR and attaching the other to a suspended proof mass. Transmission of the Fabry–Pérot interferometer is monitored by guiding the light in and out using a silicon waveguide.

1.6 Conclusions

Optical MEMS devices that integrate MEMS with optics at the micro scale have shown great potential in sensing, imaging, and communication. In this chapter, we have briefly review the MEMS technology, actuation mechanisms, and their applications in sensing, imaging, and display. Other optical MEMS devices and systems that are equally important but are less covered in this chapter include optical switches, variable attenuators, tunable lasers, tunable lenses and apertures, deformable mirrors, and many others. Many optical MEMS devices including TI's DLP are already in the market, others are rapidly developing in various research laboratories. The future of optical MEMS seems bright.

References

1. O. Brand, Fabrication Technology, in *CMOS—MEMS*, Wiley-VCH Verlag GmbH, by O. Brand, G.K. Fedder (Eds.), Weinheim, Germany, pp. 1–67, 2008.
2. V. Saile, H. H. Gatzen, J. Leuthold, *Micro and Nano Fabrication Tools and Processes*, Springer, Berlin, 2013.
3. F. Laermer and A. Schilp, Method of Anisotropic Etching of Silicon, US Patent, 2003.
4. J. A. Dziuban, *Bonding in Microsystem Technology*, Springer, Netherlands, 2006.
5. D. J. Bell, T. Lu, N. A. Fleck, and S. M. Spearing, MEMS actuators and sensors: Observations on their performance and selection for purpose, *Journal of Micromechanics and Microengineering*, vol. 15, p. S153, 2005.
6. A. Jain, H. Qu, S. Todd, and H. Xie, A thermal bimorph micromirror with large bi-directional and vertical actuation, *Sensors and Actuators A: Physical*, vol. 122, pp. 9–15, 2005.
7. S. T. Todd and H. Xie, An electrothermomechanical lumped element model of an electrothermal bimorph actuator, *Journal of Microelectromechanical Systems*, vol. 17, pp. 213–225, 2008.
8. J. Sun, S. Guo, L. Wu, L. Liu, S.-W. Choe, B. S. Sorg et al., 3D *in vivo* optical coherence tomography based on a low-voltage, large-scan-range 2D MEMS mirror, *Optics Express*, vol. 18, pp. 12065–12075, 2010.
9. S. Pal and H. Xie, Analysis and fabrication of curved multimorph transducers that undergo bending and twisting, *Journal of Microelectromechanical Systems*, vol. 21, pp. 1241–1251, 2012.
10. J. H. Comtois and V. M. Bright, Applications for surface-micromachined polysilicon thermal actuators and arrays, *Sensors and Actuators A: Physical*, vol. 58, pp. 19–25, 1997.
11. L. Li and D. Uttamchandani, Modified asymmetric micro-electrothermal actuator: Analysis and experimentation, *Journal of Micromechanics and Microengineering*, vol. 14, p. 1734, 2004.
12. H. Veladi, R. Syms, and H. Zou, A single-sided process for differentially cooled electrothermal micro-actuators, *Journal of Micromechanics and Microengineering*, vol. 18, p. 055033, 2008.

13. L. Li and D. Uttamchandani, Dynamic response modelling and characterization of a vertical electrothermal actuator, *Journal of Micromechanics and Microengineering*, vol. 19, p. 075014, 2009.
14. Y. Zhang, Y.-S. Choi, and D.-W. Lee, Monolithic micro-electro-thermal actuator integrated with a lateral displacement sensor, *Journal of Micromechanics and Microengineering*, vol. 20, p. 085031, 2010.
15. L. Que, J.-S. Park, and Y. B. Gianchandani, Bent-beam electrothermal actuators-Part I: Single beam and cascaded devices, *Journal of Microelectromechanical Systems*, vol. 10, pp. 247–254, 2001.
16. C. Lee and C.-Y. Wu, Study of electrothermal V-beam actuators and latched mechanism for optical switch, *Journal of Micromechanics and Microengineering*, vol. 15, p. 11, 2004.
17. A. M. H. Kwan, S. Song, X. Lu, L. Lu, Y.-K. Teh, Y.-F. Teh et al., Improved designs for an electrothermal in-plane microactuator, *Journal of Microelectromechanical Systems*, vol. 21, pp. 586–595, 2012.
18. R. Hokari and K. Hane, A varifocal convex micromirror driven by a bending moment, *IEEE Journal of Selected Topics in Quantum Electronics*, vol. 15, pp. 1310–1316, 2009.
19. F. Zimmer, M. Lapisa, T. Bakke, M. Bring, G. Stemme, and F. Niklaus, One-megapixel mono-crystalline-silicon micromirror array on CMOS driving electronics manufactured with very large-scale heterogeneous integration, *Journal of Microelectromechanical Systems*, vol. 20, pp. 564–572, 2011.
20. R. Syms, H. Zou, J. Stagg, and H. Veladi, Sliding-blade MEMS iris and variable optical attenuator, *Journal of Micromechanics and Microengineering*, vol. 14, p. 1700, 2004.
21. C. Lee, Arrayed variable optical attenuator using retro-reflective MEMS mirrors, *IEEE Photonics Technology Letters*, vol. 17, pp. 2640–2642, 2005.
22. T.-S. Lim, C.-H. Ji, C.-H. Oh, H. Kwon, Y. Yee, and J. U. Bu, Electrostatic MEMS variable optical attenuator with rotating folded micromirror, *IEEE Journal of Selected Topics in Quantum Electronics*, vol. 10, pp. 558–562, 2004.
23. J. A. Yeh, S.-S. Jiang, and C. Lee, MOEMS variable optical attenuators using rotary comb drive actuators, *IEEE Photonics Technology Letters*, vol. 18, pp. 1170–1172, 2006.
24. Y. Du, G. Zhou, K. L. Cheo, Q. Zhang, H. Feng, and F. S. Chau, Double-layered vibratory grating scanners for high-speed high-resolution laser scanning, *Journal of Microelectromechanical Systems*, vol. 19, pp. 1186–1196, 2010.
25. U. Krishnamoorthy, D. Lee, and O. Solgaard, Self-aligned vertical electrostatic combdrives for micromirror actuation, *Journal of Microelectromechanical Systems*, vol. 12, pp. 458–464, 2003.
26. D. Hah, S.-Y. Huang, J.-C. Tsai, H. Toshiyoshi, and M. C. Wu, Low-voltage, large-scan angle MEMS analog micromirror arrays with hidden vertical comb-drive actuators, *Journal of Microelectromechanical Systems*, vol. 13, pp. 279–289, 2004.
27. S. Kwon, V. Milanovic, and L. P. Lee, Vertical combdrive based 2-D gimbaled micromirrors with large static rotation by backside island isolation, *IEEE Journal of Selected Topics in Quantum Electronics*, vol. 10, pp. 498–504, 2004.
28. H. Ra, W. Piyawattanametha, Y. Taguchi, D. Lee, M. J. Mandella, and O. Solgaard, Two-dimensional MEMS scanner for dual-axes confocal microscopy, *Journal of Microelectromechanical Systems*, vol. 16, pp. 969–976, Aug 2007.
29. H. M. Chu and K. Hane, Design, fabrication and vacuum operation characteristics of two-dimensional comb-drive micro-scanner, *Sensors and Actuators A: Physical*, vol. 165, pp. 422–430, 2011.
30. J.-W. Jeong, S. Kim, and O. Solgaard, Split-frame gimbaled two-dimensional MEMS scanner for miniature dual-axis confocal microendoscopes fabricated by front-side processing, *Journal of Microelectromechanical Systems*, vol. 21, pp. 308–315, Apr 2012.
31. H. Xie, Y. Pan, and G. K. Fedder, A CMOS-MEMS mirror with curled-hinge comb drives, *Journal of Microelectromechanical Systems*, vol. 12, pp. 450–457, 2003.

32. D. Hah, P. R. Patterson, H. D. Nguyen, H. Toshiyoshi, and M. C. Wu, Theory and experiments of angular vertical comb-drive actuators for scanning micromirrors, *IEEE Journal of Selected Topics in Quantum Electronics*, vol. 10, pp. 505–513, 2004.
33. J. Kim, D. Christensen, and L. Lin, Monolithic 2-D scanning mirror using self-aligned angular vertical comb drives, *IEEE Photonics Technology Letters*, vol. 17, pp. 2307–2309, 2005.
34. J. Kim and L. Lin, Electrostatic scanning micromirrors using localized plastic deformation of silicon, *Journal of Micromechanics and Microengineering*, vol. 15, p. 1777, 2005.
35. W. Piyawattanametha, P. R. Patterson, D. Hah, H. Toshiyoshi, and M. C. Wu, Surface-and bulk-micromachined two-dimensional scanner driven by angular vertical comb actuators, *Journal of Microelectromechanical Systems*, vol. 14, pp. 1329–1338, 2005.
36. Y. Eun, H. Na, J. Choi, J.-i. Lee, and J. Kim, Angular vertical comb actuators assembled on-chip using in-plane electrothermal actuators and latching mechanisms, *Sensors and Actuators A: Physical*, vol. 165, pp. 94–100, 2011.
37. S. Trolier-McKinstry and P. Muralt, Thin film piezoelectrics for MEMS, *Journal of Electroceramics*, vol. 12, pp. 7–17, 2004.
38. P. Muralt, R. Polcawich, and S. Trolier-McKinstry, Piezoelectric thin films for sensors, actuators, and energy harvesting, *MRS Bulletin*, vol. 34, pp. 658–664, 2009.
39. S. Tadigadapa and K. Mateti, Piezoelectric MEMS sensors: State-of-the-art and perspectives, *Measurement Science and Technology*, vol. 20, p. 092001, 2009.
40. Y. Hishinuma and E.-H. Yang, Piezoelectric unimorph microactuator arrays for single-crystal silicon continuous-membrane deformable mirror, *Journal of Microelectromechanical Systems*, vol. 15, pp. 370–379, 2006.
41. I. Kanno, T. Kunisawa, T. Suzuki, and H. Kotera, Development of deformable mirror composed of piezoelectric thin films for adaptive optics, *IEEE Journal of Selected Topics in Quantum Electronics*, vol. 13, pp. 155–161, 2007.
42. S.-J. Kim, Y.-H. Cho, H.-J. Nam, and J. U. Bu, Piezoelectrically pushed rotational micromirrors using detached PZT actuators for wide-angle optical switch applications, *Journal of Micromechanics and Microengineering*, vol. 18, p. 125022, 2008.
43. C. Lee, F.-L. Hsiao, T. Kobayashi, K. H. Koh, P. Ramana, W. Xiang et al., A 1-V operated MEMS variable optical attenuator using piezoelectric PZT thin-film actuators, *IEEE Journal of Selected Topics in Quantum Electronics*, vol. 15, pp. 1529–1536, 2009.
44. Y. Yasuda, M. Akamatsu, M. Tani, T. Iijima, and H. Toshiyoshi, Piezoelectric 2D-optical micro scanners with PZT thick films, *Integrated Ferroelectrics*, vol. 76, pp. 81–91, 2005.
45. T. Kobayashi, R. Maeda, T. Itoh, and R. Sawada, Smart optical microscanner with piezoelectric resonator, sensor, and tuner using Pb (Zr, Ti) O 3 thin film, *Applied Physics Letters*, vol. 90, p. 183514, 2007.
46. U. Baran, D. Brown, S. Holmstrom, D. Balma, W. O. Davis, P. Muralt et al., Resonant PZT MEMS scanner for high-resolution displays, *Journal of Microelectromechanical Systems*, vol. 21, pp. 1303–1310, Dec 2012.
47. H. Miyajima, N. Asaoka, M. Arima, Y. Minamoto, K. Murakami, K. Tokuda et al., A durable, shock-resistant electromagnetic optical scanner with polyimide-based hinges, *Journal of Microelectromechanical Systems*, vol. 10, pp. 418–424, 2001.
48. T. Mitsui, Y. Takahashi, and Y. Watanabe, A 2-axis optical scanner driven nonresonantly by electromagnetic force for OCT imaging, *Journal of Micromechanics and Microengineering*, vol. 16, p. 2482, 2006.
49. A. D. Yalcinkaya, H. Urey, D. Brown, T. Montague, and R. Sprague, Two-axis electromagnetic microscanner for high resolution displays, *Journal of Microelectromechanical Systems*, vol. 15, pp. 786–794, Aug 2006.
50. C.-H. Ji, M. Choi, S.-C. Kim, K.-C. Song, J.-U. Bu, and H.-J. Nam, Electromagnetic two-dimensional scanner using radial magnetic field, *Journal of Microelectromechanical Systems*, vol. 16, p. 989, 2007.

51. H. Urey, S. Holmstrom, and A. D. Yalcinkaya, Electromagnetically actuated FR4 scanners, *IEEE Photonics Technology Letters*, vol. 20, pp. 30–32, 2008.
52. I.-J. Cho and E. Yoon, A low-voltage three-axis electromagnetically actuated micromirror for fine alignment among optical devices, *Journal of Micromechanics and Microengineering*, vol. 19, p. 085007, 2009.
53. A. D. Yalcinkaya, O. Ergeneman, and H. Urey, Polymer magnetic scanners for bar code applications, *Sensors and Actuators A: Physical*, vol. 135, pp. 236–243, 2007.
54. A. D. Yalcinkaya, H. Urey, and S. Holmstrom, NiFe plated biaxial MEMS scanner for 2-D imaging, *IEEE Photonics Technology Letters*, vol. 19, pp. 330–332, 2007.
55. H. Zeng and M. Chiao, Magnetically actuated MEMS microlens scanner for *in vivo* medical imaging, *Optics Express*, vol. 15, pp. 11154–11166, 2007.
56. T.-L. Tang, C.-P. Hsu, W.-C. Chen, and W. Fang, Design and implementation of a torque-enhancement 2-axis magnetostatic SOI optical scanner, *Journal of Micromechanics and Microengineering*, vol. 20, p. 025020, 2010.
57. N. Weber, D. Hertkorn, H. Zappe, and A. Seifert, Polymer/silicon hard magnetic micromirrors, *Journal of Microelectromechanical Systems*, vol. 21, pp. 1098–1106, 2012.
58. R. H. Webb, Confocal optical microscopy, *Reports on Progress in Physics*, vol. 59, pp. 427–471, Mar 1996.
59. D. Huang, E. A. Swanson, C. P. Lin, J. S. Schuman, W. G. Stinson, W. Chang et al., Optical Coherence Tomography, *Science*, vol. 254, pp. 1178–1181, Nov 22, 1991.
60. D. L. Dickensheets and G. S. Kino, Micromachined scanning confocal optical microscope, *Optics Letters*, vol. 21, pp. 764–766, May 15, 1996.
61. S. Kwon and L. P. Lee, Micromachined transmissive scanning confocal microscope, *Optics Letters*, vol. 29, pp. 706–708, Apr 1, 2004.
62. H. J. Shin, M. C. Pierce, D. Lee, H. Ra, O. Solgaard, and R. Richards-Kortum, Fiber-optic confocal microscope using a MEMS scanner and miniature objective lens, *Optics Express*, vol. 15, pp. 9113–9122, Jul 23, 2007.
63. K. Kumar, K. Hoshino, and X. J. Zhang, Handheld subcellular-resolution single-fiber confocal microscope using high-reflectivity two-axis vertical combdrive silicon microscanner, *Biomedical Microdevices*, vol. 10, pp. 653–660, Oct 2008.
64. C. L. Arrasmith, D. L. Dickensheets, and A. Mahadevan-Jansen, MEMS-based handheld confocal microscope for *in-vivo* skin imaging, *Optics Express*, vol. 18, pp. 3805–3819, Feb 15, 2010.
65. X. Chen, X. Y. Xu, D. T. McCormick, K. Wong, and S. T. C. Wong, Multimodal nonlinear endomicroscopy probe design for high resolution, label-free intraoperative imaging, *Biomedical Optics Express*, vol. 6, pp. 2283–2293, Jul 1, 2015.
66. W. Piyawattanametha, M. J. Mandella, H. Ra, J. T. C. Liu, E. Garai, G. S. Kino et al., MEMS based Dual-Axes Confocal Clinical Endoscope for Real Time In Vivo Imaging, in *2008 IEEE/LEOS International Conference on Optical Mems and Nanophotonics*, Freiburg, Germany, pp. 42–43, 2008.
67. S. Murugkar, B. Smith, P. Srivastava, A. Moica, M. Naji, C. Brideau et al., Miniaturized multimodal CARS microscope based on MEMS scanning and a single laser source, *Optics Express*, vol. 18, pp. 23796–23804, Nov 8, 2010.
68. C. A. Patil, C. L. Arrasmith, M. A. Mackanos, D. L. Dickensheets, and A. Mahadevan-Jansen, A handheld laser scanning confocal reflectance imaging-confocal Raman microspectroscopy system, *Biomedical Optics Express*, vol. 3, pp. 488–502, Mar 1, 2012.
69. Y. M. Wang, Y. D. Gokdel, N. Triesault, L. Y. Wang, Y. Y. Huang, and X. J. Zhang, Magnetic-Actuated Stainless Steel Scanner for Two-Photon Hyperspectral Fluorescence Microscope, *Journal of Microelectromechanical Systems*, vol. 23, pp. 1208–1218, Oct 2014.
70. W. Piyawattanametha, R. P. J. Barretto, T. H. Ko, B. A. Flusberg, E. D. Cocker, H. J. Ra et al., Fast-scanning two-photon fluorescence imaging based on a microelectromechanical systems two-dimensional scanning mirror, *Optics Letters*, vol. 31, pp. 2018–2020, Jul 1, 2006.

71. H. J. Ra, W. Piyawattanametha, M. J. Mandella, P. L. Hsiung, J. Hardy, T. D. Wang et al., Three-dimensional *in vivo* imaging by a handheld dual-axes confocal microscope, *Optics Express*, vol. 16, pp. 7224–7232, May 12, 2008.

72. W. Piyawattanametha, H. Ra, M. J. Mandella, K. Loewke, T. D. Wang, G. S. Kino et al., 3-D near-infrared fluorescence imaging using an MEMS-based miniature dual-axis confocal microscope, *IEEE Journal of Selected Topics in Quantum Electronics*, vol. 15, pp. 1344–1350, Sep-Oct 2009.

73. K. C. Maitland, H. J. Shin, H. Ra, D. Lee, O. Solgaard, and R. Richards-Kortum, Single fiber confocal microscope with a two-axis gimbaled MEMS scanner for cellular imaging, *Optics Express*, vol. 14, pp. 8604–8612, Sep 18, 2006.

74. J. T. C. Liu, M. J. Mandella, N. O. Loewke, H. Haeberle, H. Ra, and W. Piyawattanametha et al., Micromirror-scanned dual-axis confocal microscope utilizing a gradient-index relay lens for image guidance during brain surgery, *Journal of Biomedical Optics*, vol. 15, Mar–Apr 2010.

75. J. M. Schmitt, Optical coherence tomography (OCT): A review, *IEEE Journal of Selected Topics in Quantum Electronics*, vol. 5, pp. 1205–1215, Jul–Aug 1999.

76. S. M. Waldstein, H. Faatz, M. Szimacsek, A. M. Glodan, D. Podkowinski, A. Montuoro et al., Comparison of penetration depth in choroidal imaging using swept source vs spectral domain optical coherence tomography, *Eye*, vol. 29, pp. 409–415, Mar 2015.

77. J. Welzel, Optical coherence tomography in dermatology: A review, *Skin Research and Technology*, vol. 7, pp. 1–9, Feb 2001.

78. X. D. Li, S. A. Boppart, J. Van Dam, H. Mashimo, M. Mutinga, W. Drexler et al., Optical coherence tomography: Advanced technology for the endoscopic imaging of Barrett's esophagus, *Endoscopy*, vol. 32, pp. 921–930, Dec 2000.

79. X. J. Mu, G. Y. Zhou, H. B. Yu, J. M. L. Tsai, D. W. K. Neo, A. S. Kumar et al., MEMS electrostatic double T-shaped spring mechanism for circumferential scanning, *Journal of Microelectromechanical Systems*, vol. 22, pp. 1147–1157, Oct 2013.

80. X. J. Mu, G. Y. Zhou, H. B. Yu, Y. Du, H. H. Feng, J. M. L. Tsai et al., Compact MEMS-driven pyramidal polygon reflector for circumferential scanned endoscopic imaging probe, *Optics Express*, vol. 20, pp. 6325–6339, Mar 12, 2012.

81. X. M. Liu, M. J. Cobb, Y. C. Chen, M. B. Kimmey, and X. D. Li, Rapid-scanning forward-imaging miniature endoscope for real-time optical coherence tomography, *Optics Letters*, vol. 29, pp. 1763–1765, Aug 1, 2004.

82. W. Jung, D. T. McCormick, J. Zhang, L. Wang, N. C. Tien, and Z. P. Chen, Three-dimensional endoscopic optical coherence tomography by use of a two-axis microelectromechanical scanning mirror, *Applied Physics Letters*, vol. 88, April 17, 2006.

83. A. D. Aguirre, P. R. Herz, Y. Chen, J. G. Fujimoto, W. Piyawattanametha, L. Fan et al., Two-axis MEMS scanning catheter for ultrahigh resolution three-dimensional and en face imaging, *Optics Express*, vol. 15, pp. 2445–2453, Mar 5, 2007.

84. K. H. Kim, B. H. Park, G. N. Maguluri, T. W. Lee, F. J. Rogomentich, M. G. Bancu et al., Two-axis magnetically-driven MEMS scanning catheter for endoscopic high-speed optical coherence tomography, *Optics Express*, vol. 15, pp. 18130–18140, Dec 24, 2007.

85. K. H. Gilchrist, R. P. McNabb, J. A. Izatt, and S. Grego, Piezoelectric scanning mirrors for endoscopic optical coherence tomography, *Journal of Micromechanics and Microengineering*, vol. 19, p. 095012, Sep 2009.

86. Y. T. Pan, H. K. Xie, and G. K. Fedder, Endoscopic optical coherence tomography based on a microelectromechanical mirror, *Optics Letters*, vol. 26, pp. 1966–1968, Dec 15, 2001.

87. A. Jain, A. Kopa, Y. T. Pan, G. K. Fedder, and H. K. Xie, A two-axis electrothermal micromirror for endoscopic optical coherence tomography, *Ieee Journal of Selected Topics in Quantum Electronics*, vol. 10, pp. 636–642, May–Jun 2004.

88. L. Liu, L. Wu, J. J. Sun, E. Lin, and H. K. Xie, Miniature endoscopic optical coherence tomography probe employing a two-axis microelectromechanical scanning mirror with through-silicon vias, *Journal of Biomedical Optics*, vol. 16, Feb 2011.

89. C. D. Liao and J. C. Tsai, The evolution of MEMS displays, *Ieee Transactions on Industrial Electronics*, vol. 56, pp. 1057–1065, Apr 2009.

90. H. Urey, S. Madhavan, and M. Brown, MEMS microdisplays, *Handbook of Visual Display Technology*, vol. 1–4, pp. 2067–2080, 2012.

91. R. A. Conant, P. M. Hagelin, U. Krishnamoorthy, M. Hart, O. Solgaard, K. Y. Lau et al., A raster-scanning full-motion video display using polysilicon micromachined mirrors, *Sensors and Actuators a-Physical*, vol. 83, pp. 291–296, May 22, 2000.

92. U. Hofmann, J. Janes, and H. J. Quenzer, High-Q MEMS resonators for laser beam scanning displays, *Micromachines*, vol. 3, pp. 509–528, Jun 2012.

93. S. T. S. Holmstrom, U. Baran, and H. Urey, MEMS laser scanners: A review, *Journal of Microelectromechanical Systems*, vol. 23, pp. 259–275, Apr 2014.

94. K. Ikegami, T. Koyama, T. Saito, Y. Yasuda, and H. Toshiyoshi, A biaxial PZT optical scanner for pico-projector applications, *Moems and Miniaturized Systems XIV*, vol. 9375, p. 93750M, 2015.

95. O. Solgaard, F. S. A. Sandejas, and D. M. Bloom, Deformable grating optical modulator, *Optics Letters*, vol. 17, pp. 688–690, May 1, 1992.

96. D. M. Bloom, The grating light valve: Revolutionizing display technology, *Projection Displays III*, vol. 3013, pp. 165–171, 1997.

97. K. R. Kim, J. Yi, S. H. Cho, N. H. Kang, M. W. Cho, B. S. Shin et al., SLM-based maskless lithography for TFT-LCD, *Applied Surface Science*, vol. 255, pp. 7835–7840, Jun 30, 2009.

98. L. J. Hornbeck, Digital light processing(TM) for high-brightness, high-resolution applications, *Projection Displays III*, vol. 3013, pp. 27–40, 1997.

99. P. F. Van Kessel, L. J. Hornbeck, R. E. Meier, and M. R. Douglass, MEMS-based projection display, *Proceedings of the IEEE*, vol. 86, pp. 1687–1704, 1998.

100. T. Brosnihan, R. Payne, J. Gandhi, S. Lewis, L. Steyn, M. Halfman et al., Pixtronix digital micro shutter display technology—A MEMS display for low power mobile multimedia displays, *MOEMS and Miniaturized Systems IX*, vol. 7594, p. 759408, 2010.

101. D. Shim, W. Kim, and H. Choi, Fabrication of vertical moving micro-optical switch for display applications, *MOEMS and Miniaturized Systems XII*, vol. 8616, p. 86160J, 2013.

102. M. W. Miles, Interferometric modulation: MOEMS as an enabling technology for high-performance reflective displays, *Moems Display and Imaging Systems*, vol. 4985, pp. 131–139, 2003.

103. M. W. Miles, Large area fabrication of digital paper reflective displays, *ASID'04: Proceedings of the 8th Asian Symposium on Information Display*, Nanjing, China, Feb 15–17, pp. 406–408, 2004, ISBN: 978-7-81089-529-3.

104. J. Hong, E. Chan, T. Chang, T. C. Fung, B. Hong, C. Kim et al., Continuous color reflective displays using interferometric absorption, *Optica*, vol. 2, pp. 589–597, Jul 20, 2015.

105. N. V. Lavrik, M. J. Sepaniak, and P. G. Datskos, Cantilever transducers as a platform for chemical and biological sensors, *Review of Scientific Instruments*, vol. 75, pp. 2229–2253, Jul 2004.

106. H. B. Yu, G. Y. Zhou, S. K. Sinha, and F. S. Chau, Scanning grating based in-plane movement sensing, *Journal of Micromechanics and Microengineering*, vol. 20, p. 085007, Aug 2010.

107. F. Keplinger, S. Kvasnica, A. Jachimowicz, F. Kohl, J. Steurer, and H. Hauser, Lorentz force based magnetic field sensor with optical readout, *Sensors and Actuators a-Physical*, vol. 110, pp. 112–118, Feb 1, 2004.

108. S. Dante, D. Duval, B. Sepulveda, A. B. Gonzalez-Guerrero, J. R. Sendra, and L. M. Lechuga, All-optical phase modulation for integrated interferometric biosensors, *Optics Express*, vol. 20, pp. 7195–7205, Mar 26, 2012.

109. U. Wallrabe, C. Solf, J. Mohr, and J. G. Korvink, Miniaturized Fourier Transform Spectrometer for the near infrared wavelength regime incorporating an electromagnetic linear actuator, *Sensors and Actuators a-Physical*, vol. 123–124, pp. 459–467, Sep 23, 2005.

110. O. Manzardo, H. P. Herzig, C. R. Marxer, and N. F. de Rooij, Miniaturized time-scanning Fourier transform spectrometer based on silicon technology, *Optics Letters*, vol. 24, pp. 1705–1707, Dec 1 1999.

111. W. Wang, J. P. Chen, A. S. Zivkovic, Q. A. A. Tanguy, and H. K. Xie, A compact fourier transform spectrometer on a silicon optical bench with an electrothermal MEMS mirror, *Journal of Microelectromechanical Systems*, vol. 25, pp. 347–355, Apr 2016.

112. W. Wang, J. P. Chen, A. S. Zivkovic, and H. K. Xie, A Fourier transform spectrometer based on an electrothermal MEMS mirror with improved linear scan range, *Sensors*, vol. 16, Oct 2016.

113. K. Yu, D. Lee, U. Krishnamoorthy, N. Park, and A. Solgaard, Micromachined Fourier transform spectrometer on silicon optical bench platform, *Sensors and Actuators a-Physical*, vol. 130, pp. 523–530, Aug 14, 2006.

114. D. Khalil, Y. Sabry, H. Omran, M. Medhat, A. Hafez, and B. Saadany, Characterization of MEMS FTIR Spectrometer, *Moems and Miniaturized Systems X*, vol. 7930, pp. 79300J-1–79300J-10, 2011.

115. O. Manzardo, R. Michaely, F. Schadelin, W. Noell, T. Overstolz, N. De Rooij et al., Minature lamellar grating interferometer based on silicon technology, *Optics Letters*, vol. 29, pp. 1437–1439, Jul 1, 2004.

116. J. Strong and G. A. Vanasse, Lamellar grating far-infrared interferomer, *Journal of the Optical Society of America*, vol. 50, pp. 113–118, 1960.

117. C. A. Savran, T. P. Burg, J. Fritz, and S. R. Manalis, Microfabricated mechanical biosensor with inherently differential readout, *Applied Physics Letters*, vol. 83, pp. 1659–1661, Aug 25, 2003.

118. C. A. Savran, A. W. Sparks, J. Sihler, J. Li, W. C. Wu, D. E. Berlin et al., Fabrication and characterization of a micromechanical sensor for differential detection of nanoscale motions, *Journal of Microelectromechanical Systems*, vol. 11, pp. 703–708, Dec 2002.

119. G. Zhou, K. K. L. Cheo, Y. Du, and F. S. Chau, An optically interrogated microgyroscope using an out-of-plane lammelar grating, *Sensors and Actuators A: Physical*, vol. 154, pp. 269–274, 2009.

120. N. C. Loh, M. A. Schmidt, and S. R. Manalis, Sub-10 cm(3) interferometric accelerometer with nano-g resolution, *Journal of Microelectromechanical Systems*, vol. 11, pp. 182–187, Jun 2002.

121. E. B. Cooper, E. R. Post, S. Griffith, J. Levitan, S. R. Manalis, M. A. Schmidt et al., High-resolution micromachined interferometric accelerometer, *Applied Physics Letters*, vol. 76, pp. 3316–3318, May 29, 2000.

122. C. Ataman, H. Urey, and A. Wolter, A Fourier transform spectrometer using resonant vertical comb actuators, *Journal of Micromechanics and Microengineering*, vol. 16, pp. 2517–2523, Dec 2006.

123. S. R. Manalis, S. C. Minne, A. Atalar, and C. F. Quate, Interdigital cantilevers for atomic force microscopy, *Applied Physics Letters*, vol. 69, pp. 3944–3946, Dec 16, 1996.

124. M. Born, E. Wolf, Reference to the first edition, in *Principles of Optics* (Sixth (Corrected) Edition), Pergamon, 1980, ISBN 9780080264820.

125. G. C. Hill, R. Melamud, F. E. Declercq, A. A. Davenport, I. H. Chan, P. G. Hartwell, and B. L. Pruitt, SU-8 MEMS Fabry–Pérot pressure sensor, *Sensors and Actuators A: Physical*, vol. 138, pp. 52–62, 2007.

126. K. Takahashi, H. Oyama, N. Misawa, K. Okumura, M. Ishida, and K. Sawada, Surface stress sensor using MEMS-based Fabry–Pérot interferometer for label-free biosensing, *Sensors and Actuators B: Chemical*, vol. 188, pp. 393–399, 2013.

127. X. Wu, C. Jan, and O. Solgaard, Single-crystal silicon photonic-crystal fiber-tip pressure sensors, *Journal of Microelectromechanical Systems*, vol. 24, pp. 968–975, Aug 2015.

128. J. C. Xu, X. W. Wang, K. L. Cooper, and A. B. Wang, Miniature all-silica fiber optic pressure and acoustic sensors, *Optics Letters*, vol. 30, pp. 3269–3271, Dec 15, 2005.

129. C. Jan, W. Jo, M. J. F. Digonnet, and O. Solgaard, Photonic-crystal-based fiber hydrophone with sub-100 mu Pa/root Hz pressure resolution, *Ieee Photonics Technology Letters*, vol. 28, pp. 123–126, Jan 15, 2016.

130. O. Kilic, M. J. F. Digonnet, G. S. Kino, and O. Solgaard, Miniature photonic-crystal hydrophone optimized for ocean acoustics, *Journal of the Acoustical Society of America*, vol. 129, pp. 1837–1850, Apr 2011.

131. I. W. Jung, B. Park, J. Provine, R. T. Howe, and O. Solgaard, Highly sensitive monolithic silicon photonic crystal fiber tip sensor for simultaneous measurement of refractive index and temperature, *Journal of Lightwave Technology*, vol. 29, pp. 1367–1374, May 1, 2011.

132. M. C. Y. Huang, Y. Zhou, and C. J. Chang-Hasnain, A surface-emitting laser incorporating a high-index-contrast subwavelength grating, *Nature Photonics*, vol. 1, pp. 119–122, 2007.

133. B. Park, I. W. Jung, J. Provine, A. Gellineau, J. Landry, R. T. Howe et al., Double-layer silicon photonic crystal fiber-tip temperature sensors, *Ieee Photonics Technology Letters*, vol. 26, pp. 900–903, May 1, 2014.

134. R. St-Gelais, J. Masson, and Y. A. Peter, All-silicon integrated Fabry-Peacuterot cavity for volume refractive index measurement in microfluidic systems, *Applied Physics Letters*, vol. 94, Jun 15, 2009.

135. K. Zandi, J. A. Belanger, and Y. A. Peter, Design and demonstration of an in-plane silicon-on-insulator optical MEMS Fabry–Pérot based accelerometer integrated with channel waveguides, *Journal of Microelectromechanical Systems*, vol. 21, pp. 1464–1470, Dec 2012.

2

MEMS optical scanners for laser projection display

Hiroshi Toshiyoshi

Contents

2.1 MEMS for light modulation

Optical applications of MEMS (microelectromechanical systems) date back to the early stage of silicon micromachining, when a bit-map image projector was demonstrated using an array of 50 micron-size tilt mirrors in 1975 [1]. A review paper on silicon micro mechanisms published in 1982 covers a combination of a miniaturized Galvano scanner and an array of electrostatic cantilevers, both developed by the anisotropic wet etching of single crystalline silicon [2]. Later, in the 1990s, an electrostatic micromirror array was made on top of integrated circuits to construct an electrically addressable digital mirror device (DMD), which is called a digital light processing (DLP) device today [3]. A similar integrated mirror device was also reported as a pattern generator for maskless photolithography [4]. In recent years, MEMS technologies have been used to develop electrostatic tunable gratings and shutters as a light-valve array [5–7]. Various types of micro-optical scanners of both one- and two-degrees-of-freedom have been studied to construct a compact package of laser scanning displays (LSDs), which is usually referred to as a pico projector in the field of consumer electronics. Similar technology is also used to develop head-mount type displays [8].

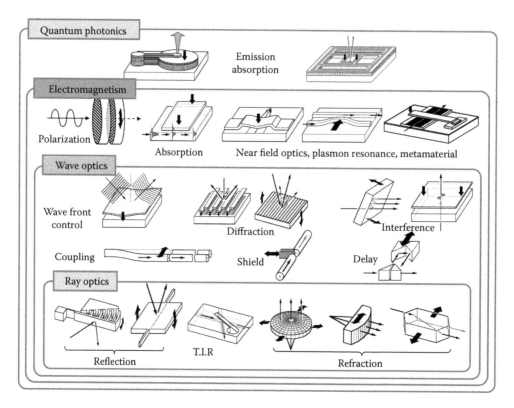

Figure 2.1 Light modulation by optical MEMS devices.

Figure 2.1 illustrates different classes of light modulation through MEMS mechanisms. The ray optics is the most classic category of spatial light modulation, where the direction of light is controlled by reflection for, for example, image displays and optical fiber cross-connects. Besides reflection, classic optical approaches such as refraction and total internal refraction are also used to alter the optical path by the mechanical motion of optical components such as lenses and prisms. In the wave optics domain, where traveling light is considered as a propagating Gaussian wave, various modulation principles are studied, including the control of the coupling efficiency between a pair of optical fibers, wave front control for aberration correction, and tunable gratings that control the direction of diffraction as well as the intensity of diffracted light. Optical interference is also a part of the wave optics. In this domain, mechanical motion as small as the wavelength of interest has an enormous effect on the optics, for example, for controlling transmission through cavity length adjustment of a Fabry–Pérot interferometer. The domain of electromagnetic wave optics provides higher degrees of light modulation through its polarization and near-field evanescent coupling. Quantum optics is the last and the highest category of optical modulation, where photons interact with electrons through mechanical motion; for instance, a wavelength tunable laser diode is developed by integrating an electrostatically controllable cantilever as a movable Bragg reflector for a vertical cavity surface-emitting laser (VCSEL) [9]. A similar mechanism can also be used as a wavelength tracking photo diode [10].

Among such a wide range of optical modulation schemes, reflection is the most popular principle as a scanning engine for laser projection displays. Figure 2.2 summarizes projection optics that is based on different types of optical scanners In Figure 2.2a, an array of digital mirrors is used to modulate the intensity of each pixel and to immediately create an image plane through a one-on-one allocation of a mirror to a pixel. Texas Instruments commercialized this type of light-valve called a DMD (digital mirror device) in 1996 for their projection displays [3]. Mirrors cannot be smaller than the physical limit of optical diffraction or the integrated electronics for individual mirror addressing; hence the chip has a limit of a minimum size, thereby leading to a higher

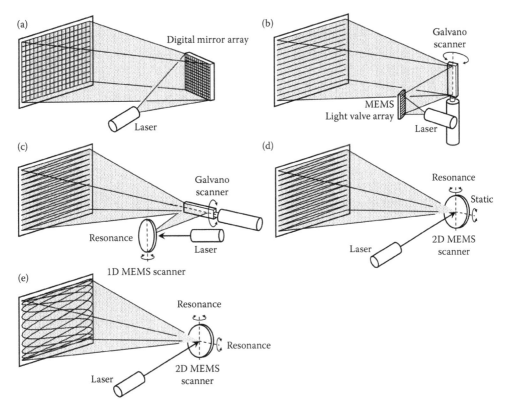

Figure 2.2 Image projection schemes based on MEMS optical scanner: (a) 1:1 bitmap projection by mirror array; (b) 1D light valve combined with a Galvano scanner; (c) a pair of resonant and non-resonant scanners; (d) 2D-MEMS scanner with a resonant and a non-resonant; and (e) 2D-MEMS resonant scanner (Lissajous scan).

production cost. On the other hand, a lower cost is expected for the scheme shown in Figure 2.2b, because the light valve occupies a smaller area due to the one-dimensional design, although there is an additional cost for another Galvano scanner to sweep the one-dimensional light intensity in the orthogonal direction to create a 2D image. For instance, Silicon Light Machines, Inc., used electrostatically addressable gratings as a 1D light valve and combined this with a scanner to produce 2D images [5]. This technology was later acquired by SONY to develop a large projection display of 2005 inches in the World Exposition in Aichi 2005 [6]. Instead of using a one-dimensional light valve, the scheme in Figure 2.2c uses a fast optical scanner for the horizontal axis, and a rather slow Galvano scanner for the vertical axis. The entire 2D space can be filled by the raster scanning, but the pixel intensity should be controlled by modulating the output of the lasers at high speed. Even smaller optics can be made as shown in Figure 2.2d, by replacing the two-mirror optics with a single MEMS optical scanner that has two-degrees-of-freedom in spatial modulation. In general, the use of a single laser source and a single optical scanner may lose the pixel-wise intensity due to the time-division control of a pixel over the two-dimensional screen. However, such optics has become more feasible these days since the emergence of relatively powerful laser diodes. In parallel with the analogy to the electron beam scan in a cathode ray tube (CRT), the scheme of Figure 2.2d usually uses a fast resonance of MEMS for the horizontal scan, while a slow off-resonance motion is used for the vertical scan, by which an image is constructed by stacking hundreds of horizontal lines in the vertical direction. The last scheme shown in Figure 2.2e is an alternative to such a method, but only mechanical resonance is used for both axes; the trajectory of the beam spot develops a Lissajous figure, and the projection system requires an image frame converter to resequence the pixel intensity data according to the position of the beam spot on the screen.

2.2 Performance index for MEMS optical scanner

The number of resolvable spots decreases when a beam is reflected off an optical scanner with a reduced mirror diameter due to the angle of diffraction θ_{diff} as shown in Figure 2.3. The diffraction angle of a Gaussian beam is written as [11]

$$\theta_{\text{diff}} = \frac{2\lambda}{\pi R}, \tag{2.1}$$

where R is the radius of the mirror, which is set to be equal to the beam waist w_0 of the propagating Gaussian beam, and λ is the wavelength of the light. Given the full optical scan angle of the spatial scanner to be $\theta_{\text{p-p}}$, the number of resolvable spots is written as

$$N = \frac{\theta_{\text{p-p}}}{\theta_{\text{diff}}} = \frac{\pi R}{2\lambda} \theta_{\text{p-p}}. \tag{2.2}$$

Here, we use a red light of wavelength $\lambda = 633$ nm reflected by a mirror of $R = 0.5$ mm, where the scan angle needed to fulfill the VGA (Video Graphic Array, 640 × 480) is

$$\theta_{\text{p-p}} = \frac{2\lambda}{\pi R} N = \frac{2 \times 630 \text{ nm}}{3.14 \times 0.5 \text{ mm}} \times 640 = 29.4°. \tag{2.3}$$

The optical angle is twice as large as the scanner's mechanical deflection when the incident light is perpendicular to the rotation axis; hence the scanner needs to deliver a mechanical full angle of 14.7° or larger. Such a scan angle is possible with recent MEMS scanners without significant difficulty, as much effort has already been given to developing scanners of large static deflection angles for free-space cross-connect applications [12–14]. For practical purposes, mirror diameter D is usually used in place of mirror radius, and $\theta_{\text{p-p}} \cdot D = 4\lambda N/\pi$ is used as a key performance index for scanners [15].

Designing a fast scanner, on the other hand, remains a technical challenge. A scanner's oscillation frequency f_{H} to draw N_{Y} lines at a frame rate of f_{FR} needs to be $f_{\text{H}} = f_{\text{FR}} \times N_{\text{Y}}$, which is 14.4 kHz for a VGA class image presented at the NTSC standard frame rate of 30 fps when only half the scan period is used for image projection (one-way scan). The scan speed can be lowered to 7.2 kHz when the laser beam is round-trip scanned. Considering a 2D raster scan, the scan speed for the vertical direction of an image to complete a frame can be as low as 30 Hz for a one-way scan (15 Hz for round-trip scan). Due to this large frequency ratio between the horizontal and vertical scans, it is generally difficult to develop a 2D scanning mechanism using the same MEMS fabrication technology.

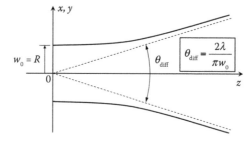

Figure 2.3 Gaussian beam model of light reflected by optical scanner.

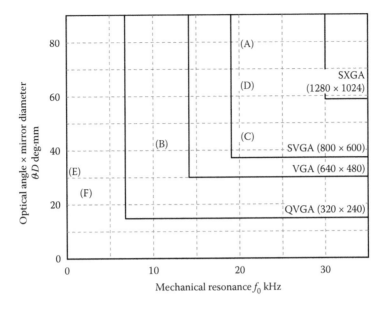

Figure 2.4 Key performance index of MEMS optical scanners for image projection. (Modified from A. D. Yalcinkaya et al. *J. Microelectromech. Syst.*, vol. 15, no. 4, pp. 789–794, 2006. Symbols—Reference: A-[15], B-[19], C-[23], D-[25], E-[28], F-[29].)

Figure 2.4 compares the performance of MEMS scanners; the vertical axis plots the key performance index $\theta_{p-p} \cdot D$, while the horizontal axis plots the scan speed [15]. In the same plot, we also show the performance needed to fulfill the QVGA (320 × 240), VGA (640 × 480), SVGA (800 × 600), and SXGA (1280 × 1024) formats. Reported MEMS scanners are capable of handling SVGA class projection. The plot in the figure mainly uses the MEMS scanners for raster scanning, where the horizontal scan lines are produced in a progressive manner at the mechanical resonance of the scanner. On the other hand, image displays based on vector scanning have also been developed [16], in which both the X- and Y-axes of the scanner are mechanically excited off the resonance, hence the beam spot trajectory produces universal patterns.

Representing a MEMS optical scanner with a quadratic oscillation system of mass m, damping coefficient c, and elastic constant k, its oscillation behavior is predictable by the equation of motion

$$m\ddot{x} + c\dot{x} + kx = F, \tag{2.4}$$

where F is the external force applied to the system. The amplitude of oscillation A at the resonant frequency is known to be $Q\left(=\sqrt{km}/c\right)$ times greater than the static displacement F/k under a force equivalent to the DC force F, hence the amplitude is written as

$$A = \frac{F}{k} \times Q = \frac{F}{k} \frac{\sqrt{km}}{c} = \frac{F}{2\pi f_0 c}, \tag{2.5}$$

where $f_0 = (1/2\pi)\sqrt{(k/m)}$ is the resonant frequency of the system when the effect of viscosity is ignored. From this governing equation, one can tell that A and f_0 cannot be tuned to greater values at the same time as long as the parameters F and c remain the same, due to the trade-off relationship as shown in Figure 2.5. In other words, either a large actuation force or a smaller loss should be used to improve the performance.

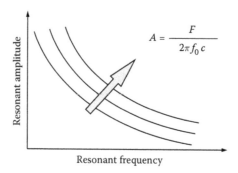

$$A = \frac{F}{2\pi f_0 c}$$

Figure 2.5 Trade-off relationship between the resonant amplitude and frequency.

As discussed, the number of resolvable beam spots is proportional to the product of the mirror diameter and the scan angle. In addition to these, the number also depends on the mirror's static and dynamic flatness and its temperature dependence. On the other hand, the scanner's resonant frequency that determines the upper limit of the scanner's response depends on the mirror diameter (or the moment of inertia) and the mechanical rigidity of the suspensions. The design parameters and the performance are closely related, and they can be presented in the cross-referenced style as shown in Figure 2.6, which is usually referred to as the parameter constellation [17]. It also tells us that the actuation mechanism for the scanner must be designed to have large force (or torque) per footprint to fulfill requirements such as a large scan angle and fast speed. For this reason, powerful actuation principles such as piezoelectric [18–20] or electromagnetic [15,21–25] are usually preferred. Thermoelectric actuation has the advantage of generating a large mechanical displacement, but the operation speed is limited by the finite heat mass [26]. An electrostatic mechanism is simple in structure, but the output force (torque) is usually smaller than that of other operation principles, hence it is usually used with mechanisms such as the vertical comb electrodes and the multi-body oscillation mechanisms [27–31] to enlarge the oscillation.

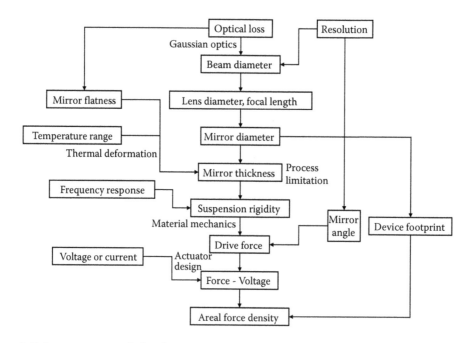

Figure 2.6 Parameter constellation for MEMS optical scanner.

2.3 Actuation principles for MEMS optical scanners

Table 2.1 compares the actuation mechanisms that could be used for MEMS optical scanners. For quantitative analysis, we use the reported representative values to estimate the output force per device footprint.

An electrostatic mechanism usually uses a pair of electrodes separated by a tiny air gap in which the electrostatic Maxwell's pressure $(1/2)\varepsilon_0 E^2$ is generated, where ε_0 is the dielectric constant of vacuum and E is the strength of the electrical field. Presuming that a DC voltage of 1–100 V is applied between a gap of 1–10 μm, the electrostatic pressure is calculated to be in a range of $10^1 \sim 10^4$ N/m^2. The electromagnetic mechanism uses the interaction between the magnetic flux density B and the electrical current I, that is, IB/w (per length) as an operation principle, where the output force is also normalized by the width w of the electrical wire by assuming a coil of large turns. Provided that a strong permanent magnet such as NdFeB (10 kOe) and electrical current of 1 mA–1 A through a coil of wire width of 10 μm are present, electromagnetic pressure is calculated to be 10^1–10^3 N/m^2, which is equivalent to that of the electrostatic mechanism.

Compared with these two mechanisms, the other actuation principles, namely piezoelectric, piezostrictive or magnetostrictive, are expected to deliver a larger force by more than four orders of magnitude. For instance, the piezoelectric strain of 0.01%–0.1% for a material of Young's modulus of 100 GPa is equivalent to a mechanical stress of 10^7–10^9 N/m^2 output. However, these device mechanisms use the extension or contraction of a rigid structure, hence the direct output displacement of a millimeter-scale body is as small as a micron level. Therefore, such a displacement is usually converted into a large angular deflection through a unimorph structure.

Electrothermal actuation uses the thermal expansion of a material through Joule heating, and the absolute value of expansion is usually written as $E\alpha\Delta T$, where E is the Young's modulus, α the linear coefficient of expansion, and ΔT the change of temperature. Thermal expansion also gives a rather large output force of 10^6–10^8 N/m^2, when α and ΔT are set to be 10 ppm and 100°C, respectively. Due to the same reason as for piezoelectricity, thermal actuation also usually uses a unimorph mechanism to amplify the output deflection.

Table 2.1 Comparison of actuation mechanisms for MEMS optical scanner

	Electrostatic	Electromagnetic	Piezoelectric	Electrothermal
Principle				
Areal Force Density	$(1/2)\varepsilon_0 E^2$	IB/w	$E\varepsilon$	$E\alpha\Delta T$
	E: Electric Field ε_0: Dielectric Constant	I: Drive Current B: Magnetic Flux w: Coil Wire Width	E: Young's Modulus ε: Strain	E: Young's Modulus α: Thermal Expansion Coefficient ΔT: Temperature Change
	10^1–10^4 N/m^2	10^1–10^3 N/m^2	10^7–10^8 N/m^2	10^6–10^8 N/m^2
Typical Design	Electrostatic Gap $g = 1$–10 μm Drive Voltage $V = 1$–100 V	Permanent Magnet $H_c \sim$10 kOe Drive Current $I = 1$ mA $\sim$1 A Coil Wire Width w–10 μm	Young's Modulus $E \sim$100 GPa Piezoelectric Strain $\varepsilon = 0.01\%$–0.1%	Young's Modulus $E \sim$100 GPa Thermal Expansion Coefficient $\alpha \sim$10 ppm Temperature Change $\Delta T = 1$–100°C

We will look further into the detailed comparison of these four actuation mechanisms by using Table 2.2. Despite its small output force, the electrostatic method is widely used in MEMS devices because of the simplicity of the structure and its compatibility with the capacitive displacement sensing that is used in gyroscopes, accelerometers, and silicon microphones. Regardless of a relatively high voltage (10–100 V) for an electrostatic drive, total power consumption is almost negligible due to the capacitive nature of the load. Electrostatic displacement is not suitable to generate a large static deflection angle for optical scanners, but it is rather widely used for raster scanning at the resonant frequency of the mechanism.

Electromagnetic actuation, on the other hand, is usually operated at a relatively low voltage (1–10 V), but power consumption is high due to the heat dissipation through the resistive load. Therefore, one may need to pay attention to the thermal drift of the scanner performance which is associated with the Joule heat. Such a problem might be solved by using the coil not on the suspended scanner structure, but on the fixed substrate; in such a case, however, the scanner mechanism may need to carry a permanent magnet for magnetic interaction, which may hamper the high frequency performance of the scanner. Hysteresis of electromagnetic operation can be avoided by using an air-core for the coil, and the mechanical deflection angle can be controlled at high repeatability. The minimum dimension of the structure is limited by the volume of the permanent magnet, but the output force is relatively large, hence it is most widely used for laser scan optics.

Besides the relatively large output force of piezoelectricity, it also has the advantage of being a smart material for properties such as strain sensing for angle detection. However, it has drawbacks in DC operation, during which the static displacement may drift when a voltage is applied for a long time and the material properties are affected by temperature as well as humidity. Such problems are partially resolved by using an electrical insulating material between the piezoelectric active layer and the metallic electrodes. The electrode for piezoelectric actuation is most effectively placed for large deformation at a place where the strain becomes large; nonetheless it sometimes results in material failure such as delamination of films after a long operation period of many cycles.

Electrothermal actuation also produces large deformation of microstructures at a potential penalty of relatively large power dissipation through Joule heating, hence it is not suitable for mobile applications driven by batteries. Operation speed could be made faster through miniaturization, but is substantially limited by the heat mass. To the best of our knowledge, no electrothermal scanner has been reported for practical operation speeds higher than 20 kHz.

2.4 Examples of MEMS optical scanners

2.4.1 Electrostatic type

An electrostatic actuator uses the Maxwell's force generated by the voltage applied in between a pair of electrodes separated by a micron-scale air gap, which has high compatibility with the semiconductor micromachining. Various types of electrostatic optical scanners have been reported, mainly in the field of optical fiber communication technology [12–14]. The electrode structure shown in Figure 2.7a is usually referred to as vertical comb electrodes that are used to generate relatively large out-of-plane motion by the electrostatic force, particularly at a mechanical resonant frequency. Vertical comb electrodes can be arranged into a 2D scanner as schematically shown in Figure 2.7b, by adopting a double-gimbal structure with orthogonal sets of torsion beams. One of the technical challenges in designing a 2D optical scanner is the layout of electrical interconnections for the inner electrodes on the gimbals because such interconnection wires are usually arranged on the thin torsion beams. A device reported in [27] and [28] uses trench etching of silicon and trench refill with an insulating material to electrically isolate the electrodes while mechanically holding the electrodes with the movable structures; on this device, relatively large mechanical angles of 30° and 8° are reported in the orthogonal directions excited at resonant frequencies of 1.45 and

Table 2.2 Detail comparison of actuation mechanisms

	Electrostatic	Electromagnetic	Piezoelectric	Electrothermal
Principle	Electrostatic attractive force between conductors	Lorentz force between current and magnetic field	Unimorph deformation by piezoelectric strain	Thermal expansion by Joule heat
Structure	A pair of electrodes with air gap	Coil and permanent magnet	Piezoelectric layer with top/bottom electrodes	Materials with different thermal expansion coefficients
Process	Silicon DRIE and metallization	Multiple layers of metal and insulator for coil	Sputtering, sol-gel, spray coating, sintering	Multiple layers of metal and insulator for heater
Output Force	Small	Large	Large	Large
Frequency Range	DC-Resonance	DC-Resonance	Low frequency-Resonance	Low Frequency-Resonance
Displacement Sensing	Capacitive, piezoresistive	Current monitor, Hall resistance	Piezoelectric	N.A.
Power Consumption	Ultra small (Capacitive)	Large (Resistive)	Small (Capacitive)	Large (Resistive)
Handling Caution	Vulnerable to electrostatic discharge	Magnetization	Moisture	Overheat
Hysteresis	Hysteresis by electrostatic charge up	No hysteresis for air-core coil	Time-dependent drift in DC mode	Temperature dependence of heater resistance
Inspection	Indistinguishable wire-breakdown	Simple short-circuit inspection	Performance degradation by electrical breakdown	Simple short-circuit inspection
Mass Producibility	Mass production compatible	Mass production compatible	Throughput limited by piezo layer formation	Mass production compatible
Temperature Dependence	Small	Small	Limited by phase transition of piezoelectric film	Bimorph structure sensitive to temperature change
Failure Mode	Particle trapping, short circuit, pull-in stiction	Over heat, migration, counter electromotive force	Dielectric breakdown, delamination, time dependent degradation	Over heat, migration
Note	Requires mechanical resonance to enhance scan angle	Size limited by permanent magnet assembly	Limited resource of foundry	Requires relatively large power

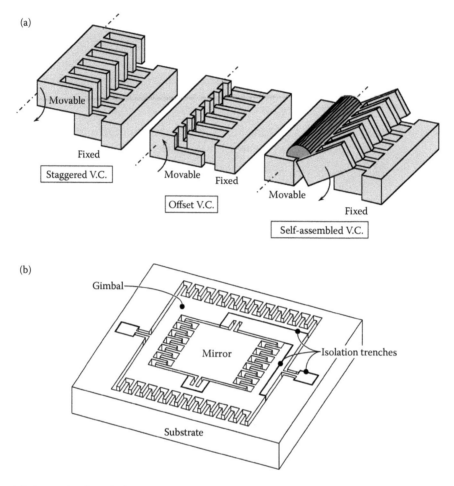

Figure 2.7 Example of optical scanner with electrostatic vertical-comb electrodes: (a) three different configurations including staggered, offset, and self-assembled vertical combs and (b) double gimbal implementation of 2D-scanner.

1.8 kHz, respectively. Due to the operation at resonance for both axes, the scanned beam spot draws Lissajous patterns as schematically shown in Figure 2.2c.

Due to the small output force per electrode pair, most electrostatic actuators are designed with a large number of comb-electrode pairs to multiply the force. Besides the output force, mechanical stroke can also be enhanced by using an amplitude magnification mechanism based on the multi-degrees-of-freedom oscillation system as shown in the SEM (scanning electron microscope) image in Figure 2.8 [31]. In this design, the actuator plate of moment of inertia J_1 is supported with a relatively thick torsion beam of elastic rigidity k_1, while the inner mirror plate of moment of inertia J_2 is held with a thinner pair of beams of rigidity k_2, where the design parameters are tuned to set the resonant frequency of the inner mirror to be equal to that of the outer gimbal (approximately $k_1/J_1 = k_2/J_2$). At the fundamental resonant mode, the oscillation amplitude of the mirror plate is magnified almost k_2/J_2 times as much as that of the actuator plate. A scanner developed for laser display [31] is reported to deliver an optical angle of 30° at 22.3 kHz with an angle magnification factor of 10.

2.4.2 Electromagnetic type

Figure 2.9 illustrates the fundamental design for an electromagnetic 2D scanner that uses the Lorentz force interaction between the microcoil integrated on the scanner gimbal and the

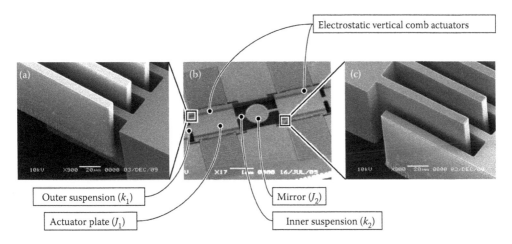

Figure 2.8 Amplitude enhancement structure based on multi-body oscillation system: (a) close-up view of the movable combs that are pulled downward by the offset angle; (b) entire view of the scanner; and (c) close-up view of the movable combs on the opposite side that are pulled upward.

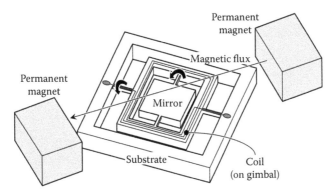

Figure 2.9 Principle of electromagnetic 2D scanner based on the interaction between coil and permanent magnet.

permanent magnets located outside of the scanner chip [15,21–23]. In contrast to this, it is also possible to have a permanent magnet on the scanner platform while having an external coil [24], but the dynamic performance of the scanner would be degraded due to the added mass of the magnet. Despite the two degrees of freedom in operation, only one coil can induce mechanical oscillations for both the X- and Y-directions owing to the fact that the magnetic flux is arranged at 45° with respect to the currents flowing through the square coil, where the Lorentz force is always perpendicular to the directions of the current and the magnetic flux. A device reported in [15] delivers relatively large optical angles of 53° in the horizontal direction at 21.3 kHz and 65° in the vertical direction at 60 Hz, which is equivalent to the SVGA class projection. Electromagnetic scanners are most widely used as a scan engine for hand-held projectors such as MicroVision's pico-projector [25].

2.4.3 Piezoelectric type

PZT or piezoelectric zirconate titanate is a piezoelectric material which is frequently used in MEMS devices due to its large piezoelectric constant compared with those of other materials such as AlN and ZnO. It is usually used in the form of a unimorph actuator as schematically shown in Figure 2.10a, where the piezoelectric layer is laminated with two metallic electrodes (usually platinum on a titanium adhesion promoting layer) on the top and bottom surfaces. Depending on the

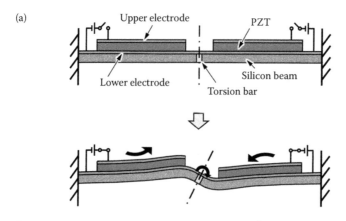

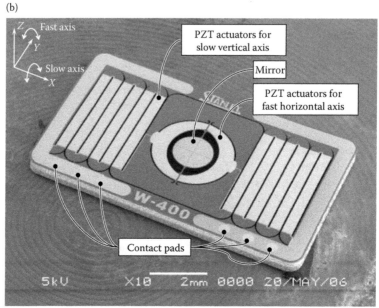

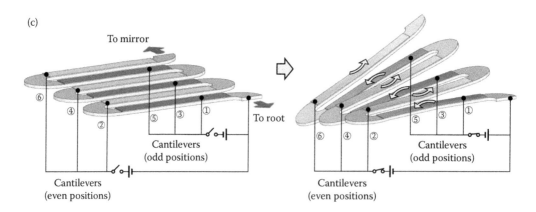

Figure 2.10 MEMS optical scanner with piezoelectric unimorph actuation mechanism: (a) layer structure of the unimorph mechanism; (b) SEM image of the 2D scanner; and (c) meandering suspension to accumulate angular displacement for the vertical axis.

polarity of the applied voltage, either tensile or compressive stress is induced on the top surface of the silicon beam, which is forced to bend up or down, respectively. Such a mechanism is used to twist the torsion beam or to deflect the gimbal for a 2D scanner mechanism [18]. The circular mirror (diameter 1 mm) in the middle part shown in Figure 2.10b is suspended with a pair of silicon beams that are 10 μm wide and 200 μm long, which are connected with an arc-shaped cantilever to induce the twisting motion to the beams. The meandering suspensions located on both sides of the gimbal are used to tilt the gimbal frame in the orthogonal direction. The electrode sets on the meandering suspensions are alternatingly grouped by even and odd orders; voltages of opposite polarities are applied to the groups of piezoelectric suspensions such that the mechanical angle is accumulated through the upward and downward deflections of the beams to deliver a large static angle to the gimbal structure, as schematically illustrated in Figure 2.10c.

One of the benefits of piezoelectric material is that the strain could be detected by the piezoelectric effect, hence it is possible to integrate an angle sensor within the scanning mechanism. The location of the strain sensor is determined by using FEM (finite element method) analysis by providing mechanical strain to the piezoelectric unimorph structure and observing the distribution of the piezoelectrically generated polarization. Piezoelectric sensors are placed to effectively pick up the strain where the polarization is calculated to be large. In the case of the structure shown in Figure 2.11a, strain in the gimbal is shown to be large at the root of the twisting suspension, and reference 19 used a sensor integrated near the suspension as shown in the SEM image of Figure 2.11b to generate a trigger signal to synchronize the control circuit for the MEMS scanner. The distribution of the strain or polarization also gives a hint for the effective layout for the piezoelectric driving electrode, as implemented in the same SEM image. Large electrodes on the beam do not always produce a large deflection angle, but it is empirically known that the electrode patterns should be confined within the piezoelectric polarization pattern of the same polarity. Electrical fields protruding outside the boundary of polarization may produce strain in the counter direction, resulting in a small deflection angle.

Figure 2.12 illustrates a block diagram of the projection system based on the piezoelectric pickup and drive of the optical scanner [19]. The image data provided from a PC are processed by the frame converter to rearrange the data sequence in the order of the laser scan, and a set of electrical currents is generated to drive the laser sources. The MEMS scanner uses a total of four voltages for the horizontal and vertical scans; the feedback system maximizes the horizontal scan angle by tuning the operation frequency. The signal from the piezoelectric sensor for the horizontal drive mechanism of the scanner is used as a synchronizing trigger for the rest of the system, so that the projected images would be clearly presented without causing blur between the scan lines projected during the forward and backward travels.

Figure 2.13 is the photograph of the projection system and the image projected onto the screen [19]. The mechanical resonant frequencies of the developed scanner are 25 and 500 Hz for the horizontal and vertical directions, respectively. The optical scan angle is 40° for the horizontal direction

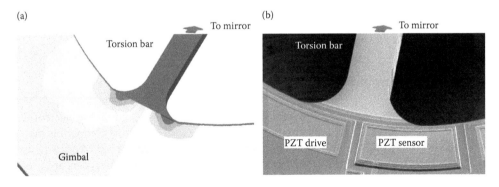

Figure 2.11 Scanner angle sensing mechanism through piezoelectric strain gauge effect: (a) FEM simulation result to show the distribution of mechanical strain when the bar is twisted and (b) SEM image of the PZT sensor, which is located at a place where the mechanical strain is large.

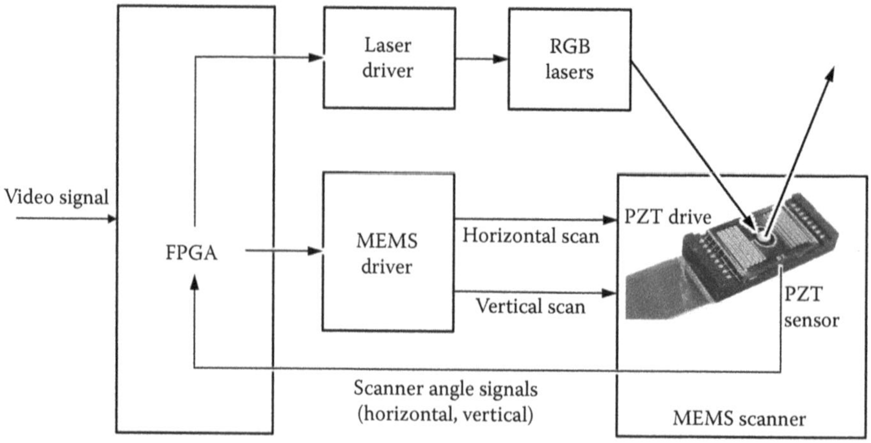

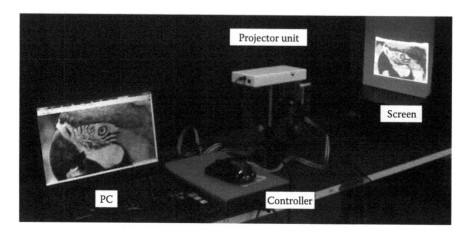

Figure 2.12 Control system for laser scan display using a piezoelectric optical scanner.

Figure 2.13 Example of projection display using MEMS optical scanner.

at 20 V at the resonance, while that for the vertical axis is 30° at 20 V with a frequency of 60 Hz. The voltages for the horizontal scan are sinusoidal waveforms while those for the vertical scan are 9:1 skewed triangular waves. The vertical scan is programmed to deflect linearly with time in 90% of the frame time and then fly back to the initial position in the remaining 10% of the time. Therefore, the image scan is performed one-way for the deflection angle, hence the frame rate is equal to the vertical scan rate.

2.4.4 Electrothermal type

Figure 2.14 shows a schematic structure of a MEMS optical scanner based on the electrothermal actuation mechanism. CMOS (complementary metal oxide semiconductor) compatible materials such as silicon oxide insulator with polysilicon and aluminum processes are frequently used to compose the mirror and the electrothermal actuators. Electrical interconnections are used as a heater to generate Joule heat in the complex materials of different thermal expansion coefficients and to induce a unimorphic deflection angle for the scanner. The device in [26] uses a voltage of 12 V DC to deliver a relatively large optical angle of 14–50°. The scan angle produced by electrothermal actuation is generally large, but the operation speed is usually limited by the heat mass.

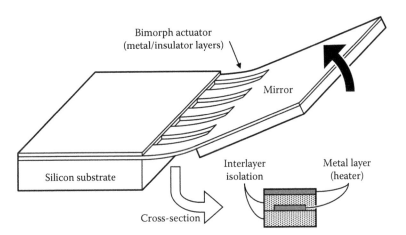

Figure 2.14 MEMS optical scanner with electrothermal actuation mechanism.

2.5 Design issues

2.5.1 Electrostatic hysteresis

Special caution needs to be paid when using a MEMS optical scanner in place of a conventional Galvano scanner based on the precision machining needed to construct optics for the image display. Electrostatic type scanners, for instance, have strong nonlinearity in their behavior due to the electrostatic force that is proportional to the square of the electric field and is inversely proportional to the square of the air gap length; hence they generally exhibit a hysteresis curve in the frequency response as shown in Figure 2.15. When the excitation frequency is increased from somewhere lower than the mechanical resonance, the amplitude of oscillation will see a small leap and then gradually decrease with the frequency. When the frequency is tuned down, on the other hand, the amplitude of oscillation monotonically increases beyond the frequency of the oscillation leap, and ends up in a chaotic oscillation before collapsing into a small oscillation. When the scanner mechanism experiences stress-hardening at a large stroke, on the other hand, the frequency response may incline toward the higher frequency side. Due to the large hysteresis, the control circuit for

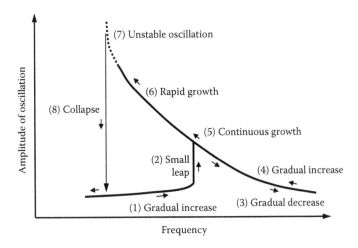

Figure 2.15 Nonlinear frequency response of electrostatic-type optical scanner.

an electrostatic scanner at resonance may need to be programmed to have a sequence to sweep the drive frequency from a higher value than the mechanical resonance.

2.5.2 Temperature dependence

Due to the finite value of the thermal expansion coefficient, optical scanners made of silicon usually have temperature dependence in their oscillation behavior, which is notable when a mirror plate is suspended with a pair of torsion bars in a straight form; in such a case, the length of the suspension increases due to the temperature elevation, leading to the mechanical buckling of the beam. After buckling, the center of gravity of the oscillation system is displaced from the initial position, which may cause a drastic change in the mechanical resonant mode, thereby resulting in a degradation of oscillation amplitude when the scanner is driven at a constant frequency. In addition to this, the elastic constant of silicon decreases slightly with the temperature rise, which causes a drift of the mechanical resonance. An automatic gain control (AGC) system for amplitude is, therefore, indispensable for operation at constant scan angles. A piezoresistive strain gauge is usually used to sense the scanner's angle [32,33].

2.5.3 Dynamic deformation of mirror

A scanning mirror fails to maintain its flat reflection surface when it is mechanically oscillated at a fast speed due to dynamic deformation, which is particularly significant when the mirror is about to turn around the peak position. The deformation leads to the unwanted expansion of the reflected beam spot, which eventually degrades the image quality by blurring near the left- and right-hand side edges of the projected image. Figure 2.16 shows the snapshot of a mirror (diameter 1 mm and thickness 50 µm) that is resonating at 20 kHz. The black and white arrows in the figure show the positions that deflect higher or lower than the mirror's average surface, respectively.

The maximum deformation of a mirror is known to be in proportion to $D^5 f^2/\theta/t^2$, where D is the mirror's diameter, t the thickness, f the operation frequency, and θ the maximum angle of deflection (peak-to-peak value) [34]. A mirror of large diameter is usually preferred for a smaller diffraction angle, but it also leads to poor flatness in dynamic motion. Mirror thickness, on the other hand, should be increased to keep the surface rigid, but it may also degrade the frequency response due to the added moment of inertia.

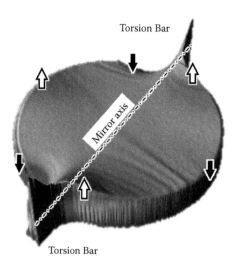

Figure 2.16 Dynamic deformation of scanning mirror in operation.

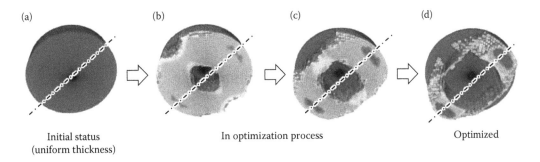

(a)   (b)   (c)   (d)

Initial status   In optimization process   Optimized
(uniform thickness)

Figure 2.17 Process of design optimization for scanning mirror: (a) Initial status; (b) and (c) in optimization process; and (d) final optimized shape.

To overcome these trade-off problems, Reference 34 used an FEM software to optimize the mass distribution on the backside of the mirror plate to keep the mirror flat while improving the frequency response. Figure 2.17 shows an example of such an optimization process; the volume of the flat disk are meshed into small grains, which are automatically removed in a trial-and-error manner to calculate the mechanical strength and the resonant frequency. As a result, masses far from the rotation axis are systematically removed to keep the resonant frequency high, while removing those in the middle part that have little contribution to the mechanical rigidity of the mirror surface [19].

2.5.4 Higher mode oscillation

An optical scanner's angle needs to have high linearity or fidelity to the operation voltage in order to project images of minimal blur, which is occasionally jeopardized by the jitter of oscillation that arises from the higher mode oscillations. The topology of a 2D scanner structure is substantially equal to that of a gyroscope of the oscillation type, hence it is susceptible to external vibration as well as intrinsic oscillation of the orthogonal axis [35]. Figure 2.18 schematically shows the mechanism of Coriolis coupling within a 2D scanner. When a scanner spinning about its Y-axis at a fast speed is forced to tilt in the orthogonal X-axis by the gimbal structure, a pair of Coriolis forces is generated on the edges of the mirror, which produces a disturbance in the rotation of the Y-axis. As a result, the projected horizontal line has a small oscillation in the Y-axis to blur the image near the

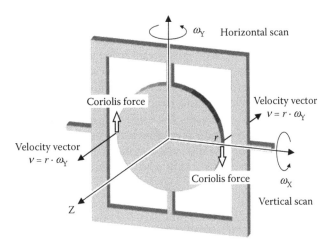

Figure 2.18 Parasitic oscillation associated with the Coriolis force.

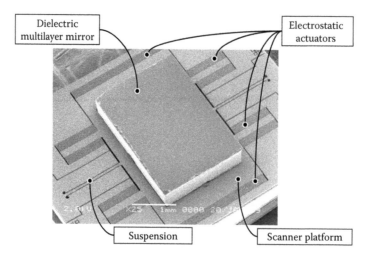

Figure 2.19 Dielectric reflector chip mounted on a scanning mechanism.

edges. Therefore, the scanner mechanism should be designed to have no resonant coupling with the mirror's rotation about the Z-axis.

2.5.5 Optical damage of mirror surface

Metallic surfaces are usually used as a reflector for a MEMS optical scanner, but their reflectivity is no greater than 96% for visible light. Therefore, the rest of the incoming energy is dissipated as heat in the mirror, which may cause thermoelastic deformation of the scanner mechanism as well as optical damage to the reflector. To avoid this problem, optical reflectors of multi-layered dielectric film of high reflectivity (>98%) can be used in place of the metallic reflector. Figure 2.19 is an example of a highly reflective mirror chip that is mounted on an electrostatic scanner platform. Owing to its reflectivity, the mirror can be used to steer a high power pulse laser (4 mJ, 20 ns) [36].

2.6 Summary

In this chapter, we looked into the practical use of MEMS optical scanners for laser projection display systems. The principles of opto-mechanical interactions were categorized into four technological layers of ray optics, wave optics, electromagnetic, and quantum photonics, among which the classic ray optics is most widely used in MEMS optical scanners. The opto-mechanical performance of scanners was studied in terms of the scan speed and the number of resolvable spots based on the Gaussian beam theory. The theoretical limits of a scanner's angle amplitude and speed were also discussed by using the mechanical equation of motion. Electromechanical actuation principles for scanners including electrostatic, electromagnetic, piezoelectric, and electrothermal were discussed to compare the pros and cons. The usual design issues, including the hysteresis in the frequency response curve, temperature dependent behavior of the scanner, dynamic deformation of the mirror surface, high mode oscillation associated with the gyroscopic effect, and optical damage of the mirror surface at an excess optical power were also discussed.

References

1. R. N. Thomas, J. Guldberg, H. C. Nathanson, P. R. Malmberg, The mirror-matrix tube: A novel light valve for projection display, *IEEE Trans. ED*, vol. 22, no. 9, pp. 765–775, 1975.

2. K. E. Petersen, Silicon as a mechanical material, *Proc. IEEE*, vol. 70, no. 5, pp. 420–457, 1982.
3. P. F. Van Kessel, L. J. Hornbeck, R. E. Meier, and M. R. Douglass, A MEMS-based projection display, *Proc. IEEE*, vol. 86, no. 8, pp. 1687–1704, 1998.
4. J. Suzuki, A. Komai, Y. Ohuchi, Y. Tezuka, H. Konishi, M. Nishiyama, Y. Suzuki, and S. Owa, Micro-mirror on Ribbon-actuator (MOR) for high speed spatial light modulator, in *Proc. IEEE Int. Conf. on Micro Electro Mechanical Systems (MEMS 2008)*, Jan. 13–17, Tucson, US, pp. 762–765, 2008.
5. D. T. Amm and R. W. Corrigan, Grating light valve technology: Update and novel applications, in *Proc. SID Symposium*, May 19, Anaheim, CA, pp. 29–32, 1998.
6. K. Saruta, H. Kasai, M. Nishida, M. Yamaguchi, Y. Ito, K. Yamashita, A. Taguchi, K. Oniki, and H. Tamada, Nanometer-order control of MEMS ribbons for blazed grating light valves, in *Proc. 19th IEEE Int. Conf. on Micro Electro Mechanical Systems (MEMS 2006)*, Jan. 22–26, Istanbul, Turkey, pp. 842–845, 2006.
7. J. L. Steyn, T. Brosnihan, J. Fijol, J. Gandhi, N. Hagood IV, M. Halfman, S. Lewis, R. Payne, J. Wu, A MEMS Digital Microshutter (DMS) for low-power high brightness displays, in *Proc. IEEE Int. Conf. on Optical MEMS and Nanophotonics*, Aug. 9–12, Sapporo Convention Center, Sapporo, Japan, pp. 73–74, 2010.
8. J. R. Lewis, In the eye of the beholder, *IEEE Spectrum*, May, pp. 24–28, 2004.
9. K. Isamoto, K. Yamashita, M. S. Khan, N. Lafitte, K. Totsuka, C. Chong, N. Nishiyama, and H. Toshiyoshi, A MEMS based electrically pumped tunable VCSEL operating at 1060 nm for SS-OCT, *SPIE Photonics West 2015—MOEMS and Miniaturized Systems XIV*, Feb. 9–12, The Moscone Center, San Francisco, CA, SPIE Vol. 9375, p. 37, 2015.
10. M. S. Wu, E. C. Vail, G. S. Li, W. Yuen, and C. J. Chang-Hasnain, Widely and continuously tunable micromachined resonant cavity detector with wavelength tracking, *IEEE Photon. Tech. Lett.*, vol. 8, no. 1, pp. 98–100, 1996.
11. J. T. Verdeyen, *Laser Electronics*, 3rd. Edition, Prentice-Hall Inc. Englewood Cliffs, New Jersey, 1995.
12. L. Y. Lin, Opportunities and challenges for MEMS in lightwave communications, *IEEE J. Selected Topics in Qaunt. Elec.*, vol. 8, no. 1, pp. 163–172, 2002.
13. J. Kim, C. J. Nuzman, B. Kumar, D. F. Lieuwen, J. S. Kraus, A. Weiss, C. P. Lichtenwalner et al., 1100×1100 port MEMS-based optical crossconnect with 4-dB maximum loss, *IEEE Photon. Tech. Lett.*, vol. 15, no. 11, pp. 1537–1539, 2003.
14. J. I. Dadap, P. B. Chu, I. Brener, C. Pu, C. D. Lee, K. Bergman, N. Bonadeo et al., Modular MEMS-based optical cross-connect with large port-count, *IEEE Photon. Tech Lett.*, vol. 15, no. 12, pp. 1773–1775, 2003.
15. A. D. Yalcinkaya, H. Urey, D. Brown, T. Montagne, and R. Sprangue, Two-axis electromagnetic microscanner for high resolution displays, *J. Microelectromech. Syst.*, vol. 15, no. 4, pp. 789–794, 2006.
16. V. Milanovic, Multilevel beam SOI-MEMS fabrication and applications, *J. Microelectromech. Syst.*, vol. 13, no. 1, pp. 19–30, 2004.
17. H. Toshiyoshi, Micro electro mechanical devices for fiber optic telecommunication, *JSME Int. Journal, Series B*, vol. 47, no. 3, pp. 439–446, 2004.
18. M. Tani, M. Akamatsu, Y. Yasuda, H. Fujita, and H. Toshiyoshi, Two-axis piezoelectric tilting micromirror with a newly developed PZT-meandering actuator, in *Proc. 20th IEEE Int. Conf. on Micro Electro Mechanical Systems (MEMS 2007)*, Jan. 21–25, Kobe Portopia Hall and Kobe Portopia Hotel, Japan, pp. 699–702, 2007.
19. K. Ikegami, T. Koyama, T. Saito, Y. Yasuda, and H. Toshiyoshi, A biaxial piezoelectric MEMS scanning mirror and its application to Pico-projectors, in *Proc. IEEE Int. Conf. on Optical MEMS and Nanophotonics (OMN 2014)*, 17–21 August, Glasgow, Scotland, pp. 95–96, 2014.
20. A. Schroth, C. Lee, S. Matsumoto, M. Tanaka, and R. Maeda, Application of sol-gel deposited thin PZT film for actuation of 1D and 2D scanners, *Sens. Actuators*, vol. 73, pp. 14–152, 1999.

21. N. Asada, M. Takeuchi, V. Vaganov, N. Belov, S. in't Hout, I. Sluchak, Silicon micro-optical scanner, *Sensors and Actuators*, vol. 83, pp. 284–290, 2000.
22. H. Urey and C. Ataman, Modeling and characterization of comb actuated resonant microscanners, *J. Micromech. Microeng.*, vol. 16, no. 1, pp. 9–16, 2006.
23. H. Miyajima, M. Nishio, Y. Kamiya, M. Ogata, Y. Sakai, and H. Inoue, Development of two dimensional scanner-on-scanner for confocal laser scanning microscope LEXT series, in *Proc. IEEE/LEOS Int. Conf. on Optical MEMS*, Aug. 1–4, Oulu, Finland, pp. 23–24, 2005.
24. K. Torashima, T. Teshima, Y. Mizoguchi, S. Yasuda, T. Kato, Y. Shimada, and T. Yagi, A micro scanner with lower power consumption using double coil layers on a permalloy film, in *Proc. IEEE/LEOS Int. Conf. on Optical MEMS*, Aug. 22–26, Takamatsu, Japan, pp. 192–193, 2004.
25. MicroVision, Inc. URL: http://www.microvision.com/
26. A. Jain and H. Xie, An electrothermal SCS micromirror for large bi-directional 2-D scanning, in *Proc. 13th Int. Conf. on Solid-State Sensors, Actuators and Microsystems (Transducers 05)*, June 5–9, Seoul, Korea, pp. 988–991, 2005.
27. H. Schenk, P. Durr, T. Haase, D. Kunze, U. Sobe, H. Lakner, and H. Kuck, Large deflection micromechanical scanning mirrors for linear scan and pattern generation, *IEEE J. Selected Topics in Quant. Elec.*, vol. 6, no. 5, pp. 715–722, 2000.
28. H. Schenk, P. Durr, D. Kunze, H. Lakner, and H. Kuck, A resonantly excited 2D-micro-scanning-mirror with large deflection, *Sens. ActuatorsA*, vol. 89, no. 1–2, pp. 104–111, 2001.
29. J. -H. Lee, Y. -C. Ko, D. -H. Kong, J. -M. Kim, K. B. Lee, and D.-Y. Jeon, Design and fabrication of scanning mirror for laser display, *Sens. Actuators A*, vol. A96, no. 2–3, pp. 223–230, 2002.
30. C. -H. Ji, S. -H. Kim, Y. Yee, M. Choi, S. -C. Kim, S. -H. Lee, and J. –U. Bu, Diamond shaped frame supported electrostatic scanning micromirror, in *Proc. 13th Int. Conf. on Solid-State Sensors, Actuators and Microsystems (Transducers 05)*, June 5–9, Seoul, Korea, pp. 992–995, 2005.
31. M. Yoda, K. Isamoto, C. Chong, H. Ito, A. Murata, S. Kamisuki, M. Atobe, and H. Toshiyoshi, A MEMS 1-D optical scanner for laser projection display using self-assembled vertical combs and scan-angle magnifying mechanism, in *Proc. 13th Int. Conf. on Solid-State Sensors, Actuators and Microsystems (Transducers 05)*, June 5–9, Seoul, Korea, pp. 968–971, 2005.
32. T. Bourouina, E. Labrasseur, G. Reyne, A. Debray, H. Fujita, A. Ludwig, E. Quandt, H. Muro, T. Oki, and A. Asaoka, Integration of two degree-of-freedom magnetostrictive actuation and piezoresistive detection: Application to a two-dimensional optical scanner, *J. Microelectromech. Syst.*, vol. 11, no. 4, pp. 355–361, 2002.
33. M. Sasaki, M. Tabata, T. Haga, and K. Hane, Piezoresistive rotation angle sensor integrated in micromirror, *Jpn. J. Appl. Phys.*, vol. 45, no. 4B, pp. 3789–3793, 2006.
34. S. Hsu, T. Klose, C. Drabe, and H. Schenk, Fabrication and characterization of a dynamically flat high resolution micro-scanner, *J. Opt. A: Pure Appl. Opt.*, vol. 10, 044005 (8p), 2008.
35. D. K. Shaeffer, MEMS inertial sensors: A tutorial overview, *IEEE Comm. Mag.*, vol. 51, no. 4, pp. 100–109, 2013.
36. A. Chekhovskiy, Y. Ohira, and H. Toshiyoshi, Laser breakdown 3D display, *IEICE Trans. Electron.*, vol. E91, no. C(10), pp. 1616–1620, 2008.

3

Optical micro-electrical-mechanical phased array

Youmin Wang and Ming-Chiang Wu

Contents

3.1 Overview of optical phased array

The beam steering and stabilization features, while very critical to many electro-optical systems, are still very much limited in performance. Systems in these related optical applications are restricted by the nature of mechanical beam steering and the stabilization that comes with it. Mechanical beam steering suffers from a lack of rapid pointing ability, requires high mechanical complexity, and demands higher cost, while having a lower reliability. Therefore, it is desirable for nonmechanical approaches to increase the pointing speed, provide random access pointing, decrease system complexity, increase overall reliability, and reduce costs. Optical phased arrays (OPAs) have generated a growing interest in many application areas regarding beam forming and beam steering [1]. Applications of OPAs range from 3D displays, printing, optical data-storage, and telecommunication to military and other industrial applications. Specifically among them, light detection and ranging (LIDAR) is a key enabling technology for target detection, surveying and mapping, self-driving cars, and other autonomous vehicles [2].

The beam steering mechanism of an optical phased array is similar to the effect of a prism. The angle of the light passing through a prism will be changed due to the different optical path differences (OPD) in different positions of the prism. However, the problem with a prism is that its OPD is fixed, and typically it requires a very large thickness. Thanks to the nature of light, which is a sinusoidal electromagnetic wave, we can fold its wavefront by 2π of phase shift, resulting in a sawtooth-like phase profile. The unfolded phase profile, which is called a modulo 2π phase profile, is shown in Figure 3.1. The advantage of this method is that the required OPD, or equivalently the device thickness, could be reduced. The maximum required OPD is roughly the same as the operating wavelength of the device. In different applications, the concept of a modulo 2π phase profile can be implemented in various designs. The phase shifters for the folded phase profile can be either actively radiating [3], or passively reflecting [4,5] or transmitting [6] the incoming light, and the OPA can consist of an array of periodic or aperiodic [7] arranged phase shifting elements in deterministic order.

The early work of phase modulation using an array of phase shifters was implemented using a number of different methods. Electro-optic effects were first investigated using different materials, including lithium tantalate [8] and lithium niobate [9] among others. Liquid crystal technologies were of particular interest due to their advantages of matured fabrication techniques, potential to achieve large steering angles, compatibility with visible wavelengths, and decent electrical and optical efficiencies [10]. Therefore, OPAs using liquid crystals approaches were investigated starting at an early age [11], and were implemented with variable blaze [12] and period [1,13] designs.

Despite the aforementioned advantages, the liquid crystal devices also suffer from several disadvantages. An important one is their slow response and steering speed. The typical steering time is on the order of milliseconds, making it impossible for fast beam steering applications such as LIDAR.

Silicon photonic waveguides with phase modulators and grating couplers have recently been reported for beam steering [3,7]. The silicon photonic phased array typically consists of several identical optical antennas. Each antenna that comprises a grating on it measures microns in length and width, and emits light with a specific amplitude and programmed phase. The amplitude and phase tuning were achieved by thermo-optically tuning a kilo-ohm level resistive heater that is connected to each pixel in the array. The power consumption of each pixel is typically a few milliwatts to achieve per π phase shift. Compared with its liquid crystal counterpart, the silicon photonics phased array offers an approach with advantages such as higher compactness, faster responding speed, and lower cost. However, its disadvantages are equally obvious. The silicon photonic phased arrays usually have much higher insertion loss, higher power consumption due to their thermo-optic tuning method, and are limited to infrared wavelengths transparent to silicon. For a wavelength-tuning approach, the beam direction is also dependent on wavelength, and it is not possible to achieve monochromatic beam steering.

A microelectromechanical system (MEMS) offers an alternative for a large-scale, compact, and highly efficient beam steering solution [14–17]. Typically, MEMS mirrors work in the reflection mode for phase modulation, where each pixel in the array is realized by a "piston" mirror that is displaced to provide the desired phase shift. The marriage of MEMS and an optical phased array started as early as more than two decades ago. D. M. Bloom introduced the Grating Light Valve™ technology using micromechanical phase gratings for display, showing a 1-μs response time and high contrast ratio [18]. Polychromix used the same basic architecture as the Grating Light Valve (GLV) with a longer ribbon for low price spectrometers [19]. However, it has not been until recently that ultra-small mirror elements have made wide-angle beam steering possible [4,20]. In this chapter, we will present an overview of the MEMS enabled optical phased array devices. We begin with the theoretical background of an optical phased array and cover various designs and fabrication considerations of MEMS OPA devices.

3.2 Physics of optical phased array

As illustrated in Figure 3.1, a MEMS OPA consists of an array of identical reflective elements in regular order. In a typical MEMS phased array, all the elements reflect coherently along a desired

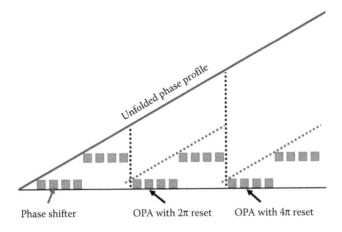

Figure 3.1 Illustration of optical phased array beam steering.

direction. In the following subsections, the OPA beam steering performances such as steering range, far-field resolution, and optical and electrical efficiencies are discussed based on these basic assumptions.

3.2.1 Beam steering range and far-field resolution

The optical performance of an OPA is determined by its pitch, aperture, and the number of elements in the array. The largest angle an OPA device can steer to using the modulo 2π approach is determined by the size of the smallest individually addressable phase element in the array. When the full array of phase shifters is illuminated with a Gaussian beam, there is an approximately uniform irradiance distribution across each individual phase shifter, and the far-field diffraction pattern is essentially the summation of the individual element fields. The full width at half maximum (FWHM) of the diffraction angle for an unobscured aperture is [21]:

$$\theta_{diff} = 1.03\frac{\lambda}{d} \tag{3.1}$$

where θ_{diff} is the beam divergence from a single element aperture of the array, λ is the operating wavelength of the device, and d is the aperture size of the element. Figure 3.2a shows the illustration of the far-field dispersion of the element. If we have n elements in the array, then the total aperture of the OPA device becomes:

$$D = n \cdot d \tag{3.2}$$

assuming there is no gap between neighboring elements. Given the total array aperture size D, the optical phased array far-field resolution is approximately given by

$$\theta_B = 1.22\frac{\lambda}{D} \tag{3.3}$$

where 1.22 is the constant indicating the position of the first diffraction minima of the Airy disk, assuming the OPA is a uniformly illuminated circular aperture. Note that different aperture shapes will change the value of this constant. The effective aperture size would also be reduced if Gaussian

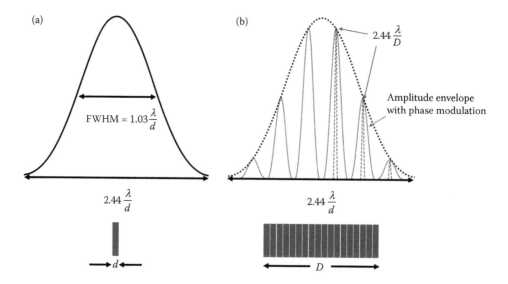

Figure 3.2 Far-field beam divergence profile of OPA's single element (a) and the entire array (b).

illumination is used. By programming the phase distribution among the elements in the array, the steered beam lobes can be arranged within the envelope of a single element dispersion profile, as shown in Figure 3.2b.

From Equations 3.1 and 3.3, we can derive the number of far-field resolvable spots of an OPA to be

$$N_{spot} = \frac{\theta_{diff}}{\theta_B} = 1.18\frac{D}{d}. \tag{3.4}$$

Therefore, a reasonable rule of thumb to calculate the number of resolvable spots in the far field is roughly the number of elements in the entire array when the OPA fill factor is high enough. However, it is also noted that the resolvable spot number given by (3.4) includes the angular positions of the diffraction side-lobes, which are typically not usable in practical applications, whereas the beam steering main lobe direction θ is restricted by the general grating equation

$$\sin\theta + \sin\theta_{inc} = \frac{\lambda}{A} \tag{3.5}$$

where θ_{inc} is the incident angle of the beam on the OPA and A is the programed grating period in a uniformly arranged array.

3.2.2 Beam forming efficiency

The beam forming efficiency is defined as the portion of light that is directed in the desired direction compared with the total input to the MEMS OPA unit. There are usually two main beam forming efficiency considerations for the MEMS OPA. The first one is the MEMS array fill factor, and the other is the quantization error caused by the discrete phase steps. The priority and complexity of these two considerations will change, depending on different beam steering mechanisms of the MEMS OPA device designs.

For example, a "piston" mirror array [20,22] as illustrated in Figure 3.1, has a fixed fill factor which is the ratio of the element width d versus array pitch A, while in a "tilting" mirror array as shown in [23], the MEMS mirror array can be fabricated to implement a variable blaze approach.

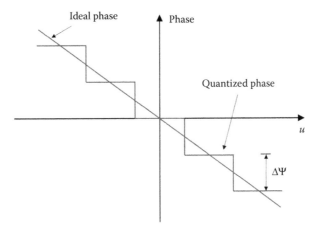

Figure 3.3 Phase quantization of a linearly phased "piston" array.

In the "tilting" array case, no phase quantization is involved whereas the fill factor will change with the increase of the mirror tilting angle.

For a "piston" mirror array in which the phase shift of each element is digitally controlled by electronic means, the number of quantized phase states is 2^N, where N is the phase shifting controlling bit number. In this case, the discrete phase interval is $\Delta\psi - 2\pi/2^N$, as shown in Figure 3.3. Since the intermediate phase values cannot be implemented by these states, in the OPA beam forming system the quantization error will affect gain and side-lobe performances. The gain loss due to phase quantization is proportional to the relative element gain at the beam peak. The diffraction efficiency caused by gain loss is given by

$$\eta = \left(\frac{\sin(\pi/2^N)}{(\pi/2^N)} \right)^2 \tag{3.6}$$

The diffraction efficiency versus the number of bits is plotted in Figure 3.4. The efficiency for a 3-bit phase shifter system is about 95.0%, and for a 4-bit system it should be around 98.7%.

The phase quantization also causes scan angle quantization, leading to a beam pointing error, as shown in Figure 3.5. In Figure 3.5, we used Equation 3.5 for OPA phase profile calculation, where the parameters used for the OPA are: $\lambda = 1.55\ \mu m$, array pitch $\Lambda = 2.5\ \mu m$, array element number $n = 256$, and bit number $N = 3$.

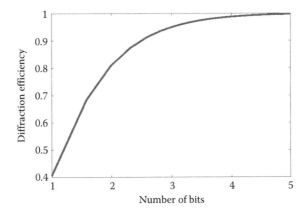

Figure 3.4 Diffraction efficiency with reference to number of bits in phase shifter.

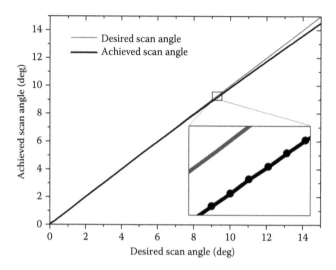

Figure 3.5 Pointing error due to quantized phase steps.

It is observed that the pointing error is more prominent when the scan angle becomes larger. It is noted that the zoomed-in inset image of Figure 3.5 also shows the discontinuity of the achieved scan angle curve, indicating that the absolute desired scan angle could not be achieved. To mitigate the problems due to the pointing error caused by quantization error, further optimization methods similar to those used in an RF antenna array such as random phasing and compensating phase feeding are required [24,25].

3.2.3 MEMS element reflectivity

The reflectivity of each MEMS OPA element is another important factor determining the overall beam forming efficiency of the OPA system. Currently, due to its simplicity and maturity within the commercial IC industry, thin-film metal coatings are still the most popular way to enhance reflectivity for most optical MEMS devices. Figure 3.6 illustrates the normal incidence reflectivity

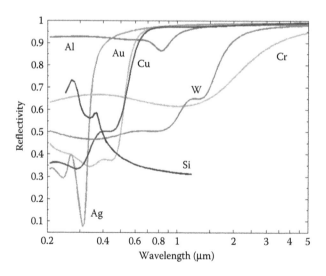

Figure 3.6 Reflectivity spectrum of aluminum, silver, gold, chromium, copper, tungsten, and silicon thin films for the wavelength range of 0.2–5 μm.

spectra of a list of common metal materials used in optical microsystems, compared with bare silicon's reflectivity. Reflection of metal films on MEMS element substrates can be calculated using the Fresnel reflection formulas extended to multilayer structures. For OPA devices which usually use perpendicular light illumination, the polarization can be ignored and the reflectivity is given by

$$R = \frac{(n-1)^2 + k^2}{(n+1)^2 + k^2} \tag{3.7}$$

where n is the material's refractive index and k is its extinction coefficient, the values of which in the calculation are referenced from [26]. Among choices of these metal types, aluminum and gold are very good reflectors in the conventional band (c-band) around 1550 nm wavelength used for fiber-optical communication. Gold has a slightly higher reflectivity in the c-band, but is much worse in visible wavelengths, as illustrated in Figure 3.6. One of the other conveniences of using a metal film reflector is that film thicknesses required to achieve bulk reflectivity are only on the order of 20–40 nm.

Meanwhile, the problems with high-reflectivity metals are power handling and temperature stability. Material and process compatibility with IC technology is another problem, for example, gold is often considered a contaminant in Si foundries. Another method for making high-reflectivity surfaces is to use Photonic Crystal reflectors [27] or subwavelength gratings for high reflectivity and power handling capability. Figure 3.7a shows a MEMS OPA with a high contrast grating (HCG) reflector. The HCG mirror comprises a thin layer of subwavelength gratings with high dielectric constants. The reflectivity and the reflection spectrum can also be tuned by varying the grating thickness, pitch, and grating bar width. Figure 3.7b shows the calculated reflection spectra of the HCG mirror using rigorous coupled wave analysis. High reflectivity can be obtained for TE-polarized light (polarization parallel to grating) over a broad wavelength band (1400–2000 nm), with a peak reflectivity of >99.9% at 1550 nm wavelength [5].

3.2.4 Addressing OPA elements

From Equations 3.3 and 3.4, we know that the far-field beam forming resolution is inversely proportional to the aperture size of the OPA, and the far-field field of view (FOV) is proportional to the inverse of the array element's pitch. Therefore, it is ideal to have an OPA that is made of a large number of tiny individual elements. Typical phased arrays use a phase array pitch size that is equal to or smaller than one-half wavelength to avoid the undesirable higher order grating lobes. For applications such as near-infrared LIDARs, the OPA aperture size should exceed a centimeter in diameter to ensure that the required detection range is greater than 200 m. In this case, the desired

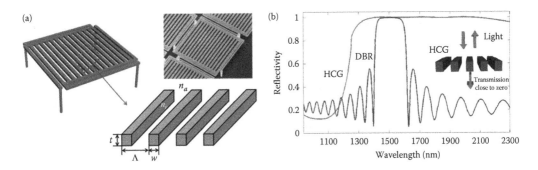

Figure 3.7 Structure schematic (a) and reflectivity spectra of the high-contrast grating (HCG) MEMS mirror in the normal-incident TE-polarized field (b).

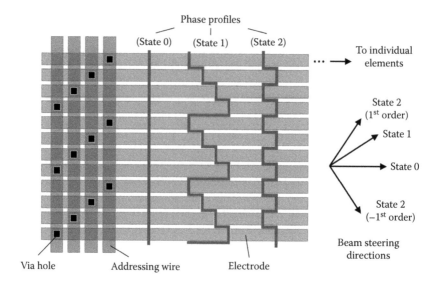

Figure. 3.8 Schematic of OPA containing subarrays to reduce the number of addressing wires.

pixel and fanout numbers of a two-dimensional (2D) OPA can easily reach the order of millions. Although recent advancements in MEMS foundries have enabled devices of similar size [28], the additional complexity of multibit driving as illustrated in Section 3.2.2 will still make the real-time driving electronics of a large size OPA overwhelming.

An alternative to circumvent this problem is to adopt a hard-wired subarray architecture to reduce the number of addressing wires. The schematic of a 2-bit OPA device that contains three subarrays is illustrated in Figure 3.8, which also shows the multiple-state beam steering concept. In a subarray architecture, the whole OPA aperture is divided into many identical subarrays. Addressing wires are used to connect in parallel the corresponding phase shifters of each of these subarrays. With this architecture, the number of addressing wires is reduced while the beam steering range and beamforming resolution are kept the same as an OPA whose elements are all individually addressable. The sacrifice, however, comes from the available number of phase profile states, which in turn limits the number of steerable angles in the far-field. A working phase profile state for subarray architecture requires the phase profile to be identical for all subarrays so as to avoid phase discontinuities between adjacent subarrays.

3.2.5 Side-lobe level suppression (SSL)

Side-lobe level is an important characteristic measuring the directivity of the OPA system. The side-lobes in a diffractive device like OPA not only come from the higher orders of diffraction lobes, but are also complicated by the element size, quantization level, gap size between elements, and array pitch. The penalty for wider spaced elements under limited quantization bit number shows up in the far-field performance where side-lobe levels are similar to that of the main lobe [3,29]. This lack of directivity could lead to the interference of crosstalk and loss of security in applications like LIDAR and optical fiber communications.

One approach to address this problem is to use an apodized array [30] where the illumination from the array has a Gaussian-like distribution. However, the results are still not ideal. Given the flexibility in a MEMS OPA design in its gap and element size, it is also possible to utilize design techniques similar to those used in radio-frequency (RF) antenna array technology that involve perturbing the sizes and locations of the array elements in order to achieve a desirable diffraction pattern. Using these methods, a much lower SLL can be obtained through nonlinear optimization methods applied to thinned arrays [31], fractals [32], and aperiodic tiling [33].

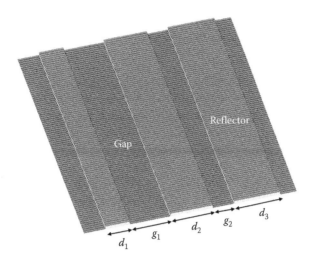

Figure 3.9 Illustration of part of an OPA whose gap sizes (d_1, d_2, d_3) and element sizes (g_1, g_2) are randomly determined.

The far-field diffraction of the arrays composed of reflector apertures with negligible mutual coupling between them can be described by the term $\Psi(\theta, \varphi)$ given by

$$\Psi(\theta,\varphi) = \sum_{n=1}^{N} \exp\left[j\frac{2\pi}{\lambda}\sin\theta \iint_{S_n} a_n(x\cos\varphi + y\sin\varphi)dS_n \right] \qquad (3.8)$$

where θ and φ are the spherical coordinates indicating the far-field direction, N is the number of elements in the array, λ is the operation wavelength, a_n and S_n are the amplitude and aperture area of the nth reflector element, and x and y are the lateral coordinates within the reflector aperture. Figure 3.9 illustrates part of an OPA whose gap and element sizes are randomly determined within a preset range. Using Equation 3.8, Figure 3.10 shows the far-field diffraction performance comparison between a uniformly distributed OPA and the random gap and location OPA as shown in

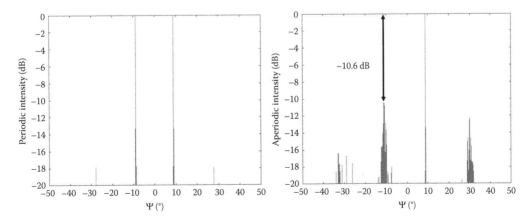

Figure 3.10 Side-lobe level (SSL) comparison of a uniform OPA and the randomly sorted aperiodic OPA. Parameters used: wavelength $\lambda = 1.55\,\mu m$; average array pitch $\Lambda = 5\,\mu m$, array elements $N = 250$, aperture size $= 1.2$ mm, and targeting beam steering angle $\theta = 9°$.

Figure 3.9. With an optimized random gap and random reflector size in the array, the SSL can be reduced from 0 dB to greater than −10 dB at the beam steering angle of 9°.

3.3 MEMS optical phased array mechanical design

We have seen the specific requirements on pitch and size for a highly compact, high resolution, and efficient optical phased array. Now we turn our attention to the mechanical structures that support the phased array elements. The mechanics is strongly dependent on the type of actuator that is used to drive the MEMS OPA element into in-plane or out-of-plane displacements, depending on the phase shifting scheme, so we need to consider the role of the actuator as we study OPA mechanical design.

3.3.1 One-dimensional (1D) OPA

One-dimensional (1D) OPA is made up of a linear array of phase shifters and can steer the illumined beam in one direction. The schematic of a 1D OPA device is shown in Figure 3.11. Since OPAs are typically working in the visible or near-infrared wavelength range, the size of the element pitch Λ should also be on the order of the wavelength length. For OPAs working in reflective mode, out-of-plane movable micromirrors can be used as phase shifters. To achieve a high fill factor while minimizing the crosstalk between adjacent mirrors, the actuators are typically designed to be hidden underneath the reflectors to save space. Figure 3.11 shows a design that integrates the vertical combdrive actuators directly under the mirror. In such a design, each actuator consists of two lower and one upper comb fingers. The mirrors are tethered to anchors through a pair of springs. The mirrors and the top combs are grounded while the bottom combs are individually addressed to produce the desired beam profiles.

In contrast to macroscopic machinery, the limitations of the fabrication technology in microsystems dictate that only simple structures be used, such as the double-clamped beam as shown in Figure 3.11.

Consider the equivalent model that has a rigid central bridge structure, is uniform in material, and is doubly clamped with spring length L, with width w, and height h as shown in Figure 3.12. The bending of the cantilever is given by the following equation when the structure is subjected to a point load F at the center

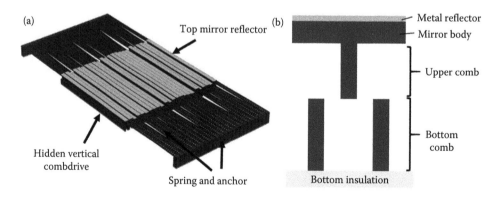

Figure 3.11 (a) Perceptual drawing of a one-dimensional (1D) MEMS OPA whose elements are suspended by pairs of vertically translational springs and (b) the close-up cross-sectional view of one element in the array showing the mirror reflector with the integrated vertical comb-drive structure.

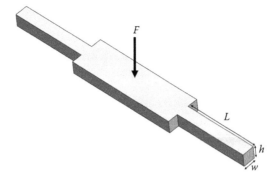

Figure 3.12 Equivalent model of a double-clamped beam.

$$y = \frac{L^3}{2Ewh^3} \cdot F \qquad (3.9)$$

where E is the Young's modulus. Therefore, the spring constant of the piston mirror structure is

$$k = F/y = \frac{2Ewh^3}{L^3} \qquad (3.10)$$

Thus, the piston mirror features a fundamental resonance in the z direction at a frequency

$$f_z = \frac{1}{2\pi}\sqrt{\frac{k}{m}} \qquad (3.11)$$

where m is the OPA micromirror element's effective mass. Higher order resonance modes should also be taken into consideration, which usually can be obtained through finite element method (FEM) simulations. Implementing the clamped boundary conditions at the fixed ends of the beam, the higher order resonance frequencies are obtained [34]. The modes described in Figure 3.13 and Table 3.1 illustrate the higher order mechanical resonances, which is an important consideration for single OPA element design. In the MEMS OPA design, the frequencies of the unwanted motions are designed to be away from the operation frequency in order to avoid the coupling of energy during operation. Clearly, the high order resonance frequencies are substantially higher than the fundamental mode frequency.

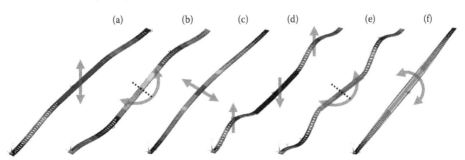

Figure 3.13 Resonant modes of the 1D piston OPA element, simulated by Finite Element Modeling (FEM) model analysis: (a) vertical out-of-plane mode, (b) tilted out-of-plane mode, (c) in-plane translational mode, (d) higher order out-of-plane mode, (e) tilted twisting mode, and (f) longitudinal twisting mode.

Table 3.1 Mode descriptions and resonance frequencies of the 1D piston OPA element as shown in Figure 3.13

Mode #	Motion	Resonance frequency
1	Out-of-plane, up-and-down linear translation, that is, fundamental mode (Figure 3.11a)	318.25 kHz
2	Rotation around an axis perpendicular to the length direction of the suspension springs (Figure 3.11b)	842.00 kHz
3	In-plane side-to-side linear translation (Figure 3.11c)	1.73 MHz
4	High order out-of-plane, up-and-down translation (Figure 3.11d)	2.83 MHz
5	Higher order rotation around an axis perpendicular to the length direction of the suspension springs (Figure 3.11e)	3.38 MHz
6	Rotation around the torsion springs (Figure 3.11e)	4.95 MHz

With the actuation force at the desired operation frequency applied, as illustrated in Figure 3.12, the spring-mass system can be described as a damped harmonic oscillator. The equation of motion for this system is

$$m\frac{d^2z}{dt^2} + b\frac{dz}{dt} + k \cdot z = F(t) \tag{3.12}$$

where m is the mass, k is the spring constant, and $F(t) = F_0 \cdot \cos(\omega t)$ is the actuation force exerted on the structure. Therefore, in the frequency domain, the equation takes the form

$$m \cdot (j\omega)^2 z + b \cdot (j\omega)z + k \cdot z = F_0 \tag{3.13}$$

with the solution given by

$$z = \frac{\frac{F_0}{m}}{-\omega^2 + j\omega \cdot \frac{b}{m} + \frac{k}{m}} = \frac{F_0}{k} \cdot \frac{1}{-\frac{\omega^2}{\omega_0^2} + j\frac{\omega}{\omega_0} \cdot \frac{1}{Q} + 1} \tag{3.14}$$

where $\omega_0 = k/m$ is the resonant frequency and $Q = (\omega_0 \cdot m)/b$ is the quality factor. The quality factor is a dimensionless parameter that describes how under-damped the MEMS OPA system is, and it has a profound effect on the mechanical response of the spring-mass system. Controlling the quality factor around 0.5 will render the system as critically damped, which in turn can avoid overshoot in the response to a step-function in the applied force. Meanwhile, it can also avoid substantially low bandwidth for an over-damped system. However, due to the limitations in MEMS designs, it is difficult to control mechanical damping in the MEMS OPA. In general, the most available approach to control damping is to control the dissipation in the gas flow around the OPA MEMS element, and meanwhile, to reduce the crosstalk from the air squeezing from adjacent OPA elements in the array.

3.3.2 Two-dimensional (2D) OPA

In the previous subsections, we have concentrated on one-dimensional OPAs. Now we extend the discussion to the two-dimensional (2D) OPA that is required by many applications. A common architecture to construct a 2D OPA is to have a reflector anchored by multiple suspension beams,

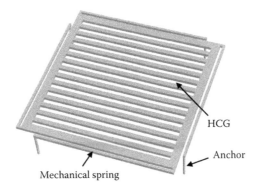

HCG

Anchor

Mechanical spring

Figure 3.14 Illustration of a single layer reflective OPA element. The high contrast grating (HCG)-based reflector is suspended with a crab-leg spring design.

as the one shown in Figure 3.14, whose reflector is made of a highly contractible grating (HCG). In the crab-leg spring suspension structure in Figure 3.14, each HCG reflector is tethered to four mechanical springs, one at each corner. It is electrostatically actuated by applying a voltage between the reflector and the substrate, both of which are electrically conductive.

For low voltage actuation with a desirable operation speed determined by a fixed resonant frequency, the light weight of a HCG is a benefit. For the aforementioned fixed resonant frequency, a lightweight HCG reflector can use a lower spring constant k than traditional mirrors. The total potential energy in this parallel plate actuation system is

$$U = -\frac{\varepsilon A}{2(g-x)}V^2 + \frac{kx^2}{2} \tag{3.15}$$

where ε is the air permittivity, A is the effective area of the HCG, V is the applied voltage, g is the initial actuation gap, and x is the HCG displacement in the vertical direction. The first term is the electrostatic potential of the HCG and the second term is the mechanical energy stored in the cantilever springs. The force acting on the movable HCG is obtained by taking the derivative of Equation 3.15 with respect to x. The equilibrium of electrostatic force and spring force provides

$$V = \sqrt{\frac{2kx}{\varepsilon A}}(g-x) \tag{3.16}$$

Though Equation 3.16 shows that the actuation voltage of the HCG-based OPA system is low thanks to the lower spring constant obtained from the light weight, critical problems such as the small size of MEMS reflector elements and the high array fill factor are still difficult to achieve for this single layer parallel plate design. This is because the crosstalk from the adjacent elements will severely affect the OPA's element performance especially in the parallel plate actuator case, and the array fill factor is limited by the width of the springs and the clearance between the structures.

One possible way to avoid these problems is to add additional layers into the structure to split the functionalities of reflection and electrical actuation. Figure 3.15 illustrates a modified design from the one shown in Figure 3.14, in which the top mirror layer is connected via a post and is resting on the spring and actuation layer below. In this way, the fill factor of the OPA element can be enhanced since the spring and anchor structures are all hidden beneath the reflector layer, and the actuator size can also be shrunk down to reduce the whole element size. The actuator in the structure shown in Figure 3.15 can still be a parallel plate design [35], though with a smaller gap size and total area, thus minimizing the fringe effects and crosstalk. It can also be made of structures such as a vertical comb drive to further enhance its actuation linearity and electrical field confinement.

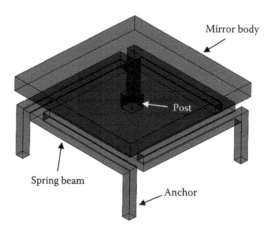

Figure 3.15 Illustration of a double layer design of the MEMS OPA element.

3.3.3 MEMS OPA examples

In this section, we give two examples of the MEMS OPAs, one is 1D and the other is 2D. These examples give different design solutions to the related problems. It should be noted that they are not necessarily the "best" solutions and they do not cover the whole MEMS OPA field. There are numerous other MEMS OPA designs, each with their own advantages and drawbacks, and their designs are highly dependent on their applications.

3.3.3.1 MEMS 1D OPA

A MEMS OPA with piston mirrors integrated on top of vertical comb drive actuators is demonstrated in the SEMs of Figure 3.16. The design is similar to the schematic depicted in Figure 3.11. The vertical comb drive design allows the MEMS 1D OPA to achieve a smaller pitch (2.4 μm) as shown in Figure 3.16, and a larger scan angle (22° at 905 nm wavelength) than all previously

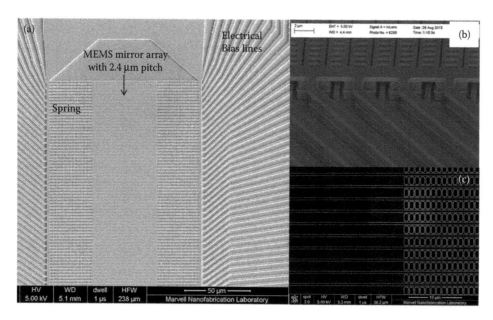

Figure 3.16 SEMs of the MEMS 1D OPA: (a) top view SEM showing an array of micromirrors between perforated springs; (b) close-up view of the anchor area where electrical bias lines pass through the tunnels between anchors; and (c) close-up view of the mirrors and perforated springs.

reported MEMS OPAs. Since the micromirror is a broadband reflective device, the OPA can operate at other wavelengths, or multiple wavelengths simultaneously. At 1550 nm, a larger scan angle of 37.7° can be realized. The vertical comb drive actuators with submicron fingers/gaps (300 nm) enable the OPA to achieve a fast time response and low operation voltage.

The vertical comb drive actuators are designed to be hidden under the mirrors. Each actuator consists of two lower and one upper comb fingers. The mirrors are tethered to anchors through a pair of springs. The mirrors and the top combs are grounded while the bottom combs are individually addressed to produce the desired beam profiles. For the double-clamped-beam spring design, the spring constant can be calculated with an analytical solution. Since the length of the central mirror region is rigidly enhanced by the T-shape upper comb structure, the OPA micromirror spring constant k can be simplified to be

$$k = \frac{32Ewt^3}{L^3_{spring}}$$

where t is the thickness of the spring layer (400 nm), w is the width of the spring beam feature size (0.5 μm), and L_{spring} is the total length of the spring beam (60 μm). With the fixed spring constant based on the dimensions above and Young's modulus ($E = 169$ GPa), $k = 0.8059$ N/m, the lightweight OPA micromirror structure has a high resonant frequency of 301.9 kHz thanks to its low mass (224 ng). The resonance frequency is designed to be 300 kHz with the actuation voltage below 10 V.

The OPA fabrication process used a 4-mask, self-aligned surface micromachining process as illustrated in Figure 3.17. Both the top and bottom combs are 1 μm high. The widths of the comb fingers are 300 nm. With such tightly spaced comb fingers, it is critical that the top and bottom comb fingers be self-aligned. The self-aligned process we used (steps 3–6 in Figure 3.17) is similar to that in Reference 36 but with tighter control of the finger widths and spacing. To protect the integrity of the bottom comb fingers, we used a thin protective oxide sidewall when thinning the lower comb fingers (steps 5–6 in Figure 3.17). After releasing in HF, the device was blanket-coated with 200 nm of gold to increase the reflectivity of the mirrors. The metal also provides bonding pads for electrical contacts. Polysilicon "gutters" were employed to block metals between addressing wires/ bonding pads to avoid electrical short circuits.

3.3.3.2 MEMS 2D OPA

Figure 3.18 demonstrates the MEMS 2D OPA that comprises a 32 × 32 array of HCG elements that is similar to the design shown in Figure 3.14. As mentioned before, for a high performance OPA, it is desirable to reduce the array pitch and increase the fill factor. However, for a given spacing between HCG reflectors determined by the minimum resolution of the fabrication process, the array fill factor decreases when element size decreases. Moreover, small MEMS elements also require a higher actuation voltage. To address these issues, the reflector area is chosen to be 20 × 20 μm². The spacing between the HCG reflectors is designed to be 2 μm, resulting in a fill factor of 85% and a TFOV of ±2° at 1550 nm wavelength.

The spring constant of the HCG suspension, based on the dimensions given above and silicon's Young's modulus set at $E = 169$ GPa, is estimated to be $k = 1.185$ N/m. The lightweight grating structure can be designed to obtain a high resonant frequency of 420 kHz thanks to its low mass of 139 pg. The FEM simulation result using COMSOL software reveals the fundamental mode to be around 420 kHz, with the mode shape shown in Figure 3.19. The specific design parameters of the HCG reflector for our MEMS OPA are shown in Table 3.2: grating period = 1250 nm, grating beam width = 570 nm, and thickness = 400 nm. The reflector, 20 × 20 μm², is tethered to four anchors with four springs. The widths (w) and thicknesses (t) of the springs are designed to be 350 and 400 nm, respectively. The length of each spring is 18.7 μm.

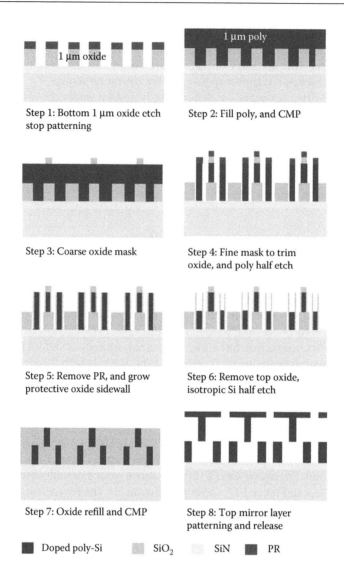

Figure 3.17 Fabrication process of the MEMS 1D OPA.

This developed 2D MEMS OPA successfully demonstrated the beam steering in one dimension [5]. Thanks to the light mirror weight, the achieved response time is less than 4 µs. This unique HCG design for the 2D MEMS OPA enabled a fast tuning speed, high power handling capability, and ultra-high reflectivity in the fiber communication wavelength range.

3.4 Summary of MEMS optical phased array

Optical phased arrays are ideal for monochromatic laser beam steering and dynamic hologram generation in general. They will greatly simplify beam scanning as well as pointing, resulting in a much more compact, more efficient, and lower cost overall system. Among all their potential applications, light detection and ranging (LIDAR) is a key enabling technology for self-driving cars and other autonomous vehicles. Most current LIDARs employ mechanical scanning units such as motors, which are bulky and often intrusive in their deployment. An OPA system will not only provide fast scanning in a compact form, but also enable more sophisticated beam forming such as simultaneous scanning and tracking of multiple objects, or even direct line-of-sight communications.

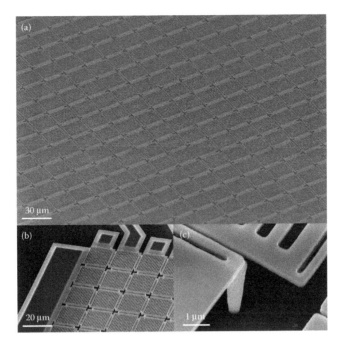

Figure 3.18 SEMs of the MEMS HCG-based 2D OPA. (a) Isometric view of the 32 × 32 array of MEMS HCG reflectors. (b) A corner of the array showing the anchor of the array and the bias lines. (c) Close-up view of one HCG element's anchor and spring area.

Figure 3.19 Simulated mode shape of a single HCG reflector element. The grayscale contour demonstrates the relative displacement in vertical direction.

Table 3.2 Specifications of the MEMS 2D OPA HCG mirror element

HCG dimension (14 silicon beams)		Spring dimension	
HCG beam width	570 nm	Length	18.7 μm
HCG period	1250 nm	Width	350 μm
Thickness:		400 nm	
Resonant frequency:		420 kHz	

Even with the limited options for development, a MEMS OPA shows significant advantages over various other types of phase modulators including liquid crystal, lithium niobate, and current silicon photonics devices. With state-of-the-art design and improved IC foundry fabrication techniques, MEMS OPA has overcome the conventional limitations imposed on it, and achieved a high fill factor and small element size. With the element size approaching half of the operation wavelength and a fill factor greater than 85%, MEMS OPAs have been designed in both 1D and 2D configurations and are revolutionizing the way OPAs are developed. For performance improvements of existing MEMS OPA designs as well as for the enabling of entire beam steering systems, novel MEMS structure designs are expected to solve problems, including inter-element crosstalk, array uniformity, large-array electrical addressing, and potentially the dispersion from the formed grating phase profile, etc. Future growth areas, including integration with adaptive optics, ultra-sensitive photodetectors, and enhanced optical pattern recognition methods, will enable more potential applications, which in turn may further drive the development of MEMS OPAs.

References

1. P. F. McManamon et al., Optical phased array technology, *Proc. IEEE.*, vol. 84, no. 2, pp. 268–298, Feb. 1996.
2. J. Levinson et al., Towards fully autonomous driving: Systems and algorithms, *2011 IEEE Intelligent Vehicles Symposium (IV)*, 2011, pp. 163–168.
3. J. Sun, E. Timurdogan, A. Yaacobi, E. S. Hosseini, and M. R. Watts, Large-scale nanophotonic phased array, *Nature*, vol. 493, no. 7431, pp. 195–199, Jan. 2013.
4. Y. Wang and M. C. Wu, An optical phased array for LIDAR, *In Journal of Physics: Conference Series*, vol. 772, no. 1, p. 012004. IOP Publishing, 2016.
5. B.-W. Yoo et al., A 32 × 32 optical phased array using polysilicon sub-wavelength high-contrast-grating mirrors, *Opt. Express*, vol. 22, no. 16, pp. 19029–19039, Aug. 2014.
6. N. R. Smith, D. C. Abeysinghe, J. W. Haus, and J. Heikenfeld, Agile wide-angle beam steering with electrowetting microprisms, *Opt. Express*, vol. 14, no. 14, pp. 6557–6563, Jul. 2006.
7. D. N. Hutchison et al., High-resolution aliasing-free optical beam steering, *Optica*, vol. 3, no. 8, p. 887, Aug. 2016.
8. R. A. Meyer, Optical beam steering using a multichannel lithium tantalate crystal," *Appl. Opt.*, vol. 11, no. 3, pp. 613–616, Mar. 1972.
9. Y. Ninomiya, Ultrahigh resolving electrooptic prism array light deflectors, *IEEE J. Quantum Electron.*, vol. 9, no. 8, pp. 791–795, Aug. 1973.
10. R. A. Soref and M. J. Rafuse, Electrically controlled birefringence of thin nematic films, *J. Appl. Phys.*, vol. 43, no. 5, pp. 2029–2037, May 1972.
11. A. F. Fray and D. Jones, Liquid crystal light deflector, US4066334 A, Jan. 03 1978.
12. R. M. Matic, *Blazed phase liquid crystal beam steering*, vol. 2120, pp. 194–205, 1994.
13. D. P. Resler, D. S. Hobbs, R. C. Sharp, L. J. Friedman, and T. A. Dorschner, High-efficiency liquid-crystal optical phased-array beam steering, *Opt. Lett.*, vol. 21, no. 9, pp. 689–691, May 1996.
14. P. F. V. Kessel, L. J. Hornbeck, R. E. Meier, and M. R. Douglass, A MEMS-based projection display, *Proc. IEEE*, vol. 86, no. 8, pp. 1687–1704, Aug. 1998.
15. R. S. Muller and K. Y. Lau, Surface-micromachined microoptical elements and systems, *Proc. IEEE*, vol. 86, no. 8, pp. 1705–1720, Aug. 1998.
16. R. T. Howe, R. S. Muller, K. J. Gabriel, and W. S. N. Trimmer, Silicon micromechanics: sensors and actuators on a chip, *IEEE Spectr.*, vol. 27, no. 7, pp. 29–31, Jul. 1990.
17. U. Krishnamoorthy, K. Li, K. Yu, D. Lee, J. P. Heritage, and O. Solgaard, Dual-mode micromirrors for optical phased array applications, *Sens. Actuators Phys.*, vol. 97–98, pp. 21–26, Apr. 2002.
18. O. Solgaard, F. S. A. Sandejas, and D. M. Bloom, Deformable grating optical modulator, *Opt. Lett.*, vol. 17, no. 9, pp. 688–690, May 1992.
19. S. D. Senturia, D. R. Day, M. A. Butler, and M. C. Smith, Programmable diffraction gratings and their uses in displays, spectroscopy, and communications, *J. MicroNanolithography MEMS MOEMS*, vol. 4, no. 4, pp. 41401–41401–6, 2005.

20. Y. Wang and M. C. Wu, Micromirror based optical phased array for wide-angle beamsteering, *In Micro Electro Mechanical Systems (MEMS), 2017 IEEE 30th International Conference on,* (pp. 897-900). IEEE, Jan. 2017.
21. J. W. Goodman, *Introduction to Fourier optics.* 2005.
22. T. G. Bifano, J. Perreault, R. K. Mali, and M. N. Horenstein, Microelectromechanical deformable mirrors, *IEEE J. Sel. Top. Quantum Electron.,* vol. 5, no. 1, pp. 83–89, Jan. 1999.
23. D. Hah, S. T. Y. Huang, J.-C. Tsai, H. Toshiyoshi, and M. C. Wu, Low-voltage, large-scan angle MEMS analog micromirror arrays with hidden vertical comb-drive actuators, *J. Microelectromechanical Syst.,* vol. 13, no. 2, pp. 279–289, Apr. 2004.
24. R. C. Hansen, *Phased Array Antennas.* John Wiley & Sons, vol. 213, 2009.
25. A. Toshev, Suppression of phaser quantization lobes for large flat phased array antennas, *IEEE Antennas and Propagation Society International Symposium. Digest. Held in Conjunction with: USNC/CNC/URSI North American Radio Sci. Meeting (Cat. No.03CH37450),* vol. 2, pp. 488–491, 2003.
26. A. D. Rakic, A. B. Djurisic, J. M. Elazar, and M. L. Majewski, Optical properties of metallic films for vertical-cavity optoelectronic devices, *Appl. Opt.,* vol. 37, no. 22, pp. 5271–5283, 1998.
27. I. W. Jung, S. Kim, and O. Solgaard, High-reflectivity broadband photonic crystal mirror MEMS scanner with low dependence on incident angle and polarization, *J. Microelectromechanical Syst.,* vol. 18, no. 4, pp. 924–932, Aug. 2009.
28. TI's Digital Light Processor Shrinks to Glasses Size | EE Times, EETimes. [Online]. Available: http://www.eetimes.com/document.asp?doc_id=1324839
29. B.-W. Yoo et al., Optical phased array using high contrast gratings for two dimensional beamforming and beamsteering, *Opt. Express,* vol. 21, no. 10, p. 12238, May 2013.
30. J. Sun et al., Two-dimensional apodized silicon photonic phased arrays, *Opt. Lett.,* vol. 39, no. 2, p. 367, Jan. 2014.
31. R. L. Haupt, Thinned arrays using genetic algorithms, *IEEE Trans. Antennas Propag.,* vol. 42, no. 7, pp. 993–999, Jul. 1994.
32. V. Pierro, V. Galdi, G. Castaldi, I. M. Pinto, and L. B. Felsen, Radiation properties of planar antenna arrays based on certain categories of aperiodic tilings, *IEEE Trans. Antennas Propag.,* vol. 53, no. 2, pp. 635–644, Feb. 2005.
33. T. G. Spence and D. H. Werner, Design of broadband planar arrays based on the optimization of aperiodic tilings, *IEEE Trans. Antennas Propag.,* vol. 56, no. 1, pp. 76–86, Jan. 2008.
34. O. Brand and H. Baltes, Micromachined resonant sensors—An overview, *Sens. Update,* vol. 4, no. 1, pp. 3–51, Aug. 1998.
35. D. López et al., Two-dimensional MEMS array for maskless lithography and wavefront modulation, *In Proceeding Microtechnologies New Millennium,* pp. 65890S1–65890S, 2007.
36. U. Krishnamoorthy, D. Lee, and O. Solgaard, Self-aligned vertical electrostatic combdrives for micromirror actuation, *J. Microelectromechanical Syst.,* vol. 12, no. 4, pp. 458–464, Aug. 2003.

4

Optical MEMS for space
Design, characterization, and applications

Frederic Zamkotsian

Contents

4.1 Introduction

Micro-optoelectromechanical systems (MOEMS) could be key components in the future generations of space instruments. In Earth observation, universe observation, and planet exploration, the return of the scientific instruments must be optimized in future missions. MOEMS devices are based on mature micro-electronics technology and in addition to their compactness, scalability, and specific task customization, they could generate new functions not available with current technologies. National Aeronautics and Space Administration (NASA), Japan Aerospace Exploration Agency (JAXA), European Space Agency (ESA), and Centre National d'Etudes Spatiales (CNES), that is, the US, Japanese, European, and French space agencies, have initiated several studies developing new MOEMS-based instruments.

Innovative instruments could be based on this new family of components. Replacing classical components by MOEMS will improve performances, but in order to develop breakthrough instruments, we have to identify and list new functions associated with several types of microelectromechanical systems (MEMS), and develop new ideas for instruments.

Designing components able to operate properly in space and survive launch conditions is very specific. In addition to the design rules of conventional MOEMS, specific parameters must drive their concepts, including harsh conditions, such as vacuum, cryogenic temperatures, radiation, vibrations, shocks, and acoustic loads. Low power consumption and compact electronics drivers are also required. Several MEMS devices are considered in terms of their performance and abilities for different functions; among them, we can cite:

- Programmable slits
- Programmable micro-diffraction gratings
- Micro-deformable mirrors

For the programmable slits, three devices are promising for developing new applications; two of them have been developed by institutes while the last one is commercially available. The first ones are the micro-shutter arrays (MSA) selected by NASA to be on board the Near Infrared Spectrograph (NIRSpec) instrument in James Webb Space Telescope (JWST). They have been designed, realized, and characterized by NASA's Goddard Space Flight Center. They use a combination of magnetic effect for shutter actuation, and electrostatic effect for shutter latching in the open position. They have been tested at 30 K and integrated in the instrument. JWST will be launched in 2018 and this device will be the first MOEMS device fully operational in a space instrument. A European development is under way between Laboratoire d'Astrophysique de Marseille (LAM, France), Ecole Polytechnique Fédérale de Lausanne (EPFL, Switzerland), and Centre Suisse d'Electronique et de Microtechnologies (CSEM, Switzerland) in order to develop micromirror arrays for generating reflective slit masks in future multi-object spectrographs. These programmable reflective slit masks are composed of 2048 individually addressable $100 \times 200 \ \mu m^2$ micromirrors in a 32×64 array. Each silicon micromirror is electrostatically tilted by a precise angle of at least 20° for an actuation voltage of 130 V. These micromirrors have demonstrated very good surface quality with a deformation below 10 nm, individual addressing using a line-column scheme, and they are working in a cryogenic environment at 162 K. Finally, the commercial array is the popular Digital Micromirror Device (DMD) from Texas Instruments (TI). The DMD features 2048×1080 mirrors on a 13.68-μm pitch and has been tested for space applications.

With the development of programmable slits, several functions have been listed and possible applications have been found. The functions involve the selection of objects or parts of an FOV, feeding instruments in two directions, selecting two pointing directions, modulation of the light intensity by zone on a temporal basis, pupil shape configuration, spectral selection, and slit masks generation.

Programmable diffraction gratings are a new class of MOEMS devices. Piston-motion parallel moving ribbons could be set locally as a grating and then diffract the light. If the incoming light spectrum is dispersed along the device, any wavelength could be selected or removed by tuning the device. Two devices have been developed: a silicon-based device designed and realized by laboratories (CSEM, LAM, EPFL) and a commercial array from Silicon Light Machines.

For programmable diffraction gratings, several functions have been listed and possible applications have been found. The functions include wavelength selection for spectroscopy or spectral tailoring, optical beam selection, and attenuation. The former two functions include temporal behavior and beam addressing by using the diffraction orders.

Micro-deformable mirrors are used for wavefront correction, mainly in adaptive optics systems. Boston Micromachines Corporation (BMC) produces the most advanced MEMS deformable mirrors. One prototype has been tested in space, and several others are under test in space environment facilities.

Several functions have been listed and possible applications have been found for micro-deformable mirrors. The functions include wavefront correction (active or adaptive modes depending on loop frequency), optical beam tailoring, and focal plane curvature compensation.

In this chapter, we will list all the new functions available for MOEMS devices on board space instruments, the device requirements to fulfill the harsh space environment, review the different MOEMS candidates for space, discuss the characterization needs through several examples, develop the different types of applications/missions including universe observation and Earth observation, and finally, give some perspectives for further developments.

4.2 Instrumental functions

For each MOEMS type, functions have been listed below. The position of the MOEMS device in the optical design is noted as "i" if the device is in the image plane, or "p" if the device is in the pupil plane.

Table 4.1 Functions with micromirror arrays

Function	Position	Description
F1	i	Selection of objects or part of field of view (FOV)
	p	Pupil configuration
F2	i	Flux orientation in two directions (feeding two instruments with
	p	the same FOV)
F3	i	Observation along two directions and possible combinations (two
	p	FOV observed with a single instrument)
F4	i	Temporal modulation of the flux within FOV or part of FOV
	p	
F5	i	Spectral selection, large FOV programmable spectrograph

4.2.1 Functions with micromirror arrays

In Table 4.1, new functions using micromirror arrays are listed and described. Full descriptions of these functions are given in the following paragraphs.

4.2.1.1 Function F1: Selection of objects or part of FOV and pupil configuration

Using micromirror arrays located in the image plane of an optical system, objects or part of an FOV could be selected. This function is one of the first investigated functions with this type of MEMS in astronomy. Multi-object spectroscopy (MOS) allows measuring the infrared spectra of faint astronomical objects that provide information on the evolution of the universe. By placing the programmable slit mask in the focal plane of the telescope, the light from selected objects is directed toward the spectrograph, while the light from other objects and from the sky background is blocked. This application will be developed in Section 4.6. A schematic view of this function is shown in Figure 4.1a.

Any windowing (open and/or close) application can be foreseen in this function, including coronography or the masking of any part of the FOV, following any shape. This application is limited by the "resolution" of the device, that is, the number of micro-elements in the device versus the FOV. If the micromirror array is located in a pupil plane, the pupil could be tailored as desired. Application

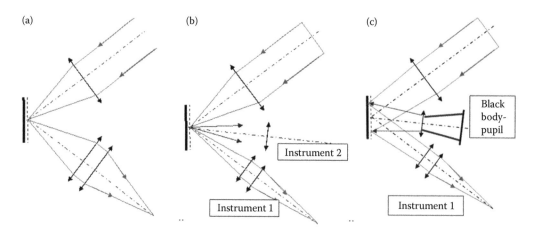

Figure 4.1 (a) Function F1: Selection of objects or part of FOV. (b) Function F2: Flux orientation in two directions. (c) Function F3: Observation along two directions and possible combination.

in the apodization or in the masking part of the pupil is possible. Removing wavefront defects or simulating a diaphragm is also possible. However, the size of the pupil is limited to the size of the array if a monolithic pupil is considered; buttable components may lead to larger pupils.

4.2.1.2 Function F2: Flux orientation in two directions

Flux incoming on the micromirror array located in an image plane could be oriented in two directions, for example, for feeding two instruments with the same FOV (or part of the same FOV). This function could be applied simultaneously or sequentially. Then the collecting optics is unique and is fully shared by the instruments. For the MOEMS device requirements, the tilt angle positions of the device must be precisely determined in order to properly feed the instruments. This function could also be used as an achromatic beamsplitter. A schematic view of this function is shown in Figure 4.1b.

If the micromirror array is located in a pupil plane, the function is identical. However, the size of the pupil is limited to the size of the array if a monolithic pupil is considered; buttable components (see, for example, MIRA component in Section 4.3) may lead to larger pupils.

4.2.1.3 Function F3: Observation along two directions and possible combinations

This original function permits to use and/or combine flux coming from two directions in the same instrument, simultaneously or sequentially. A schematic view of this function is shown in Figure 4.1c. For example, this function could be used in an instrument where a fast calibration is needed after each exposure, on the sky or on a black body inside the instrument. The tilting speed is a key requirement in this application. Instrument design as well as system feasibility have to be investigated.

4.2.1.4 Function F4: Temporal modulation of the flux within FOV or part of FOV

Temporal modulation of the optical beam is possible with the micromirror arrays thanks to their high resonance frequency. Within each exposure, the position of each micromirror could be changed from one stable position to the other stable position, modulating the incoming flux on this mirror, that is, on the associated detector pixel(s). This function is commonly used with these devices (e.g., DMD components from Texas Instruments) for display applications. In our case, saturation of the signal could be avoided on the FOV or part(s) of the FOV. The incoming signal could also be modulated for application in heterodyne or synchronous detection. The signal-to-noise ratio could then be increased by orders of magnitude. This concept could be used in the infrared or in the visible for discriminating the signal and the noise.

Dynamical masking of the FOV or part of the FOV is also possible with applications of faint objects' detection, "programmable neutral density filter" generation or tailoring the incoming beam according to detector dynamic range. Limitation of this function is linked to the resonance frequency of the micro-elements as well as the contrast of the device. A schematic view of this function is identical to function 1, and is shown in Figure 4.1a.

4.2.1.5 Function F5: Spectral selection, large FOV programmable spectrograph

For this concept, the principle is to use a MOEMS component to select the wavelengths. Two MOEMS components could be considered, such as programmable micro-diffraction gratings (PMDG) and micromirror arrays like the DMD from TI. Indeed, this component is placed in the focal plane of a first diffracting stage (using a grating for instance) and is used as a wavelength selector by reflecting or switching on/off the light (by diffraction for the PMDG or by deflection for the DMD). Then the light is combined at the second diffraction stage as shown in Figure 4.2. It then becomes possible to realize a programmable and adjustable filter in both wavelength (λ) and bandwidth ($\Delta\lambda$).

For a point-like object, a 1D MOEMS device is required. A 1D demonstration has been already done using a PMDG device [1,2]. For a 1D FOV, a 2D device is mandatory, limiting the MOEMS choice to a micromirror array (Figure 4.2). Then, on the DMD surface, the spatial dimension is

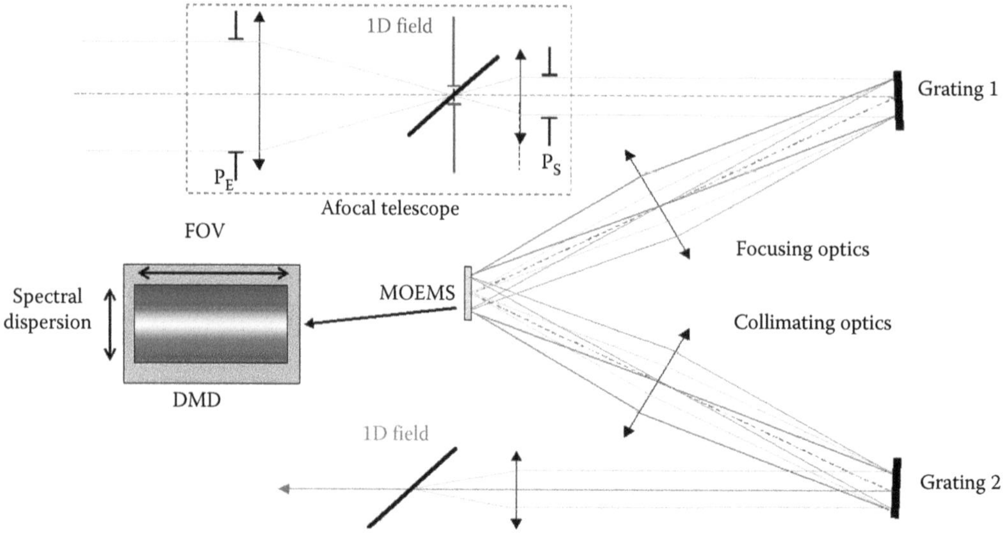

Figure 4.2 Function F5: Spectral selection, large FOV programmable spectrograph. (Copyright SPIE. Reproduced with permission.)

along one side of the device and the spectrum for each spatial point is displayed along the perpendicular direction. Each spatial and spectral feature of the 1D FOV is fully adjustable and/or programmable for each exposure or even during an exposure by using the temporal behavior of the component.

4.2.2 Functions with programmable micro-diffraction gratings

In Table 4.2, new functions using programmable micro-diffraction gratings (PMDG) are listed and described. Full descriptions of these functions are given in the following paragraphs.

4.2.2.1 Function F1: Spectral selection

Using a PMDG located in the image plane of an optical system, the spectrum of the source could be tailored. This function is one of the first investigated functions with this type of MEMS in astronomy [1,3]. The spectral resolution is directly linked to the number of micro-elements in the device. A schematic view of this function is identical to Figure 4.2, where the DMD device is replaced by PMDG, and the input and output fields are point-like.

4.2.2.2 Function F2: Flux selection

When used as a flux selector, a PMDG is based on the diffraction efficiency of the device. By placing a stop in the Fourier plane, the diffracted light could be blocked (OFF position) while the

Table 4.2 Functions with programmable micro-diffraction gratings

Function	Position	Description
F1	i	Spectral selection
F2	i	Flux selection
F3	i	Temporal modulation of the flux
F4	i	Flux reorientation by diffraction effect

nondiffracted light goes out of the optical system (ON position). In a more complex optical design, the diffracted light could be the ON position.

4.2.2.3 Function F3: Temporal modulation of the flux

Temporal modulation of the optical beam is possible with a PMDG, thanks to the ribbons' high resonance frequency. Within each exposure, the position of each ribbon could be changed in an analog or digital way, modulating the incoming flux on each PMDG pixel. This function is commonly used with these devices for display applications. The image is obtained by columns, and the lines are populated using an additional scanning mirror [4,5].

4.2.2.4 Function F4: Flux reorientation by diffraction effect

This potential function could be very efficient if the diffraction efficiency of the grating could be adjusted by a blazed grating configuration. However, the realization of blazed gratings at the scale of the PMDG ribbons, that is, around 4 µm, is a challenge.

4.2.3 Functions with micro-deformable mirrors

In Table 4.3, new functions using micro-deformable mirrors are listed and described. Full descriptions of these functions are given in the following paragraphs.

4.2.3.1 Function F1: Wavefront correction

Wavefront correction like adaptive optics systems is based on a combination of three elements, namely the wavefront sensor for the measurement of the shape of the wavefront arriving in the telescope, the deformable mirror as the wavefront correcting element, and, finally, the real time computer closing the loop of the system at a frequency ranging from 0.5 to 2 kHz in order to follow the evolution of the atmospheric fluctuations (Figure 4.3).

Using a micro-deformable mirror located in the pupil plane of an optical system, a distorted wavefront could be corrected. This function is one of the first investigated functions with this type of MEMS in astronomy. Active or adaptive modes are used depending on the loop frequency. Atmosphere turbulence, optical system aberrations, alignment residuals, and dynamic effects (thermal, gravity) could be corrected. The wavefront correction is directly linked to the number of actuators, their stroke, and their resonance frequency. The actual size of the micro-deformable mirror also limits the FOV due to the Lagrange invariant of an optical system.

4.2.3.2 Function F2: Optical beam tailoring

Optical beam tailoring/shaping is of prime interest for high power laser beams in order to avoid unexpected peak intensities. Designing MOEMS deformable mirrors to enable the system to operate under high optical fluxes is mandatory.

4.2.3.3 Function F3: Focal plane curvature compensation

If the micro-deformable mirror is located in an image plane, the focal plane curvature could be corrected. This function is limited by the stroke of the component, usually from a few microns to tens of microns.

Table 4.3 Functions with micro-deformable mirrors

Function	Position	Description
F1	p	Wavefront correction (active or adaptive modes depending on loop frequency)
F2	p	Optical beam tailoring
F3	i	Focal plane curvature compensation

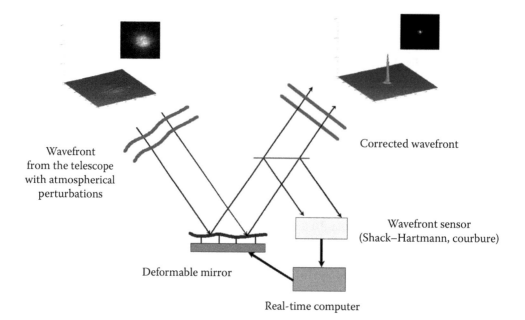

Wavefront
from the telescope
with atmospherical
perturbations

Corrected wavefront

Wavefront sensor
(Shack–Hartmann, courbure)

Deformable mirror

Real-time computer

Figure 4.3 Function F1: Wavefront correction.

4.3 MOEMS requirements for space

Space environment could affect MOEMS due to their design, their materials or their packaging. The main failure modes of MOEMS devices are [6]:

- Mechanical failure modes, including stiction and wear, fatigue, plastic deformation, delamination, curvature change, shock, and vibration-induced fracture
- Electrical failure modes, including short and open circuits, arcing across small gaps, electrostatic discharge, dielectric charging, and corrosion.

Specific parameters must drive space-oriented MOEMS design. Most of them are linked to the harsh conditions surrounding the devices while in operation. These specific parameters are

- Vacuum
- Cryogenic temperatures
- Radiation
- Vibrations and shocks
- Acoustic loads

4.3.1 Vacuum

If the devices are not packaged in sealed mounts, they must be operated in ultra-vacuum conditions. This leads to the consideration of only non- or low-outgassing materials. In terms of operation, electrostatic actuators may suffer from sticking effects or fast degradation on contact surfaces. In vacuum, the damping effect from the air is removed, increasing the speed of moving elements and adding bumping effects of the mobile parts. Contact parts must be designed accordingly; new driving schemes could also be considered.

4.3.2 Cryogenic temperatures

This is a common condition for all components in space if they are not placed in a dedicated temperature-regulated area. Cryogenic temperatures cause deformation of the optical surfaces, degradation/delamination of the coatings, or possibly block the devices: this could be a cause of single point failure. Thus, these effects must be considered in the design and realization phases, and characterized carefully in order to guarantee the proper operation of the MOEMS device when in orbit. This domain is still very new and there is not yet a dedicated and official procedure established by the agencies. In any case, the learning curve is incremental and the technology readiness level (TRL) chart has to be followed with care for convincing the space agencies and placing MOEMS devices on board future instruments in space.

Cold temperatures could range from 100 to 150 K in low orbit missions down to 30 K for instruments passively cooled at Lagrange point L2, which is the case for JWST instruments. These temperatures correspond to instruments working at wavelengths up to 30 μm; for instruments working at longer wavelengths, the devices must operate at even lower temperatures.

4.3.3 Radiation

The mechanical properties of silicon and metals are mostly unchanged when considering radiation conditions in space. We could then claim that silicon as a structural material is intrinsically radiation hard. But the control-driving electronics may suffer from radiation. Radiation damage typically causes latch-up or single event upsets for electronics; these damages could be curable or permanent. The devices and their electronics must be designed to be radiation hard as much as possible, or at least designed to slow down device degradation. Additional radiation hardening could be done by shielding the components. Two types of radiation are usually evaluated: Total ionizing dose (TID) test for cumulative degradation measurement and proton single event effects (SEE) radiation test. Classical TID values must exceed 40 to 50 krad for classical 5-year missions and the SEE have to be avoided or at least detected and recorded in the instruments.

4.3.4 Vibrations and shocks

The launch will generate very high figures in terms of vibrations and shocks, which can cause heavy damage from the satellite level down to the component level. Thus, this issue must be studied in detail for all devices. The small sizes and masses of elementary elements in the MOEMS devices lead to high resonance frequencies, thus resulting in very low to nonsensitivity of the MOEMS architecture to vibrations and shocks. But the packaging might still be affected. In some cases, latching the moving parts of the devices may be a solution if powered components are accepted during launch.

4.3.5 Acoustic loads

Acoustic loads may also cause dramatic damage to the components during the launch of the satellite. Proper design and packaging of the devices are mandatory. As considered for vibrations and shocks, latching the moving parts of the devices may be also be a solution.

More conventional parameters must also be adapted for the space design conditions, such as power consumption and device volume.

4.3.6 Power consumption

In satellites, power consumption must be minimized for all subsystems. MOEMS devices might also fulfill this requirement and are designed to be within the allocated power budget. This condition may push for electrostatic actuation (voltage-driven), instead of thermal or magnetic actuation (current driven).

4.3.7 Device volume and mass

While MOEMS' volume and mass are usually small, their driving electronics as well as their mounts must be optimized for minimizing these two parameters.

4.4 MOEMS for space

MOEMS devices have been developed or successfully used in a wide range of ground-based commercial applications, from telecom to life science, and from imaging to spectroscopy. MOEMS are not yet widely used for space applications, but this technology will most likely provide the key to new science and instrumentation in space. Several potential MOEMS devices for future space missions are described. The first major application will be the micro-shutters array in near infrared multi-object spectrograph (NIRSpec) for the James Webb Space Telescope (JWST).

4.4.1 Micro-shutters

Next generation MOS for space, such as the NIRSpec for the JWST will use a programmable multi-slit mask [7], which will be used for selecting objects. The programmable multi-slit mask requires remote control of the multi-slit configuration in real time. Micro-shutter arrays (MSA) have been selected for generating transmissive slits of NIRSpec, with the ability to record more than 100 targets in a single exposure.

NASA's Goddard Space Flight Center (GSFC) has developed the MSA. They use a combination of a magnetic effect for shutter actuation and an electrostatic effect for shutter latching in the open position. The individual size of a micro-shutter is 100×200 μm^2. The shutter is composed of several layers, that is, a silicon nitride blade coated with a cobalt-iron layer for the magnetic actuation. Light shields are realized in each cell, surrounding each shutter in order to prevent light leakage. Submicron bumps and micron-scaled ribs are fabricated on the light shields and the side walls to prevent stiction. The front side of an individual 100×200 μm^2 micro-shutter, without the light shields, is shown in Figure 4.4a [8]. MSAs have been successfully tested at cryogenic temperatures and results will be presented in the next paragraph.

The micro-shutter assembly located at the focal plane of NIRSpec is composed of 4 quadrants. Each quadrant is a 365×171 shutters array (Figure 4.4b). A macroscale magnet is sweeping above the assembly for opening the shutter; a line-column addressing technique permits the latching of the desired combination of shutters.

Two failure modes are possible, in the open and close positions. The first one is the more problematic and these failed open shutters are manually and mechanically closed by local opaque patches.

A new generation of MSA is currently under study at GSFC, based only on the electrostatic effect for opening and latching the shutters. This new design will permit avoiding the complex and heavy magnet system moving in front of the array needed in the previous design. Cryogenic and vacuum tests will be done on these new MSAs.

(a) 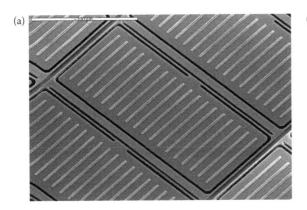 (b)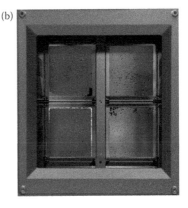

Figure 4.4 Micro-shutters array developed at NASA's Goddard Space Flight Center. (a) Individual $100 \times 200\ \mu m^2$ micro-shutter, front side, without the light shields; (b) packaged 4 quadrants micro-shutter assembly for JWST NIRSpec instrument. (Copyright NASA.)

4.4.2 Micromirror arrays: DMD

Digital micromirror devices (DMD) from Texas Instruments are the most popular MOEMS devices used in image display applications [9]. DMDs could act also as an objects' selection reconfigurable mask. Let us consider the largest DMD chip developed by TI for covering the largest FOV in the instruments; it features 2048×1080 mirrors on a 13.68 μm pitch, where each mirror can be independently switched between an ON (+12°) position and an OFF (−12°) position. The MOEMS architecture is built on top of a CMOS driving circuit, and it is based on metallic alloys and realized by surface micromachining. Each 13-μm-square individual mirror is made of aluminum (Figure 4.5a). The DC2k packaged chip is shown in Figure 4.5b and is primarily used for displaying movies in movie theaters.

This component has been extensively studied in the framework of an ESA technical assessment using this DMD DC2k component for space applications, for example, in the EUCLID mission. The ESA's EUCLID mission is a cosmology mission devoted to a dark matter and dark energy search in the universe by mapping the galaxies' distribution in 3D. Redshift values of 100 million galaxies will be revealed by the measurement of their spectrum analysis. During the early phase studies, a MOS instrument based on a DMD device has been assessed. Due to complexity and cost reasons, slitless spectroscopy was chosen for EUCLID despite a much higher efficiency with slit spectroscopy.

(a) 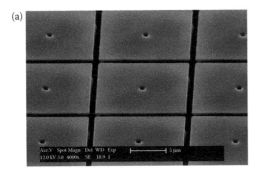 (b)

Figure 4.5 Digital micromirror device (DMD) from Texas Instruments. (a) Individual $13 \times 13\ \mu m^2$ micromirror, front side; (b) packaged DC2k device featuring 2048×1080 micromirrors.

For a first space evaluation of this device, ESA has engaged a study led by LAM. A specialized driving electronics and a cold temperature test set-up have been developed. Our tests reveal that the DMD remains fully operational at −40°C and in vacuum. A 1038-hours life test in space survey conditions (−40°C and vacuum) has been successfully completed. TID and SEE radiation tests, thermal cycling (over 500 cycles between room temperature and cold temperature, on a non-operating device), and vibration and shock tests have also been done; no degradation has been observed from the optical measurements. *These results do not reveal any concerns regarding the ability of the DMD to meet environmental space requirements* [10]. The characterization tests will be extensively presented later.

The DMD device cannot be customized and cannot operate at cryogenic temperatures. An alternative solution specifically designed for space applications is under development in Europe.

4.4.3 Micromirror arrays: MIRA

In Europe, an effort is currently under way to develop single-crystalline silicon micromirror arrays for future generation infrared multi-object spectroscopy [11]. A collaboration within Laboratoire d'Astrophysique de Marseille (LAM, France), Ecole Polytechnique Federale de Lausanne (EPFL, Switzerland), and Centre Suisse d'Electronique et de Microtechnologies (CSEM, Switzerland) has proposed to develop a European programmable MMA that can be used as a reflective slit mask for MOS; the project is called *MIRA*. The requirements for our MMA were determined from previous simulation results and measurements [12]. It has to achieve a high optical contrast of 1000:1 (goal: 3000:1), a fill factor of more than 90%, and a mechanical tilt angle greater than 20°. Furthermore, the performance must be uniform over the whole device; the mirror surface must remain flat in operation throughout a large temperature range and it has to work at cryogenic temperatures.

This MMA concept is based on the electrostatic double plate actuator. A micromirror is suspended by two polysilicon flexion hinges, which were attached to a sustaining frame (Figure 4.6a). To generate an electrostatic force, an electrode is placed underneath the micromirror and pillars are placed to set a precise electrostatic gap. Two landing beams were placed under the micromirror to keep it from touching the electrode and generating short circuits during the actuation. A stopper beam is placed under the frame to precisely set the tilt angle of the micromirror after actuation and electrostatically lock it in this position.

When a voltage higher than the pull-in voltage was applied on the electrode, the micromirror was attracted toward the electrode by an electrostatic force. During this motion, it touched its stopper beam (Figure 4.6b) and landed on its landing pads. Therefore, after pull-in, the micromirror is electrostatically clamped at a precise tilt angle (Figure 4.6c). When the voltage is decreased, the micromirror takes off from its stopper beam. When the restoring force of the flexure beams is higher than the electrostatic force, the micromirror returns to its rest position.

The $100 \times 200~\mu m^2$ micromirrors were made of single-crystal silicon, assuring flat optical surfaces. Because silicon is transparent in the infrared range, a gold thin-film coating was deposited on the topside of the mirrors. For MMA realization, a combination of bulk and surface silicon

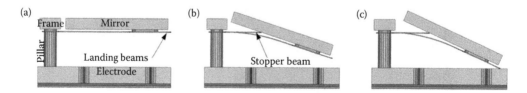

Figure 4.6 Schematic of MIRA micromirror (a). When a voltage is applied on the electrode, the micromirror is attracted toward its electrode by the action of an electrostatic force. During this motion, the mirror touches its stopper beam (b) and at the end of the motion, it is clamped at a precise tilt angle thanks to its contacts with its stopper beam and its landing pads (c).

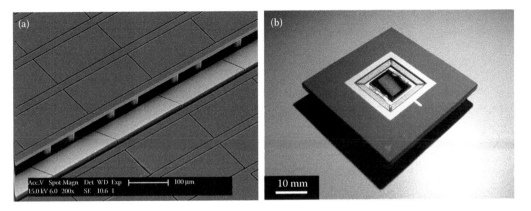

Figure 4.7 64 × 32 micromirror array. (a) Close view of 100 × 200 μm² micromirrors. (b) Mounted MMA in a pin grid array (PGA).

micromachining was used. They were made of two wafers: one for the mirrors and one for the electrodes, which were processed separately and assembled by wafer level bonding. Prototypes of MMA with 2048 individually addressable micromirrors (64 × 32 mirrors) have been successfully realized (Figure 4.7) [13].

MMA have been tested at LAM on bench set-ups dedicated to the characterization of MOEMS devices. The surface quality of the micromirror is measured by phase-shifting interferometry, and a total deformation of 10 nm peak-to-valley is measured, with 1 nm roughness. These mirrors can be electrostatically tilted by 24° at an actuation voltage of 130 V. In many MOS observations, astronomers need to have the spectrum of the background near the studied object ("long slit" mode, see Figure 4.7a). Our locking mechanism is designed to ensure this goal and an angle of a few arcminutes' difference has been obtained. The fill factor is 83% for the mirror surface and 98% in the direction along the micromirror lines. The MMA is then mounted in a pin grid array (PGA) for connection to the electronics (Figure 4.7b). Individual addressing of the mirrors is based on a line-column scheme. As a proof of concept, a 2 × 2 subpart of an MMA of 32 × 64 micromirrors was successfully actuated. The contrast of a micromirror was characterized on a dedicated optical bench at LAM: a contrast ratio of 1000:1 was obtained.

In order to avoid spoiling the astronomical objects' spectra by the thermal emission of the instrument, the micromirror array has to work in a cryogenic environment. *The micromirrors can be successfully actuated before, during, and after cryogenic cooling at 162 K.* We can measure the surface quality of the gold-coated micromirrors at room temperature and at 162 K without a large deformation difference. *A 9.8 nm peak-to-valley (PTV) surface deformation was measured at 293 K, which increased to 27.2 nm PTV at 162 K. When coming back to room temperature, we again measured the mirror surface deformation and obtained a value of 9.9 nm PTV, identical to the value measured before cooling of the array.* The deformation is due to the coefficient of thermal expansion (CTE) mismatch between the thick silicon micromirror and the thin gold coating layer on top. However, the surface deformation stays within the limit of 50 nm. Cryogenic characterization of the MIRA is described in the next paragraph.

In future space instruments for universe and Earth observation, FOVs will increase to larger scales than that of the current FOVs. In order to not degrade the device yield too much, the limit of the component size is fixed at around 20 × 20 mm². For reaching larger actual areas, we need to develop a mosaicking strategy as was done for paving large FOVs with the CCDs. For MIRA's next goals, we focus our efforts in two directions:

1. Increasing the size of the arrays up to 20 × 20 mm², that is, 100 × 200 micromirrors.
2. Develop and assemble integrated electronics on board for device control-command at each actuator level.

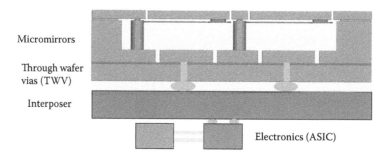

Figure 4.8 Schematic view of MIRA integrated with a hardened electronics (ASIC).

For the first item, MIRA design will be scaled up to the given size. Thanks to its original design, this step will be done homothetically, taking care, however, to maintain the quality of the wafer level bonding on pillars between the mirror- and the electrode-wafers. The yield of the realized arrays will be a key parameter. For the second item, a schematic view of MIRA is given in Figure 4.8. We are currently developing a through wafer vias (TWV) technology for moving the addressing lines from the top of the electrode wafer to its back side. This will increase the electrode areas on the top side and will ease the routing of the command lines on the back side. Below the MOEMS architecture, an integrated electronics (ASIC) will be attached. An interposer may be needed in order to adapt the pitch of the MOEMS actuators to the pitch of the contacts on the ASIC. This high voltage hardened ASIC will be designed for a space environment. Existing chips already developed by micro-electronics foundries could be used or newly designed chips might be developed. This is under evaluation.

MIRA with its integrated electronics is scheduled to be available in the next three years.

4.4.4 Programmable micro-diffraction gratings

Programmable MEMS diffraction gratings are used for spectroscopic applications because of their potential in tailoring visible and infrared spectra. Piston-motion parallel moving ribbons could be set locally as a grating and then diffract the light. If the incoming light spectrum is dispersed along the device, any wavelength could be selected or removed by tuning the device. Two devices have been studied: a commercial array from Silicon Light Machines (SLM) and a silicon-based device designed and realized by laboratories (CSEM, LAM, EPFL). While the first device is not designed for space application, the second one is potentially designed and developed for space instruments.

The PMDG device made by SLM consists of 1086 "pixels," each with 6 ribbons per pixel (3 fixed, 3 variable in Z location) [4]. The width of the ribbons is 3.775 μm, and the gap width is 0.475 μm, leading to a pitch of 4.250 μm; the length of the ribbons is 220 μm supported between two posts. A window with an antireflecting coating in the visible protects the device (Figure 4.9). The printed circuit board (PCB) is connected to an external electronics box through a flex cable, and the electronics is linked to the computer with a serial cable. We have developed the serial link configuration and the driving software in MATLAB®, in order to directly drive the PMDG device via the serial link. Our measurements were conducted in the zeroth order with a spatial filter to remove the higher orders in the Fourier plane. When the ribbons are in the rest position, they are all in the same plane, generating a flat mirror-like surface. When the ribbons are actuated and set at the λ/4 position, these pixels diffract the incoming light, and while this light is blocked in the Fourier plane, they appear dark on the charge coupled device (CCD) camera. A close-up view focused on a few PMDG pixels is shown in Figure 4.9, where all ON and one ON—one OFF pixels are displayed. Note that this feature is obtained only in the active area of the PMDG, that is, within 150 μm located at the center of the ribbons. Outside the active area, no diffraction occurs and the Al-coated surfaces reflect light. For spectrum tailoring, the incoming light is dispersed

(a)

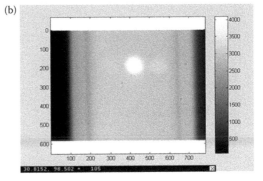
(b)

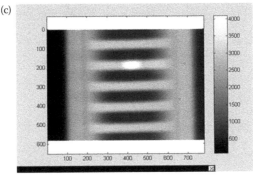
(c)

Figure 4.9 PMDG component by SLM (a). Close-up view of a few PMDG pixels: with all pixels ON (b), and with a serial of one ON and one OFF pixels (c).

and then imaged on the PMDG surface. According to the wavelength λ arriving on each pixel, the OFF state is obtained when the stroke of the ribbons is exactly equal to $\lambda/4$. A calibration step is required for determining the right voltage to be applied for obtaining this condition. We can then implement the optimal OFF position for any wavelength on any pixel. These values have been measured for all PMDG pixels in the field of view. The contrast is determined as the light flux ratio between ON and OFF states. The measured contrast is better than 30 over our entire wavelength band (550–650 nm) [1].

A silicon-based device has been designed and realized by European laboratories (CSEM, LAM, EPFL) [14]. A fully programmable MEMS diffraction grating (FPMDG), where every micromirror can move independently in a range $0-\lambda/2$, where λ is the wavelength of light, leads to a better control of the intensity for each wavelength in the synthetized spectrum—the intensity can take any value from 0 (micromirror $\lambda/4$-condition) to the maximum (no micromirror displacement). The FPMDG chip contains 64 micromirrors which are electrostatically actuated. Rigid Si micromirrors are connected to side electrodes via linkage arms, permitting the micromirror to follow a pure vertical displacement, and reducing micromirror bending throughout actuation (Figure 4.10a). Microfabrication is based on a 4-mask photolithography process, using SOI and Pyrex wafers. The FPMDG is then mounted in a PGA (Figure 4.10b).

Each micromirror of the FPMDG chip can move by 1.25 µm at voltages below 100 V. Two families of micromirrors, 50 or 80 µm wide, show negligible cross-talk during actuation. The bowing is as small as 0.14 over a 700-µm long micromirror and remains unchanged throughout actuation. Extinction ratios of up to 100 have been achieved by actuating only three adjacent micromirrors. The measurements have shown high stability and good reproducibility over time. Finally, FPMDGs are used to demonstrate shaping of the input spectrum: the intensity in a particular wavelength region is controlled through independent actuation of a set of adjacent micromirrors. The result is attenuation or cancellation of the corresponding wavelengths (Figure 4.11).

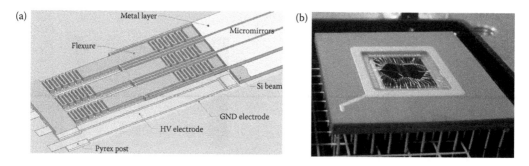

Figure 4.10 (a) FPMDG schematic view; (b) FPMDG mounted in a PGA.

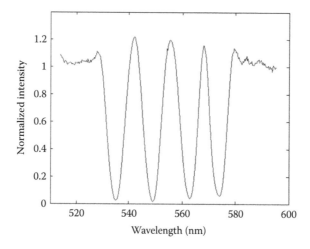

Figure 4.11 Spectrum of the input light dispersed over the FPMDG actuated at 4 spectral positions; the response is normalized by the spectral response of non-actuated FPMDG.

4.4.5 Micro-deformable mirror

Boston Micromachines Corporation (BMC) produces the most advanced MEMS deformable mirrors [15]. The concept is based on an array of electrostatic actuators linked one by one to a continuous top mirror (Figure 4.12), or a segmented mirror. Their main parameters are approaching the requirement values, that is, a large number of actuators (up to 4096, see Figure 4.12), a large

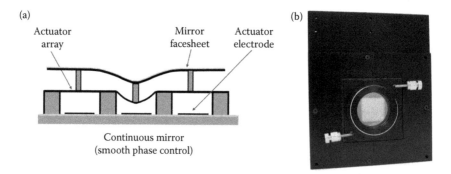

Figure 4.12 (a) Micro-deformable mirror principle; (b) 4095 actuator BMC deformable mirror. (Courtesy BMC company.)

stroke (up to 5.5 μm), and good surface quality, but they still need large voltages for their actuation (150–250 V).

Space qualification is an issue and to our knowledge, tests have been conducted for that purpose. NASA has selected BMC for two Phase 1 contracts in order to develop devices and electronics suited for space applications (main goal: wavefront control in space-based high contrast imaging instruments).

Device space characterization results are presented in the next section.

4.5 MOEMS characterization

MOEMS characterization in a space environment is the key to increase the TRL level of these devices in order to be considered first in future instrumental concepts and then to be actually integrated into space instruments. We will develop four examples of MOEMS device testing, showing the wide range of measurements to be made.

4.5.1 Micro-shutter characterization

MSA space qualification has been fully and successfully done by the GSFC-NASA team involved in this project [8]. This work has been conducted both at the device level and at the subsystem quadrant level.

At the MOEMS level, the characterization includes prescreening, open/close test, electrical test, and bowing test. The micro-shutters have been designed in order to be flat at 30 K, leading to curved shutters at room temperature. Thus, characterization of the MSA could only be done at cryogenic temperatures to avoid degradation of the light shields if the curved shutters are actuated at ambient temperature. A bowing test has been conducted, but no precise data on the actual bow at cryogenic or ambient temperatures are available. The open/close test has two failure modes: failed open and failed close. In the flight-format MSA, 150 failed closed shutters and 3 failed open shutters have been revealed, well below the requirements.

At the subsystem level, that is, the MSA quadrant level, the tests include 2D-addressing, life cycling, optical tests, and environmental tests. Two-dimensional-addressing is done by a line-column addressing scheme where the sweeping magnet, moving over the array, "pushes" the shutters and a 40 V voltage is applied to latch the shutters; 20 V is used for keeping the shutters open. The shutters are released by synchronizing the switching-off voltage and a damping magnetic force from the sweeping magnet in order to avoid shutter impacts on the light shields. Two-dimensional-addressing has been successfully demonstrated on the whole array. Life cycling (1 million cycles) at cryogenic temperatures as well as radiation, vibration, and acoustic tests were successful. Optical testing is a key issue as the shutters have to block the unwanted light in the FOV (background light and nonselected sources light) with a required contrast exceeding 2000. The contrast is defined by the ratio of the light passing through an open shutter with the light leaking behind a closed shutter. Contrast >10,000 has been measured, well above the requirement.

Finally, complete MSA quadrant assembly has been functionally tested and delivered to the NIRSpec consortium.

4.5.2 MIRA characterization

In order to avoid spoiling of the astronomical objects' spectra by thermal emission of the instrument, the array has to work in a cryogenic environment. MIRA is conceived such that all structural elements (single and polycrystalline silicon) have a matched CTE in order to avoid deformation within the device when cooling down to the operating temperature.

Figure 4.13 Cryogenic chamber installed on our interferometric setup for characterizing our MMA in space environment. Micromirrors could be successfully actuated before, during, and after cryogenic cooling at 162 K.

For characterizing the surface quality and the performance of these MMAs at low temperature, a cryogenic chamber optically coupled to a high-resolution Twyman–Green interferometer has been developed [16]. The interferometer provides a subnanometer accuracy, and the cryogenic chamber allows pressure down to 10^{-6} mbar and cryogenic temperatures (Figure 4.13). In order to get such temperatures, the chamber is equipped with an internal screen that radiatively insulates the sample from the chamber. Control of the environment is obtained by means of temperature sensors and local heaters. They are wired to the outside environment through a Dutch connector and connected to custom control electronics [17].

The MMA device is packaged in a pin grid array (PGA) chip carrier. The PGA is inserted in a zero insertion force (ZIF)-holder integrated on a PCB board. Large copper surfaces on the PCB facilitate cooling down the system; renouncing the solder-stop layer eases outgassing of the PCB FR4 base material during evacuation of the chamber. The PCB itself is mounted via a fix-point-plane-plane attachment system to a solid aluminum block, the latter being interconnected to the cryogenic generator. Thick copper wires between the PCB and the aluminum block further enhance thermal transport between the sample chip and the cryostat. A 100-pins feed-through connector links the chip with a custom built MMA control electronics. Temperature sensors are connected to the aluminum block and to the grid zip connector adjacent to the sample chip.

The micromirrors can be successfully actuated before, during, and after cryogenic cooling at 162 K. Figure 4.14 displays part of the mirror array in the direction of the spectrograph for different mirrors and at different temperatures. In Figures 4.14b and c, micromirrors are tilted when 130 V is applied both at room temperature and at 162 K. However, although all mirrors are tilted at room temperature, at 162 K, mirrors 1 and 3 in the first column are not fully tilted. This voltage is not large enough for tilting all of the mirrors; by increasing the voltage to 148 V, an additional mirror (mirror 1 of the first column) could be tilted (Figure 4.14d). This effect is due to a combination of a low doping level in the flexion hinges supporting the mirrors and a decrease of carrier speed at low temperature, thus decreasing the applied electric field at the actuator level at cryogenic temperatures with respect to the electric field at ambient temperature. Although our electronics are limited to 150 V, a higher voltage could permit to successfully tilt all mirrors.

We can measure the surface quality of the gold-coated micromirrors at room temperature and at 162 K without a large deformation difference. Thanks to the use of a reference plate in the reference arm of the interferometer, identical to the chamber window, we could get a high contrast in our measurements. Interference fringes are clearly visible on the first and last columns of actuated

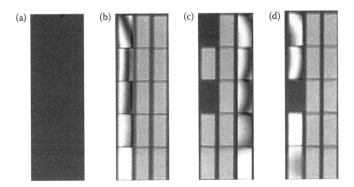

Figure 4.14 Interferometric observations of the tilted lines of micromirrors during the cryogenic experiment at 162K: (a) all micromirrors in OFF state at room temperature (RT); (b) all micromirrors in ON state at RT, 130 V applied; (c) all micromirrors in ON state at 162 K, 130 V applied, except mirrors 1 and 3 of the first column; (d) all micromirrors in ON state at 162 K, 148 V applied, except mirror 3 of the first column.

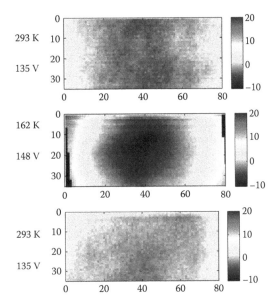

Figure 4.15 Surface quality of the gold-coated micromirrors in actuation, at room temperature before cooling, at 162 K, and at room temperature after cooling. Deformation from 10 nm to 30 nm peak-to-valley is measured.

mirrors in Figure 4.14, and we could then measure the mirror surface deformation when the device was actuated at room temperature and at the cryogenic temperature.

A 9.8 nm PTV surface deformation was measured at 293 K, increasing up to 27.2 nm PTV at 162 K. When coming back to room temperature, we again measured the mirror surface deformation and obtained a value of 9.9 nm PTV, identical to the value measured before the cooling of the array (Figure 4.15). The deformation is due to the CTE mismatch between the thick silicon micromirror and the thin gold coating layer on top. However, the surface deformation stays within the limit of 50 nm.

4.5.3 DMD characterization

DMD DC2k chip characterization in a space environment is the most advanced and detailed MOEMS device space evaluation done in Europe. This study has been engaged by ESA and lead by LAM.

The philosophy of the whole test campaign was the validation of any test through a photometric evaluation of each micromirror of the array before and after the test.

For this study, a specialized driving electronics and a cold temperature test set-up have been developed. The tests include cold temperature and vacuum tests, life tests in space survey conditions, radiation (TID and SEE) tests, thermal cycling, and vibration and shock tests, as well as MOS-like tests. No degradation is observed from the optical measurements. *These results do not reveal any concerns regarding the ability of the DMD to meet environmental space requirements.*

Over the last few years, the Laboratoire d'Astrophysique de Marseille has developed an expertise in the characterization of micro-optical components [12,16,18]. This expertise in small-scale surface deformation characterization of micro-optical components as well as operational testing of MOEMS components has been used for the development of a dedicated cold temperature test set-up for DMD measurements in EUCLID operating conditions.

4.5.3.1 Cryostat and optical bench

For environmental testing (vacuum and low temperature), a cryostat has been developed at LAM. The bench (Figure 4.16) is used as a photometric bench. In order to get enough resolution on each micromirror, the FOV images approximately 200 × 200 micromirrors onto a 1k × 1k camera. To inspect the complete DMD, a stitching procedure is carried out by means of motorized stages. For the sake of test accuracy and efficiency, the characterization set-up is automated as much as possible. Three computers are used for managing the tests. The thermal chamber enables tests in a vacuum environment with a temperature adjustment in the range of −60°C to +20°C. Temperature change is obtained through a liquid cooler. The cross-shaped chamber is made of stainless steel. Each side is devoted to a specific task. The first side includes a window in order to view the DMD sample; the second side holds the feed-through connectors for driving the DMD board; the third side permits all sensing wires (for temperature sensors) to pass into the chamber. Finally, the fourth side receives the pipes from the liquid cooler device. The cryostat is monitored with temperature sensors and vacuum sensors driven by a computer (*cryostat computer* in Figure 4.16). Illumination of the DMD array is made by a collimated beam. Optical imaging is made by two doublets (200–400 mm) mounted on rails. The system is diffraction-limited on the detector, leading to an optimized photometric measurement. The device is divided into *50 zones*. The FOV to be imaged by the CCD camera covers one zone, equivalent to *205 × 216 micromirrors* (44,280 micromirrors). The scale on the 1k × 1k camera is exactly 4.07 × 4.07 detector pixels per micromirror. For complete DMD testing, the stitching procedure is done by means of motorized stages in three directions (XYZ), moving the whole imaging optical train with a travelling range of 100 mm and a resolution of 0.1 μm. The

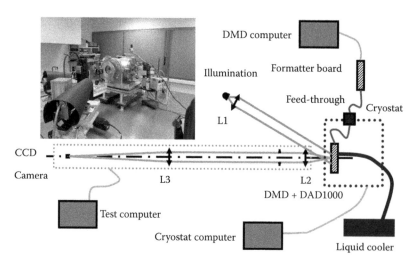

Figure 4.16 DMD cold temperature test set-up. (Copyright SPIE. Reproduced with permission.)

(a)   (b)

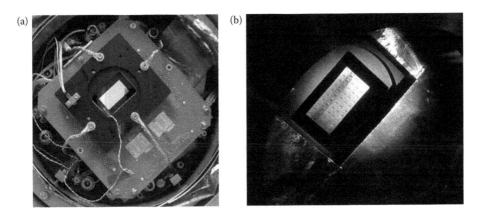

Figure 4.17 Pictures of the device during life test: (a) cryostat open and (b) cryostat closed.

optical test equipment (stages + camera) are monitored by a computer (*test computer* in Figure 4.16). All software is developed in MATLAB.

The DMD board is mounted on the thermal interface through a point-plane-plane mounting scheme in order to avoid any additional stress on the board when the temperature is changing. The thermal link between the thermal interface and the DMD board as well as the DMD itself is done through copper wires connected to the DMD board, the DMD heat sink, and the front mounting surface of the DMD device (Figure 4.17, left-hand image). The DMD board is linked by 300 wires through the chamber to the formatter board. Specialized feed-throughs for such a high number of wires have been realized, tested, and have been mounted on our chamber. All materials used in the chamber are vacuum compatible except for the wires leading to the DMD board and some parts on the DMD board. After DMD board integration, the thermal screen with a multilayer insulator (MLI) cover is mounted around the device, and the optical window is closed on the chamber. We decided to align and fix the optical input and output beams as a very precise and reliable reference. Then all optical alignments are performed in the chamber, using the tip-tilt and rotation mounting under the cryogenic chamber. When the DMD chip and the detector are in parallel planes, the stitching of the images can be done much easier with the motorized stages moving the optical train. A focused image across the whole device is maintained without the need for Z stage adjustment during stitching with the X and Y stages. In this way, the optical input and output beams stay fixed with respect to the DMD.

4.5.3.2 Device driving

Hardware and software were developed by Visitech (Norway) and LAM for driving the DMD boards. The hardware is controlled by an RS-232 serial link and a digital visual interface (DVI) port is used for loading an image onto the DMD. The software was developed in MATLAB for driving the DMD chip by a computer (*DMD computer* in Figure 4.16). In addition to the extreme conditions of a EUCLID environment, this DMD-based instrument also has a very nontypical DMD operation: during data capture, each DMD micromirror will be held in one position for 1500 seconds before changing state. This is quite a challenge to accomplish with TI proprietary chips to format the DMD signal.

The two major patterns used during the tests, pattern 1 and pattern 2, are designed as "positive" and "negative" to each other (Figure 4.18a). Each pattern row is divided in 10 zones for a total of 50 zones, and each zone is imaged on the CCD camera. Each zone includes specific patterns that are identical from zone to zone and numbers at the edge of each zone are set for easy processing and archiving (central picture of Figure 4.18). The individual patterns show lines with different widths and orientations, chessboard features, and MOS-like patterns. In addition to pattern 1 and pattern 2, all ON and all OFF mirrors will also be displayed and measured at each measurement step. The background will be measured and subtracted from the images.

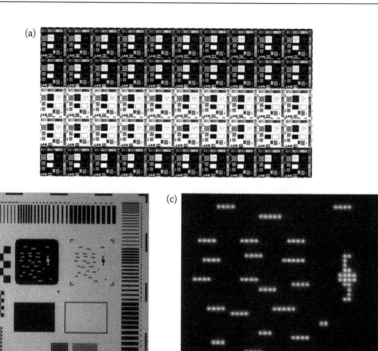

Figure 4.18 (a) LAM designed pattern (pattern 1) applied to the DMD; (b) image of a zone taken with the characterization set-up CCD camera; and (c) close-up view on a MOS-like pattern.

Each micromirror is imaged on the CCD camera on about 4×4 detector pixels, which is enough for the monitoring and detection of failures (if any) during the tests. A zoom-in the area, simulating a MOS-like pattern with multi-slits, is shown in the right-hand side of Figure 4.18. Any slit location and shape can be generated. It has to be noticed that the OFF mirrors cannot be imaged on the detector due to the high contrast performance of the DMD. Thus, the contrast cannot be measured in this test. An evaluation of the contrast will be done during the MOS-like tests, using a high dynamic range camera.

4.5.3.3 Analysis procedure

A data pipeline for data reduction has been developed using MATLAB software. Photometric measurements are done before, during, and after each test, and compared to the reference measurements (taken before the test, at room temperature). Any degradation in performance of a mirror will be revealed. Differences between mirrors as well as tests on different patterns applied on the same device (patterns in space domain and time domain) are analyzed. The analysis procedure comprises background subtraction, pattern recentering, mirror envelope detection, mirror photometric measurement, and comparison with the nominal mirror photometric measurement, with local variations of the illumination beam taken into account. We have adopted three mirror degradation definitions: —the *blocked mirror* when the mirror is non-responsive (e.g., "stuck"), —the *lossy mirror* when the optical throughput is decreased by more than 20%, and—the *weak mirror* when the optical throughput is decreased between 10% and 20%.

We have developed software for an automatic analysis of all data recorded during our measurements. All failure types are searched and detected. Results are displayed as maps and graphs. All fifty zones from the DMD are shown in a matrix pattern (zone 00 to zone 49), and the number of affected mirrors is assigned to each zone. Comparison of measurements before and after testing shows the evolution of the failure rate.

4.5.3.4 Cold temperature test

Four sets of cold temperature tests have been completed: cold temperature step stress test, nominal cold temperature test, thermal cycling, and life test. Pictures of the device mounted in the chamber during the life test in LAM's cold chamber are shown in Figure 4.17.

4.5.3.5 Cold temperature step stress test and nominal cold temperature test

The cold temperature step stress test has been done from room temperature down to $-60°C$. *This test shows that permanent failure, that is, stuck mirrors, is appearing on some mirrors when the device is taken down to $-55°C$. The failure rate increases at $-60°C$. Based on these results, the nominal temperature condition for EUCLID was set to $-40°C$.* In order to confirm that this is an acceptable nominal operating condition for the device, a single DMD was tested. The DMD was mounted in the cold temperature chamber and tested three times at this temperature. For minimizing the stress on the device when cooling down, a fast ($-40°C/hour$) and homogeneous cooling was provided. Using the definitions of blocked, lossy, and weak mirrors, we have analyzed all (over 2 million) mirrors from the DMD for each measurement. No blocked mirror was revealed, while only 12 mirrors were defined as lossy, and between 3 and 7 were defined as weak mirrors. We highlight the fact that these effects have no impact on a typical DMD display mode, but they have to be taken into consideration and calibrated for MOS application. These numbers are very low when compared to the total of 2 million mirrors operating in the device. *This test revealed no degradation of the device when three consecutive cycles in EUCLID conditions are applied.*

4.5.3.6 Thermal cycling

Thermal cycling has been done at Istituto Nazionale di Astrofisica/Istituto di Astrofisica Spaziale e Fisica Cosmica, INAF/IASF's facility in Bologna, Italy, and the first five cycles as well as the optical characterizations have been performed at LAM's facility. Two sets of cycles (249 and 313 cycles) have been applied, with intermediate and final optical characterization. No degradation of device performance was observed.

4.5.3.7 Life test

The device was tested in EUCLID conditions, which means in vacuum and at $-40°C$, with the device operating with the following test cycle: pattern 1 is applied for 1500 s, the whole device is switching between pattern 1 and pattern 2 for 60 s, and pattern 2 is applied for 1500 s. In this way, there is an identical duty cycle for all mirrors. *The life test lasted for 1038 hours.* Full optical tests were done during the whole life test. Measurements were done at room temperature (reference measurement), a first test at cold temperature was done at the starting temperature, 11 intermediate measurements were done, and finally, a last measurement was done at the end of the life test, after *1038 h.* We can note that when the DMD is turned off, a drop of *2–3°C* in temperature is revealed, showing the heating capability of the DMD chip due to its power consumption. After recording all data, we used the previously described analysis procedure for extracting the number and locations of affected mirrors. Three blocked mirrors and five lossy mirrors have been observed, but they are detected at ambient temperature, at cold temperature at the beginning of the test, and at cold temperature at the end of the test; they are not a consequence of the space conditions applied to the device. Other affected mirrors are only weak mirrors and their number is very low, with no increase while the life test was running. *This shows that space conditions did not degrade the device performances within this life test period.*

4.5.3.8 Radiation

Radiation and vibrations tests have been conducted on DMD devices [19]. Total ionizing dose (TID) radiation tests and proton single event effects (SEE) radiation tests have also been done.

4.5.3.9 Total ionizing dose (TID) test

A complete test vehicle has to be in operation during the radiation testing, but by carefully shielding the test vehicle with lead bricks, only selected devices receive a significant radiation dose. The devices to be irradiated were selected in such a way that the artifacts of a failing device would help in pinpointing which device had failed. The devices to be irradiated were the DMD and the electronics. During the radiation testing, the DMD displayed full black and full white test patterns that alternated every 1500 s. This is similar to the expected EUCLID duty cycle. A data logger measured the current consumption of major power supplies of the DMD chip set during radiation and provided the opportunity to monitor if and when any of the devices were degrading. *Total ionizing dose (TID) radiation tests established a tolerance level of 10–15 krads for the DMD; at the mission level, this limitation could likely be overcome by shielding the device.*

Optical characterizations were done on the samples before and the day after the radiation. An additional test was carried out one week later in order to measure longer term effects. No blocked mirror and five lossy mirrors have been observed before and after TID radiation; all lossy mirrors were at the same location. Other affected mirrors are weak mirrors and their number is low (around 30 mirrors), with no increase after the test. *We can then conclude that these results show that space conditions did not degrade the device performances, within this TID radiation test condition.*

4.5.3.10 Proton single event effects (SEE) radiation test

The proton radiation test has been performed at the KVI facility in the Netherlands. During proton radiation, a live optical characterization has been conducted on a limited FOV around the center of the radiation beam hitting the device. The proton beam has an angle of around 45° with respect to the DMD surface, and the beam is located precisely on the DMD surface thanks to a laser guiding beam provided on the KVI equipment. The optical set-up is based on the set-up used in Marseille with a collimated input beam from a light emitting diode (LED) and an imaging system with two doublets, which gives the right magnification on the camera for observing the individual micromirrors with enough spatial resolution.

The condition for the first run was a 48 MeV beam from the accelerator, leading to 34.7 MeV on the DMD after crossing the DMD window at 45°. The radiation started at low flux (6×10^5 p/cm²/s for 300 s) and was to be increased to a higher flux (5×10^7 p/cm²/s for 896 s) in order to reach a total dose of 10 krads on the DMD. The proton radiation testing was unfortunately ended prematurely because of a breakdown in the accelerator after 120 seconds of radiation at low flux. *The lack of observed single events upsets is promising, but considering the length of the test, no conclusions can be reached on the tolerance of the test vehicle in regards to single events upsets.*

4.5.3.11 Vibrations and shocks

Vibration test conditions using the standards MIL-STD-883F, methods 2005 (vibration fatigue) and 2007 (vibration at variable frequency) condition A, and shock test condition B of the MIL-STD-883F Method 2002 were applied. It has been demonstrated that space conditions did not degrade device performances [20].

4.5.3.12 Micromirrors throughput analysis

From the recorded measurements, we have engaged a detailed micromirror throughput analysis. Blocked, lossy, and weak mirrors have been defined for mirrors exhibiting losses larger than 10%. In this paragraph, we consider micromirror throughputs on the whole range. Calculation has been made on 12 experiments realized during the life test. The analysis is done on a typical zone, zone # 25, considering 42,177 micromirrors; the number of samples is large enough for a good statistical analysis.

In order to define the accuracy of our measurements, standard deviation calculation is done for each set of measurements. These standard deviation values include:

- Measurement precision
- Throughput calculation method precision
- Local mirror throughput variation

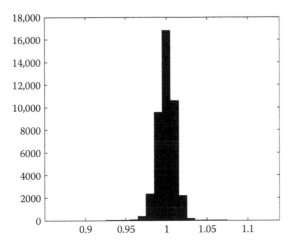

Figure 4.19 Histogram for a single experiment during the DMD life test (No. of mirrors versus through-put). (Copyright SPIE. Reproduced with permission.)

A typical histogram for a single experiment is shown in Figure 4.19. The figure is very sharp with nearly all micromirrors included in the ±5% range.

The throughput evolution has been detected only on one mirror for each type: one mirror is always blocked and one weak mirror is back to normal at the end of the life test. All other mirrors show only temporary modification of their throughput. Throughput variations are due to slight differences in the tilting angle and possible slight deformations of the mirror surface [10].

4.5.3.13 MOS-like tests

In order to evaluate the capability of a DMD device to select objects in a field of view, a dedicated test set-up has been developed, demonstrating the concept of a DMD-based MOS spectrograph. From a field of view simulator, objects could be selected with the DMD device. Thus, device performances have been measured and the object selection procedure has been evaluated.

The bench is used as a photometric bench, and the FOV is imaged on a 1k × 1k camera in order to get sufficient resolution on each micromirror. Each micromirror is imaged on about 9 × 9 detector pixels. In comparison with the set-up developed for vacuum and low temperature testing (4 × 4 detector pixels/micromirror), the optical magnification is higher in this new set-up; we want to get a higher photometric accuracy in DMD performance parameters, as well as a higher spatial resolution on each micromirror. A 24° angle is set between the input and output beams, and both input and output beams have been set to F/3. In front of the camera, a neutral density filter is inserted in order to increase the dynamic range of the bench. This feature is very important for precise DMD contrast determination. An FOV containing three objects was imaged on the DMD device surface (Figure 4.20a). In Figure 4.20b, the same FOV is presented, but the DMD is programmed for selecting the left-hand object (the mirrors are ON only on the object to be selected, and the remainder of the mirrors are OFF). *This picture shows the full capability of the DMD device to generate any slit pattern (reflective slit) sending the light toward the spectrograph, when all other sources as well as the background are hidden by the OFF micromirrors.* The contrast of a micromirror is defined as the ratio of the throughput when the mirror is in the ON position with respect to the mirrors in the OFF position. In order to be as accurate as possible, the throughput is integrated within a mask applied on a whole micromirror. The background light has been removed. *This gives a final value for the contrast of 2250.* Contrast has been measured on several mirrors and they exhibit identical values.

In conclusion, *these results do not reveal any show-stopper concerning the ability of the DMD to meet environmental space requirements.* Insertion of such devices into final flight hardware would still require additional efforts such as development of space compatible electronics, and original

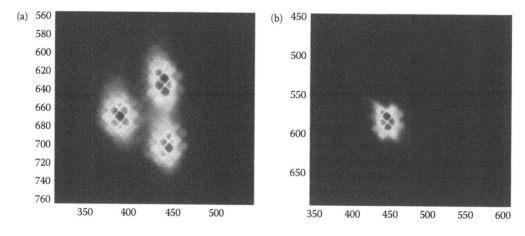

Figure 4.20 (a) FOV with three objects imaged on the DMD device surface. (b) Same FOV when the DMD is programmed in order to select the left-hand side object (the mirrors are ON only on the object to be selected, and the rest of the mirrors are OFF).

opto-mechanical design of the instrument. From an ESA perspective, the micromirror arrays have achieved a reasonable TRL (Technology Readiness Level). Insertion of such devices into final flight hardware would still require additional effort (estimation is approximately two years) in terms of change of the window coating and redevelopment of space compatible electronics as well as a different package interface compatible with spacecraft launch conditions.

In the US, a team from Space Telescope Science Institute (STScI) and University of Rochester is also considering DMDs in astronomical instruments. They are involved in instrument design [21] and DMD testing [22]. They plan to use the DMD device and its associated electronics developed in Europe for ground-based and possibly for space instruments.

4.5.4 DM characterization

BMC is currently conducting two types of actions for developing DMs devoted to space applications. As early as 2011, a classical DM was tested with a sounding rocket in the US, in the Planetary Imaging Concept Testbed Using a Rocket Experiment (PICTURE). PICTURE-B was launched in November 2015. This experiment directly measures optical light scattered by the debris disk around Epsilon-Eri star.

In parallel, several studies have been engaged to modify the DM architecture and make it compatible with a space environment [23]. Topography of the mirror surface has been improved by reducing the print-through from 13 nm RMS down to 2.5 nm RMS, and the "scalloping" from 100 PTV to 20 nm PTV. In order to enhance the reliability of the actuators, especially to avoid an actuator sticking because of electrostatic over stress (EOS) damage, hard stops have been added below the actuator membrane and grounded landing pads have been added in the electrode plane, leading to the elimination of the sticking effect even if a snap-in occurs. Finally, for reaching 10k actuators DMs, the wire bonding technique has been replaced by through wafer via (TWV) technology in order to connect the driving lines on the back of the DM wafer to the drive electronics. In terms of new electronics, Dr. E. Bendek from NASA Ames facility is developing electronics boards directly integrated with DM packaging, eliminating the need for cables. In this way, the volume, power, and mass of the driving electronics are greatly reduced; this is called a kilo-driver and works at 1 kHz with a 14 bits resolution.

Within the NASA Technology Demonstration for Exoplanet Missions program, testing is conducted extensively at NASA facilities for vibration, acoustic loads, and shocks, increasing the TRL in flight-like conditions. Characterizations also include thermal testing at 95 K under vacuum and radiation exposure testing. The measured parameters are the mirror surface quality (at rest and actuated), voltage versus deflection response, influence function, frequency response, and actuator yield.

4.6 Applications 1: Universe observation

Multi-object spectroscopy (MOS) is a key technique for large field of view surveys. MOEMS programmable slit masks could be next generation devices for selecting objects in future infrared astronomical instrumentation for space telescopes. MOS is used extensively to investigate astronomical objects by optimizing the signal-to-noise ratio (SNR): high precision spectra are obtained and the problems of spectral confusion and background level confusion occurring in slitless spectroscopy are cancelled. Fainter limiting fluxes are reached and the scientific return is maximized both in cosmology and in legacy science. Major telescopes around the world are equipped with MOS in order to simultaneously record several hundred spectra in a single observation run. Conventional masks or complex fiber-optics-based mechanisms are not attractive for space. The programmable multislit mask requires remote control of the multislit configuration in real time. Next generation MOS for space, such as the NIRSpec instrument for the JWST telescope, will use micro-shutter arrays as object selectors. During the early phase studies of ESA's EUCLID mission, a MOS instrument based on a DMD device has been studied. Complexity and cost reasons have forced ESA to choose slitless spectroscopy, despite a much higher efficiency with slit spectroscopy.

In future space MOS, the use of MOEMS devices such as micromirror arrays (MMA) [7,21,24] or micro-shutter arrays (MSA) [8] is scheduled. MMAs are designed for generating reflecting slits, while MSAs generate transmissive slits. By placing the programmable slit mask in the focal plane of the telescope, the light from selected objects is directed toward the spectrograph, while the light from other objects and from the sky background is blocked or sent toward an imager (Figure 4.21).

BATMAN flies is proposed by a European consortium gathering, LAM and Thales Alenia Space (TAS) in France, CSEM in Switzerland, IMTEK—University of Freiburg in Germany, and Trieste Observatory in Italy. The concept is based on a spectrograph and an imager working in parallel, fed by a tiltable micromirror array (MOEMS-MIRA). To dynamically correct the in-coming wavefront, a MOEMS deformable mirror (MOEMS-DM) placed at a pupil before this configuration

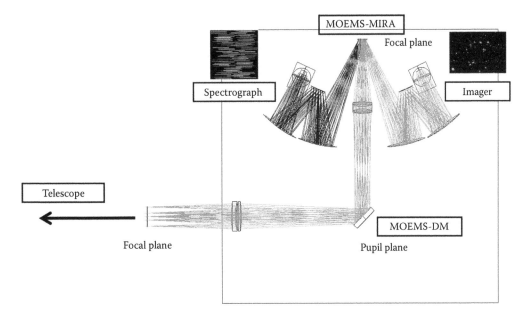

Figure 4.21 *BATMAN flies* conceptual design.

provides an optimal efficiency on both instruments, regardless of the telescope at the entrance and the type of mission, by dynamically correcting the in-coming wavefront. The complex system control has to be built in order to control all subsystems, as well as provide new operational modes not feasible with current technologies. A schematic view of the instrument we are proposing is given in Figure 4.21.

In this concept, one point not yet addressed is wavefront sensing. However, we see several sensing possibilities within three classes: measurement of the wavefront in a closed loop (or open loop), measurement of the scientific image (phase diversity, e.g.), and internal metrology for open-loop correction.

BATMAN flies will provide unprecedented versatile, programmable, optimized in mass, volume, and cost, and efficient spectro-imagers for space missions.

ESA and NASA roadmaps include, in terms of missions, next generation ultraviolet (UV) to infrared (IR) observatories including spectrographs for universe observations, next generation hyperspectral imagers for Earth observations, and laser detection and ranging systems or LIDARS and spectrographs for planetology observations. ESA's roadmap in new technologies includes, as a priority, the digital micromirror array (DMA) for space optical instruments. *BATMAN flies* is following these roadmaps and proposing breakthrough solutions.

In universe observation, *BATMAN flies* is a deep multisurvey mission in the infrared with a multi-object spectrograph based on a reconfigurable slit mask using MOEMS devices. Unique science case space observations are reachable with this instrument:

- Deep survey of high-z galaxies: large sample of 200,000 galaxies down to H = 25 on 5 deg^2, and all z > 7 candidates at H = 26.2 over 5 deg^2
- Deep survey of nearby galaxies: characterization of the initial mass function (IMF) in several thousands of young stellar clusters in a large sample of nearby galaxies
- Deep survey of the Kuiper Belt: spectroscopic survey of *all* known objects down to H = 22 (700 objects, current sample multiplied by 10).

Pathfinder toward *BATMAN flies* is already running. *BATMAN*, the new generation MOEMS-based spectro-imager of the sky, is of prime importance for characterizing the actual performance of this new family of MOS instruments as well as for investigating the operational procedures on astronomical objects (Figure 4.22a). Thanks to a French-Italian collaboration, this instrument will be placed on a ground-based 4-m-class telescope, the Telescopio Nazionale Galileo (TNG) in the Canary Islands at the nasmyth focus in 2017. ROBIN, a *BATMAN* demonstrator on an optical bench, has been built and has already delivered images and spectra in parallel, allowing us to validate all expected performances (Figure 4.22b). We have tested the instrument's abilities in terms

(a) 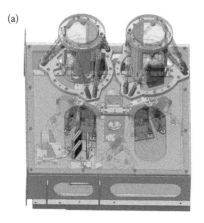 (b)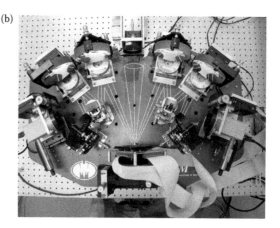

Figure 4.22 (a) 3D general design view of *BATMAN*; (b) integrated ROBIN picture.

of variable spatial bin and variable spectral resolution and any combination of the above modes over the whole FOV. Integral field unit (IFU) can be generated by using the scanning slit technique. MOS and IFU modes have been studied with any slit mask configurations (any shape, including long slit) as well as real time reconfiguration. [25]

4.7 Applications 2: Earth observation

Two promising concepts of space instruments for Earth observation (EO) are presented. In EO instruments, bright sources in the observed scenes degrade the recorded signal. Our first concept, called *Smart Slit*, consists of an active row of MOEMS for removing the bright sources. Experimental demonstration has been conducted and the stray light in the instrument has been removed almost completely. The second concept is a *programmable wide-field spectrograph* where both the FOV and the spectrum could be tailored thanks to 2D micromirror arrays: A demonstrator has been designed, built, and tested.

4.7.1 Smart Slit

Earth observation instruments aim at observing the ocean and the land. But the sea observed in infrared wavelengths is very dark when a very bright signal is reflected by the clouds. In order to prevent CCD saturation, a very high dynamical range is required for the detector. The stray light in the spectrometer (detector backscattering) in the IR band is so important that each time a cloud is present in the field of view, the image is considered to be lost. In consequence, a new concept called *Smart Slit* is proposed in order to enhance spectrometer performances by placing MOEMS in the entrance slit to dynamically filter out the light coming from clouds.

According to the luminance of the clouds and sunglints, the stray light level in the instrument is around 60% in the near IR. This stray light is detrimental to the Earth observation signal. The scatter within the spectrometer clearly and largely dominates the overall stray light, that is, after the slit, and more precisely by the detector backscatter. Placing a MOEMS in the entrance slit of the spectrometer to filter out the light coming from clouds and the light coming from the sea should improve the stray light performance of this instrument.

In the new design, the MOEMS is a micromirror array located at the entrance of the spectrometer. All other elements are kept such as the scrambling window unit, the catadioptric objective, the concave grating, and the focal plane [26]. The results of end-to-end stray light simulations have been postprocessed. The maximum value in *spectral band 1020 nm is around 60%*, definitely higher than the values in other bands. This stray light is detrimental to the Earth observation signal. The scattering within the spectrometer clearly and largely dominates the overall stray light, that is, after the slit. This stray light is inherent to all back-illuminated thinned CCD detectors which are used with wavelength close to or larger than 1 micron.

Stopping the light coming from the cloud before entering the spectrometer should decrease these values. Simulations are postprocessed for an ideal cloud remover, filtering a cloud spread on one side of the slit. In the presence of the concept we propose, making the assumption that we can predict the cloud location in the entrance slit, the stray light introduced by the spectrometer can be fully cancelled. In consequence, only the stray light introduced by the ground imager remains. This simulated level is definitely lower, *around 5% for the 1020-nm band*.

4.7.1.1 First experimental demonstration

Experimental demonstration of this concept has been conducted on a dedicated bench at LAM (Figure 4.23). A scene (dark area simulated by the neutral density filters) with a contiguous bright area mimicking the clouds (factor 10^2) has been focused on a micromirror array and imaged on a CCD detector. The micromirror array is a DMD device from Texas Instruments made of 13.5-μm

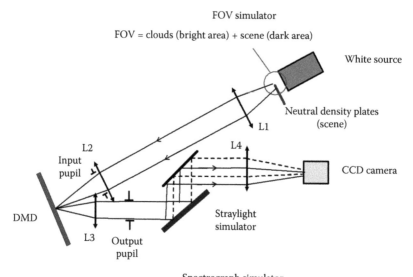

Figure 4.23 Experimental set-up for cloud removal demonstration in the Smart Slit concept. (Copyright SPIE. Reproduced with permission.)

mirrors. After the programmable slit, the stray light issued from the bright zone is set to the right level, that is, equal to the scene signal level, and pollutes the scene [27].

The resulting signal with the clouds and the polluted signal is recorded on a CCD camera. In order to restore the signal, the micromirrors located on the bright area are switched off, almost completely removing the stray light in the instrument. In Figure 4.24a), the profile of the FOV including the scene (left part) and the clouds area (right part) is shown, where the green curve represents the FOV when the programmable slit is all ON, and the blue curve represents the FOV when the micromirrors located in the bright area (clouds) are switched OFF. A close-up view (Figure 4.24b) permits to see the benefit of optically removing the polluting source where the stray light produced by the bright area is nearly completely removed; the light blue curve represents a perfect scene, without clouds. *This successful demonstration shows the high potential of this new concept in future spectro-imagers for Earth observation.*

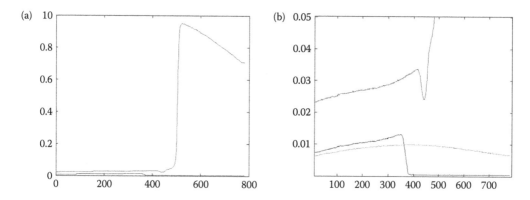

Figure 4.24 Cloud removal experiment: (a) profile of the FOV including the scene (left part) and the clouds (right part) when the programmable slit is all ON (upper curve/green curve) and when the micromirrors located on the bright area are switched OFF (lower curve/blue curve); (b) close-up view of the transition area; the lowest curve (light blue curve) represents the perfect scene, without clouds. (Copyright SPIE. Reproduced with permission.)

A good candidate for realizing this dynamic programmable slit at the entrance of the spectrometer is a MIRA-like device (see description in Section 4.4).

4.7.2 Programmable wide-field spectrograph

Our innovative reconfigurable instrument concept, a programmable wide-field spectrograph, is based on a 2D micromirror array (MMA) where both the FOV and the spectrum could be tailored. For a linear 1D field-of-view (FOV), the principle is to use an MMA to select the wavelengths (Figure 4.2). This component is placed in the focal plane of a first grating. On the MMA surface, the spatial dimension is along one side of the device and for each spatial point, its spectrum is displayed along the perpendicular direction. Each spatial and spectral feature of the 1D FOV is then fully adjustable dynamically and/or is programmable. A second stage with an identical grating recomposes the beam after wavelength selection, leading to an output tailored 1D image.

As for the programmable slits, two devices are promising for developing new applications: MIRA and a DMD, described in Section 4.4.

A mock-up has been designed, fabricated, and tested. The design has been developed from ideas proposed many years ago for the JWST near-infrared multi-object spectrograph [24], and developed more recently for the *BATMAN* project [25]. The micromirror array is the largest DMD described above. Our optical design for both spectrographs, mounted before and after the DMD plane, is a compact design, all-reflective with F/4 on the DMD component and robust 1:1 Offner relays (Figure 4.25a). The most critical component of the system is the convex grating, with 150 gr/mm line density, 140 mm radius of curvature, and 40 mm in diameter. This makes the system simple and efficient, not suffering from chromatic aberrations. The image quality delivered onto the DMD and onto the detector is also high enough to not degrade the resolving power and spatial resolution. Typical monochromatic RMS spot diameters are 20 μm at the intermediate focal plane (DMD surface) and 10 μm at the focal plane over the whole FOV for wavelengths between 450 and 750 nm.

The general mechanical design of the mock-up consists of a main optical bench with two spectrographs in serial (Figure 4.25b). An organic light emitting diode (OLED) at the entrance simulates the field-of-view. Earth images are displayed using their red-green-blue (RGB) values. We have developed software in MATLAB in order to drive the input beam with an adjustable image size on the OLED, programmable slits (shape, width), and colors' selection as well as the possibility to have a scanning slit-FOV, mimicking a push-broom type Earth observation.

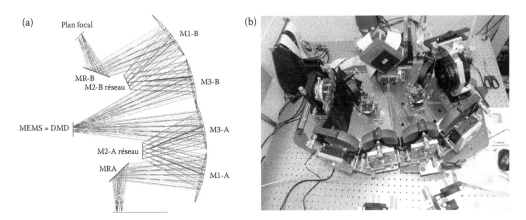

Figure 4.25 (a) Optical design of the programmable wide-field spectrograph demonstrator. Light from the telescope is coming from below (1D entrance slit), passing through two Offner-type spectrographs in serial separated by the DMD, up to the detector at the top of the drawing; (b) integrated demonstrator picture. Input/OLED is on the left, the DMD at the top, and the output/detector on the right.

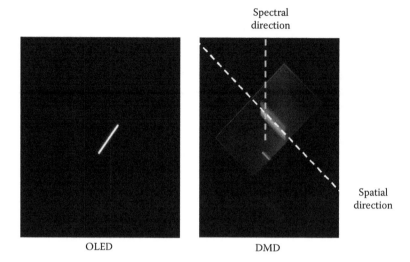

Figure 4.26 Images of the input (OLED) plane and intermediate focal plane (DMD surface). Dispersed spectra (1st order) are displayed on the DMD surface, while 0th-order and −1 order are also recorded on the image. Spatial and spectral directions are labeled.

The alignment strategy uses a combination of mechanical and optical measurements. Figure 4.26 shows images of the input (OLED) plane and intermediate focal plane (DMD surface). The OLED displays a white slit at 45° with respect to the horizontal (and vertical) axis. After passing through the first spectrograph, the slit is dispersed on the DMD surface. Spatial and spectral directions are labeled in Figure 4.26. The zeroth order is partly imaged on the DMD surface as a narrow white stripe (image of the entrance slit), and the first order is faintly displayed at the bottom-center of this picture. Using these images and exact wavelength locations, thanks to interference bandpass filters, we have determined *the exact spectral dispersion* to be 200 nm wavelength range dispersed along 353 micromirrors, that is, *0.57 nm/micromirror*.

A synthetic linear FOV mimicking Earth views is generated and typical images have been recorded at the output focal plane of the instrument. The FOV exhibits different features including bright objects and "colored" elements. By tailoring the DMD, we could successfully modify each pixel of the input image: for example, it is possible to remove bright objects or, for each spatial pixel, modify the spectral signature [28].

The very promising results obtained on the mock-up of the programmable wide-field spectrograph reveal the efficiency of this new instrument concept for Earth observation.

4.8 Conclusions and perspectives

In future generations of space instruments for Earth observation, Universe observation, and planet exploration, micro-optoelectromechanical systems (MOEMS) could be key components for generating new functions not available with current technologies. New functions associated with several types of MEMS (programmable slits, micro-deformable mirrors, programmable micro-diffraction gratings) are foreseen for intensity/field selection, wavefront control, or spectral tailoring.

However, the number of MOEMS designed, realized, and tested for space applications is limited: three main devices are specifically announced for these goals: the micro-shutter array from NASA, the micromirror array, MIRA, from LAM-EPFL-CSEM, and the micro-deformable mirror from Boston Micromachines. Testing in vacuum and at cryogenic temperature has been successfully done for all of them, and the micro-shutters will be the first fully operational devices flying in space on-board JWST in 2018.

These devices will be used in many more applications if they can pave larger FOVs. For this goal, they must fulfill two additional conditions:

- Enable mosaicking strategies with a buttable architecture on maximized array sizes.
- Develop and assemble integrated electronics (ASIC) on board for device control-command at each actuator level.

The next step is to realize a 3D-assembly of these components and increase their TRL to be proposed as standard building blocks of new space missions.

In parallel, at the instrument level, their architecture must be originally designed for integrating these components with specific behaviors, and then optimizing the instrumental efficiency. The spectro-imager family named *BATMAN* is an example of this new class of instruments designed for taking advantage of these MOEMS devices; they are also scalable to be adapted to a wide range of missions, from large satellites to compact instruments on board planetary exploration missions.

Finally, MOEMS devices developed for space may benefit many other space and non-space fields, in physics (LIDARS, beam-shaping, microscopy) and in biology (cell-trapping, wide-field, and ultra-high-resolution microscopy, 3D imagery).

New MOEMS devices could and will be developed in the future for (new) space applications. For faster access to the market, they will have to be designed from the beginning for this purpose; they will need a robust test strategy for confirming their ability to operate in harsh environments; and they must also be designed and optimized together with their foreseen instrumentation to get the highest performance. Lessons learned from the first candidates presented in this chapter will facilitate this bright future.

References

1. F. Zamkotsian, P. Lanzoni, T. Viard, Convolution spectrometer demonstration using programmable diffraction grating, in *Proceedings of the Optical MEMS Conference 2011*, Istanbul, Turkey, 2011.
2. T. Viard, C. Buisset, F. Zamkotsian, V. Costes, L. Venancio, MOEMS for prospective space applications, invited paper, in *Proceedings of the SPIE Conference on MOEMS 2011, Proc. SPIE 7928*, San Francisco, USA, 2011.
3. F. Zamkotsian, P. Lanzoni, T. Viard, C. Buisset, New astronomical instrument using MOEMS-based programmable diffraction gratings, in *Proceedings of the SPIE Conference on MOEMS 2009, Proc. SPIE 7208*, San Jose, USA, 2009.
4. J. Trisnadi, C. Carlisle, R. Monteverde, Overview and applications of Grating Light Valve based optical write engines for high-speed digital imaging, in *Proceedings of the SPIE Conference on MOEMS 2004, Proc. SPIE 5348*, San Jose, USA, 2004.
5. S. Senturia, Programmable diffraction gratings and their uses in displays, spectroscopy, and communications, in *Proceedings of the SPIE Conference on MOEMS 2004, Proc. SPIE 5348*, San Jose, USA, 2004.
6. H. R. Shea, Reliability of MEMS for space applications, in *Proceedings of the SPIE Conference on MOEMS 2006, Proc. SPIE 6111*, San Jose, USA, 2006.
7. R. Burg, P.Y. Bely, B. Woodruff, J. MacKenty, M. Stiavelli, S. Casertano, C. McCreight, A. Hoffman, Yardstick integrated science instrument module concept for NGST, in *Proceedings of the SPIE Conference on Space Telescope and Instruments V, SPIE 3356*, 98–105, Kona, Hawaii, 1998.
8. M. J. Li, A. D. Brown, A. S. Kutyrev, H. S. Moseley, V. Mikula, JWST microshutter array system and beyond, in *Proc. SPIE 7594*, San Francisco, USA, 2010.
9. L. J. Hornbeck, Digital light processing and MEMS: Timely convergence for a bright future, in *Proceedings of Micromachining and Microfabrication Process Technology*, K.W. Markus, ed., *SPIE 2639*, 2, Austin, USA, 1995.

10. F. Zamkotsian, P. Lanzoni, E. Grassi, R. Barette, C. Fabron, K. Tangen, L. Valenziano, L. Marchand, L. Duvet, Successful evaluation for space applications of the 2048×1080 DMD, in *Proceedings of the SPIE Conference on MOEMS 2011, Proc. SPIE 7932*, San Francisco, USA, 2011.

11. S. Waldis, F. Zamkotsian, P.-A. Clerc, W. Noell, M. Zickar, N. De Rooij, Arrays of high tilt-angle micromirrors for multiobject spectroscopy, *IEEE Journal of Selected Topics in Quantum Electronics* 13, pp. 168–176, 2007.

12. F. Zamkotsian, J. Gautier, P. Lanzoni, Characterization of MOEMS devices for the instrumentation of Next Generation Space Telescope, in *Proceedings of the SPIE Conference on MOEMS 2003, Proc. SPIE 4980*, San Jose, USA, 2003.

13. M. Canonica, F. Zamkotsian, P. Lanzoni, W. Noell, N. de Rooij, The two-dimensional array of 2048 tilting micromirrors for astronomical spectroscopy, *Journal of Micromechanics and Microengineering*, 23, 055009, 2013.

14. F. Zamkotsian, B. Timotijevic, R. Lockhart, R. P. Stanley, P. Lanzoni, M. Luetzelschwab, M. Canonica, W. Noell, M. Tormen, Optical characterization of fully programmable MEMS diffraction gratings, *Optics Express*, 20(23), 25267–25274, 2012.

15. S. Cornelissen, T. G. Bifano, Advances in MEMS deformable mirror development for astronomical adaptive optics, in *Proceedings of the SPIE Conference on MOEMS 2012, Proc. SPIE 8253*, San Francisco, USA, 2012.

16. A. Liotard, F. Zamkotsian, Static and dynamic micro-deformable mirror characterization by phase-shifting and time-averaged interferometry, in *Proceedings of the SPIE Conference on Astronomical Telescopes and Instrumentation 2004, Proc. SPIE 5494*, Glasgow, UK, 2004.

17. F. Zamkotsian, E. Grassi, S. Waldis, R. Barette, P. Lanzoni, C. Fabron, W. Noell, N. de Rooij, Interferometric characterization of MOEMS devices in cryogenic environment for astronomical instrumentation, *Proc. SPIE 6884*, San Jose, USA, 2008.

18. F. Zamkotsian, K. Dohlen, Surface characterization of micro-optical components by Foucault's knife-edge method: The case of a micro-mirror array, *Applied Optics*, 38(31), 6532–6539, 1999.

19. F. Zamkotsian, E. Grassi, P. Lanzoni, R. Barette, C. Fabron, K. Tangen, L. Marchand, L. Duvet DMD chip space evaluation for ESA EUCLID mission, in *Proceedings of the SPIE Conference on MOEMS 2010, Proc. SPIE 7596*, San Francisco, USA, 2010.

20. F. Zamkotsian, P. Lanzoni, E. Grassi, R. Barette, C. Fabron, K. Tangen, L. Valenziano, L. Marchand, L. Duvet Space evaluation of 2048×1080 mirrors DMD chip for ESA EUCLID mission, in *Proceedings of the SPIE Conference on Astronomical Telescopes and Instrumentation, Proc. SPIE 7731*, San Diego, USA, 2010.

21. M. Robberto, A. Cimatti, A. Jacobsen, F. Zamkotsian, F. M. Zerbi, Applications of DMDs for Astrophysical Research, in *Proceedings of the SPIE Conference on MOEMS 2009, Proc. SPIE 7210*, San Jose, USA, 2009.

22. A. Travinsky, D. Vorobiev, Z. Ninkov, Heavy ion radiation testing of a digital micromirror device for performance in space, in *Proceedings of the SPIE Conference on MOEMS 2016, Proc. SPIE 9761*, San Francisco, USA, 2016.

23. P. Bierden, MEMS Deformable Mirrors for space imaging, in *ESA Workshop on Innovative Technologies for Space Optics*, Noordwijk, Netherlands, 2015.

24. F. Zamkotsian, K. Dohlen, D. Burgarella, V. Buat, Aspects of MMA for MOS: Optical modeling and surface characterization, spectrograph optical design, in *Proceedings of the NASA Conference on NGST Science and Technology Exposition, ASP Conf. Ser.* 207, 218–224, Hyannis, USA, 1999.

25. F. Zamkotsian, H. Ramarijaona, M. Moschetti, P. Lanzoni, M. Riva, N. Tchoubaklian, M. Jaquet et al., Building BATMAN: A new generation spectro-imager on TNG telescope, in *Proceedings of the SPIE Conference on Astronomical Instrumentation 2016, Proc. SPIE 9908*, Edinburgh, UK, 2016.

26. A. Liotard, F. Zamkotsian, W. Noell, T. Viard, M. Freire, B. Guldimann, S. Kraft, Optical MEMS for space spectro-imagers, in *Proceedings of the SPIE Conference on Astronomical Instrumentation 2012, Proc. SPIE 8450*, Amsterdam, The Netherlands, 2012.
27. F. Zamkotsian, A. Liotard, P. Lanzoni, T. Viard, Optical MEMS in space instruments for Earth Observation and Astronomy, in *Proceedings of the SPIE Conference on MOEMS 2013, Proc. SPIE 8616*, San Francisco, 2013.
28. F. Zamkotsian, P. Lanzoni, A. Liotard, T. Viard, V. Costes, P.-J. Hebert, DMD-based programmable wide field spectrograph for Earth Observation, in *Proceedings of the SPIE Conference on MOEMS 2015, Proc. SPIE 9376*, San Francisco, 2015.

5

MEMS vibratory grating scanners and their applications

Guangya Zhou

Contents

5.1 Introduction

Light scanning has numerous applications spanning a wide range of fields including manufacturing, life science and healthcare, defense and homeland security, information and communication systems, and consumer electronics. These scanning systems can mainly be categorized into one-dimensional (1D) and two-dimensional (2D) scanning, as shown in Figure 5.1a and b, respectively. In 1D scanning systems, light scans only in one dimension. Two-dimensional imaging or projection can also be achieved by moving the target or optical system in a direction perpendicular to the scan line. One-dimensional scanning applications include laser barcode readers, laser printers for the computer industry, terrestrial laser scanning systems for structural deformation and health monitoring. In 2D scanning systems, light scans in two mutually perpendicular directions as shown in the figure. Two-dimensional scanning applications include laser scanning confocal microscopes for biomedical and healthcare applications, raster-scanning laser projection displays/TVs for consumer markets, laser scanning 3D imaging ladar (laser detection and ranging) for unmanned ground vehicles, laser material processing for marking, cutting, welding, drilling, and so on, and laser cameras.

Scan frequency and optical resolution are two major performance characteristics for optical scanners (Urey et al., 2000). Scan frequency is defined as the number of scan lines generated per second, while optical resolution is defined as the number of resolvable spots/pixels per unidirectional scan, or in other words, the ratio of the optical scan angle to the optical beam divergence. Theoretically, the scan resolution is proportional to the product of the maximum optical scan angle θ and beam diameter (or the scanner's aperture size) D, since the beam divergence is inversely proportional to D due to the diffraction effect.

(a)

(b)

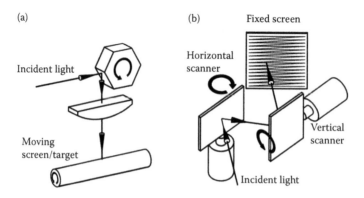

Figure 5.1 Schematics showing the applications of (a) 1D optical scanning (target is moving in an orthogonal direction) and (b) 2D optical scanning using two orthogonal scanners (target/screen is stationary).

A brief overview of the optical scanning technology (Beiser, 1992, 2003, Marshall and Stutz, 2012) is listed in Figure 5.2; it can be classified into two main principle categories, namely mechanical and nonmechanical, as shown in the figure. Mechanical optical scanners can be further categorized according to their motions, for example, continuously rotating and oscillatory. Without mechanical moving parts, nonmechanical scanners such as acousto-optic and electro-optic scanners are capable of very fast scan frequencies (MHz to GHz). However, the high cost, small scan angle, and limited scan resolution are among the major drawbacks of these types of scanners. Mechanical scanners, on the other hand, are attractive for their effectiveness, moderate scan frequency (kHz),

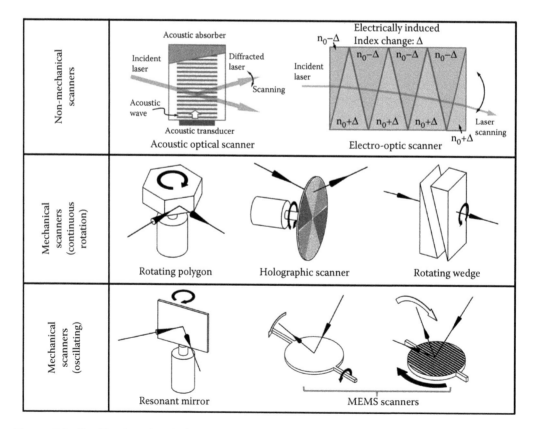

Figure 5.2 Classification of optical scanners.

and high resolution. Among those mechanical scanners using continuous rotation, polygonal scanners and holographic scanners (rotating diffraction gratings) (Beiser, 1988) are the most commercially successful ones. They can perform a kHz-rate simple linear sawtooth optical scan pattern with very high optical resolution. However, the relatively large inertias of polygons and gratings and the need for high-speed motors and pneumatic rather than mechanical bearings are among the major drawbacks. It is also noted that rotating grating scanners, although they are dispersive in nature (wavelength dependent), offer additional advantages in Bragg angle wobble correction and low aerodynamic loading. Hence, when properly designed, the holographic scanners can outperform the rotating polygons in speed. Oscillatory mechanical scanners, also known as low-inertia scanners, are usually operated in resonance to gain the advantages of both high-speed scanning and a large optical scan angle (hence, a high scan resolution). A typical design of such scanners consists of a suspended mirror that can be sent into resonant torsional oscillation by introducing a periodic signal into coils around a magnetic core. As compared with continuously rotating mechanical scanners, oscillatory scanners have certain advantages, including being optically simple, relatively compact, having an adjustable scan angle over a range to suit different applications, and having a flexure suspension (mechanical motion does not cause any long-term wear). The disadvantages of the resonant scanners mainly originate from the sinusoidal nature of the scan, for example, nonuniform scan velocity and restricted duty cycle.

In recent years, a new type of mechanical oscillatory optical scanner, namely microelectromechanical systems (MEMS)-based scanners (Muller and Lau, 1998), has been rapidly developed. Fabricated through silicon micromachining technology, these scanners have outstanding advantages including having an extremely low inertia thus being capable of fast scanning up to tens of kHz, ultra-compact, lightweight, and having a low per-unit-cost through IC-like batch fabrication. MEMS optical scanning technology can not only provide significant performance enhancements to existing applications (e.g., faster, smaller, and portable optical systems), but can also form a technological basis for a wide range of new applications including scanning endoscopic probes for healthcare applications (Sun et al., 2010) and raster-scanning retinal projection wearable displays (Urey, 2000). The majority of the research and development in the area of MEMS-based optical scanners is based on oscillatory micromirrors suspended by flexural suspensions and driven by MEMS actuators. Significant progress has been made in recent years (Yalcinkaya et al., 2006, Holmstrom et al., 2014, Ikegami et al., 2015).

Due to the nature of the microfabrication process, a micromirror is usually very thin; therefore, at high scan frequencies, the micromirror plate may lose its rigidity and tend to dynamically deform during scanning due to the high out-of-plane acceleration/deceleration forces. This introduces dynamic aberrations into the optical system and degrades the system performance (Hart et al., 2000). Engineering such MEMS mirrors to operate at high speed with high optical resolution is possible, but challenging. In this chapter, we introduce a new type of MEMS optical scanners, termed vibratory diffraction grating scanners (Zhou et al., 2004a, b), which originated from rotating holographic grating scanners. These scanners utilize rotational in-plane motion to change the orientation of the grating lines, hence causing the diffracted laser beams to scan. It is well known from elastodynamics that the dynamic deformation of a thin plate under in-plane excitation (i.e., movement perpendicular to the surface normal) is much smaller than that for out-of-plane excitation. Hence, MEMS-based vibratory grating scanners have the potential to scan at high frequencies without optical performance degradation resulting from the dynamic non-rigid-body deformation prevalent in conventional high speed out-of-plane torsional micromirror-based scanners.

5.2 Operation principle of the diffraction grating scanner

Let us consider the grating scanner shown in Figure 5.3. As shown, the grating, having a period of Λ, lies in the xy plane and rotates about the z-axis. A collimated monochromatic beam of light

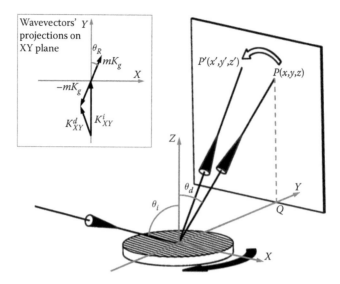

Figure 5.3 Schematic of an in-plane rotational diffraction grating for optical scanning applications.

(wavelength λ) in the yz plane illuminates the grating with an incident angle of θ_i. At the grating's rest position, the grooves are parallel to the x-axis, hence all diffracted beams also lie in the yz plane. Let us further consider a selected mth-order diffracted beam for scanning. This beam leaves the grating with a diffraction angle θ_d and projects to a spot centered at P on a screen. Here, the right-hand rule is used for all angle signs, that is, a right-hand rotation around an axis is defined as positive. To simplify the analysis, we assume the projection screen is parallel to the xz plane and intercepts the y-axis at $Q(0,l,0)$, as shown in the figure. It is noted that l is negative if the diffraction angle θ_d is positive.

As shown in the inset of Figure 5.3, the wave-vectors of the incident and diffracted beams when projected onto the plane of the grating surface (k^i_{xy} and k^d_{xy}) must fulfill the following relation

$$k^d_{xy} = k^i_{xy} - mk_g \tag{5.1}$$

where m is the diffraction order and k_g is the grating-vector with its magnitude defined as $2\pi/\Lambda$ and its direction perpendicular to the grating lines. Thus, it is easy to obtain the x- and y-components of the diffracted beam wave-vector as follows:

$$k^d_x = k^i_x + m\frac{2\pi}{\Lambda}\sin\theta_R = m\frac{2\pi}{\Lambda}\sin\theta_R \tag{5.2}$$

$$k^d_y = k^i_y - m\frac{2\pi}{\Lambda}\cos\theta_R = \frac{2\pi}{\lambda}\sin\theta_i - m\frac{2\pi}{\Lambda}\cos\theta_R \tag{5.3}$$

where θ_R is the rotation angle of the grating around the z axis. Taking the signs of the angles θ_i and θ_d into consideration, it can also be easily verified that the above equations are equivalent to the commonly used diffraction grating equation when $\theta_R = 0$

$$\sin\theta_i + \sin\theta_d = \frac{m\lambda}{\Lambda} \tag{5.4}$$

Since the wavelength is unchanged during diffraction, the wave number is preserved and the z-component of the diffracted beam wave-vector can then be obtained by

$$k_z^d = \left[\left(\frac{2\pi}{\lambda} \right)^2 - \left(k_x^d \right)^2 - \left(k_y^d \right)^2 \right]^{1/2} \tag{5.5}$$

Subsequently, the unit vector (normalized wave-vector) pointing to the direction of the outgoing diffracted beam can be expressed as

$$\begin{aligned} \boldsymbol{r} = (G\sin\theta_R)\boldsymbol{e}_x + (\sin\theta_i - G\cos\theta_R)\boldsymbol{e}_y \\ + [1 - G^2 - \sin^2\theta_i + 2G\sin\theta_i\cos\theta_R]^{1/2}\boldsymbol{e}_z \end{aligned} \tag{5.6}$$

where G is defined as $G = (m\lambda/\Lambda)$, and $\boldsymbol{e}_x$, $\boldsymbol{e}_y$, and $\boldsymbol{e}_z$ are the unit vectors along the x, y, and z axes, respectively.

With the direction of the diffracted beam known, the coordinates of the projected light spot $P'(x', y', z')$ on the screen can be calculated as

$$x' = \left(\frac{G\sin\theta_R}{\sin\theta_i - G\cos\theta_R} \right) l \tag{5.7}$$

$$y' = l \tag{5.8}$$

$$z' = \left[\frac{(1 - G^2 - \sin^2\theta_i + 2G\sin\theta_i\cos\theta_R)^{1/2}}{\sin\theta_i - G\cos\theta_R} \right] l \tag{5.9}$$

It can be easily seen from Equations 5.7 through 5.9 that the scan trajectory on the screen is usually a bow-like curve instead of a straight line for an arbitrary incidence angle θ_i. For example, the black and blue curves in Figure 5.4 show, respectively, the scan trajectories of a

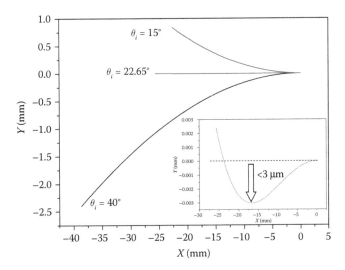

Figure 5.4 Various scan trajectories when the incident angle θ_i changes, where G is set at 1.2 and the grating rotates from 0° to 10°.

diffraction grating scanner with $G = 1.2$ and $\theta_R = 10°$ when the incident angle is 40° and 15°. Now the question is whether there exists an optimal angle θ_i that minimizes the scan bow. To answer this question, let us consider the difference between the z-components of two selected points on the scan trajectory, that is, P' (with grating rotation θ_R) and P (with grating rotation angle 0).

$$\Delta = \left| \left[\frac{(1 - G^2 - \sin^2\theta_i + 2G\sin\theta_i \cos\theta_R)^{1/2}}{\sin\theta_i - G\cos\theta_R} - \frac{(1 - G^2 - \sin^2\theta_i + 2G\sin\theta_i)^{1/2}}{\sin\theta_i - G} \right] l \right| \quad (5.10)$$

For simplicity, we assume θ_R is small so that $\cos\theta_R = 1 - \theta_R^2/2$, and further define

$$a = 1 - G^2 - \sin^2\theta_i + 2G\sin\theta_i \quad (5.11)$$

$$b = \sin\theta_i - G \quad (5.12)$$

The difference Δ can then be written as

$$\Delta = \left\{ \left[\frac{\left[a - (G\sin\theta_i)\theta_R^2\right]^{1/2}}{b + (G/2)\theta_R^2} - \frac{a^{1/2}}{b} \right] l \right\} \quad (5.13)$$

Further expanding the above equation into a power series of θ_R^2 using Taylor expansion and neglecting higher-order terms, we obtain

$$\Delta \simeq -\frac{1}{2} \left(\frac{bG\sin\theta_i + aG}{b^2 a^{1/2}} \right) \cdot \theta_R^2 \cdot l \quad (5.14)$$

Hence, the absolute value of Δ will be minimized if the following condition is fulfilled

$$bG\sin\theta_i + aG = 0 \quad (5.15)$$

Inserting Equations 5.11 and 5.12 into Equation 5.15 above, we obtain

$$\sin\theta_i = G - \frac{1}{G} = \frac{m\lambda}{\Lambda} - \frac{\Lambda}{m\lambda} \quad (5.16)$$

With the diffraction grating equation, that is, Equation 5.4, it can also be deduced that

$$\sin\theta_d = \frac{1}{G} = \frac{\Lambda}{m\lambda} \quad (5.17)$$

This is a well-known condition for a diffraction grating to perform a bow-free scanning (Beiser, 1988). It is noted that $-1 < \sin\theta_i < 1$ and $-1 < \sin\theta_d < 1$, hence we can conclude that G must fulfill the condition that $1 < G < 1.618$ in order for the bow-free scanning to exist. Taking $G = 1.2$ as an example, when the incident angle θ_i is around 22.65° as computed from Equation 5.16, the scan trajectory on the screen at this bow-free incident angle for a grating rotation angle of 10° is also plotted as a red curve in Figure 5.4 for comparison. An enlarged view in the inset of Figure 5.4

shows that the maximum deviation of the scan trajectory from a perfectly straight line is less than 3 μm throughout the 25-mm scan length. Clearly, the scan bow is now minimized to a negligible level.

5.3 Diffraction efficiency of the grating scanner

If the first diffraction order ($m = 1$) is selected for scanning, it is obvious that the grating period Λ must be greater than 0.618λ and less than λ in order for the bow-free scanning to exist. This grating is then a subwavelength grating and its diffraction efficiency (i.e., the percentage of incident light energy that is diffracted to the first order for scanning) can be optimized through the design of the surface profile within a unit grating period. While scalar diffraction theory may not be sufficient to determine the diffraction efficiency of the subwavelength gratings, rigorous coupled-wave analysis (RCWA) (Moharam and Gaylord, 1983) can be employed for this optimization task.

An example of using the RCWA approach to optimize the grating profile to achieve high diffraction efficiency is shown in Figure 5.5 (Zhou et al., 2008). The diffraction grating considered here has a period of 500 nm with a duty cycle of 50% and is made of a highly reflective metal such as gold. The RCWA calculates the diffraction efficiency of the grating as a function of grating groove depth. Considering the scanning application, we also fixed the light incident angle to fulfill the bow-free scanning condition from Equation 5.16. For a 632.8-nm wavelength laser, this incident angle is 28.4°. The RCWA results reveal a few interesting facts. First, the grating is polarization sensitive, that is, TE- (electric field vector parallel to the grating grooves) and TM-polarized (electric field vector perpendicular to the grating grooves) laser beams have different diffraction efficiencies. Second, the TM-polarized efficiently oscillates up and down as the groove depth increases and can be as high as 80% at the peaks. Third, the TE-polarized efficiency is generally low and increases monotonically as the groove depth increases. These RCWA results have been verified by experiments with microfabricated gratings. These gratings are fabricated on a silicon substrate and coated

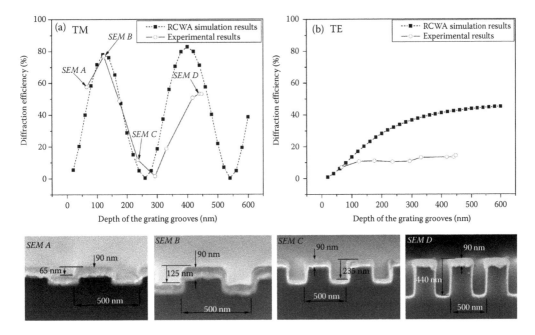

Figure 5.5 Diffraction efficiency as a function of the grating groove depth for (a) TM-polarized incident light and (b) TE-polarized incident light. Insets show the cross-sections of the various diffraction gratings with different groove depths imaged under an SEM. (Copyright IOP Publishing. Reproduced with permission. All rights reserved.)

with a 90-nm thick gold layer to enhance the reflectivity. Some of the selected cross-sectional views are also provided in Figure 5.5, indicated as SEM A to D. The results show conclusively that high efficiency laser scanning can be achieved using TM-polarized light incident on a simple binary subwavelength grating at an optimized groove depth.

5.4 MEMS-driven mechanisms

In the previous sections, we showed that high-efficiency bow-free optical scanning can be obtained by rotating a diffraction grating through an axis perpendicular to its surface. Such an in-plane motion has less mechanical dynamic deformation, thus is preferred for high-speed scanning with thin platforms at a microscale. There are several MEMS mechanisms available to drive a platform (having a diffraction grating) to execute an in-plane rotation, among which the use of micromotors is the most direct one. In fact, as shown schematically in Figure 5.6a, the diffraction grating scanner was first implemented at the microscale using surface-micromachined continuously rotating electrostatic poly-silicon micromotors (Smith et al., 1995, Yasseen et al., 1999). This can be considered the MEMS version of the rotating holographic scanners. However, a slow scanning rate due to the lack of a sufficiently good bearing system for high-speed rotation, a short lifetime resulting from the friction, and wear and tear issues arising from MEMS moving surfaces are disadvantages.

Another method is to drive a suspended platform/grating into an in-plane rotational oscillation (about the axis perpendicular to the grating surface) instead of continuous rotation or spinning. This can be considered the resonant version of the holographic scanners. As the platform is suspended by flexures and the motion is oscillatory in nature and is supported by the elastic deformation of the flexural springs, there are no contact surfaces that move relative to each other and hence

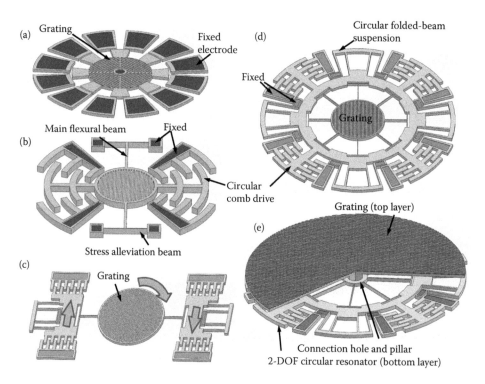

Figure 5.6 MEMS driving mechanisms for vibratory grating scanners. The in-plane rotation of the grating is driven by (a) micro motor, (b) 1-DOF vibration system with circular comb drives, (c) 2-DOF vibration system with linear comb drives, (d) 2-DOF vibration system with circular comb drives, and (e) double-layered MEMS structure.

no friction or wear and tear issues. In fact, similar to conventional low-inertia scanning mirrors, such MEMS grating designs can also be operated in their structural resonances to take the advantages of both high-speed scanning and a large scan amplitude. These devices are called MEMS vibratory diffraction grating scanners (Zhou et al., 2004a).

To implement a MEMS vibratory grating scanner, the simplest mechanism is a single degree-of-freedom (DOF) spring-mass vibration design as shown in Figure 5.6b, utilizing circular electrostatic comb drives (Tang et al., 1990), where the fixed and movable comb-like electrodes are concentric. The platform containing the diffraction grating is suspended with T-shaped flexure springs. The bending of the main flexural beams supports a rotational motion about the grating normal, and the deflection of the stress alleviation beams reduces the stresses in the main flexural beams and supports their large bending deformations. However, a severe problem of such a driving mechanism is that it is difficult for the circular comb drives to achieve a large rotation angle, especially when the grating platform is large (long finger electrodes and large finger overlaps coupled with the large required displacements for the finger electrodes at a large radius lead to an increased chance of electrostatic pull-in, i.e., short circuit through a contact between stationary and mobile electrodes).

Hence, a practical mechanism to drive a large in-plane rotational motion is to use two linear electrostatic comb drive resonators as shown in Figure 5.6c (Zhou et al., 2004a, 2008). The two resonators are identical and are operated at the same amplitude and frequency, but they are 180° out-of-phase. This mechanism is equivalent to a two-DOF (2-DOF) vibration system. If the masses of the platform and the comb drive resonators and spring constants of their respective suspensions are carefully designed, one can obtain a suitable vibration mode at its resonant frequency such that small linear displacements of the comb drive resonators can drive a large rotational vibration of the platform. T-shaped flexural suspensions that support large rotation angles are again necessary to connect the platform to the driving comb drive resonators. Indeed, this driving mechanism was utilized in our demonstration of the first prototype MEMS vibratory grating scanner. It is also noted that by adding additional T-shaped suspensions connecting the platform to some fixed supports (not the driving resonators), one can further greatly increase the operation frequency of the scanner to above 50 kHz while maintaining a large scan angle of 14° (Du et al., 2010a).

To further enhance the performance of the scanner, one has to perfectly balance the two driving resonators such that they are identical, which is difficult to achieve when the microfabrication imperfections are taken into consideration. Hence, the driving mechanism is modified into that shown in Figure 5.6d (Du et al., 2009). Instead of two linear comb drive resonators, a single circular comb drive resonator is utilized to drive the grating platform to eliminate the balancing problem. A major difference between the mechanisms shown in Figure 5.6d and b is that the former is a 2-DOF system that one can again design to optimize the masses and springs of the platform and driving resonator to achieve a suitable vibration mode that drives a large platform rotation with a small angular vibration of the circular comb drive resonator. Here, circular folded-beam suspensions are used to support the circular comb drive resonator and T-shaped flexural springs are used to connect the grating platform to the resonator. An additional advantage of such a mechanism is that a large number of T-shaped flexural springs can be placed to connect the grating to the driving resonator, as shown in the schematic. This effectively reduces the maximum stress level in each individual suspension beam while maintaining a high frequency and large scan amplitude optical scanning, which in turn is highly desirable for linear, stable, reliable, and long lifetime operation of the scanner.

As we discussed previously, a high scan resolution requires both a large optical scan angle and a large device aperture size (or grating size here). For a single-layered design (the grating and its driving mechanism are made of the same structure layer) as shown in Figure 5.6d, there is an inevitable trade-off between the radius of the grating platform and the maximum possible platform rotation angle. To overcome this trade-off/limitation, a double-layered driving mechanism design is proposed and demonstrated as shown in Figure 5.6e (Du et al., 2010b). In this case, the grating platform and MEMS driving mechanism are fabricated separately and then assembled together to form the scanner. As shown, the driving mechanism is similar to that shown in Figure 5.6d, however, the center platform without a grating has significantly shrunk to a small connection hole that is used to mechanically connect the diffraction grating located in a separate layer above the MEMS driving mechanism. The connection is established by inserting the connection pillar located underneath the grating into the connection hole and gluing them together. In this way, both a large grating aperture size (2 ∼ 3 mm) and a large optical scan angle (∼30°) can be achieved simultaneously, leading to a very high optical scan resolution at a high operation speed (20–30 kHz) (Du et al., 2011).

5.5 Fabrication and testing of the double-layered MEMS vibratory grating scanners

We choose to use the double-layered grating scanner as an example to elaborate the microfabrication process. The other scanners are fabricated in a similar way. As shown in Figure 5.7a and b, respectively, the grating platform and the MEMS driving mechanism are fabricated separately. The grating platform fabrication starts from a single-crystal silicon wafer. First, the subwavelength diffraction grating is patterned on the top surface of the wafer using deep-UV lithography and a timed plasma etching process that results in a grating groove depth of about 120–150 nm. Next, the grating platform is defined by another lithography and a deep reactive ion etch (DRIE) process with the etch depth controlled to be around 10 μm, which roughly determines the final thickness of the platform. The wafer is then thinned down through etching the backside using KOH. Finally, a set of masking, lithography patterning, and DRIE processes are conducted to etch the wafer from the

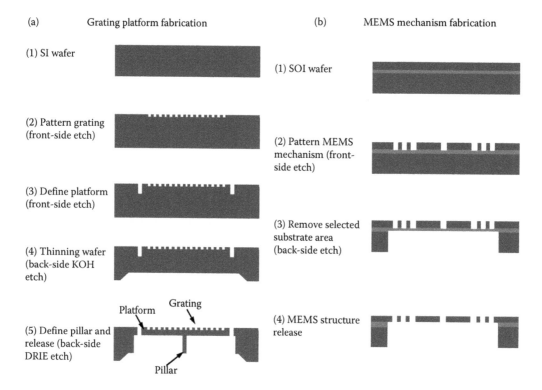

Figure 5.7 Microfabrication process flow for the double-layered vibratory grating scanners, with (a) showing the grating fabrication process and (b) showing the MEMS driving mechanism fabrication process.

backside to fabricate the connection pillar and at the same time release the grating platform from the wafer substrate.

The fabrication of the MEMS driving mechanism, however, starts from a silicon-on-insulator (SOI) wafer. First, the MEMS structures, including the circular comb drive resonator, T-shaped suspensions, center platform, and connection hole are patterned using lithography and etched into the silicon device layer (about 80 μm thick) using DRIE stopping on the buried oxide (BOX) layer. Next, the SOI wafer is patterned on the backside and another DRIE process is used to completely remove the silicon substrate in areas that are under the movable MEMS structures. Finally, a hydrofluoric wet etching process is conducted to remove the exposed BOX layer to release the MEMS structures. After both the grating platform and its MEMS driving mechanism are fabricated, they are assembled under an optical microscope by carefully inserting the connection pillar into the connection hole and fixing them with an epoxy. The final assembled MEMS grating scanner is shown in Figure 5.8a. This scanner has a grating diameter of 2 mm and is capable of scanning an optical angle of 33.5° at its resonant frequency of 21.6 kHz when operated in a low pressure vacuum chamber (0.12 mTorr) with a 65 V DC bias voltage and 130 V peak-to-peak AC voltage (Du et al., 2011). The experimental results showing the scanning performance of the device are given in Figure 5.8b.

In order to demonstrate the low dynamic deformation of the grating platform during such a high-speed scanning, a stroboscopic method needs to be used. A testing setup is built in-house, the schematic of which is illustrated in Figure 5.9a. As shown, a CW helium-neon (He-Ne) laser at 632.8 nm wavelength is first focused to a Bragg cell to generate a pulsed laser beam with a pulse duration around 25 ns using the acousto-optic effect. The laser pulse frequency is perfectly synchronized with the MEMS scanner's operation frequency, such that the motion of the scanning beam is effectively "frozen" due to the stroboscopic effect. The beam can be "frozen" at any location along the scan trajectory by adjusting the phase between the laser pulse and the device's driving signal. With this technique to freeze the motion, we can then construct laser interferometers to examine the wavefront distortions of the outgoing strobed laser beams from the MEMS scanning device. An example is illustrated in Figure 5.9a. We first expand the strobed laser beam using a beam expander and then input the beam to a lateral shearing interferometer (Malacara, 2007). The amount of lateral shearing can be controlled by moving one of the mirrors translationally while maintaining its orientation. The interferogram from the lateral shearing interferometer is monitored with a CCD camera. Figure 5.9b shows the typical experimental results—four interferograms recorded at four different scan locations. In our experiments, we purposely introduce a small amount of defocus such that an aberration-free wavefront will result in uniform straight fringes

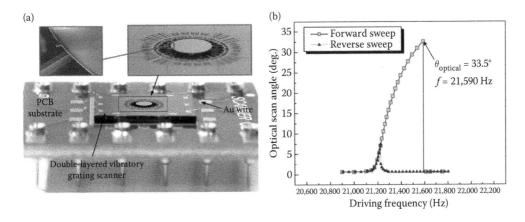

Figure 5.8 (a) Images of the fabricated double-layered diffraction grating scanner and (b) measured optical scan angle as a function of the driving frequency. (Figures reproduced with permission from Elsevier.)

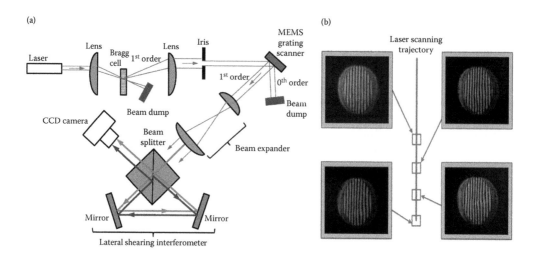

Figure 5.9 (a) Schematic showing the experimental setup to test the MEMS vibratory grating scanners, which involves stroboscopy and lateral shearing interferometry. (b) The recorded lateral shearing interferograms at different scan locations. (Figures reproduced with permission from Elsevier.)

with equal gaps. Judging from the recorded interferograms, we can conclude that no significant wavefront distortions are observed at all locations during the high speed scanning. We also use a similar stroboscopic method to measure the spot sizes at different scan locations. Combining the measured average spot size with the length of the scan line, a scan resolution of about 1450 pixels per scan line is experimentally obtained for the MEMS grating scanner shown in Figure 5.8 (Du et al., 2011).

5.6 Applications of the MEMS vibratory grating scanners

The MEMS vibratory grating scanners' outstanding performances at high speed, including low dynamic deformation, large aperture size, and high scan resolution, make them attractive in a wide range of applications. These include laser printers, laser marking systems, laser scanning confocal microscopes, light detection and ranging (LiDAR) systems, laser cameras, laser projection displays, and many others. It should be noted that the MEMS grating scanners are dispersive and this characteristic is inherently different from the MEMS scanning mirrors. For monochromatic (or narrowband) applications, MEMS vibratory grating scanners can be used in the same way as the MEMS mirrors without changing the system configuration. However, for broadband applications, special considerations need to be taken into account at the system level. Taking the laser projection color display, for example, the display has to project the red (R), green (G), and blue (B) images on the screen at the same time. A possible configuration of the laser project display using MEMS vibratory grating scanners is highlighted in Figure 5.10. As shown, beams from three laser sources with different wavelengths (R, G, and B) are scanned, respectively, by three MEMS vibratory grating scanners. The beams are then combined to produce a white spot on the screen. The MEMS grating scanners are synchronized such that the RGB laser spots are always at the same location during scanning. Alternatively, synchronization can also be implemented using three different gratings integrated on the same MEMS vibration platform (Zhou and Chau, 2006). Fast MEMS grating scanners are operated at resonance, and the slow scanner (micromirror) is driven by a sawtooth waveform. The combined motions of the fast and slow scanners create a 2D sinusoidal raster. Images are created by modulating the intensities of the RGB laser sources according to the spot position on the screen.

Another application of the MEMS vibratory grating scanners is hyperspectral imaging (Sellar and Boreman, 2005, Mouroulis et al., 2007), which utilizes both its light scanning and dispersion

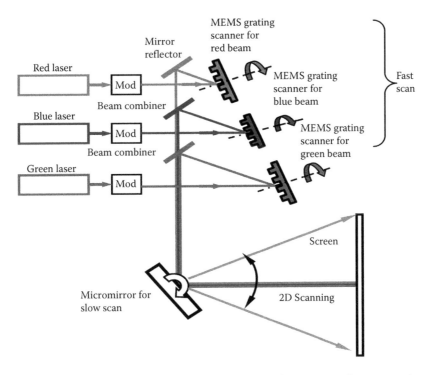

Figure 5.10 Schematic of a laser projection color display using three MEMS vibratory grating scanners.

characteristics. To illustrate its working principle (Zhou et al., 2009), let us first consider light scanning using a grating scanner and a mirror, as shown in Figure 5.11. A broadband light beam from a light source is dispersed by the grating and then reflected by the mirror, and after the mirror reflection it is projected onto a screen. When the grating rotates in-plane as shown in the figure, light with different wavelengths scans at different locations. We assume, for simplicity at this moment, that all these scan lines are straight and equal in length for all wavelengths. Next, let us consider the hyperspectral imaging application by reversing the ray direction and replacing the light source with a photodetector (single-pixel) and the screen with a slit. As shown in the figure, due to the

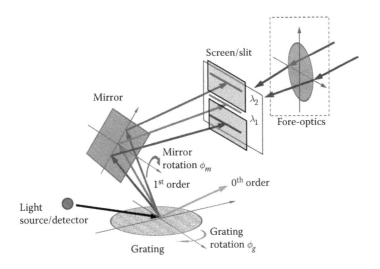

Figure 5.11 Schematic of a hyperspectral imaging system utilizing a MEMS vibratory grating scanner and a scanning mirror.

reciprocity in light transport, only the wavelength λ_1 selected by the slit opening can reach the detector. Furthermore, when the grating rotates, the light scans and the detector records the narrow-band light intensities at different locations along the slit. In other words, when the grating scanner completes one scan, the image of the slit at wavelength λ_1 is captured. The slit images at different wavelengths can be captured by simply rotating the mirror, for example, by moving the scan line of wavelength λ_2 into the slit as shown in the figure. Hence, a hyperspectral image of the slit can be obtained through spectral scanning with the mirror and spatial scanning with the grating. We can further configure the system to work in a push-broom mode (Sellar and Boreman, 2005) by attaching a fore-optics in front of the slit as shown in a dashed box in Figure 5.11. A three-dimensional (3D) hyperspectral data cube can then be obtained by moving/scanning the imaging system along a direction perpendicular to the slit direction.

Such a hyperspectral system possesses a few interesting characteristics. With optical scanning and a single-pixel-based photodetector, large-scale integration of electronics on a sensor chip is not required. Thus, the imager can be operated at wavelengths currently unavailable or prohibitively expensive for detector arrays (an example is the expensive IR focal plane arrays). This is especially attractive for IR imaging. In addition, single-pixel-based imaging also simplifies the sensor calibration process, as it is inherently free of array uniformity errors. With the MEMS technology, the hyperspectral imagers can be potentially constructed in miniature form as lightweight and low cost imagers.

In a practical hyperspectral imaging system using the above-mentioned working principle, the MEMS grating scanner and the mirror are in constant motion—the mirror at a slower rate in a sawtooth wave form and the grating at a higher rate in a sinusoidal resonance form—in order to fulfill the frame rate requirement of the system. In addition, the scan trajectories are not always straight lines with equal lengths for different wavelengths as the grating rotates. Therefore, one might ask: at an instant of time t during sampling, when the grating has an orientation angle of $\phi_g(t)$ and the mirror has its orientation angle of $\phi_m(t)$, what is the instantaneous field of view along the slit seen by the single-pixel detector? And what is the wavelength it recorded? The answers to the above questions are critical to the image reconstruction. The problem is difficult to solve analytically, but fortunately it can be easily computed numerically (Du et al., 2012).

We use a reverse ray-tracing method as illustrated in Figure 5.12. As shown in Figure 5.12a, the ray connecting the centers of the pinhole and the focusing lens is incident on the diffraction grating with a rotation angle $\phi_g(t)$. The wavelength is initially chosen to be at the center of the operational spectral band $(\lambda_{min}, \lambda_{max})$, that is, $\lambda = (\lambda_{min} + \lambda_{max})/2$. The direction of the first order diffracted ray is computed using Equation 5.6, which is then further incident on the reflecting mirror rotated with an angle $\phi_m(t)$. The reflected ray is computed using Snell's law. The outgoing ray is then focused by the collimating lens and intersects the imaging slit at $P(x,y)$. Since there is a narrow slit placed along the Y_s axis as shown, if the computed coordinate x of point P is large (i.e., $|x| \geq \varepsilon$), the ray is blocked by the imaging slit. In this case, the above ray-tracing process needs to be iterated with a new wavelength. As shown in the flowchart in Figure 5.12b, either λ_{min} or λ_{max} is replaced with the current wavelength λ depending on the sign of x, and a new wavelength is computed again using $\lambda = (\lambda_{min} + \lambda_{max})/2$. The same process is then carried out again until the x coordinate of P on the slit is sufficiently small such that $|x| < \varepsilon$. We can then conclude that this ray can pass through the slit. The final values of wavelength and height (y coordinate of P) are recorded as λ_t and y_t, respectively. The method is similar to the numerical bisection method to find the root of an equation. Based on the reciprocity principle in light transport, by changing the ray direction, we can also conclude that a ray with wavelength λ_t coming from a height of y_t on the slit can pass through the system and reach the photodetector. In other words, when the grating has an orientation angle of $\phi_g(t)$ and the mirror has its orientation angle of $\phi_m(t)$, the instantaneous wavelength and field of view seen by the detector are λ_t and y_t, respectively. Thus, the intensity measured by the photodetector is assigned to the data point (λ_t, y_t). In this way, when all the spectral and spatial dimensions of a slit image are sampled, a distorted hyperspectral image of the slit is created. In our system, we use a galvano mirror operated at 100 Hz for a slow sawtooth scan and a MEMS vibratory grating

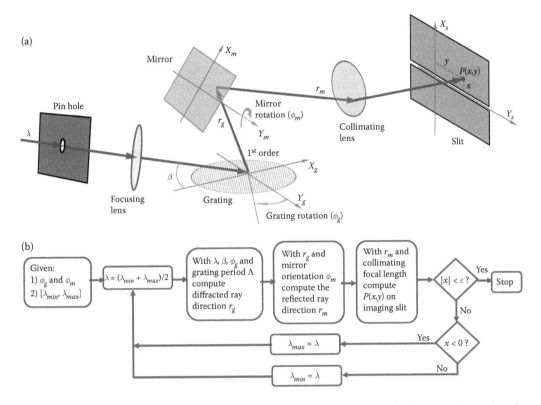

Figure 5.12 (a) Schematic and (b) flow chart of the inverse ray-tracing algorithm to determine the instantaneous wavelength and field of view recorded by the single-pixel photodetector.

scanner operated at 21.3 KHz for a fast resonant scan. Thus, the image system has a frame rate of about 200 Hz with a large number of data points sampled in each frame.

To illustrate how the distortion correction algorithm works, we schematically show a portion of the distorted data points (blue dots) here in Figure 5.13a for clear visualization. These represent the raw sampled spatial and spectral data points with the photodetector intensity reading at each

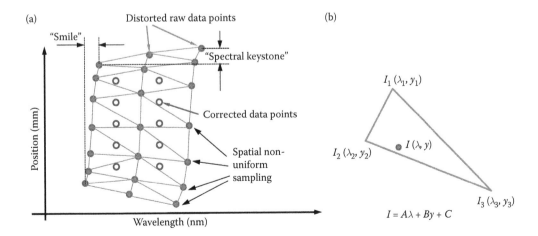

Figure 5.13 (a) Schematic showing the major distortions in the recorded hyperspectral image and the meshed network using Delaunay Triangulation. (b) Schematic showing the linear interpolation method used to compute the intensity value at the corrected data point.

point removed. It can be seen that due to the constant motions of the two scanners, the curved scan trajectory and the wavelength-dependent scan length of the grating scanner, the recorded image is distorted. As shown, there are mainly smile, spectral keystone, and nonuniform spatial sampling distortions. To correct these distortions, the data points are first meshed into triangular areas. Delaunay Triangulation (DT) (Zhao et al., 2005) is a good method to use for meshing. Then, a new undistorted grid of data sampling points is created and these are indicated as the red open dots, as shown in the figure. The intensity value at each undistorted data point is calculated through a linear interpolation using the measured intensities at the three vertexes of the triangle within which it is located, as shown in Figure 5.13b. In other words, the three coefficients A, B, C are first determined using the three measured intensities I_1, I_2, and I_3. Next, the unknown intensity at (λ, y) is then computed using the linear equation shown in the figure.

The hyperspectral imaging system is developed. The system is configured to operate in a spectral band from 450 to 650 nm with a spatial field of view over 40 mm total in length. A test target, as shown in Figure 5.14a, consisting of two light emitting diodes (LEDs), is employed to evaluate its performance. The upper LED emits blue light at a peak wavelength of 462 nm, while the lower LED emits green light at a peak wavelength of 528 nm. Their centers are spatially separated around 20.5 mm apart. Figure 5.14b shows an obtained hyperspectral image of the test target along a vertical line running through the two centers of the LEDs. The intensity data are color-coded. This is the final distortion-corrected image using the above-described method. A 3D plot showing the intensity, spatial, and spectral data of the captured hyperspectral image is also provided in Figure 5.14c. It is quite clear from the experimental results that the hyperspectral imaging system developed using a MEMS scanning grating and a mirror can capture not only the image of the target, but also its spectral features.

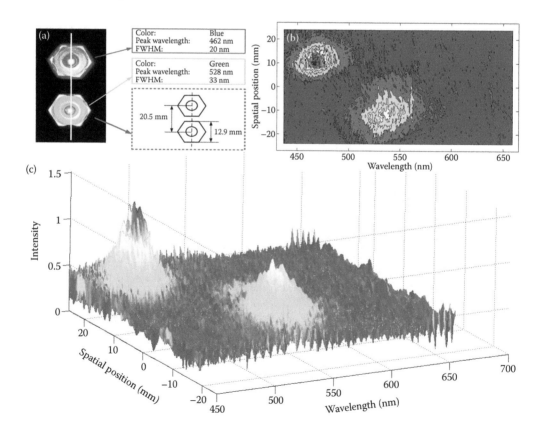

Figure 5.14 (a) LED test target, (b) corrected hyperspectral image of the LED target with intensities coded in colors, and (c) 3D plot of the hyperspectral image.

5.7 Summary

In this chapter, we discuss the operation principle of MEMS vibratory grating scanners, which rely on the in-plane rotation of diffraction gratings to change the orientation of the grating lines, thus causing the diffracted light to scan. When properly designed, such scanners can achieve high diffraction efficiency and bow-free scanning. The in-plane motion also greatly reduces the dynamic structural deformation during the scanning operation making these scanners capable of high resolution scanning at very high frequencies. Several MEMS driving mechanisms are discussed in detail, including 2-DOF vibration systems and double-layered structures. Microfabrication processes and scanner performance testing methods are also highlighted. The applications of MEMS vibratory scanners are broad and include a range of laser and no-laser applications. In particular, we highlight a novel hyperspectral imaging system using such MEMS vibratory grating scanners. In this system, a MEMS grating scanner is used for fast spatial scanning and a mirror is used for slow spectral scanning. The synchronized operation of those two scanners allows the hyperspectral imaging system to work with a single-pixel photodetector, which is attractive for IR imaging where arrayed detectors are bulky and expensive.

References

Beiser, L. 1988. *Holographic Scanning*, New York, Wiley.

Beiser, L. 1992. *Laser Scanning Notebook*, Bellingham, Wash, SPIE Optical Engineering Press.

Beiser, L. 2003. *Unified Optical Scanning Technology*, New York, Wiley-Interscience.

Du, Y., Cheo, K. K. L., Zhou, G. G., and Chau, F. S. 2012. A spectral line imager based on a MEMS vibratory grating scanner. *2012 International Conference on Optical Mems and Nanophotonics (Omn)*, Banff, Alberta, Canada, 210–211.

Du, Y., Zhou, G. Y., Cheo, K. L., Zhang, Q. X., Feng, H. H., and Chau, F. S. 2009. A 2-DOF circular-resonator-driven in-plane vibratory grating laser scanner. *Journal of Microelectromechanical Systems*, 18, 892–904.

Du, Y., Zhou, G. Y., Cheo, K. K. L., Zhang, Q. X., Feng, H. H., and Chau, F. S. 2010a. A high-speed MEMS grating laser scanner with a backside thinned grating platform fabricated using a single mask delay etching technique. *Journal of Micromechanics and Microengineering*, 20(11), 115028.

Du, Y., Zhou, G. Y., Cheo, K. L., Zhang, Q. X., Feng, H. H., and Chau, F. S. 2010b. Double-layered vibratory grating scanners for high-speed high-resolution laser scanning. *Journal of Microelectromechanical Systems*, 19, 1186–1196.

Du, Y., Zhou, G. Y., Cheo, K. K. L., Zhang, Q. X., Feng, H. H., and Chau, F. S. 2011. A 21.5 kHz high optical resolution electrostatic double-layered vibratory grating laser scanner. *Sensors and Actuators a-Physical*, 168, 253–261.

Hart, M. R., Conant, R. A., Lau, K. Y., and Muller, R. S. 2000. Stroboscopic interferometer system for dynamic MEMS characterization. *Journal of Microelectromechanical Systems*, 9, 409–418.

Holmstrom, S. T. S., Baran, U., and Urey, H. 2014. MEMS laser scanners: A review. *Journal of Microelectromechanical Systems*, 23, 259–275.

Ikegami, K., Koyama, T., Saito, T., Yasuda, Y., and Toshiyoshi, H. 2015. A Biaxial PZT Optical Scanner for Pico-Projector Applications. *Moems and Miniaturized Systems Xiv*, 9375.

Malacara, D. 2007. *Optical Shop Testing*, Hoboken, NJ, Wiley-Interscience.

Marshall, G. F. and Stutz, G. E. 2012. *Handbook of Optical and Laser Scanning*, Boca Raton, FL, CRC Press.

Moharam, M. G. and Gaylord, T. K. 1983. Rigorous coupled-wave analysis of grating diffraction—E-mode polarization and losses. *Journal of the Optical Society of America*, 73, 451–455.

Mouroulis, P., Sellar, R. G., Wilson, D. W., Shea, J. J., and Green, R. O. 2007. Optical design of a compact imaging spectrometer for planetary mineralogy. *Optical Engineering*, 46(6), 063001.

Muller, R. S. and Lau, K. Y. 1998. Surface-micromachined microoptical elements and systems. *Proceedings of the IEEE*, 86, 1705–1720.

Sellar, R. G. and Boreman, G. D. 2005. Classification of imaging spectrometers for remote sensing applications. *Optical Engineering*, 44(1), 013602-1.

Smith, S. W., Yasseen, A. A., Mehregany, M., and Merat, F. L. 1995. Micromotor grating optical switch. *Optics Letters*, 20, 1734–1736.

Sun, J. J., Guo, S. G., Wu, L., Liu, L., Choe, S. W., Sorg, B. S., and Xie, H. K. 2010. 3D In Vivo optical coherence tomography based on a low-voltage, large-scan-range 2D MEMS mirror. *Optics Express*, 18(12), 12065–12075. doi: 10.1364/Oe.18.012065

Tang, W. C., Nguyen, T. C. H., Judy, M. W., and Howe, R. T. 1990. Electrostatic-comb drive of lateral polysilicon resonators. *Sensors and Actuators a-Physical*, 21, 328–331.

Urey, H. 2000. Optical advantages of retinal scanning displays. *Helmet- and Head-Mounted Displays V*, 4021, 20–26.

Urey, H., Wine, D. W., and Osborn, T. D. 2000. Optical performance requirements for MEMS-scanner based microdisplays. *Moems and Miniaturized Systems*, 4178, 176–185.

Yalcinkaya, A. D., Urey, H., Brown, D., Montague, T., and Sprague, R. 2006. Two-axis electromagnetic microscanner for high resolution displays. *Journal of Microelectromechanical Systems*, 15, 786–794.

Yasseen, A. A., Mitchell, J. N., Smith, D. A., and Mehregany, M. 1999. High-aspect-ratio rotary polygon micromotor scanners. *Sensors and Actuators a-Physical*, 77, 73–79.

Zhao, Y., Tay, F. E. H., Chan, F. S., and Zhou, G. Y. 2005. A nonlinearity compensation approach based on Delaunay triangulation to linearize the scanning field of dual-axis micromirror. *Journal of Micromechanics and Microengineering*, 15, 1972–1978.

Zhou, G. Y. and Chau, F. S. 2006. Micromachined vibratory diffraction grating scanner for multiwavelength collinear laser scanning. *Journal of Microelectromechanical Systems*, 15, 1777–1788.

Zhou, G. Y., Cheo, K. K. L., Du, Y., Chau, F. S., Feng, H. H., and Zhang, Q. X. 2009. Hyperspectral imaging using a microelectricalmechanical-systems-based in-plane vibratory grating scanner with a single photodetector. *Optics Letters*, 34, 764–766.

Zhou, G. Y., Du, Y., Zhang, Q. X., Feng, H. H., and Chau, F. S. 2008. High-speed, high-optical-efficiency laser scanning using a MEMS-based in-plane vibratory sub-wavelength diffraction grating. *Journal of Micromechanics and Microengineering*, 18(8), 085013.

Zhou, G. Y., Vj, L., Chau, F. S., and Tay, F. E. H. 2004a. Micromachined in-plane vibrating diffraction grating laser scanner. *Ieee Photonics Technology Letters*, 16, 2293–2295.

Zhou, G. Y., Vj, L., Tay, F. E. H., and Chau, F. S. 2004b. Diffraction grating scanner using a micromachined resonator. *MEMS 2004: 17th IEEE International Conference on Micro Electro Mechanical Systems, Technical Digest*, Maastricht, Netherlands, 45–48.

6

F–P filters and applications in spectrometers and gas sensing

Chong Pei Ho and Chengkuo Lee

Contents

6.1 Background of Fabry–Pérot filter

The Fabry–Pérot Filter (FPF) is comprised of two flat and highly reflective mirrors with an air gap between them. Thus, the FPF can be considered to be a system with three media. The individual media can first be considered as shown in Figure 6.1. For an incoming beam from Medium 1 striking on the interface between Medium 1 and Medium 2, the beam can be characterized with complex reflection amplitude, r_1, and complex transmission amplitude t_1. Likewise, for an incoming beam from Medium 2 incident on the interface between Medium 1 and Medium 2, the beam is undergoing reflection back into Medium 2 with complex reflection amplitude, r_1' and transmission into Medium 1 with complex transmission amplitude t_1'. Finally, when a beam from Medium 2 strikes the interface between Medium 2 and Medium 3, the complex reflection amplitude is r_2 and the complex transmission amplitude is t_2.

The FPF is characterized by its overall reflectance, R and transmittance, T. Both R and T are Poynting vectors of the portion of energy reflected and transmitted, respectively. In order to obtain R and T of the FPF, we consider that the incoming beam from Medium 1 strikes the interface between Medium 1 and Medium 2 and resonates within Medium 2 before being

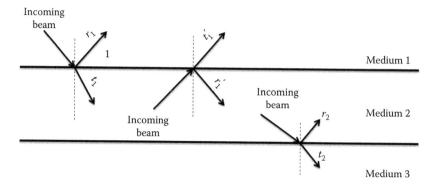

Figure 6.1 The three individual mediums of the FPF system with the various complex reflection amplitudes and complex transmission amplitudes when the incoming beam originated from different mediums.

transmitted into Medium 3. This causes it to incident repeatedly within the two interfaces (n times) as shown in Figure 6.2. The distance between the interfaces is fixed at **d**. At every incidence onto the interfaces, a portion of the complex amplitudes travel into either Medium 1 or Medium 3.

Since Medium 1 and Medium 3 are the same, we can simplify the amplitude components to

$$r_1' = r_2 = r$$

$$r_1' = -r_1$$

$$r_2 = -r_1$$

$$t_1' = t_1 = t$$

$$t_1 t_1' = 1 - r^2$$

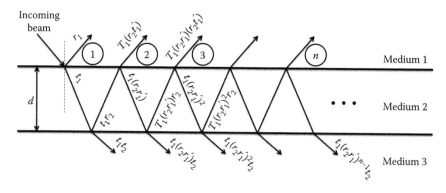

Figure 6.2 The FPF system with an incoming beam from Medium 1 and resonating within Medium 2. The overall reflection amplitude of the FPF can be obtained by summing all the complex reflection amplitudes into Medium 1 and the overall transmittance amplitude is obtained by summing all the complex transmittance amplitudes into Medium 3.

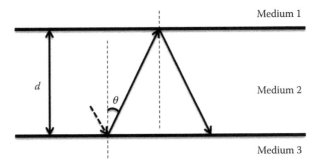

Figure 6.3 Geometry of the system to determine the phase difference between two successive transmitted waves.

As shown in Figure 6.3, based on the geometry of the system, the phase difference between two successive transmitted waves can be defined as

$$\delta = \frac{2d\cos\theta}{\lambda_0/n_0}2\pi$$

In order to obtain the overall reflection amplitude of the FPF, r_{FPF}, we sum all the individual complex reflection amplitudes travelling into Medium 1 when the incoming beam strikes the interface between Medium 1 and Medium 2. The overall reflection amplitude is given as follows:

$$
\begin{aligned}
r_{FPF} &= r_1 + t_1\left(r_2 t_1'\right)e^{-i\delta} + t_1\left(r_2 r_1'\right)\left(r_2 t_1'\right)e^{-i2\delta} + \cdots + t_1\left(r_2 r_1'\right)^{n-2}\left(r_2 t_1'\right)e^{-i(n-1)\delta} + \cdots \\
&= r_1 + \frac{t_1\left(r_2 t_1'\right)e^{-i\delta}}{1 - r_2 r_1' e^{-i\delta}} \\
&= \frac{r_1 + r_2 e^{-i\delta}\left(-r_1 r_1' + t_1 t_1'\right)}{1 - r_2 r_1' e^{-i\delta}} \\
&= r_1 + \frac{r_1 + r_2 e^{-i\delta}}{1 + r_1 r_2 e^{-i\delta}}
\end{aligned}
$$

The overall reflectance, R of the FPF can be calculated as follows:

$$
\begin{aligned}
R = r_{FPF}^2 &= \frac{\left(r_1 + r_2 e^{-i\delta}\right)\left(r_1 + r_2 e^{+i\delta}\right)}{\left(1 + r_1 r_2 e^{-i\delta}\right)\left(1 + r_1 r_2 e^{+i\delta}\right)} \\
&= \frac{(r_1 + r_2)^2 - 4 r_1 r_2 \sin^2(\delta/2)}{(1 + r_1 r_2)^2 - 4 r_1 r_2 \sin^2(\delta/2)} \\
&= \frac{4|r|^2 \sin^2(\delta/2)}{(1 - |r|^2)^2 + 4|r|^2 \sin^2(\delta/2)}
\end{aligned}
$$

Likewise, in order to obtain the overall transmission amplitude of the FPF, t_{FPF}, we sum all the individual complex transmission amplitudes travelling into Medium 3 when the incoming beam strikes the interface between Medium 2 and Medium 3. The overall transmission amplitude is given as follows:

$$
\begin{aligned}
t_{FPF} &= t_1 t_2 + t_1\left(r_2 r_1'\right)t_2 e^{-i\delta} + t_1\left(r_2 r_1'\right)t_2 e^{-i2\delta} + \cdots + t_1\left(r_2 r_1'\right)^{n-1} t_2 e^{-i(n-1)\delta} + \cdots \\
&= \frac{t_1 t_2}{1 - r_2 r_1' e^{-i\delta}}
\end{aligned}
$$

The overall transmittance, T of the FPF can be calculated as follows:

$$T = t_{FPF}t^*_{FPF} = \frac{t_1^2 t_2^2}{\left(1 - r_2 r_1' e^{-i\delta}\right)\left(1 - r_2 r_1' e^{+i\delta}\right)}$$

$$= \frac{t_1^2 t_2^2}{\left(1 - r_2 r_1'\right)^2 + 4 r_2 r_1' \sin^2(\delta/2)}$$

$$= \frac{t_1^2 t_2^2}{\left(1 - r_2 r_1'\right)^2 + 4 r_2 r_1' \sin^2(\delta/2)}$$

$$= \frac{t_1^2 t_2^2}{\left(1 - r_2 r_1'\right)^2} \frac{1}{1 + F \sin^2(\delta/2)}$$

where

$$F = \frac{4 r_2 r_1'}{\left(1 - r_2 r_1'\right)^2}$$

F is also known as the coefficient of finesse and is an indication of the Q-factor of the filtered peak. When the reflectance of the mirrors is high, the finesse of the FPF increases and the high Q-factor of the filtered peak can be obtained. When assuming normal incidence of the incoming beam, that is, $\theta = 0°$, maximum overall transmittance occurs when d is set to half of an integer multiple of the wavelength.

6.2 Current state of the art

FPF has being researched and developed intensely by industry across various wavelengths over a number of years. One such major company is VTT Technical Research Centre of Finland. The general approach to form the reflector is through the use of Bragg reflectors. The Bragg reflector is formed by depositing thin layers of two alternating materials with high and low refractive indices. The larger the difference between the refractive indices of the two materials is, the larger the reflection will be from the Bragg reflector. The thickness of each layer is designed to be a quarter of the designed wavelength divided by the refractive index of that layer. In other words, the longer the designed wavelength, the thicker the layers have to be and this might pose some fabrication issues.

Based on the analysis above, it can be seen that as the cavity distance between the two reflectors changes, the filtered output of the FPF can be tuned. In order to achieve this actuation, the typical methodology used is electrostatic actuation. By applying a voltage difference between the top and bottom reflectors, the movable top reflector will be actuated toward the bottom reflector. When voltage is applied to the mirrors and electrostatic actuation is induced, the top mirror deforms and there is slight bulking of the mirror due to the lack of any spring design, as shown in Figure 6.4a [1]. This causes the distance between the top and bottom reflectors to vary across the FPF, hence resulting in ineffective filtering of the input radiation. In order to avoid this, the reflectors are designed to be tensile-stressed, hence pulling the released reflectors toward the anchors. This results in a relatively flat surface in the middle of the reflector. To ensure that the FPF only uses this region of the reflector for the filtering process, an aperture is used and the opening is aligned to the center of the reflector as shown in (b) [2].

Across different desired wavelengths, the techniques and materials used to form the Bragg mirror and the FPF are vastly different. For an FPF that works in the visible light region as shown in Figure 6.5a, the layers forming the Bragg mirrors are typically very thin. In the design [2], the Bragg

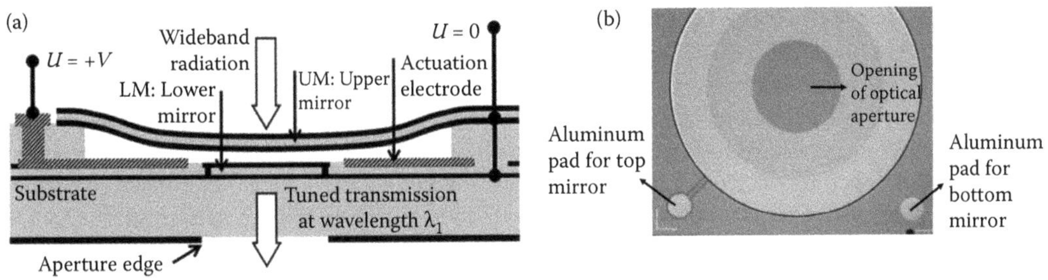

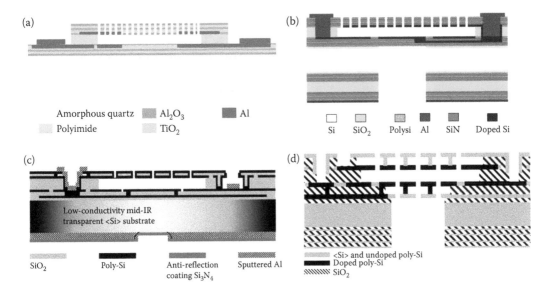

Figure 6.4 (a) Cross-sectional view of the FPF when actuated. (b) Optical image of the FPI structure with the optical aperture opening in the center of the device reflecting the designed wavelength.

Figure 6.5 Design of various FPF which work in (a) visible light region, (b) near infrared region, (c) mid-infrared region, and (d) far infrared region.

mirror is formed by layers of TiO_2 and Al_2O_3 with the thicknesses of TiO_2 and Al_2O_3 being typically around 50–75 nm. In order to achieve high quality films and accurate thicknesses, the method employed for the deposition is atomic layer deposition (ALD). An additional advantage of using ALD is that the deposition is performed at a low temperature ($T = 300°C$) which allows the use of sacrificial materials such as polyimide.

As the designed wavelength of an FPF increases to the near infrared region, the materials used to form the Bragg mirror are SiN and polycrystalline Si as shown in Figure 6.5b [3]. In order to form thicker films, a low pressure chemical vapor deposition (LPCVD) is used. As an LPCVD typically involves a higher temperature, the sacrificial material used for the cavity is SiO_2, which can be removed using vaporous hydrofluoric acid (VHF).

In the mid-infrared region, the FPF is formed by polycrystalline Si and SiO_2 which are deposited using chemical vapor deposition (CVD), as shown in Figure 6.5c [4]. The sacrificial material used for the cavity is also SiO_2. Hence, in order to ensure that the SiO_2 in the Bragg reflectors is not removed when the sacrificial SiO_2 is removed, an analysis for the amount of undercut into the Bragg mirror is performed.

When the wavelength of the FPF is long, such as in the far infrared region, the material of choice for the Bragg mirror is polycrystalline Si and air as shown in Figure 6.5d. As the wavelength extends beyond 6 μm, the optical constants of SiO_2 make it impractical to use in the Bragg mirror [5]. SiO_2

is used instead as the sacrificial material. The use of polycrystalline Si and air also enables better resolution in the output and more uniform tuning over a wider spectral range.

Measurements of the FPF in the various wavelength regions show a wide tuning range achievable through electrostatic actuation, and the voltage for actuation is kept relatively low, at a maximum of around 30 V [6]. As shown in Figure 6.6a, the tunable wavelengths of the FPF range from 550 to 700 nm which covers colors from blue to red [6]. For the FPF working in the mid-infrared region, the measured results show a tunability from 3950 to 4400 nm under an actuation voltage of 13 V as shown in Figure 6.6b. The significance of this result is that it covers the wavelength of 4260 nm, which is the absorption wavelength of carbon dioxide, making this FPF applicable to devices for carbon dioxide sensing [7].

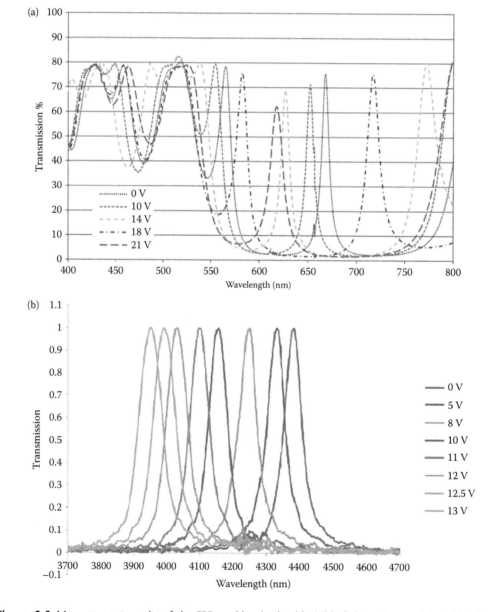

Figure 6.6 Measurement results of the FPF working in the (a) visible light region and (b) mid-infrared region.

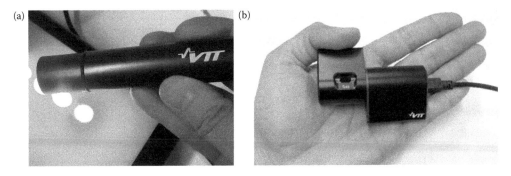

Figure 6.7 Devices using FPF for demonstration as (a) miniature imaging spectrometers and (b) gas sensors.

By combining the FPF with other components such as infrared emitters and optical lenses and filters, the device can be used for various applications. For instance, in the visible light region, it was demonstrated that the device is able to perform as a miniature imaging spectrometer about the size of a pen as depicted in Figure 6.7a [6]. When the FPF is working in the mid-infrared region, the main application is in gas sensing, as gases such as hydrocarbons and carbon dioxide have their signature absorption peaks in this range. The gas sensor can be a size that is smaller than the palm as shown in Figure 6.7b, and more recently, a hydrocarbon sensor has been integrated into a cell phone for portability [4].

6.3 Single slab photonic crystal-based reflector

A high performance reflector is the most important part of an FPF. In this section, we introduce the design, fabrication, and characterization of a single slab photonic crystal-(PhC)-based reflector. Two designs of PhC membranes with circular and square air holes were investigated and characterized to show greater than 90% reflectivity in the MIR wavelength range around 3.45 μm, which is an important wavelength for applications of the detection of gases such as hydrocarbons. Compared to circular air holes, PhC with square air holes maintains high reflectivity in the MIR region. However, square air holes when compared to circular air holes have a lower filling factor, which is defined as the volume of air holes over the volume of Si within a unit cell. This makes the membrane less brittle while keeping the reflectance at more than 90% around 3.45 μm.

6.3.1 Plane wave expansion (PWE) and finite-difference time-domain (FDTD) simulation

It is commonly known that PhC-based resonators provide very low bending loss due to good optical confinement. They are different from an Si slab waveguide, which is based on the total internal reflection principle as the confinement method for the light. Instead, confinement of light within PhCs is based on the photonic band gap (PBG) which is created by their unique structures. In PhC, patterns typically include the formation of air holes in Si or Si pillars. This induces an alternating variation of dielectric properties which is responsible for the formation of the PBG. To characterize the frequency response of the PhC structure, it was found that calculation is only needed for the first Brillouin zone. The results are then applicable to the whole structure since it is periodic. This is the basis for the calculation of the frequency response using the method known as plane wave expansion (PWE) [8]. Several commercial programs, such as Rsoft and MIT Photonic Bands, are available for such simulation work.

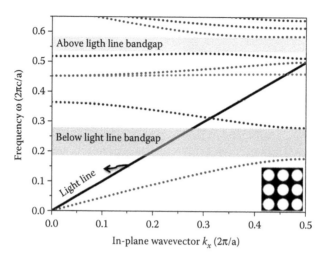

Figure 6.8 Band structure of PhC design.

For a full characterization of the electromagnetic wave propagation in the Si slab, the finite-difference time-domain (FDTD) method is typically used in modelling. For such simulations, commercial programs such as Lumerical are also available. FDTD calculation, while being accurate, is very time-consuming and memory intensive. In order to reduce the computation time while not reducing the accuracy, two-dimensional (2D) FDTD calculation is often employed rather than three-dimensional (3D) calculation. M. Kitamura et al. and M. Belotti et al. have both reported that the cavity modes obtained by the PWE method and the FDTD method in PhCs' slab structure are in good agreement with the measured results [9,10].

For example, using the PWE method, the band structure of a silicon PhC slab with a square lattice of air holes is derived and shown in Figure 6.8. The ratio between the radii of the holes (r) and lattice constant (a) is selected as 0.395. According to the derived photonic band structure diagram, there are two photonic band gaps which range from the normalized frequency range of photonic $0.192(2\pi c/a)$ to $0.275(2\pi c/a)$ and $0.528(2\pi c/a)$ to $0.587(2\pi c/a)$. An important element of the band structure is the light line. For bands and bandgaps below the light line, they are referred to as in-plane guided modes, which are completely confined by the slab without any coupling to external radiations. The bandgap region as shaded in blue indicates the frequency range that is not allowed to propagate within the PhC design. This concept is used in in-plane photonic designs such as for PhC waveguides. For bands and bandgaps above the light line, the PhC is in radiation mode, which allows or restricts coupling from free-space illumination. The band gap region as shaded in red indicates the bandwidth within an out-of-plane illumination that will be reflected as it cannot couple into the PhC slab.

6.3.2 CMOS-compatible fabrication process flow

The schematic of the design with air holes is shown in Figure 6.9a [11]. The radius of the air hole is indicated as *r*, the lattice constant is defined as *a*, and the thickness of the PhC membrane is *t*. In our proposed PhC device, the ratio of *r/a* and *t/a* are set to be $0.395(2\pi c/a)$ and $0.513(2\pi c/a)$, respectively.

Figure 6.9b shows the corresponding band structure of the PhC design that was calculated based on PWE. As can be seen, a band gap (shaded in red) is found from $0.528(2\pi c/a)$ to $0.587(2\pi c/a)$. In order to have a high reflectivity at 3.45 µm, the lattice constant is determined to be 1.92 µm. The radius is calculated to be 0.758 µm and the thickness of the Si membrane is 0.985 µm.

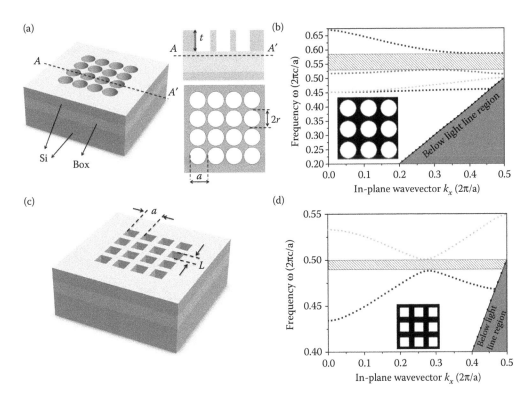

Figure 6.9 Schematic of the structure with circular air holes and (b) band structure of the proposed PhC with circular air holes with the band gap region shaded. (c) Schematic of the structure with square air holes and (d) band structure of the proposed PhC with square air holes with the band gap region shaded.

Fabrication of the PhC began with a bare Si wafer and a 1 μm thick silicon oxide (BOX) layer that was grown using thermal oxidation (Figure 6.10a). This was followed by a 1 μm thick LPCVD Si layer which acts as the device layer (Figure 6.10b). In order to ensure that the Si device layer is polycrystalline and to reduce the residual stress of the eventual suspended membrane, a high temperature anneal of 1000°C was done for 30 minutes (Figure 6.10c). The surface was then patterned by using deep-UV lithography and deep reactive ion etching (DRIE) (Figure 6.10d). The photoresist (PR) on the wafer is then stripped and the wafer is cleaned. Finally, the PhC is released using vaporous hydrofluoric acid. The importance of the VHF release will be highlighted in a later section.

In Figure 6.11a, the SEM photograph is shown. Due to fabrication uncertainties, the radii of the air holes are 0.77 μm and the lattice constant is 1.95 μm. The thickness of the device layer is measured as 1 μm. Simulation is done using FDTD methodology to examine the performance of the PhC membranes. The refractive index of the Si is assumed to be 3.464 and the boundary conditions of the unit cell are set to periodic. The incidence angle of the input light beam is set to 45°, which is consistent with the experimental setup that is used for measurement. As shown in Figure 6.11b, the simulated reflectivity displays a peak around 3.60 μm and more than 90% reflectivity over a 286 nm range. In addition, it also matches well with the band gap region from the band structure calculated in Figure 6.9b, where the reflectance is high within the band gap region and experiences a drop once outside the band gap region. Generally, the simulated result agrees well with the measurement result except for two dips in reflectivity at 3.31 and 3.45 μm. The higher reflectivity measured after the 3.70 μm wavelength is due to the reflection caused by the presence of the substrate.

A similar approach is adopted for the design of the square air holes as shown in Figure 6.9c. The length of the square air hole, L, is designed to be $0.618a$ and the thickness, t, is set to be $0.588a$. Based on the band gap region from $0.488(2\pi c/a)$ to $0.501(2\pi c/a)$ as shown in Figure 6.9d, the lattice

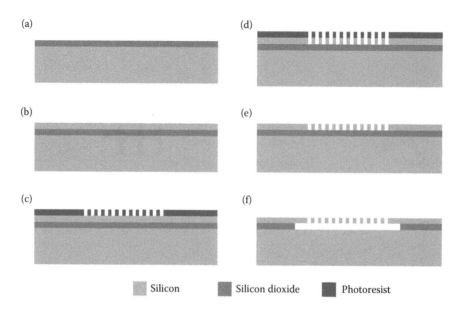

Silicon   Silicon dioxide   Photoresist

Figure 6.10 Process flow for PhC-based reflector. (a) 1 μm thermal SiO$_2$ is grown on a bare 8″ Si wafer. (b) 1 μm Si is then grown using LPCVD and annealing is done at 1000°C. (c) Coat and photolithography of PR to define air holes. (d) DRIE of Si to pattern the air holes. (e) PR strip and cleaning of wafer. (f) Device release in VHF.

constant is determined to be 1.70 μm. This equates to an **L** of 1.05 μm and t of 1 μm. The SEM photograph of the fabricated PhC membrane is shown in Figure 6.12a. Similar to the circular air holes, fabrication uncertainties caused the measured length of the square air holes to be 1.05 μm and the lattice constant to be 1.75 μm. The device layer of the square air holes is still 1 μm. From the fabricated parameters, it can be calculated that the filling factor of the square air holes in the membrane is 36%. In contrast, the filling factor for circular air holes is 49%. The lower filling factor allows the membrane to be mechanically stronger, hence less brittle. This allows more flexibility in design and fabrication, especially in MEMS applications where such membranes are released and are freestanding, as demonstrated in our PhC design. Simulation is also performed using FDTD calculation. The simulated result shows a high peak around 3.55 μm as shown in Figure 6.12b, and this matches well with the measurement data. It can be observed that the square air holes are rounded

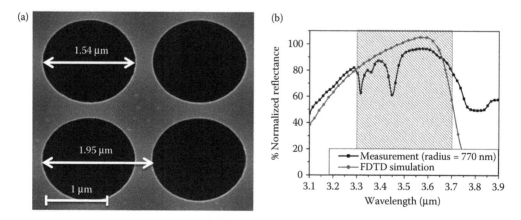

Figure 6.11 (a) SEM image of the fabricated device and (b) the simulation of the reflectance based on the fabricated PhC with circular air holes overlaid with the measurement result.

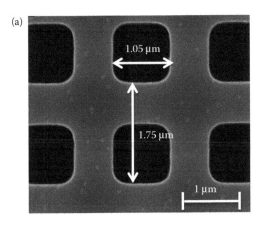
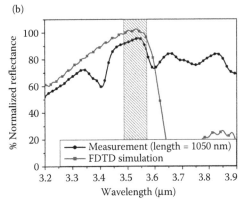

Figure 6.12 (a) SEM image of the fabricated device and (b) the simulation of the reflectance based on the fabricated PhC with square air holes overlaid with the measurement result.

and the radius of the curvature of the edges is estimated to be around 200 nm. In order to examine the effect of the rounded edges of the square holes, we have simulated the structures using FDTD. The curvature of the edge is related to the radius of the circle as shown in Figure 6.13a. At a radius of 525 nm, the air hole becomes circular in shape rather than the intended square hole. The simulated reflectance is summarized in Figure 6.13b. As the edges get more rounded, the reflectance starts to red shift and the maximum reflectance decreases. If we benchmark the high reflectance to be more than 90% at the desired wavelength of 3.55 μm, the tolerance is close to a rounded edge of radius 200 nm. Based on the fabricated device, the curvature radius is around 200 nm and this is within the tolerance that is allowed for a high reflectance.

6.3.3 Characterization of photonic crystal-based reflector

The experimental results of circular air holes are shown in Figure 6.14a. The PhC reflectors are designed to be 300 by 300 μm² in order to ensure that the input beam is illuminated only on the PhC patterns. The measured reflectance of the circular air holes of radius 770 nm shows a peak of 95.5% at wavelengths around 3.61 μm and a reflectance greater than 90% is present from 3.56 to

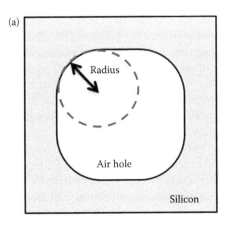
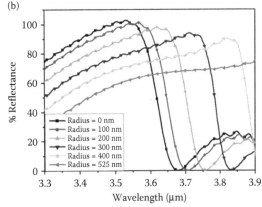

Figure 6.13 (a) Definition of radius with respect to the curvature of the edge, (b) FDTD simulation of reflectance with various radii of curvature at the edge.

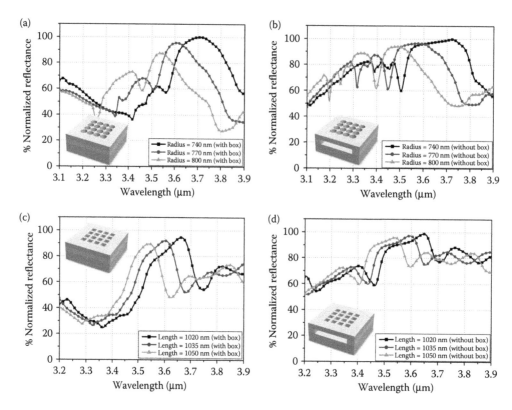

Figure 6.14 Experimental results of circular air hole design (with BOX) and (b) released circular air hole design (without BOX). (c) Experimental results of stationary square air hole design (with BOX) and (d) released square air hole design (without BOX).

3.68 μm. As the radius of the circular air hole changes, it is observed that the reflected bandwidth experiences a blue shift as the radius increases. This can be attributed to the slight shift in the band gap region toward higher frequencies as the ratio of *r/a* increases. Outside the band gap region, low reflectance values are measured.

For an increased performance of the PhC membrane with circular air holes, the BOX layer is removed by isotropic etching using VHF. The schematic of the released structure is shown in the inset of Figure 6.14b. The measured reflectance of the PhC membrane with different radii for the circular air holes is shown as well. A peak of 96.5% reflectance is observed at 3.58 μm for circular air holes with a radius of 770 nm. Generally, the spectra measured for the released PhC membrane with circular air holes show a distinct blue shift when compared to the unreleased circular air hole design. As the BOX layer is removed, the refractive index of the cladding below the PhC membrane decreased from 1.44 (SiO_2) to 1 (air). This reduces the effective refractive index which leads to the shift of the spectra toward lower wavelengths. Sharp dips in reflection are also found at 3.31 and 3.45 μm, and such dips in reflection are attributed to a non-zero angle of incidence according to K. B. Crozier [12]. In this case, the incidence angle is set to 45°.

Figure 6.14c shows the experimental results of a 300 by 300 μm² unreleased PhC membrane with square air holes. For square air holes with a length of 1035 nm, a reflectance peak of 92% is present at a 3.60 μm wavelength. The high reflectance region spans across a much smaller bandwidth from 3.58 to 3.61 μm. Similar to the PhC membrane with circular air holes, when the length of the square air hole increases, the reflectance peak shows a blue shift because more area within the membrane becomes air, which has a lower refractive index. This causes the bandgap region to move toward higher frequencies. After the BOX layer is removed using isotropic etching by VHF, the reflectance is enhanced as shown in Figure 6.14d. The measured reflectance for the square air holes with length

1035 nm is 97.2% at a 3.59 μm wavelength. A similar blue shift can be seen when the length of the square air holes is increased.

6.4 Photonic crystal-based Fabry–Pérot filter

In the previous section, we described our work in realizing a high performance PhC. By using a PhC reflector in the formation of the FPF, the Q-factor obtained can be much higher than that with conventional methods. In this section, we will demonstrate the design, fabrication, and characterization of an FPF which is fabricated using a CMOS-compatible monolithic fabrication process which is highly desirable [13]. In this section, we will introduce the optimization steps that were taken in the fabrication process in order to realize the FPF. Several failures were experienced before the fabrication process was fixed. In our previous section, where we fabricated the PhC mirror, the process used involved a high temperature anneal of 1000°C. The annealing step induces very high thermal stress and this causes cracking of the silicon layers and wafer breakage. A single etch process was also attempted to define both PhC mirrors, but this was also found to be unsuccessful. Details of these fabrication attempts will be discussed in the following section. In the end, we alleviated the thermal stress issue through the use of low stress epitaxial polycrystalline Si, which is deposited at the lower temperature of 610°C, and a bottom-up approach in the realization of the FPF. The fabricated FPF shows a transmission peak centered at 3.51 μm with a Q-factor of around 300 [14]. While this is lower than those in the simulations, it is still significantly higher than existing works which typically have Q-factors of a few tens [15–17]. This opens the possibility of utilizing such a PhC FPF for high-resolution applications like gas sensing [18,19] and hyperspectral imaging [15,20].

6.4.1 Fabrication process flow

In order to fabricate the PhC-based FPF, we propose a monolithic approach to the fabrication process in order to avoid any physical bonding of the mirrors, which is usually present in existing works. Such a monolithic fabrication approach provides simplicity and low-risk fabrication which can be achieved across the whole wafer. The schematic of the proposed PhC FPF to be fabricated is shown in Figure 6.15 [14].

The first fabrication process proposed was a continuation of the steps used in the previous section in this chapter, where we used LPCVD Si and annealing at 1000°C. We also attempted to use a single etch step to define both PhC mirrors in order to prevent the use of chemical mechanical

Figure 6.15 Schematic of the PhC-based FPF to be fabricated.

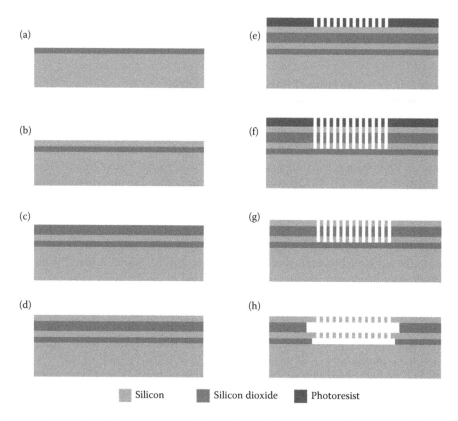

Figure 6.16 Process flow for PhC FPF. (a) 1 μm PECVD SiO₂ is grown on a bare 8″ Si wafer. (b) 1 μm Si is then grown using LPCVD and annealing is done at 1000°C. (c) 1.75 μm SiO₂ is deposited using PECVD to define the cavity length. (d) The 1 μm Si top reflector is deposited using LPCVD. (e) Coat and photolithography of PR to define air holes. (f) DRIE of Si/SiO₂/Si layers to pattern the air holes in both Si layers. (g) PR strip and cleaning of wafer. (h) VHF to realize PhC FPF.

polishing (CMP) of SiO₂. The effect of the CMP process will be highlighted in a subsequent section. The detailed fabrication steps are shown in Figure 6.16.

Fabrication of the FPF starts with a 1 μm PECVD SiO₂ on a bare Si wafer as shown in Figure 6.16a. Then a 1 μm LPCVD Si (Figure 6.16b) and 1.72 μm PECVD SiO₂ are deposited (Figure 6.16c). This 1.72 μm SiO₂ defines the Fabry–Pérot cavity length between the two PhC mirrors. To form the upper PhC mirror, another 1 μm LPCVD Si is deposited and the wafer is annealed at 1000°C (Figure 6.16d). Photolithography is then done to define the air holes (Figure 6.16e) and DRIE is done to etch both the Si mirror layer and the SiO₂ layer (Figure 6.16f). The PR is then stripped and the wafer is cleaned (Figure 6.16g). Finally, VHF is used to remove the SiO₂ in order to form the FPF (Figure 6.16h). When attempting the proposed fabrication, two critical problems were faced. The first is the thermal stress that occurs during the annealing step in Figure 6.16d. Unlike the previous fabrication flow which includes only one layer of Si, the SiO₂/Si/SiO₂/Si stack in this case cracks more readily at high temperatures as shown in Figure 6.17. While not all the wafers crack after annealing, the high temperature anneal is a high risk process for the fabrication of the FPF.

In order to avoid CMP of the SiO₂ which defines the Fabry–Pérot cavity so as to obtain maximum accuracy of its thickness, we used a single etch step to define the FPF as shown in Figure 6.16f. In Figure 6.18a, when only the top Si layer is etched, the etched holes are still well defined. However, after etching the Si/SiO₂/Si stack as shown in Figure 6.18b, it was observed that the top Si experiences over-etching, which causes the size of the air hole to enlarge. In the etching of SiO₂, it is also observed that the sidewall profile is sloping. This causes the air hole pattern on the bottom Si mirror to be much smaller than was designed. With different air hole sizes between the top and

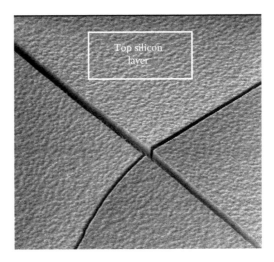

Figure 6.17 Cracking of the wafer after the 1000°C anneal step.

bottom Si layers, based on the analysis in the previous section, it is expected that they will display vastly different reflection spectra. Without working at a common wavelength, the efficiency of the Fabry–Pérot cavity to filter unwanted wavelengths will be greatly reduced. Due to the inability to control the etching profile of the Si/SiO$_2$/Si stack, the fabrication process has to be revised.

In order to reduce the thermal budget of the fabrication flow, we have to replace the high temperature steps with lower temperature steps. First, the LPCVD Si which forms the Si reflectors is grown in a furnace at 540°C. However, the Si is amorphous and requires an annealing at 1000°C to form polycrystalline Si. The high temperature anneal induces very high thermal stress and causes the wafers to crack as shown in Figure 6.17. In order to avoid the high temperature anneal, epitaxial Si is used instead. Although the epitaxial Si is deposited at 610°C, the Si deposited is polycrystalline. This means that there is no need for the wafers to undergo the 1000°C anneal step. The next issue faced is the inability to control the sidewall profile of the Si/SiO$_2$/Si stack after etching. The proposed method to overcome this is the use of a bottom-up approach in the fabrication. Instead of depositing the Si/SiO$_2$/Si stack before etching the air holes, the reflectors in the FPF are etched layer by layer. The detailed fabrication process is shown in Figure 6.19.

The revised fabrication process flow of the FPF starts with a 1 μm PECVD SiO$_2$ on a bare Si wafer as shown in Figure 6.19a. The device layer of 1 μm thick polycrystalline Si of the bottom PhC mirror is then deposited using epitaxy (Figure 6.19b). Photolithography followed by DRIE of the

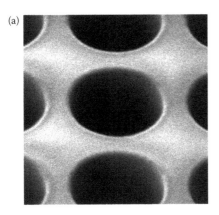

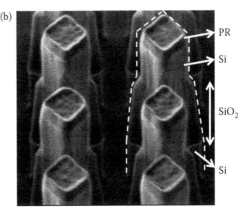

Figure 6.18 (a) SEM image of the wafer when only the top Si is etched and (b) sloping sidewall profile after the Si/SiO$_2$/Si stack is etched.

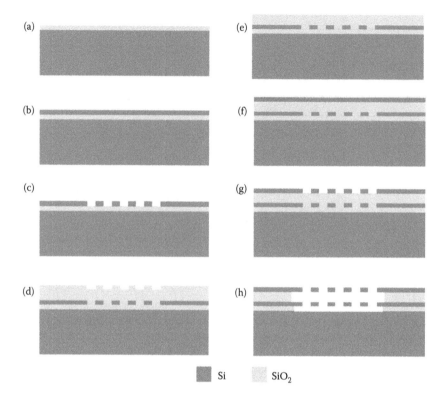

Figure 6.19 Revised process flow for PhC FPF. (a) 1 μm PECVD SiO$_2$ is grown on a bare 8″ Si wafer. (b) 1 μm epitaxial Si is then deposited. (c) Photolithography of PR and DRIE of Si to define air holes. (d) 2 μm SiO$_2$ is deposited using PECVD to define the cavity length. (e) CMP of the SiO$_2$ layer to remove topology of the top wafer. (f) Deposition of epitaxial Si for top reflector. (g) DRIE of Si to pattern the air holes in top Si layers. (h) VHF to realize PhC FPF.

polycrystalline Si layer to form the air holes is then performed to define the bottom PhC mirror (Figure 6.19c). A 2 μm PECVD SiO$_2$ is deposited (Figure 6.19d) before a CMP of 0.2 μm of SiO$_2$ is done to remove topology issues and also to define the cavity length of the FPF (Figure 6.19e). The 1 μm Si device layer of the top PhC mirror is then deposited using epitaxy (Figure 6.19f) and the air holes are defined through photolithography and DRIE (Figure 6.19g). Finally, the FPF is released by using VHF (Figure 6.19g).

The CMP step in Figure 6.19e is needed due to the topology issue. When SiO$_2$ is deposited after the etching of the bottom Si reflector (Figure 6.19d), it was found that the top surface of the SiO$_2$ is not flat, as shown in Figure 6.20. If this topology issue on the top surface of the SiO$_2$ is not removed, such an unevenness will be transferred to the top Si layer when it is deposited. In order to prevent this, we deposit an additional 0.2 μm of SiO$_2$ and CMP is used to thin it back to achieve the desired 1.80 μm thickness of SiO$_2$. The compromise that has to be made is the incapability to exactly control the thickness of SiO$_2$ left to define the Fabry–Pérot cavity. This is because the CMP of SiO$_2$ is time-based and variation cannot be avoided. The variation can, however, be minimized by using multiple test wafers to determine the CMP rate before proceeding with the FPF wafers.

The fabricated device is shown in Figure 6.21a. In order to reveal the bottom PhC reflector, the fabricated device is cleaved along the dotted line. It is also noted that, although the air holes were etched after different photolithography steps, the misalignment between the top Si reflector and bottom Si reflector air holes is minimal. In addition, the dimensions of the air holes are also well controlled. The cross-sectional view of the FPF before release is shown in Figure 6.21b. The bottom SiO$_2$ layer shown is the BOX while the top SiO$_2$ defines the Fabry–Perot cavity; it is measured to be 1.70 μm. This will have an impact on the FPF performance, which will be discussed later.

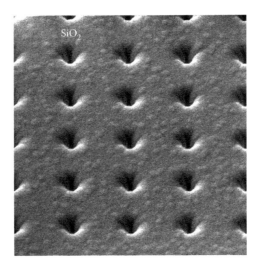

Figure 6.20 SEM image of the wafer after the SiO_2 is deposited to define the cavity length. Topology issue is apparent at the top surface and this is removed by performing CMP of SiO_2.

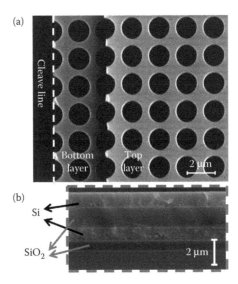

Figure 6.21 SEM image of the (a) top and (b) cross-sectional view of the fabricated FPF before VHF release.

6.4.2 Characterization of photonic crystal-based Fabry–Pérot filter

In this section, the feasibility of implementing 2D PhC designs as the high reflectivity mirrors in FPF is explored using FDTD simulations. As discussed, such high reflectivity mirrors are commonly implemented using multi-layer structures which face the problem of high residual stress in the layers. In contrast, 2D PhC designs are able to achieve high reflectivity using a thin Si membrane which is lightweight and easy to fabricate. We have also verified in previous sections that the PhC reflector is able to produce extremely high reflection at any desired wavelength just by altering the geometry of the design. The schematic of our designed FPF is shown in Figure 6.22a. The PhC membranes are used as the mirrors and the top Si PhC membrane is supported using springs. The cavity length can be made tunable by using MEMS techniques, such as applying a voltage between the top and bottom Si slabs. This induces an attractive electrostatic force which will pull the top

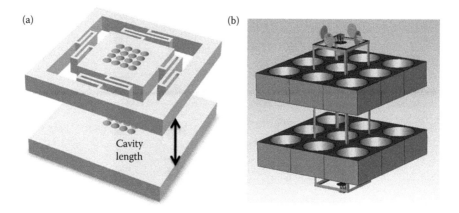

Figure 6.22 (a) Schematic of FPI using PhC reflector as mirrors and (b) model of PhC-based FPF used in simulation program, CST MWS.

and bottom slabs toward each other, hence reducing the gap between them. More discussion on the design of the spring will be done in a later chapter.

The transmission output of the FPF as the cavity length changes is simulated. The simulation model is shown in Figure 6.22b. Only the unit cell indicated by the red box is simulated and a periodic boundary condition is set on all four sides. This helps to reduce computation time while maintaining high accuracy in the simulated output. In the previous sections, we have shown the possibility of using both circular and square holes in the PhC reflector design.

In Figure 6.23a, the simulated spectrum of the FPF using circular air holes is presented. In the simulations, the incident light is propagating along the z direction and the electric field is along the x direction. The radius of the circular air hole is 770 nm and the lattice constant is 1.95 μm. The thickness of the top and bottom Si slabs are maintained at 1 μm. The FWHM of the transmittance peak corresponding to a cavity length of 2.00 μm is 0.07 nm, which equates to a Q-factor of about 52,000. The tuning range of the FPF is 28 nm from 3.68 to 3.71 μm when the cavity length is changed to 1.95–2.05 μm. This is around an order of magnitude higher than the simulated Q-factor of an FPF using multi-layered structures [16,17,21].

In order to fully assess the practicability of using square air holes in an FPF, an FPF with square air hole reflectors is also simulated for comparison using a similar method as that mentioned above. The values of L, a, and t are 1.05, 1.70, and 1 μm, respectively. The simulated results are presented in

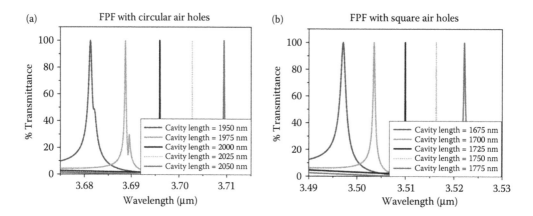

Figure 6.23 (a) Simulated transmittance of the FPI with various cavity length using circular air holes and (b) using square air holes.

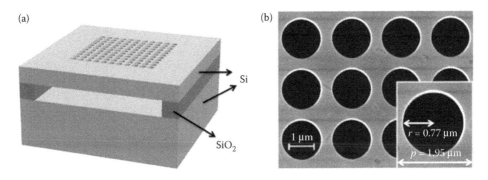

Figure 6.24 (a) Schematic of PhC reflector and (b) SEM of fabricated chip.

Figure 6.23b. The FPI shows a tuning range of 25 nm from 3.50 to 3.52 μm when the cavity length is changed to 1.675–1.775 μm. When the cavity length is 1.725 μm, the FWHM of the simulated peak is 0.08 nm, which corresponds to a Q-factor of 43,800. Based on the simulations of the FPF using both circular air holes and square air holes, it has been shown the proposed PhC designs are ideal candidates for the realization of the high reflectivity mirrors. At the same time, the use of PhC with square air holes in an FPF also displays a good optical performance that is comparable to the PhC with circular air holes. As described in a previous section, using PhC with square air holes offers higher mechanical strength, which is vital in applications such as an FPF where the PhC reflectors are free-standing.

As mentioned above, the PhC mirrors are now fabricated using epitaxial Si instead of the previously utilized LPCVD Si which requires high temperature annealing. Hence, there is a need to characterize the epitaxial Si PhC mirror to ensure that the reflection peak is still around 3.55 μm. The PhC reflector is suspended in order to make it compatible with the subsequent fabrication of the FPF as shown in Figure 6.24a. Fabrication of the PhC reflector starts by growing a 1 μm PECVD SiO$_2$ on a bare 8″ Si wafer. The device layer of 1 μm thick polycrystalline Si is then deposited using epitaxy. The air holes are patterned using deep-UV lithography and etched using DRIE. The PhC reflector is finally released using VHF. The SEM of the fabricated PhC reflector is shown in (b).

Measurement of the PhC mirror is also done using the FTIR microscope from 2 to 8 μm. Due to the experimental setup, the angle of incidence for the reflection measurement is limited to 45°. The reflection measurement of the PhC mirror is shown in Figure 6.25. A high reflection of 96.4% is measured at 3.60 μm with a bandwidth of 160 nm for wavelengths that have reflection of more than 90%. The dips in reflection observed at 3.29 and 3.43 μm were looked into in a previous section

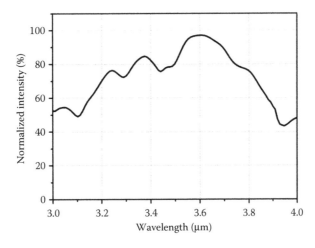

Figure 6.25 Reflection measurement of the fabricated PhC reflector.

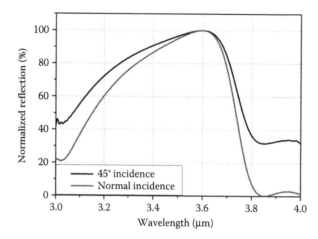

Figure 6.26 3D FDTD simulation of the PhC reflector at 45° incidence and normal incidence.

where we attributed them to the 45° angle of incidence. It is noted that the peak reflection wavelength of 3.60 µm of the PhC reflector using epitaxial Si is slightly higher than that of the PhC reflector using LPCVD Si. This is due to a slight increase in the refractive index of the epitaxial Si over the LPCVD Si.

As the subsequent measurement of the FPF is based on transmission which has an angle of incidence fixed to normal incidence, the dependence of the performance of the PhC mirror with an incident angle is investigated. Based on FDTD simulations of the PhC mirror, the wavelengths of high reflection when the input IR light is incident at 45° and normal incidence remain the same at 3.60 µm, as shown in Figure 6.26. Based on the measurement results from the PhC mirror where the peak reflection is at 3.60 µm, the cavity gap between the mirrors is designed to be around 1.80 µm, which is half of the peak reflection wavelength. However, due to the inability to control the thickness variation induced by the CMP process on the top SiO_2 layer, it was found that the cavity length of the fabricated FPF is measured to be 1.70 µm. This has an impact on the Q-factor of the eventual transmitted Fabry–Pérot filtered peak.

Simulation of the FPF is done using 3D FDTD as well where two identical 1 µm thick Si slabs with an air hole radius of 0.77 µm are drawn with a separation defined by the cavity length. Similar to the PhC reflector simulation, the unit cell consists of both the Si slabs with lengths of 1.95 µm on each side and the refractive index of Si is set to 3.464. The boundary conditions are also set to be periodic boundary conditions with perfectly matched layers. In order to obtain the theoretical Q-factor of the FPF, unlike the PhC reflector, the expected high Q-factor is determined by the slope of the envelope of the decaying signal in the simulation. This is because the energy within the cavity cannot completely decay in a time that can be reasonably simulated and the maximum Q-factor that can be simulated scales with the simulation time. The simulated Q-factor in an optimized case of a cavity length of 1.80 µm is found to be in excess of 45,000 at a wavelength of 3.59 µm (ideal case) as shown in Figure 6.27. However, after taking fabrication variations into account, the simulated Q-factor drops to around 540 at 3.52 µm (non-ideal case). The enhanced Q-factor over that of existing works is attributed to the additional filtering effects of the PhC reflector. The first filtering effect is within the Fabry–Pérot cavity where the undesired wavelengths are attenuated due to destructive interference. The second filtering effect is due to the intrinsic wavelength selective reflectivity of the PhC reflector, which has a bandwidth of around 160 nm with more than 90% reflection. In contrast, a multilayer Bragg reflector has a bandwidth of more than 3 µm in the MIR wavelengths with a reflection of more than 90% [22–24]. This makes the PhC reflector able to filter unwanted wavelengths more efficiently than the multilayer Bragg reflector, hence resulting in a higher Q-factor. Measurement of the FPF is also done by Agilent Cary 620 FTIR Microscope from 2 to 8 µm. Similarly, the size of the FPF is designed to be 200 by 200 µm². In the case of transmission measurement, the incidence angle is

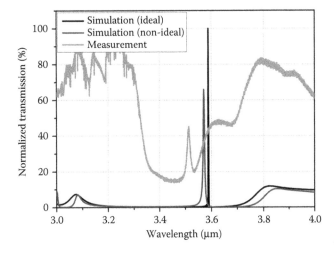

Figure 6.27 Simulation of ideal and non-ideal case of the FPF and measurement of the fabricated FPF.

normal to the sample. The measurement result is shown in Figure 6.27. Measurement of the fabricated device reveals a Q-factor of around 300 at a wavelength of 3.51 μm. The lower transmission intensity for the wavelength range below 3.35 μm and that above 3.55 μm in the simulations can be attributed to the simulation methods adopted for the high Q-factor simulations, as the transmission intensity of wavelengths where there are no resonances are suppressed. While the measured Q-factor is lower than the simulated Q-factor, it is still around an order of magnitude higher than that of existing works where the Q-factor is typically a few tens. The shift in the transmission wavelength and the drop in the Q-factor can be attributed to the variation in the cavity length. When the cavity length is not at the optimal distance, the MIR light is unable to be confined within the cavity due to its higher transmission. This reduces the efficiency of the constructive interference of the desired wavelength, which causes a drop in output intensity as well as a broadening of the transmission peak. Both these factors result in a much lower Q-factor. In addition, the presence of the Si substrate in the FPF causes a drop in the transmitted intensity. Based on measurements of bare Si, the transmitted intensity is reduced to around 60% for wavelengths around 3.60 μm. The effect of the Si substrate will be removed in future iterations by performing a DRIE etch of the Si substrate. In order to alleviate the fabrication variations introduced by the CMP process on the cavity length, MEMS technology can be incorporated into the design. With MEMS technology, actuation of the PhC reflector will be enabled, hence the cavity length will be tunable. This not only reduces the impact of cavity length variation in the fabrication process, it also offers the possibility of realizing a tunable Fabry–Pérot interferometer.

6.4.3 Gas sensing using a Fabry–Pérot filter

One major application of highly reflective surfaces is in the formation of a Fabry–Pérot filter (FPF) where two highly reflective surfaces are placed parallel to each other. The air gap between the mirrors is designed to be $n\lambda/2$, where n is an integer and λ is the desired filtered wavelength, in order to achieve constructive interference of the desired wavelength. The gap between the two highly reflective surfaces can be tuned by incorporating MEMS technology. Such tunable filters are extremely important in applications such as gas sensing [16,19] and hyperspectral imaging [20,25]. Many attempts in recent works use Bragg reflectors with a high reflectivity to realize an FPF [15–17,20,21]. However, the Bragg reflectors are typically heavy and require a high actuation voltage to achieve tunability. In addition, as mentioned before, when working at longer wavelengths, the thicknesses of the layers have to be increased and this complicates fabrication. Suh et al. introduced the idea of using 2D PhC as the reflective mirror and theoretically proved the high performance of such an FPF [26].

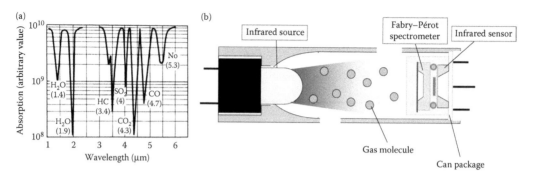

Figure 6.28 Infrared absorption dips for common gases of environmental concern.

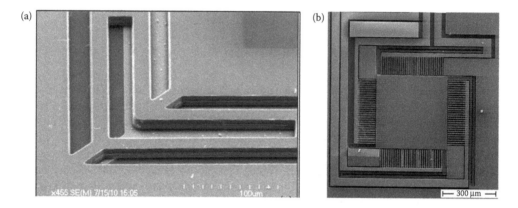

Figure 6.29 (a) An FPF that is formed through the use of bonding of the top and bottom mirror and (b) shows an FPF that uses more exotic materials.

MIR photonics has recently attracted tremendous attention due to its potential for industrial applications as well as the rapid improvement in MIR laser technologies. MIR spectroscopy in particular is an extremely powerful technique that can be used for chemical and biological sensing in environmental monitoring and medical diagnosis [27–29]. For example, as shown in Figure 6.28, for gas sensing, MIR wavelengths from 3 to 6 μm are particularly important [30]. In order to sense these gases through spectroscopy, it is important that the detected light only has the wavelength that the gas absorbs, as shown in (b) [31]. This is usually realized through the output of the FPF. By using 2D PhC reflectors in the FPF, their high Q-factor output would allow for a more sensitive and higher resolution sensing.

Current efforts to realize FPF also typically involve bonding of the mirrors, as shown in Figure 6.29a, or the use of non-CMOS compatible materials such as polyimide as the sacrificial material to define the cavity length [15–17,20,25]. In addition, more exotic materials such as lead chalcogenide and europium(II) telluride might be needed for the formation of the mirror used in the FPF as shown in (b) [32]. The Q-factors of the filtered peak by current reports of FPF are typically a few tens and using PhC reflectors as mirrors is deemed to be a method to increase the Q-factor for more sensitive and higher resolution applications.

6.5 Prospective of photonic crystal-based Fabry–Pérot filter

In previous work, we have fabricated FPF that displays a high Q-factor. However, due to the CMP process of the SiO_2 in between the two Si layers, the Fabry–Pérot cavity gap is not well controlled. In order to overcome such variations, MEMS technology can be incorporated to implement actuators

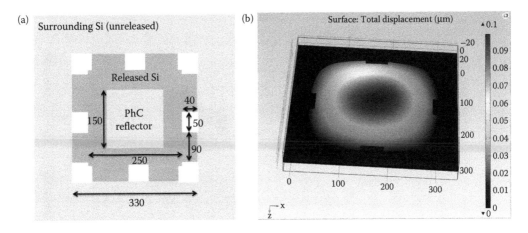

Figure 6.30 (a) Spring design of the Si membrane and (b) simulation of the Si membrane when undergoing electrostatic actuation. All dimensions indicated are in µm.

into the design. This enables electrostatic actuation of one of the PhC reflectors, and hence achieves tunability of the cavity length. This not only reduces the impact of cavity length variation in the fabrication process, it also offers the possibility of realizing a tunable Fabry–Pérot interferometer. The design of the actuator plays a critical role as the Si layer has to be made less stiff in order to introduce displacement. This will create issues such as buckling of the Si reflector when actuated, which will reduce the performance of the tunable FPF. While buckling can be reduced by having softer springs, it will introduce other issues such as the Si membranes sticking together after release. Hence, an optimization process has to be undertaken.

One simple spring design is to have eight fixed-fixed beams connected to the released Si membrane which includes the PhC reflector, as shown in Figure 6.30a. We maintained a PhC reflector area of 150 by 150 µm² and eight fixed-fixed beams of 40 by 90 µm². The thickness of the Si is still 1 µm. However, when actuated, as shown in Figure 6.30b, it was realized that the buckling effect of the Si membrane is severe, especially in the middle. The height difference between the edge and the middle of the PhC reflector is around 60 nm and this causes the Fabry–Pérot length to vary. In order to maintain a good performance for the FPF, the Si membrane should be made to be as flat as possible when actuated.

In order to reduce the buckling effect of the Si membrane when actuated, an enhanced version of the spring design can be implemented. One of the better designs is the meander-shaped spring. As in the previous design, the PhC reflector area is 150 by 150 µm² and the thickness of the Si is still 1 µm. The parameters are shown in Figure 6.31a. When the Si membrane is actuated, as shown in Figure 6.31b, the deformation in the released Si membrane is very flat in the middle. The height difference between the edge of the PhC reflector and the middle is also minimized to 1.7 nm. This amount of height variation is negligible.

The fabrication process flow of a tunable FPF has to be changed as the two Si layers have to be doped and metal vias created for the contact pads. The proposed fabrication flow is shown in Figure 6.32. The proposed process flow is very similar to that of the FPF described earlier. The main additional steps are the inclusion of selective area doping so that the Si membranes are conductive and can be actuated through electrostatic attraction. 1 µm of SiO_2 is first grown on the bare Si wafer and 1 µm of epitaxial Si is then deposited (Figure 6.32a). The PhC reflector design is then etched on the bottom layer (Figure 6.32b) and the area of the released Si is doped through ion implantation (Figure 6.32c). Care has to be taken to ensure that the area of the PhC reflector is not doped as it will cause the refractive index of the Si to change, hence affecting the performance of the reflector. The frame of the bottom Si reflector is then defined (Figure 6.32d) and the wafer is covered with 2 µm of SiO_2 (Figure 6.32e). CMP is then done to remove 0.2 µm of SiO_2 to reduce the topology (Figure 6.32f). The top 1 µm epitaxial Si is then grown (Figure 6.32g) and the PhC reflector is defined (Figure 6.32h).

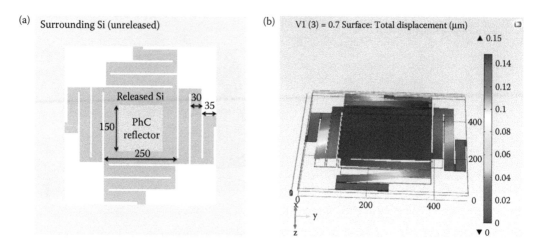

Figure 6.31 (a) Enhanced spring design of the Si membrane and (b) simulation of the Si membrane when undergoing electrostatic actuation. All dimensions indicated are in μm.

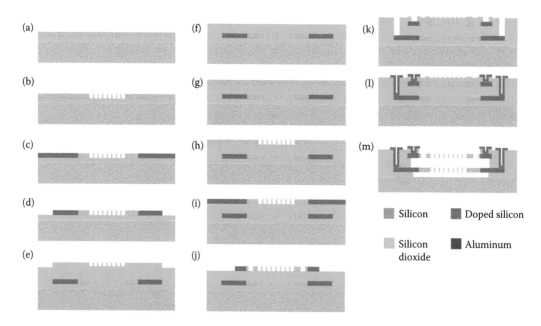

Figure 6.32 Fabrication process flow for tunable FPF. See Section 6.5 text for description of individual panels.

The top Si is then doped (Figure 6.32i) and the frame and springs designs are defined through DRIE (Figure 6.32j). The wafer is then covered with SiO_2 (Figure 6.32k) before vias are etched into the SiO_2 for metal contacts. Aluminum is then deposited and etched to form the metal pads (Figure 6.32l), and the device undergoes VHF release to realize the tunable FPF (Figure 6.32m).

References

1. Tuohiniemi, M. et al., MOEMS Fabry–Pérot interferometer with point-anchored Si-air mirrors for middle infrared. *Journal of Micromechanics and Microengineering*, 2014. 24(9): p. 095019.

2. Rissanen, A. and R.L. Puurunen, Use of ALD thin film Bragg mirror stacks in tuneable visible light MEMS Fabry-Perot interferometers. In *SPIE MOEMS-MEMS*. 2012. International Society for Optics and Photonics, San Francisco, California, Vol. 8249, p. 82491A.

3. Rissanen, A. et al., Tunable MOEMS Fabry-Perot interferometer for miniaturized spectral sensing in near-infrared. In *SPIE MOEMS-MEMS*. 2014. International Society for Optics and Photonics, San Francisco, California, pp. 89770X–89770X.

4. Mannila, R. et al., Hydrocarbon gas detection with microelectromechanical Fabry-Perot interferometer. In *Proc. SPIE*. Baltimore, Maryland, 2013. 8726: p. 872608.

5. Tuohiniemi, M. et al., Micro-machined Fabry-Pérot interferometer for thermal infrared. In *SENSORS, 2013 IEEE*. 2013. IEEE, Baltimore, Maryland, pp. 1–4.

6. Rissanen, A. et al., MOEMS miniature spectrometers using tuneable Fabry-Perot interferometers. *Journal of Micro/Nanolithography, MEMS, and MOEMS*, 2012. 11(2): p. 023003-1–023003-6.

7. Mannila, R. et al., Gas detection with microelectromechanical Fabry-Perot interferometer technology in cell phone. In *SPIE Sensing Technology + Applications*. 2015. International Society for Optics and Photonics. Baltimore, Maryland, pp. 94820P–94820P.

8. Johnson, S. and J. Joannopoulos, Block-iterative frequency-domain methods for Maxwell's equations in a planewave basis. *Optics Express*, 2001. 8(3): p. 173–190.

9. Kitamura, M., S. Iwamoto, and Y. Arakawa, Enhanced light emission from an organic photonic crystal with a nanocavity. *Applied Physics Letters*, 2005. 87(15): p. 151119.

10. Belotti, M. et al., All-optical switching in 2D silicon photonic crystals with low loss waveguides and optical cavities. *Optics Express*, 2008. 16(15): p. 11624–11636.

11. Ho, C.P. et al., Development of polycrystalline silicon based photonic crystal membrane for mid-infrared applications. *IEEE Journal of Selected Topics in Quantum Electronics*, 2014. 20(4): p. 94–100.

12. Crozier, K.B. et al., Air-bridged photonic crystal slabs at visible and near-infrared wavelengths. *Physical Review B*, 2006. 73(11): p. 115126.

13. Wu, X., C. Jan, and O. Solgaard, Single-crystal silicon photonic-crystal fiber-tip pressure sensors. *Journal of Microelectromechanical Systems*, 2015. 24(4): 968–975.

14. Ho, C.P. et al., Two-dimensional photonic-crystal-based Fabry-Perot etalon. *Optics Letters*, 2015. 40(12): p. 2743–2746.

15. Russin, T.J. et al., Fabrication and analysis of a MEMS NIR Fabry Perot interferometer. *Journal of Microelectromechanical Systems*, 2012. 21(1): p. 181–189.

16. Mayrwöger, J. et al., Fabry-Perot-based thin film structure used as IR-emitter of an NDIR gas sensor: Ray tracing simulations and measurements. SPIE Microtechnologies, Prague, Czech Republic, 2011. p. 80660K–80660K.

17. Milne, J.S. et al., Widely tunable MEMS-based Fabry Perot filter. *Journal of Microelectromechanical Systems*, 2009. 18(4): p. 905–913.

18. Noro, M. et al., CO_2/H_2O gas sensor using a tunable Fabry-Perot filter with wide wavelength range. In *Micro Electro Mechanical Systems, 2003. MEMS-03 Kyoto. IEEE The Sixteenth Annual International Conference on*. 2003. IEEE, Kyoto, Japan, Japan, pp. 319–322.

19. Wöllenstein, J. et al., Miniaturized multi channel infrared optical gas sensor system. SPIE Microtechnologies, Prague, Czech Republic, 2011. p. 80660Q–80660Q.

20. Keating, A.J. et al., Design and characterization of Fabry–Pérot MEMS-based short-wave infrared microspectrometers. *Journal of Electronic Materials*, 2008. 37(12): p. 1811–1820.

21. Malak, M. et al., Cylindrical surfaces enable wavelength-selective extinction and sub-0.2 nm linewidth in 250 um-gap silicon Fabry-Perot cavities. *Journal of Microelectromechanical Systems*, 2012. 21(1): p. 171–180.

22. Ebermann, M. et al., Resolution and speed improvements of mid-infrared Fabry-Perot microspectrometers for the analysis of hydrocarbon gases. In *SPIE MOEMS-MEMS*. 2014. International Society for Optics and Photonics, San Francisco, California, pp. 89770T–89770T.

23. Tuohiniemi, M. and M. Blomberg, Surface-micromachined silicon air-gap Bragg reflector for thermal infrared. *Journal of Micromechanics and Microengineering*, 2011. 21(7): p. 075014.
24. Yichen, S. et al., Fano-resonance photonic crystal membrane reflectors at mid- and far-infrared. *IEEE Photonics Journal*, 2013. 5(1): p. 4700206–4700206.
25. Musca, C.A. et al., Monolithic integration of an infrared photon detector with a MEMS-based tunable filter. *IEEE Electron Device Letters*, 2005. 26(12): p. 888–890.
26. Suh, W. et al., Displacement-sensitive photonic crystal structures based on guided resonance in photonic crystal slabs. *Applied Physics Letters*, 2003. 82(13): p. 1999–2001.
27. Rossel, R.V. et al., Visible, near infrared, mid infrared or combined diffuse reflectance spectroscopy for simultaneous assessment of various soil properties. *Geoderma*, 2006. 131(1): p. 59–75.
28. Rochat, N. et al., Multiple internal reflection spectroscopy for quantitative infrared analysis of thin-film surface coating for biological environment. *Materials Science and Engineering: C*, 2003. 23(1): p. 99–103.
29. Artioushenko, V.G. et al., Medical applications of MIR-fiber spectroscopic probes. In *Europto Biomedical Optics' 93*. 1994. International Society for Optics and Photonics, Beijing, China, pp. 758–761.
30. Puscasu, I. et al., Photonic crystals enable infrared gas sensors. In *Optical Science and Technology, the SPIE 49th Annual Meeting*. 2004. International Society for Optics and Photonics, Denver, Colorado, Vol. 5515, p. 59.
31. Enomoto, T. et al., Infrared absorption sensor for multiple gas sensing. Development of a Fabry–Perot spectrometer with ultrawide wavelength range. *Electronics and Communications in Japan*, 2013. 96(5): p. 50–57.
32. Quack, N. et al., Mid-Infrared tunable resonant cavity enhanced detectors. *Sensors*, 2008. 8(9): p. 5466–5478.

7

Electrothermally actuated MEMS mirrors
Design, modeling, and applications

Huikai Xie, Xiaoyang Zhang, Liang Zhou, and Sagnik Pal

Contents

7.1 Introduction

Mirrors are the fundamental building blocks for a wide range of optical systems, such as lasers, cameras, and optical microscopes. They may be stationary or scanning. Scanning mirrors manipulate light in free space and can be used to modulate, steer, deflect, or switch optical beams; they are the critical components in many active optical systems for optical imaging, ranging, switching, display, and laser cutting. Scanning mirrors can be categorized into two types: angular scanning mirrors and linear scanning mirrors. Most commonly used angular scanning mirrors are galvanometers, while linear scanning mirrors are often based on linear motors or piezoelectric actuators.

With the rapid growth of the Internet of Things, more and more tiny sensor nodes must be wirelessly (either RF or optical) connected. The fascinating virtual reality (VR) and augmented reality (AR) applications suddenly seem to be within our reach and are demanding small, fast optical scanners. Robots are also becoming more and more intelligent, but require miniature laser radars or microlidars for better agility. High resolution endoscopic optical imaging is being applied to detect early cancer inside the human body. All these latest technology movements point to one common device – optical microscanners. Galvanometers or linear motors are too bulky, and/or too slow, and/or too expensive for these applications. Fortunately, the optical microscanners in demand can be realized using microelectromechanical systems (MEMS) technology.

7.1.1 MEMS micromirrors

MEMS technology leverages integrated circuits (IC) technology to manufacture microscale devices and systems [1,2], and thus is the natural choice to make scanning microdevices, that is, MEMS micromirrors. Micromirrors can be actuated electrothermally, electrostatically, piezoelectrically, or electromagnetically. Among the four actuation mechanisms, electrostatic actuation is the most popular. Electrostatic mirrors are fast, but the scan range is small, and the required driving voltage is high, often on the order of 100 V. In addition to that, the fill factor, defined as the ratio between the size of the mirror aperture and the footprint of the chip, is small for electrostatic comb-drive micromirrors. Electromagnetic micromirrors have larger scanning ranges and lower driving voltages, but the power consumption is high and the fill factor is typically low. Another limiting factor of electromagnetic mirrors is the requirement of packaging external magnets or depositing magnetic materials. Piezoelectric MEMS mirrors have the advantages of high speed and low power consumption, but there exist the hysteresis effect and charge leakage problem. In contrast, electrothermal MEMS mirrors have a high fill factor, large scan range, and low driving voltage, but they are relatively small due to the long thermal response time. All four actuation mechanisms can be used to make either angular scanning micromirrors or linear scanning micromirrors.

7.1.2 Scope and organization of this chapter

This chapter introduces a new class of thermally actuated micromirrors that can generate large angular scans and large linear piston scans at low driving voltages. The key innovation is the thermal bimorph actuator design, which has evolved from the single bimorph actuator, double bimorph actuator, and triple bimorph actuator all the way to the quadruple bimorph actuator. A series of bimorph material pairs has also been tested and compared from Al/SiO$_2$ to Cu/W. Electrothermomechanical models have been developed to predict the thermal distribution and responses of the bimorph micromirrors.

This chapter is organized as follows: The bimorph actuation principle and basic micromirror structure design are introduced in Section 7.2. The several generations of bimorph micromirrors and their fabrication processes are described in Section 7.3. Analytical and lumped-element modeling are the focus of Section 7.4. In Section 7.5, some applications of these electrothermal micromirrors are presented, including laser scanning confocal microscopy and Fourier transform spectroscopy.

7.2 Principle of electrothermal bimorph actuation

An electrothermal bimorph consists of two layers of materials with different coefficients of thermal expansion (CTEs), as illustrated in Figure 7.1. When there is a temperature change across the

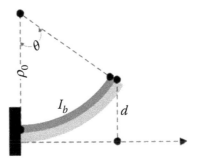

Figure 7.1 Side view of the cantilevered bimorph actuator with curling.

bimorph, the strain generated in each layer is different, resulting in a curling of the bimorph. In this section, a cantilevered bimorph actuator is used to introduce the working principle of electro-thermal bimorph actuation.

7.2.1 Stress and curvature

There are intrinsic stresses and extrinsic stresses in thin films [3,4]. Intrinsic stresses may be caused by the coalescence of grain boundaries, annihilation of excess vacancies, impurity incor-poration, and structure damage. Thus, intrinsic stresses are developed during deposition and are strongly dependent on the deposition process parameters. Extrinsic stresses, on the other hand, may be induced by material phase change, external forces, or temperature changes. For a cantile-vered bimorph as depicted in Figure 7.1, both types of stress will make the bimorph bend: intrin-sic stresses are used to generate an initial curvature while temperature change-induced extrinsic stresses are used for actuation. The curvature of the bimorph is given by [4]

$$\frac{1}{\rho} = \frac{\beta}{t_1 + t_2} \cdot (\varepsilon_{in} + \varepsilon_{ex}), \tag{7.1}$$

where ρ is the radius of curvature, t_1 and t_2 are the thicknesses of the top and bottom layers, respec-tively, ε_{in} is the intrinsic strain, and ε_{ex} is the extrinsic strain induced by Joule heating. Here, ε_{in} and ε_{ex} are either positive or negative, which will make the cantilevered bimorph curl upward or downward. β is a bimorph curvature coefficient given by [5]

$$\beta = 6 \cdot \frac{\left(1 + \frac{t_1}{t_2}\right)^2}{\frac{E_1'}{E_2'} \cdot \frac{t_1^3}{t_2^3} + 4 \cdot \frac{t_1^2}{t_2^2} + 6 \cdot \frac{t_1}{t_2} + \left(\frac{E_1'}{E_2'} \cdot \frac{t_1}{t_2}\right)^{-1}}, \tag{7.2}$$

where E_1' and E_2' are the biaxial elastic moduli for the two materials in the bimorph. The biaxial modulus E' is related to the Young's Modulus E and Poisson's ratio v by the equation $E' = E/(1 - v)$.

From Equation 7.1, the initial curvature after the release of the cantilevered bimorph actuator can be written as

$$\frac{1}{\rho_0} = \frac{\beta}{t_1 + t_2} \cdot \varepsilon_{in}. \tag{7.3}$$

This initial curvature is constant during the activation of bimorph actuation unless there are any material property changes or bimorph structural changes. Similarly, the Joule heating induced curvature is given by

$$\frac{1}{\rho_\Delta} = \frac{\beta}{t_1 + t_2} \cdot \varepsilon_{ex},$$

(7.4)

The thermally induced strain for each material can be expressed as

$$\varepsilon_i = \alpha_i \cdot \Delta T_i, \quad i = 1,2$$

(7.5)

where α and ΔT are, respectively, the CTE and the temperature change for each layer of the bimorph. As the two layers are stacked together and the thicknesses are small, the temperature difference between the two layers along the thickness direction is usually neglected, while only the temperature distribution along the length of the bimorph is taken into consideration. So, for an infinitesimal length dx at a position x on the bimorph, its extrinsic strain is given by $\varepsilon_{ex}(x) = (\alpha_1 - \alpha_2) \Delta T(x) = \Delta\alpha\Delta T(x)$. Thus, Equation 7.4 can be rewritten as

$$\frac{1}{\rho_\Delta} = \frac{\beta}{t_1 + t_2} \cdot \frac{\int_0^{L_b} \varepsilon_{ex}(x)dx}{L_b} = \frac{\beta}{t_1 + t_2} \cdot \frac{\Delta\alpha}{L_b} \cdot \int_0^{L_b} \Delta T(x)dx = \frac{\beta}{t_1 + t_2} \cdot \Delta\alpha \cdot \overline{\Delta T}$$

(7.6)

where $\overline{\Delta T} = (1/L_b)\int_0^{L_b} \Delta T(x)dx$ is the average temperature change along the cantilevered bimorph beam, and L_b is the length of the bimorph.

The tilt angle at the bimorph tip, which is equal to the arc angle of the curved beam, can be expressed as

$$\theta = \frac{L_b}{\rho} = \theta_0 + \theta_\Delta = \theta_0 + \frac{\beta}{t_b} \cdot L_b \cdot \Delta\alpha \cdot \overline{\Delta T},$$

(7.7)

where t_b is the total thickness of the bimorph, θ_0 is the initial arc angle of the curved bimorph, and θ_Δ is the thermally induced arc angle change.

Taking the initial curling case in Figure 7.1 as an example, the bimorph curls upward after structure releasing. If the material with a higher CTE is set as the upper layer, the thermally induced curvature will be negative, which means the Joule heating will make the bimorph bend downward; if the upper layer is set with a lower CTE material, then the bimorph bends upward upon heating.

7.2.2 Angular responsivity

The angular responsivity of the cantilevered bimorph beam can be defined as the bending angle divided by the temperature change, that is,

$$S_T = \frac{\theta_\Delta}{\Delta T} = \frac{\beta L_b}{t_b} \cdot \Delta\alpha$$

(7.8)

From the above equation, it can be concluded that S_T can be increased simply by increasing the bimorph length and decreasing the bimorph thickness. However, that will also reduce the structural stiffness, which leads to mechanical robustness issues.

For a given total thickness of a bimorph beam, the condition to reach the maximum value of β can be easily found [5]. By defining the thickness ratio as $\mu = t_1/t_2$, and the biaxial elastic modulus ratio as $\gamma = E_1'/E_2'$, Equation 7.2 can be rewritten as

$$\beta = 6 \cdot \frac{(1+\mu)^2}{\gamma\mu^3 + 4\mu^2 + 6 \cdot \mu + (\mu\gamma)^{-1}} \tag{7.9}$$

β reaches its maximum value of 1.5 when

$$\mu^2 = 1/\gamma \quad \text{or} \quad \frac{t_1}{t_2} = \sqrt{\frac{E_2'}{E_1'}} \tag{7.10}$$

This means for any selected material combinations, by setting the thickness ratio based on their biaxial elastic moduli, the same value of β can always be obtained. Thus, to maximize the angular responsivity, the most efficient way is to select materials with a large CTE difference.

7.2.3 Steady state thermal response

To study the steady state thermal response of the cantilever bimorph actuator, not only the actuator itself is the target of analysis, but the substrate, mirror plate or stage connected to the bimorph actuator, as well as the bridges connecting them and the ambient environment are also required to be taken into consideration. Together, these will determine the thermal response. As shown in Figure 7.2, a steady-state LEM model of a 1D MEMS mirror based on the cantilevered bimorph actuator has been developed by S. Todd [6]. The entire bimorph actuator is lumped as a single current source at the point where the temperature is highest and splits the bimorph into two sections. The thermal flux flows from the highest temperature to these lower temperature parts. The definitions of the thermal resistors in Figure 7.2 are given in Table 7.1.

The heat is generated by a resistive heating source. The electrical resistance of the heater is given by

$$R_{E,h} = \int_0^{L_b} \frac{\rho_e}{wt}[1+\gamma_h \cdot \Delta T(x)]dx = R_0(1+\gamma_h\overline{\Delta T}), \tag{7.11}$$

where $R_{E,h}$ is the overall resistance of the heater, ρ_e is the resistivity of the heater at room temperature, w and t are the width and thickness of the heater, respectively, γ_h represents the thermal

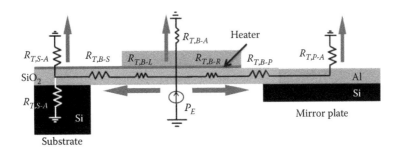

Figure 7.2 The steady-state LEM model of a MEMS mirror based on the cantilevered bimorph actuator. Heat paths are also shown.

Table 7.1 Definitions of the thermal resistors shown in Figure 7.2

$R_{T,S-A}$	Equivalent thermal resistance from the substrate to ambient
$R_{T,B-S}$	Conduction thermal resistance from the left end of the bimorph to the substrate
$R_{T,B-L}$	Conduction thermal resistance of bimorph from highest temperature point to the left end of the bimorph
$R_{T,B-R}$	Conduction thermal resistance of bimorph from highest temperature point to the right end of the bimorph
$R_{T,B-A}$	Convection thermal resistance from bimorph to ambient
$R_{T,B-P}$	Conduction thermal resistance from the right end of the bimorph to the mirror plate
$R_{T,P-A}$	Equivalent thermal resistance from the mirror plate to ambient

coefficient of resistivity (TCR) of the heater material, and $\overline{\Delta T}$ is the average temperature change along the bimorph. Thus, with a voltage signal V applied to the heater, the Joule heating generated by the heater is

$$P_E = \frac{V^2}{R_{E,h}} = \frac{V^2}{R_0(1+\gamma_h \Delta T)}. \tag{7.12}$$

In the thermal domain, by considering the conduction thermal resistance and convection thermal resistance and ignoring the radiation, the temperature change of the cantilevered bimorph at the substrate end, T_L, can be expressed as

$$T_L = f \cdot P_E \cdot (R_{T,B-S} + R_{T,S-A}), \tag{7.13}$$

where f is a balancing factor having a value between 0 and 1 which is the percentage of the total heat flux P_E flowing into the substrate. f is determined by the thermal resistances in Figure 7.2, and is expressed as [7]

$$f = \frac{\dfrac{R_{T,b}}{2} + R_{T,B-P} + R_{T,P-A}}{R_{T,b} + R_{T,B-P} + R_{T,P-A} + R_{T,B-S} + R_{T,S-A}}, \tag{7.14}$$

where $R_{T,b}$ is the total thermal resistance of the cantilever bimorph. Thus, by combining Equations 7.12 through 7.14, the temperature change of the bimorph at the substrate end is

$$T_L(\overline{\Delta T}) = \frac{\left(\dfrac{R_{T,b}}{2} + R_{T,B-P} + R_{T,P-A}\right) \cdot (R_{T,B-S} + R_{T,S-A})}{R_{T,b} + R_{T,B-P} + R_{T,P-A} + R_{T,B-S} + R_{T,S-A}} \cdot \frac{V^2}{R_0(1+\gamma_h \overline{\Delta T})}. \tag{7.15}$$

Similarly, the temperature change at the mirror plate end can be expressed as

$$T_R(\overline{\Delta T}) = \frac{\left(\dfrac{R_{T,b}}{2} + R_{T,B-S} + R_{T,S-A}\right) \cdot (R_{T,B-P} + R_{T,P-A})}{R_{T,b} + R_{T,B-P} + R_{T,P-A} + R_{T,B-S} + R_{T,S-A}} \cdot \frac{V^2}{R_0(1+\gamma_h \overline{\Delta T})}. \tag{7.16}$$

By comparing Equations 7.15 and 7.16, in order to obtain a uniform and even temperature distribution along the entire bimorph actuator, the thermal resistances from both ends of the bimorph

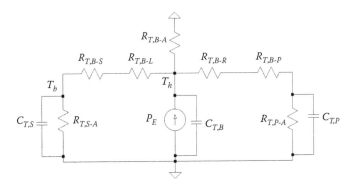

Figure 7.3 Simplified LEM thermal model of the cantilever bimorph-based MEMS mirror. R's are thermal resistances and C's are thermal capacitances.

are required to be close to each other, and should both be much larger than the thermal resistance of the bimorph itself.

7.2.4 Transient thermal response

To evaluate the rise time, a simplified lumped element model is built as shown in Figure 7.3 [6,8]. The highest temperature point T_h along the bimorph actuator is taken into consideration for convenience, and the directions of the heat flow are shown. There are three parts of the heat flow: the heat flowing to the mirror-plate, the heat flowing to the substrate, and the heat dissipating to the surrounding ambient environment.

If we look at the point of T_h, the heat transfer equation of the bimorph actuator can be simplified as

$$C_T \frac{d\Delta T_h}{dt} + \frac{\Delta T_h}{R_T} = P_E. \tag{7.17}$$

Here, the R_T and C_T are the equivalent thermal resistance and equivalent thermal capacitance between the node T_h and the substrate T_b, and PE is the input electrical power. The step response is then given by

$$\Delta T_h = \frac{P_E}{C_T}\left[1 - e^{-\frac{t}{R_T C_T}}\right]. \tag{7.18}$$

The thermal response time is approximately

$$t_r = 2.2\tau = 2.2R_T C_T. \tag{7.19}$$

To evaluate the equivalent thermal resistance and capacitance, the position of the highest temperature point must be located. As mentioned above, a balancing factor f is defined, which represents the percentage of the total heat flux flowing into the substrate. Thus the heat flux flowing into the substrate and the mirror plate are $P_{B-S} = f \times P_E$ and $P_{B-P} = (1 - f) \times P_E$, which means at point $x = f \times L_b$, the temperature is highest. Here, the convection effect on f is not considered, which may cause some error, but is in the acceptable range.

From the LEM circuit in Figure 7.3, the equivalent thermal resistance can be expressed as

$$R_T \approx (R_{T,B-S} + fR_{T,b})//(R_{T,B-P} + (1-f)R_{T,b} + R_{T,P-A})//R_{T,B-A}, \tag{7.20}$$

where $R_{T,B-S}$, $R_{T,b}$ and $R_{T,B-P}$ are thermal conduction resistances, which are determined by the equation below

$$R_{cond} = \frac{L}{A \times \gamma},$$ (7.21)

where L is the length of the thermal path, A is the cross-section area of the thermal path, and γ is the thermal conductivity of the material. $R_{T,B-A}$ and $R_{T,P-A}$ are thermal convection resistances, which are given by

$$R_{conv} = \frac{1}{hS},$$ (7.22)

where h is the convection coefficient and S is the surface area.

The thermal capacitance is mainly determined by the bimorph beam itself and is given by

$$C_{T,b} = \sum_i m_i * C_{T,i}$$ (7.23)

where m_i and $C_{T,i}$ are the mass and specific heat of the active bimorph layers. The thermal response time can be calculated by combining Equations 7.19, 7.20, and 7.23.

Also note that the conduction thermal resistance is inversely proportional to the average thermal conductivity of the bimorph, γ_{avg}, and the thermal capacitance is the average volumetric specific heat capacity of the bimorph, $C_{T,v-avg}$. Thus, the conductive thermal response time $t_{r,conductive}$ is inversely proportional to the average thermal diffusivity, that is,

$$t_{r,conductive} \propto \frac{C_{T,v-avg}}{\gamma_{avg}} = \frac{1}{d_{avg}},$$ (7.24)

where d_{avg} is the average thermal diffusivity of the bimorph. Thus, for the bimorph design with the same dimension settings, a high equivalent thermal diffusivity will help improve the response speed of the bimorph actuator.

From the calculation above, when the thermal isolation between the substrate and bimorph and the thermal isolation between the bimorph and mirror plate are fixed, the thermal response time will generally decrease with the decrease of $R_{T,b}C_{T,b}$. In order to achieve electrothermal bimorph-based actuators with a fast thermal response, there are mainly two feasible methods: (a) to use materials with high thermal diffusivity and (b) to reduce the thermal convection resistance by improving the convection coefficient through packaging or improving the effective surface area.

7.2.5 Materials selection

The materials selected for the two layers of the bimorph will definitely affect the performance of the bimorph actuator in multiple aspects, including the actuation range, responsivity, stiffness, and response speed. The yield strength and ultimate strength will determine the repeatable actuation range and robustness of the bimorph actuator. Table 7.2 shows a summary of commonly used MEMS materials with their thermal and mechanical properties.

To achieve high angular responsivity, as studied in Section 7.2.2, the ideal material combination is two materials with a large CTE difference. From Table 7.2, Al and SiO_2 have one of the largest possible CTE differences. This is the main reason a series of Al/SiO_2 bimorph actuator designs

Table 7.2 Overview of properties of popular MEMS materials

Material	CTE (10⁻⁶/K)	Thermal conductivity (W/mK)	Young's modulus (GPa)	Poisson ratio	Melting point (°C)	Yield strength (MPa)	Ultimate strength (MPa)
Si	3.0	150.0	179	0.27	1414	–	300
SiO$_2$	0.4	1.4	70	0.17	1700	–	690–1380
Si$_3$N$_4$	3.3	30.0	310	0.24	1900	–	–
Al	23.6	237.0	70	0.35	660	124	176
Au	14.5	318.0	78	0.44	1064	–	127
Cu	16.9	401.0	120	0.34	1083	262	310
Pt	8.9	71.6	168	0.38	1768	185	240
Pb	28.7	35	160	0.42	327	12	12
Cr	5.0	93.9	140	0.21	1907	200	282
Ti	8.6	21.9	116	0.32	1668	100	240
W	4.5	173	410	0.28	3410	550	620
Ni	12.8	90.9	200	0.31	1455	35	140

have been developed and high angular responsivity has been achieved. Al/SiO$_2$ bimorph actuator designs will be introduced in the following section.

In addition to high angular responsivity, it is desired to obtain bimorph actuators with higher mechanical stiffness, which will result in higher mechanical resonance and make the actuators more resistant to environmental vibrations. With the same dimensions, bimorphs with greater Young's modulus materials can ensure higher stiffness, which means the selection requires that both layers have a high Young's modulus. A bimorph design based on this idea will also be discussed in the next section.

Furthermore, yield strength and ultimate strength are responsible for the repeatability of the actuation and the robustness of the bimorph actuator. With both mechanical design optimization and proper material selection, robust bimorph actuators with large repeatable actuation ranges can be achieved.

7.3 Electrothermal bimorph-based micromirrors

Various electrothermal bimorph-based actuators and micromirrors have been developed. Depending on the number of bimorph segments contained in the mechanical structure of the actuator, the electrothermal bimorph actuators are divided into four types: single beam design, double beam design, triple beam design, and quadruple beam design. In this section, we will introduce these actuator designs and the micromirrors based on these actuators. The corresponding fabrication processes are also described.

7.3.1 Single beam design

A 1D electrothermal micromirror design consisting of an array of single bimorph beams and a single-crystal silicon (SCS) supported mirror plate was developed by Xie et al. [9]. The schematic and SEM of the fabricated micromirror is shown in Figure 7.4, where the SCS mirror plate with Al coated on the top surface is connected to the Al/SiO$_2$ bimorph array. The bimorph array is linked to form a mesh to enhance the mechanical robustness. A layer of polysilicon is encapsulated within

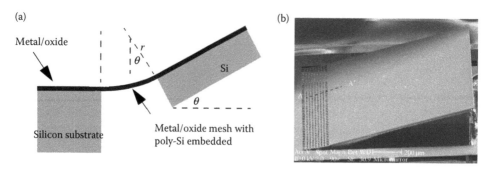

Figure 7.4 An electrothermal bimorph micromirror based on the single beam design. (a) The conceptual design. (b) An SEM of a released micromirror. (Adapted from Xie, H. et al. *Sensors and Actuators A: Physical* 103, no. 1, 2003: 237–241.)

the SiO_2 as the heater. The micromirror can scan about one axis with a maximum scan angle up to 17° and the mechanical resonance frequency is 165 Hz.

The micromirror is fabricated with a DRIE CMOS-MEMS process as shown in Figure 7.5. The front side layers are formed by a standard foundry CMOS process. The MEMS process starts from the backside deep silicon etch, which leaves a 40 μm thick SCS membrane to be used as the mechanical support of the mirror. Then a dielectric etch is performed from the front side, followed by an anisotropic deep silicon etch. Finally, the silicon underneath the bimorph mesh is fully removed by an isotropic silicon etch. The release process is maskless and only dry etch is used, which is completely compatible with the commercial CMOS processes. The mirror plate of the released micromirror has an initial upward tilt due to the intrinsic stresses in the bimorph thin-film layers.

Based on the same process in Figure 7.5, a 2D electrothermal bimorph micromirror was developed by Jain et al. [10], where two sets of single beam bimorphs are arranged at orthogonal directions to realize 2-axis area scanning, as shown in Figure 7.6. The mirror has a rotation angle of 40° at 15 V and the frame rotates by 25° at 17 V, with power consumptions of 95 and 135 mW, respectively. The resonant frequencies of the mirror and frame actuators are 445 and 259 Hz, respectively.

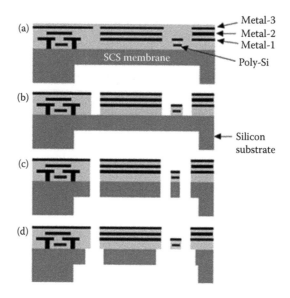

Figure 7.5 DRIE CMOS-MEMS process flow of the 1D micromirror using polysilicon for heaters: (a) backside etch, (b) oxide etch, (c) anisotropic deep Si etch, and (d) Si undercut. (Adapted from Xie, H. et al. *Sensors and Actuators A: Physical* 103, no. 1, 2003: 237–241.)

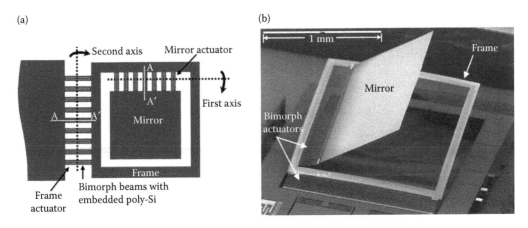

Figure 7.6 2D micromirror based on the single beam design. (a) Schematic of the micromirror design showing the axes of rotation. (b) An SEM of a fabricated 2D micromirror. (Adapted from Jain, A. et al. *Selected Topics in Quantum Electronics, IEEE Journal of* 10, no. 3, 2004: 636–642.)

Although large rotation angles are achieved with single beam bimorphs, the single beam design-based micromirrors have a large initial tilt angle and a non-stationary center of rotation of the mirror plate. The large initial tilt angle makes the packaging process complex when the micromirror is employed in optical systems. A shifting center of rotation axis will introduce image distortion or phase errors.

7.3.2 Double beam design

To overcome the problem of the large initial tilt of the mirror plate in the single beam design-based micromirror and to achieve axial scanning, a double beam design electrothermal bimorph actuator was proposed by Jain et al. [11]. This double beam design employs two complementary bimorph actuators oriented in a folded structure to keep the mirror plate parallel to the substrate, as shown in Figure 7.7. The initial elevation of the mirror plate is above the substrate, which comes from the frame connecting in the longitudinal direction of the curling bimorph. The mirror actuator and frame actuator rotate the mirror in opposite angular directions and can be controlled individually. Bi-directional scanning can be generated by alternatively applying voltage to the mirror and frame actuators. Another mode of operation is the large piston motion, which is achieved by simultaneously applying the same voltage to both actuators, which can generate equal curling angles. As the tilt angles of the two actuators can be compensated, a pure vertical motion can be achieved.

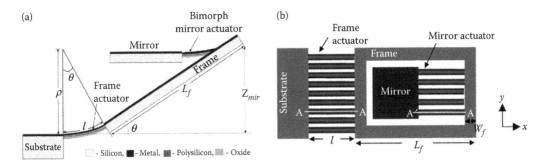

Figure 7.7 (a) Schematic of the double beam design-based bimorph actuator. (b) The topology design of a double beam-based LVD micromirror. (Adapted from Jain, A. et al. *Sensors and Actuators A* 122, no. 1, 2005: 9–15.)

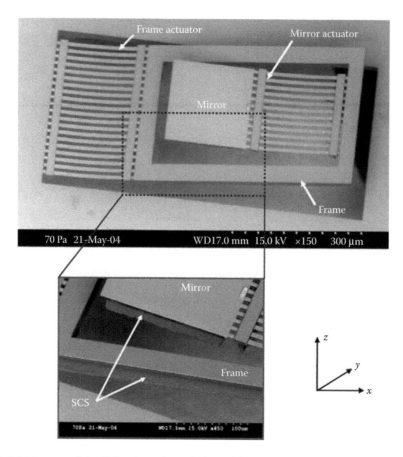

Figure 7.8 SEM images of the LVD micromirror. (Adapted from Jain, A. et al. *Sensors and Actuators A* 122, no. 1, 2005: 9–15.)

The micromirror is fabricated using a DRIE CMOS-MEMS process similar to that shown in Figure 7.5. SEMs of a released device are shown in Figure 7.8. The mirror plate has a maximum rotation angle of 26.5° when 3 V is applied to the mirror actuator, with a power consumption of 21 mW. The frame with the mirror plate has a maximum deflection of −16.5° when 5.5 V is applied to the frame actuator and the power consumption is 50 mW. The maximum piston displacement that can be achieved is 200 μm with the angular tilt compensation of the mirror and frame actuators, thus the double beam actuator design is also called a large-vertical-displacement (LVD) actuator.

As an extension of this double beam LVD actuator design, a 2D LVD micromirror was developed by Jain et al. [12]. Similar to the single beam design-based 2D micromirror, two LVDs are cascaded in two orthogonal directions, as shown in Figure 7.9. The mirror plate can be actuated to rotate about both x and y axes, and a piston motion can be obtained as well by proper simultaneous actuation of the four bimorph actuators. The optical scan angles of the 2D LVD micromirror are up to ±40° in the x axis and ±30° in the y axis at DC actuation voltages less than 12 V. The maximum piston displacement achieved is 0.5 mm at about 15 V. The fabrication process is the same as the one shown in Figure 7.5.

Although the double beam LVD actuator design solves the problem of a large initial tilt, the rotation center of the mirror plate is still not fixed during scanning. This also produces a new problem of lateral shift, as illustrated in Figure 7.10. The vertical displacement Z and the lateral shift Ls of the mirror center can be expressed as

$$Z = L_f \times \sin\theta \tag{7.25}$$

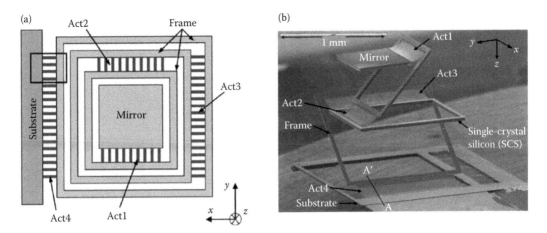

Figure 7.9 2D LVD micromirror: (a) schematic of the device, and (b) SEM image of the fabricated device. (Adapted from Jain, A., and H. Xie. *Sensors and Actuators A: Physical* 130, 2006: 454–460.)

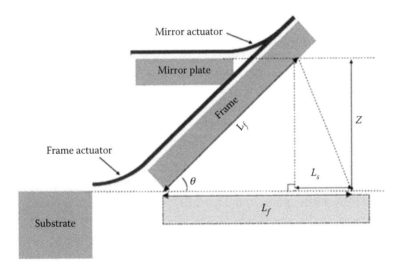

Figure 7.10 Cross-section of the LVD actuator showing lateral shift. (Adapted from Wu, L. and H. Xie. *Sensors and Actuators A: Physical* 145, 2008: 371–379.)

$$L_s = L_f \times (1 - \cos\theta). \tag{7.26}$$

From Equations 7.25 and 7.26, we can conclude that the lateral shift is proportional to the vertical displacement, which means the lateral shift problem becomes more serious with the increase of the vertical range. The lateral shift of the LVD actuator will greatly distort the scan patterns and also reduces the effective aperture size of the micromirror.

Another problem is that the fill factor of the LVD micromirror is very low due to the large area occupied by the frames and the bimorph beams. For the 2D LVD micromirror, the aperture size is 0.5 by 0.5 mm² while the device footprint is 2.7 by 1.9 mm², which means the fill factor is only 4.8%.

7.3.3 Triple beam design

To continue to solve the problems such as a non-stationary rotation center, lateral shift, and small fill factor, a novel LVD electorthermal actuator design with cancelation of both angular tilt and

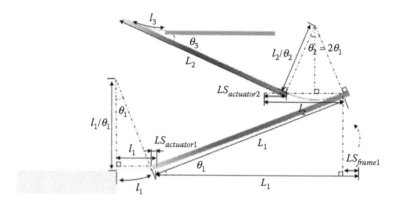

Figure 7.11 Concept of LSF-LVD actuator design. (Adapted from Wu, L., and H. Xie. *Sensors and Actuators A: Physical* 145, 2008: 371–379.)

lateral shift was proposed by Wu et al. [13], which is the lateral-shift-free-large-vertical-displacement (LSF-LVD) actuator.

The LSF-LVD actuator is a triple beam design that employs three bimorph segments to cancel both the angular tilt at the bimorph tip and the lateral shift at the end of the actuator. It can still obtain a large vertical displacement. The LSF-LVD actuator design consists of three sets of Al/SiO$_2$ bimorphs and two frames connected in series at the end of the corresponding sets of bimorphs, as shown in Figure 7.11, where l_1, l_2, and l_3 denote the respective lengths of the three bimorph segments, L_1 and L_2 denote the lengths of the two frames, and θ_1, θ_2, and θ_3 are the arc angles of the three bimorph segments. The mirror plate is connected at the end of the upper bimorph. To cancel the initial tilt of the mirror plate, the arc angles of the three bimorph segments need to satisfy the relation $\theta_2 = \theta_1 + \theta_3$. As the arc angle of each bimorph is proportional to its length, the condition for the initial tilt cancelation becomes $l_2 = l_1 + l_3$.

The lateral shift of the mirror plate, from Figure 7.11, can be expressed as

$$LS = LS_{bimorph1} + LS_{frames1} - LS_{bimorph2} - LS_{frames2} + LS_{bimorph3} \tag{7.27}$$

For simplicity, l_1 and l_3 are chosen as half of l_2, yielding

$$LS = (L_1 - L_2)(1 - \cos\theta_1). \tag{7.28}$$

With the tilt angle canceled, to further eliminate the lateral shift, simply choosing the same length for the two frames can achieve the goal. Assuming $L_2 = L_1 = L$ and $\theta_3 = \theta_1 = \theta$, the pure vertical displacement generated at the end of the upper bimorph is given by

$$Z = 2L\sin\theta. \tag{7.29}$$

The schematic of a 2D micromirror employing the LSF-LVD actuator is shown in Figure 7.12, where four identical LSF-LVD actuators are placed, respectively, on the four sides of a square mirror plate to control the relative displacements to achieve either piston motion or angular scanning.

The device is fabricated based on a combined surface- and bulk-micromaching process using an SOI wafer [13], as shown in Figure 7.13. The process starts from (a) a PECVD SiO$_2$ layer deposition and sputtering and lift-off of a Pt layer to form the heater. (b) A dielectric layer is deposited by PECVD on top of the Pt layer as the electrical insulation, followed by (c) removing the SiO$_2$ on top of the mirror plate. (d) Al evaporation and lift-off is performed to form the bimorph and the reflective

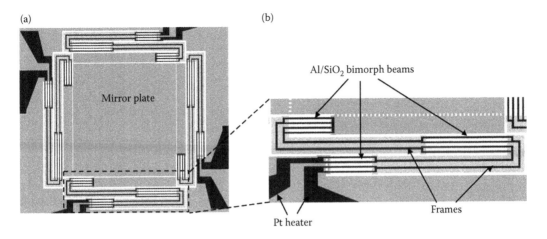

Figure 7.12 Schematic of the LSF-LVD 2D micromirror with four identical actuators at four sides of the mirror plate. (Adapted from Wu, L. and H. Xie. *Sensors and Actuators A: Physical* 145, 2008: 371–379.)

mirror surface. (e) Front-side SiO_2 etching is done to define the mirror plate and the release area. The backside process starts from (f) a deep silicon etch to form the backside chamber and mirror plate, followed by (g) a front-side anisotropic silicon etch and (h) front-side isotropic silicon etch to release the bimorphs. Instead of using polysilicon as the heater material, Pt is employed in this process. Pt is more stable and has a smaller resistivity, which means that a lower driving voltage is needed. Pt can also be used for temperature sensing.

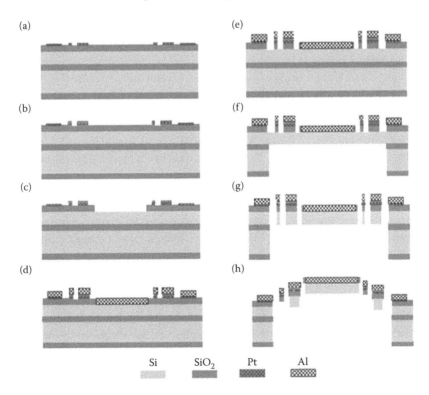

Figure 7.13 Fabrication process of the Al/SiO_2 LSF-LVD 2D micromirror. (a) Pt sputtering and lift-off. (b) PECVD dielectric SiO_2. (c) Mirror SiO_2 etch. (d) Al evaporation and lift-off. (e) Front-side SiO_2 etch. (f) Backside SiO_2 and Si etch. (g) Front-side Si anisotropic etch. (h) Front-side Si isotropic etch. (Adapted from Wu, L. and H. Xie. *Sensors and Actuators A: Physical* 145, 2008: 371–379.)

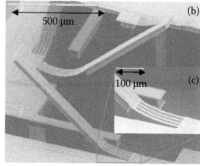

Figure 7.14 SEMs of the LSF-LVD 2D micromirror using Al/SiO$_2$ bimorphs. (Adapted from Sun, J. et al. *Optics Express* 18, no. 12, 2010: 12065–12075.)

The SEMs of the fabricated devices are shown in Figure 7.14 [14]. The frames have a thick Si layer underneath to ensure rigidity. The footprint of the device is 2 by 2 mm^2 and the mirror plate is 1 by 1 mm^2. Thus, the fill factor is about 25%, which is five times that of the original LVD 2D micromirror. A static vertical displacement of more than 600 μm has been achieved at only 5.5 V and 86 mW. The maximum optical angle can be up to $\pm 31°$.

The LSF-LVD actuator design solves the problem of lateral shift from the LVD actuator. Meanwhile, when differential voltages are applied on opposite actuators on one axis, the rotation axis will be along the center of the mirror plate, so the center shift problem is also overcome. Thus, together with the zero initial tilt, the optical alignment is simplified and optical system design flexibility is greatly improved. However, due to the small Young's moduli of Al and SiO$_2$, the frames contain a thick Si layer to ensure the rigidity, which will introduce a large thermal capacitance and slow the thermal response. For the micromirror employing the LSF-LVD actuator for large range actuation, the thermal response time is in the order of 100 ms [15], which means a thermal cutoff frequency of less than 3 Hz.

To improve the thermal response speed of the LSF-LVD actuator, a new LSF-LVD actuator with high Young's modulus materials, Cu and W, as the two bimorph layers was proposed by Zhang et al. [16]. In this new actuator design, the frames are made of Cu/W/Cu thin-film multimorphs instead of thick silicon, which can greatly reduce the thermal capacitance of the LSF-LVD actuator. Although Cu/W has a smaller CTE than Al/SiO$_2$, Cu/W can work under a much higher temperature range than Al. In other words, Cu/W has a relatively lower mechanical responsivity, but can stand a much higher temperature change, so a Cu/W-based LSF-LVD actuator can still produce a large actuation range compared to its Al/SiO$_2$ counterpart. As the thermal diffusivity of the Cu/W combination is larger than that of Al/SiO$_2$ due to the small thermal diffusivity of SiO$_2$, the thermal conductive response time will also be improved [17]. Moreover, W itself is used as both one of the active bimorph layers and the heater, which will also lower the cost compared to using Pt as the heater.

The Cu/W LSF-LVD 2D micromirror is fabricated using a combined surface- and bulk-micromachining process [16], which is illustrated in Figure 7.15. Different from the previous fabrication processes, the device release is done from the backside, which has the advantage of keeping the front side intact. The process starts from the front side, including the following main steps: (a) PECVD SiO$_2$ deposition and wet etch to form the thermal isolation bridges, followed by the deposition of a SiO$_2$/Si$_3$N$_4$ layer; (b) sputtering and lift-off of the first Cu layer to form the bimorphs and multimorph frames, followed by the deposition of a thin Si$_3$N$_4$ layer as the electrical isolation layer; (c) W sputtering and lift-off for the bimorphs and multimorph frames, followed by the deposition of another electrical isolation Si$_3$N$_4$ layer; (d) contact via etching, followed by the sputtering and lift-off of the second Cu layer to form the multimorph frames, and then the deposition and patterning of the top protective SiO$_2$/Si$_3$N$_4$ layer; and (e) sputtering and lift-off of a thin Al layer to form the mirror surface and bonding pads. With this step, the front side process is completed and now the process

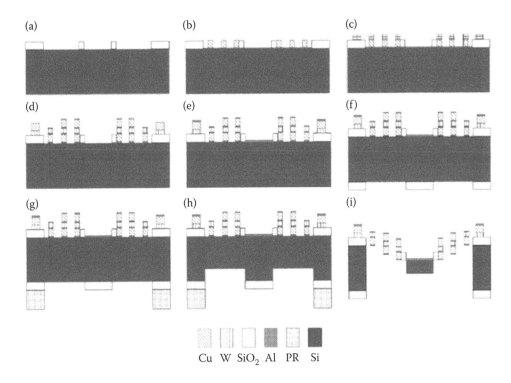

Cu   W   SiO$_2$ Al   PR   Si

Figure 7.15 Fabrication process of the Cu/W LSF-LVD 2D micromirror. See text for description of (a)–(i). (Adapted from Zhang, X. et al. *Micromachines* 6, no. 12, 2015: 1876–1889.)

is switched to the backside. The backside process includes the following steps: (f) SiO$_2$ PECVD and spattering for the opening areas of the LSF-LVD actuator regions; (g) thick photoresist patterning for the backside chamber; (h) silicon anisotropic DRIE to form a silicon trench corresponding to the actuator regions, followed by RIE removal of SiO$_2$; and (i) silicon anisotropic DRIE to remove all the silicon in the actuator regions. Compared with the front side release process, the backside release can keep the front side intact during the release and ensure a better mirror surface quality.

SEMs of a Cu/W LSF-LVD 2D micromirror are shown in Figure 7.16, where the device footprint is 2.5 by 2.5 mm^2 and the mirror plate size is 0.9 by 0.9 mm^2. The symmetric Cu/W/Cu multimorph frames show a negligible deformation and well balanced thin-film stresses, which are also temperature insensitive during actuation of the actuator. The micromirror can perform both piston and tip-tilt scanning. A static piston displacement of 320 μm and angular optical scan of ±18° have

Figure 7.16 SEMs of the LSF-LVD 2D micromirror using Cu/W bimorphs. (Adapted from Zhang, X. et al. *Micromachines* 6, no. 12, 2015: 1876–1889.)

been achieved with a voltage of only 3 V DC and a power consumption of 56 mW. The thermal response time is 15.4 ms and the mechanical resonances of the first piston mode and the second tilt mode are 550 and 832 Hz, respectively. The effective stiffness of the Cu/W bimorph is about three times greater than that of the Al/SiO$_2$ bimorph due to the large Young's moduli of Cu and W. Thus, the Cu/W LSF-LVD actuator design has a larger force and faster response compared with the Al/SiO$_2$ design.

7.3.4 Quadruple beam design

The triple beam LSF-LVD actuator design solves the problem of lateral shift existing in the double beam LVD actuator design. The 2D micromirror employing the triple beam design achieves a large piston with a very small lateral shift, and large a tip-tilt scan with a stationary rotation center. However, due to the structural asymmetry, the device still suffers from a slightly skewed scan motion, resulting in an in-plane rotation of the mirror plate. The slow thermal response of the LSF-LVD actuator can be improved by the Cu/W/Cu multimorph frame design, but still needs to be further improved by modification in the actuator structure design. Moreover, the area taken by the bimorphs and the frames is still large, so the fill factor is not high.

To further solve these existing problems, Jia et al. [18] proposed the quadruple beam design, which utilizes four bimorph beams and a unique structure design to compensate the angular tilt and lateral shift at the tip of the actuators. The principle of the quadruple beam actuator design, which is also called an inverted-series-connected (ISC) actuator, is illustrated in Figure 7.17. For a single beam bimorph, the tip will have both tangential tip-tilt and lateral shift during actuation, as shown in Figure 7.17a. If a second bimorph with the same length but an inverted layer setting is connected at the tip of the first bimorph to form an S-shaped bimorph beam, the tangential tip-tilt will be cancelled, as shown in Figure 7.17b; however, a lateral shift is still present at the tip of the second bimorph. To compensate the lateral shift, a second S-shaped bimorph beam is connected with the first one end-to-end but in a folded fashion, as shown in Figure 7.17c, where the tip at the end of the second S-shaped bimorph beam, to the first order, has no lateral shift or tilt. Thus, by using this quadruple beam ISC actuator design, both tip-tilt and lateral shift are cancelled at the tip of the ISC actuator, and pure vertical motion can be achieved. As no frames are needed, the thermal response is faster compared to the LSF-LVD actuator designs. The area occupied by this ISC actuator is also smaller, leading to a higher fill factor.

A detailed design of an ISC bimorph is illustrated in Figure 7.18a, where the lengths of the inverted (IV) and non-inverted (NI) segments are denoted as L_{IV} and L_{NI}, respectively. According to Equation 7.7, to compensate the tangential tilt angle, the optimal length ratio L_{IV}/L_{NI} is chosen to be 2.1 according to FEM simulation [18]. As the stress will be concentrated at the joint point of the NI bimorph and IV bimorph where they are directly connected, a sandwiched multimorph overlap portion is added between the NI and IV segments to improve the robustness of the ISC actuator. SEMs of a fabricated

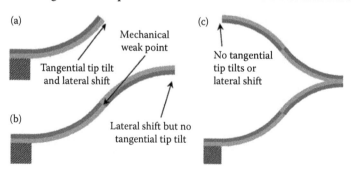

Figure 7.17 Concept of the ISC bimorph actuator. (Adapted from Jia, K. et al. *Microelectromechanical Systems, Journal of* 18, no. 5, 2009: 1004–1015.)

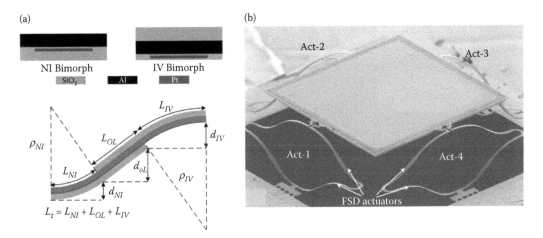

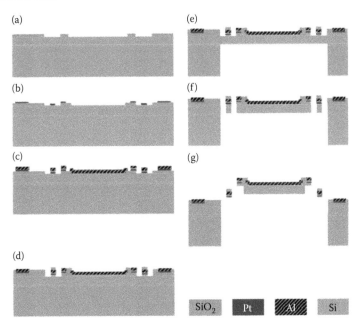

Figure 7.18 (a) Cross-sectional view of the S shaped half-ISC actuator. (b) An SEM of the ISC 2D micromirror. (Adapted from Jia, K. et al. *Microelectromechanical Systems, Journal of* 18, no. 5, 2009: 1004–1015.)

ISC 2D micromirror are shown in Figure 7.18b. The fabrication process, as shown in Figure 7.19, is a combined surface- and bulk-micromachining fabrication process similar to the one shown in Figure 7.13. The main difference is that ISC actuators require two SiO_2 layers, respectively deposited before and after the patterning of the Al layer to form both non-inverted and inverted bimorphs.

The device has 3-DOF actuations, including 2-axis rotation and piston mode. The maximum optical scan range is up to $\pm 30°$, and the maximum piston displacement is 480 μm at a DC voltage less than 8 V. The thermal response time is about 10 ms. The resonance frequencies of the piston and rotation modes are 336 and 488 Hz, respectively. The unique actuator design and the symmetrical structure of the micromirror ensure a large piston-free lateral shift and tip-tilt scan with a stationary rotation axis.

Figure 7.19 Fabrication process of Al/SiO_2 ISC 2D micromirror. (a) PECVD bottom SiO_2 and BOE. (b) Pt sputtering and lift-off. (c) Al sputtering and lift-off. (d) PECVD top SiO_2 and RIE. (e) Backside chamber DRIE. (f) Frontside anisotropic DRIE. (g) Frontside isotropic DRIE. (Adapted from Jia, K. et al. *Microelectromechanical Systems, Journal of* 18, no. 5, 2009: 1004–1015.)

Figure 7.20 SEMs of the Cu/W ISC 2D micromirror. (Adapted from Zhang, X. et al. In *Solid-State Sensors, Actuators and Microsystems (TRANSDUCERS), 2015 Transducers-2015 18th International Conference on*, Shanghai, China, pp. 912–915. IEEE, 2015.)

As mentioned in Section 7.3.3, the Al/SiO$_2$ bimorph has a relatively small stiffness due to the small Young's moduli of Al and SiO$_2$. SiO$_2$ limits the thermal response speed and its brittle nature also causes poor robustness of the actuator. By replacing Al/SiO$_2$ bimorphs with Cu/W bimorphs, the mechanical stiffness, thermal response speed, and robustness can all be greatly improved. Moreover, as the W layer can act as both the bimorph layer and the heater, no extra heater layer is needed, which makes it possible to make asymmetrical layer settings for the NI and IV bimorphs. Thus, the ISC concept shown in Figure 7.17 can be easily realized by simply choosing the same length instead of tuning the length ratio between the NI and IV bimorphs based on the material properties that may vary with process conditions.

Zhang et al. reported a 2D micromirror using a Cu/W ISC actuator [19], and the SEMs of the micromirror are shown in Figure 7.20. The fabrication process is shown in Figure 7.21, and is similar to the one illustrated in Figure 7.15. A maximum piston displacement of 169 μm and a maximum optical angle of ±13° are achieved with a DC driving voltage of only 2.3 V. The micromirror gives a fast thermal response of less than 7.5 ms. The resonances of the piston and tip-tilt modes are as high as 1.58 and 2.74 kHz, respectively. The actuation range is limited by the average temperature change the actuator can achieve. Currently, the maximum temperature change is only 380 K which is much below the melting point of Cu or W due to the oxidation problem of Cu. More study on the protection of Cu from oxidation during thermal actuation is ongoing, and is expected to extend the actuation range of the Cu/W ISC actuator.

7.4 Applications

7.4.1 Laser scanning confocal microscopy

Confocal microscopy, which was first introduced by Marvin Minsky in 1957 [20], has revolutionized biological research by enabling noninvasive imaging with high resolution and depths of hundreds of micrometers into tissue samples. It is one of the most widely used technologies for *in vivo* imaging of tissues [21]. It uses the linear light-tissue interactions to generate a signal and utilizes a pinhole to reject out of focus light to increase signal contrast. The resolution is in the order of a single micron; thus, confocal microscopy can delineate subcellular details and detect cancers at an early stage. Laser scanning confocal microscopy (LSCM) acquires 3D images via scanning in all three dimensions. The 3D scanning presents a big challenge for applying LSCM for endoscopic applications because of the difficulty of making miniature scanners that can fit into the human body.

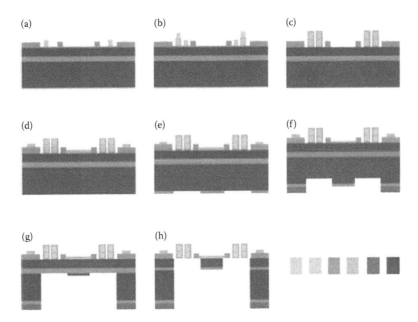

Figure 7.21 Fabrication process of Cu/W ISC 2D micromirror. (a) Cu-1 sputtering and lift-off. (b) W sputtering and lift-off. (c) Cu-2 sputtering and lift-off. (d) Al sputtering and lift-off. (e) Backside SiO_2-1 PECVD and pattering. (f) Backside SiO_2-2 PECVD and pattering, followed by "stage-down". (g) DRIE "stage down" to BOX. (h) BOX removal and final bimorph release. (Adapted from Zhang, X. et al. In *Solid-State Sensors, Actuators and Microsystems (TRANSDUCERS), 2015 Transducers-2015 18th International Conference on*, Shanghai, China, pp. 912–915. IEEE, 2015.)

Based on the large scan range of MEMS electrothermal actuators and micromirrors reported in Section 7.3, Liu et al. demonstrated a MEMS-based 3D LSCM endoscopic system [22]. The schematic of the MEMS LSCM system is shown in Figure 7.22, including a 638 nm laser (CLAS2-635-100C, Blue Sky Research), a spatial filter, a collimating lens, a polarizing beam splitter (PBS), an avalanche photodiode (APD), and a MEMS-based fiber-optic endoscopic probe. The laser is first guided into the MEMS probe through a fiber to a GRIN lens where the light beam is collimated. The collimated light passes through a quarter-wave plate (QWP) to convert the linearly polarized light into a circularly polarized light and then is incident on a MEMS mirror placed on a 45° slope. A fixed mirror that is also on a 45° slope is used to steer the light beam back to the probe length direction. The light beam continues to propagate through a beam expander before reaching a MEMS lens scanner. In this probe, the MEMS 2D mirror performs a two-axis transverse scan while the MEMS lens scanner provides the depth (z-axis) scan.

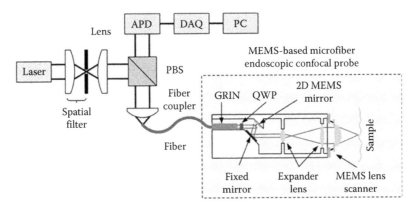

Figure 7.22 Schematic of a 3D confocal microendoscope.

Figure 7.23 The MEMS 2-axis scanning mirror and z-axis tunable lens. (a) SEM of a MEMS mirror. (b) Picture of an assembled MEMS lens scanner. (c) SEM of a MEMS scanning platform. (Adapted from Liu, L. et al. *Sensors and Actuators A: Physical* 215, 2014: 89–95.)

An SEM of the LSF-LVD MEMS mirror is shown in Figure 7.23a, where the mirror plate is 1 by 1 mm² and the device footprint is 2 by 2 mm². The MEMS mirror scans ±20° at a DC driving voltage of 4.5 V and its first resonant frequency is 240 Hz. The MEMS lens scanner, shown in Figure 7.23b, is actually a miniature aspheric lens assembled on a MEMS scanning platform (Figure 7.23c). The center of the MEMS platform is hollow to allow light to pass. An Al/SiO₂ LSF-LVD bimorph actuator design is employed here. Note that the actuator for the MEMS lens scanner has a much wider bimorph width than those of the MEMS mirror. This is to provide enough stiffness to support the relatively heavy lens. The aspheric lens has a diameter of 2.4 mm, an NA of 0.55, and a back focal length of 0.88 mm. The aspheric lens is glued on the MEMS scanning platform using UV optical glue. At a driving voltage of 3.2 V, the MEMS lens scanner generates a 515 μm vertical displacement, which is sufficiently large for the depth scan of confocal imaging.

Figure 7.24 shows a picture of the MEMS probe, which is assembled from three separate precision-machined metal pieces. The first piece (Part I) is used to hold the fiber, GRIN lens, and QWP. The second piece (Part II) is used to house the MEMS mirror, the fixed mirror, and the first expander lens. The third piece (Part III) is used to house the second expander lens and the MEMS lens scanner. The probe diameter is 7 mm. The measured axial and lateral resolutions are 9.0 and 1.2 μm, respectively [22].

7.4.2 Fourier transform microspectrometers

Spectrometers are mostly based on dispersive gratings, which require photodetecctor arrays or rotating gratings. In either case, to obtain a high spectral resolution, some of the signal-to-noise ratio (SNR) must be sacrificed. Photodetector arrays for wavelengths greater than 1 μm are very expensive; and the cost almost linearly increases with the photodetector array size. On the other hand, Fourier transform spectrometers (FTS) do not require dispersive components, and use only a single photodetector that collects the optical signal of the entire spectral range. Therefore, FTS has a much higher SNR and much more affordable photodetectors can be used.

To obtain FT spectroscopy, a Michelson interferometer is typically used. As shown in Figure 7.25, the light is split into two beams which are directed to two mirrors. One mirror is fixed and the other mirror is moveable. The two optical beams from the beam splitter (BS) are reflected back by both mirrors then the merged optical beam passes through the sample before it reaches the photodetector. When the movable mirror scans in the z-axis, the photodetector picks up an interferogram, $I(z)$, which can be expressed as

$$I(Z) = \frac{1}{2}I(0) + \frac{1}{2}\int_{-\infty}^{\infty} G(k)e^{ikz}dk, \tag{7.30}$$

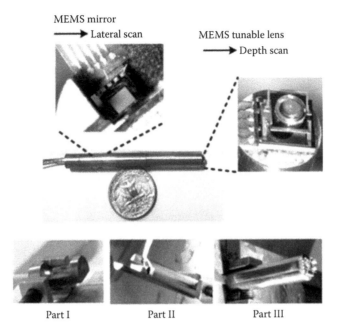

MEMS mirror
Lateral scan

MEMS tunable lens
Depth scan

Part I   Part II   Part III

Figure 7.24 Photographs of an assembled endoscopic probe and its parts. (Adapted from Liu, L. et al. *Sensors and Actuators A: Physical* 215, 2014: 89–95.)

where z is the position of the movable mirror, $I(0)$ is the intensity for a zero path difference, k $(=2\pi/\lambda)$ is the angular wave number, and $G(k)$ is the spectral power density. Since $I(z)$ and $G(k)$ are a Fourier transform pair, the spectral density can be directly recovered through a reverse Fourier transform, that is,

$$G(k) = \frac{1}{\sqrt{2\pi}} \int_{-\infty}^{\infty} [2 \cdot I(z) - I(0)] e^{-ikz} dz \qquad (7.31)$$

The spectral resolution of an FTS is inversely proportional to the scan range of the movable mirror; the larger the mirror scan range, the higher the resolution.

FT spectroscopy has extensive applications for detecting various biological and chemical agents. However, conventional FTS systems are bulky and expensive, and are mostly only for lab use. To meet the requirements of on-site, real-time analysis applications, portable or even handheld miniature FTS must be developed. MEMS provides a solution for miniaturization, but the scan range of MEMS mirrors is typically small, in the order of 10 microns. Fortunately, as introduced in Section

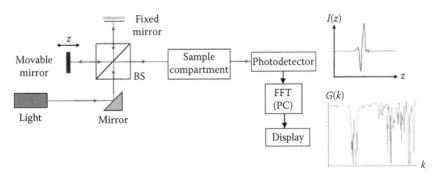

Figure 7.25 Simplified block diagram of an FTS system. BS: Beamsplitter.

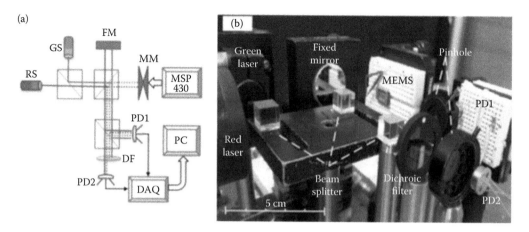

Figure 7.26 An MEMS FTS design. (a) Schematic. (b) Benchtop set-up. (Adapted from Wang, W. et al. *Photonics Technology Letters, IEEE* 27, no. 13, 2015: 1418–1421.)

7.3, the LVD electrothermal bimorph actuators can provide hundreds of microns linear piston motion, which is very suitable for making MEMS FTS. Several such MEMS FTS have been demonstrated [23–25].

Wang et al. reported an LSF-LVD MEMS mirror-based FTS system [24]. The schematic of the MEMS FTS system is shown in Figure 7.26a. The first beam splitter (BS) combines an unknown light (GS) and a reference laser (RS) into a single beam and directs it to the second BS that splits the light beam onto a fixed mirror (FM) and a MEMS movable mirror (MM). Then the reflected light beams from both FM and MM are combined into the third BS, which directs half of the light to the first photodiode (PD1) and the other half to a dichroic filter (DF) which only passes the reference laser to the second photodiode (PD2). The MM is controlled by a microprocessor MSP430 (Texas Instruments). A picture of the implemented FTS is shown in Figure 7.26b. The MEMS mirror employed is shown in Figure 7.27, which is a meshed ISC design with four double-lined ISC bimorph pairs oriented in a redundant configuration. This meshed ISC actuator design can lead to a more repeatable vertical scan response.

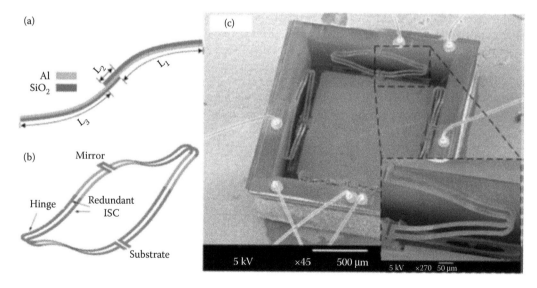

Figure 7.27 The MEMS mirror employed in the MEMS FTS system. (a) S-shaped bimorph. (b) A full mesh ISC actuator structure. (c) SEM of a fabricated mesh ISC MEMS mirror. (Adapted from Wang, W. et al. *Photonics Technology Letters, IEEE* 27, no. 13, 2015: 1418–1421.)

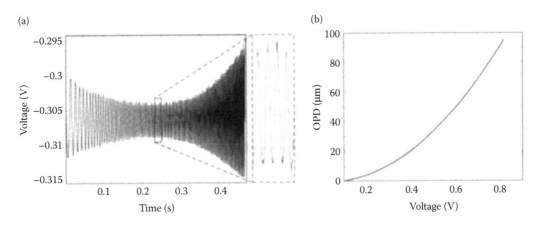

Figure 7.28 (a) The raw interferogram of a reference signal. (b) Mirror displacement versus applied voltage. (Adapted from Wang, W. et al. *Photonics Technology Letters, IEEE* 27, no. 13, 2015: 1418–1421.)

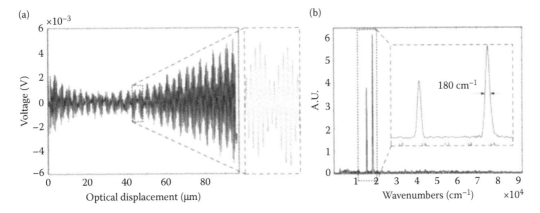

Figure 7.29 (a) The corrected interferogram of the unknown source. (b) The spectrum of the unknown source. (Adapted from Wang, W. et al. *Photonics Technology Letters, IEEE* 27, no. 13, 2015: 1418–1421.)

Just for demonstration purposes, a He-Ne laser ($\lambda = 632.8$ nm, 1125P, JDS Uniphase) is used as the reference light and a green laser diode ($\lambda = 532$ nm, C01440, Lightvision Tech.) is combined with the He-Ne laser as the unknown source. The reference interferogram, whose envelope reflects the mirror tilting, picked up the PD and is plotted in Figure 7.28a. Note that the fringe spacing is not uniform over the scan range, which is caused by the varying velocity of the mirror plate. Since the wavelength of the reference light is known, the vertical displacement of the mirror plate can be precisely determined from the interferogram, as plotted in Figure 7.28b. At 0.81 V, a total 47.5 μm usable mirror displacement is achieved, which leads to a 95 μm optical path difference (OPD). This displacement versus voltage relation in Figure 7.28b is then used to obtain the corrected spatial interferogram of the unknown light source, which is shown in Figure 7.29a. In this case, the unknown source is the reference red laser combined with a green laser diode. The corresponding spectrum curve of the combined source is shown in Figure 7.29b. There are two spectral peaks, 15,800 and 18790 cm^{-1}, or 632.9 and 532.2 nm in wavelength, which represent the center wavelengths of the He-Ne laser and the green laser, respectively. The measured spectral resolution is 180 cm^{-1}, or 5 nm for the 532 nm green light. Later, Wang et al. improved the spectral resolution to 40 cm^{-1}, corresponding to 1.1 nm for the green laser [26]. Recently, this electrothermal MEMS FTS technology has been developed into a portable FTS device and even a handheld device by WiO Tech. (Wuxi, China), as shown in Figure 7.30 [25]. The uniformity and repeatability are much improved due to the foundry fabrication of the MEMS mirrors.

(a) (b) (c)

Figure 7.30 MEMS FTS products. (a) Foundry-fabricated ISC MEMS mirror. (b) Portable MEMS FTS. (Adapted from Xie, H. et al. In *SENSORS, 2015 IEEE*, pp. 1–4. IEEE, 2015.) (c) Handheld MEMS FTS. (Courtesy of WiO Tech.)

7.5 Summary

Based on a simple biomaterial thermal effect, a new class of multi-axis actuators with a large scan range at a low drive voltage has been developed. It has evolved from a simple single beam design, to a double beam LVD design, to a triple beam LSF-LVD design, and eventually to a quadruple beam ISC design. With these innovative actuator designs, unrivalled tip-tilt piston MEMS mirrors with large apertures, high fill factors, large scan ranges, and low drive voltages have been demonstrated and applied to a variety of applications. The laser scanning confocal microscopy with MEMS scanning in all three axes is a full utilization of the large piston and large tip-tilt scanning capability of the LSF-LVD actuator design. The unparalleled, record-setting 1-mm vertical displacement of the LSF-LVD actuator design finds a perfect niche in Fourier transform spectrometers, leading to portable and handheld microspectrometers; there is even a great potential to achieve chip-size Fourier transform spectrometers. In addition to the laser scanning confocal microscopy and Fourier transform microspectrometers, these MEMS mirrors have been applied to various MEMS-based endoscopic microscopy, including endoscopic Optical Coherence Tomography (OCT) [27], two-photon microscopy [28], photoacoustic microscopy (PAM) [29], dual-mode OCT and PAM [30], and dual-mode OCT and DOT (diffuse optical tomography) [31].

Acknowledgments

The authors would like to thank the US National Science Foundation for financial support under the award numbers 1512531, 1002209, and 0901711.

References

1. Bustillo, J. M., R. T. Howe, and R. S. Muller. Surface micromachining for microelectromechanical systems. *Proceedings of the IEEE* 86, no. 8, 1998: 1552–1574.
2. Kovacs, G. T. A., N. I. Maluf, and K. E. Petersen. Bulk micromachining of silicon. *Proceedings of the IEEE* 86, no. 8, 1998: 1536–1551.
3. Freund, L. B. and S. Suresh. *Thin Film Materials: Stress, Defect Formation and Surface Evolution.* Cambridge: Cambridge University Press, 2004.
4. Thornton, J. A. and D. W. Hoffman. Stress-related effects in thin films. *Thin Solid Films* 171, no. 1, 1989: 5–31.

5. Chu, W.-H., M. Mehregany, and R. L. Mullen. Analysis of tip deflection and force of a bimetallic cantilever microactuator. *Journal of Micromechanics and Microengineering* 3, no. 1, 1993: 4.
6. Peng, W., Z. Xiao, and K. R. Farmer. Optimization of thermally actuated bimorph cantilevers for maximum deflection. *Nanotech Proceedings* 1, 2003: 376–379.
7. Todd, S. T. and H. Xie. Steady-state 1D electrothermal modeling of an electrothermal transducer. *Journal of Micromechanics and Microengineering* 15, no. 12, 2005: 2264.
8. Pal, S. and H. Xie. A parametric dynamic compact thermal model of an electrothermally actuated micromirror. *Journal of Micromechanics and Microengineering* 19, no. 6, 2009: 065007.
9. Xie, H., Y. Pan, and G. K. Fedder. Endoscopic optical coherence tomographic imaging with a CMOS-MEMS micromirror. *Sensors and Actuators A: Physical* 103, no. 1, 2003: 237–241.
10. Jain, A., A. Kopa, Y. Pan, G. K. Fedder, and H. Xie. A two-axis electrothermal micromirror for endoscopic optical coherence tomography. *Selected Topics in Quantum Electronics, IEEE Journal of* 10, no. 3, 2004: 636–642.
11. Jain, A., H. Qu, S. Todd, and H. Xie. A thermal bimorph micromirror with large bi-directional and vertical actuation. *Sensors and Actuators A* 122, no. 1, 2005: 9–15.
12. Jain, A. and H. Xie. A single-crystal silicon micromirror for large bi-directional 2D scanning applications. *Sensors and Actuators A: Physical* 130, 2006: 454–460.
13. Wu, L. and H. Xie. A large vertical displacement electrothermal bimorph microactuator with very small lateral shift. *Sensors and Actuators A: Physical* 145, 2008: 371–379.
14. Sun, J., S. Guo, L. Wu, L. Liu, S.-W. Choe, B. S. Sorg, and H. Xie. 3D *in vivo* optical coherence tomography based on a low-voltage, large-scan-range 2D MEMS mirror. *Optics Express* 18, no. 12, 2010: 12065–12075.
15. Zhang, X., R. Zhang, S. Koppal, L. Butler, X. Cheng, and H. Xie. MEMS mirrors submerged in liquid for wide-angle scanning. In *Solid-State Sensors, Actuators and Microsystems (TRANSDUCERS), 2015 Transducers-2015 18th International Conference on*, Anchorage, Alaska, pp. 847–850. IEEE, 2015.
16. Zhang, X., L. Zhou, and H. Xie. A fast, large-stroke electrothermal MEMS mirror based on Cu/W bimorph. *Micromachines* 6, no. 12, 2015: 1876–1889.
17. Pal, S. and H. Xie. Fabrication of robust electrothermal MEMS devices using aluminum–tungsten bimorphs and polyimide thermal isolation. *Journal of Micromechanics and Microengineering* 22, no. 11, 2012: 115036.
18. Jia, K., S. Pal, and H. Xie. An electrothermal tip–tilt–piston micromirror based on folded dual S-shaped bimorphs. *Journal of Microelectromechanical Systems* 18, no. 5, 2009: 1004–1015.
19. Zhang, X., B. Li, X. Li, and H. Xie. A robust, fast electrothermal micromirror with symmetric bimorph actuators made of copper/tungsten. In *Solid-State Sensors, Actuators and Microsystems (TRANSDUCERS), 2015 Transducers-2015 18th International Conference on*, Shanghai, China, pp. 912–915. IEEE, 2015.
20. Minsky, M. Microscopy apparatus. U.S. Patent 3,013,467, issued December 19, 1961.
21. Stephens, D. J. and V. J. Allan. Light microscopy techniques for live cell imaging. *Science* 300, no. 5616, 2003: 82–86.
22. Liu, L., E. Wang, X. Zhang, W. Liang, X. Li, and H. Xie. MEMS-based 3D confocal scanning microendoscope using MEMS scanners for both lateral and axial scan. *Sensors and Actuators A: Physical* 215, 2014: 89–95.
23. Wu, L., A. Pais, S. R. Samuelson, S. Guo, and H. Xie. A miniature Fourier transform spectrometer by a large-vertical-displacement microelectromechanical mirror. In *Fourier Transform Spectroscopy*, p. FWD4. Optical Society of America, Vancouver, Canada, 2009.
24. Wang, W., S. R. Samuelson, J. Chen, and H. Xie. Miniaturizing Fourier Transform Spectrometer With an Electrothermal Micromirror. *Photonics Technology Letters, IEEE* 27, no. 13, 2015: 1418–1421.
25. Xie, H., S. Lan, D. Wang, W. Wang, J. Sun, H. Liu, J. Cheng et al. Miniature fourier transform spectrometers based on electrothermal MEMS mirrors with large piston scan range. In *SENSORS, 2015 IEEE*, Busan, South Korea, pp. 1–4. IEEE, 2015.

26. Wang, W., J. Chen, A. S. Zivkovic, C. Duan, and H. Xie. A silicon based Fourier transform spectrometer base on an open-loop controlled electrothermal MEMS mirror. In *Solid-State Sensors, Actuators and Microsystems (TRANSDUCERS), 2015 Transducers-2015 18th International Conference on*, Anchorage, Alaska, pp. 212–215. IEEE, 2015.
27. Samuelson, S. R., L. Wu, J. Sun, S.-W. Choe, B. S. Sorg, and H. Xie. A 2.8-mm imaging probe based on a high-fill-factor mems mirror and wire-bonding-free packaging for endoscopic optical coherence tomography. *Microelectromechanical Systems, Journal of* 21, no. 6, 2012: 1291–1302.
28. Fu, L., A. Jain, H. Xie, C. Cranfield, and M. Gu. Nonlinear optical endoscopy based on a double-clad photonic crystal fiber and a MEMS mirror. *Optics Express* 14, no. 3, 2006: 1027–1032.
29. Xi, L., J. Sun, Y. Zhu, L. Wu, H. Xie, and H. Jiang. Photoacoustic imaging based on MEMS mirror scanning. *Biomedical Optics Express* 1, no. 5, 2010: 1278–1283.
30. Xi, L., C. Duan, H. Xie, and H. Jiang. Miniature probe combining optical-resolution photoacoustic microscopy and optical coherence tomography for in vivomicrocirculation study. *Applied Optics* 52, no., 9, 2013: 1928–1931.
31. Yang, H., L. Xi, S. Samuelson, H. Xie, L. Yang, and H. Jiang. Handheld miniature probe integrating diffuse optical tomography with photoacoustic imaging through a MEMS scanning mirror. *Biomedical Optics Express* 4, no. 3, 2013: 427–432.

MEMS tunable optics
Liquid and solid methods

Yongchao Zou and Guangya Zhou

Contents

8.1 Introduction of MEMS tunable lenses and apertures

As the key components in optical systems, lenses and apertures (sometimes called "diaphragms" or "irises") have never stepped out of the spotlight in both academia and industrial research and development areas since their first appearance. Much effort is spent to not only improve the performances of these devices, but also scale down their dimensions and further equip them with tunability in order to meet the increasing demand for miniature camera modules integrated in various portable devices, including mobile phones, medical imaging systems, and surveillance systems. It is apparent that compact tunable lenses and apertures can provide these imaging systems with optical zooming/autofocus functions and improved image quality without sacrificing their slim profiles, which were never possible before with conventional optical devices.

To distinguish these from conventional bulky zoom lenses which achieve optical power variation by mechanically moving the lens elements (or lens groups) along the optical axis, MEMS tunable lenses here refer to those compact ones which realize optical power tuning by modulating their geometric shapes or refractive indices with the assistance of various MEMS actuation mechanisms. Thanks to the rapid development of microelectronics as well as micro-/nano-machining technologies during the past years, various compact tunable lens designs have become possible, as summarized in Figure 8.1. Based on the materials used for the lens implementation, compact MEMS tunable lenses are divided into two categories, namely solid tunable lenses and liquid tunable lenses. The former type can be further classified into two groups according to their different working principles, which are based on the modulation of overall thicknesses and the tuning of geometric shapes, respectively. The examples include Alvarez lenses and thermal-actuated tunable solid lenses. The liquid tunable lenses also consist of two different types of tuning principles, one of which is to modulate the refractive index or index distribution of the lens (such as liquid crystal lenses), and the other one is to alternate the geometric shape of the lens (such as various MEMS-based membrane lenses). Currently, there are extensive technical and review papers on liquid tunable lenses, but few on the solid ones.

The optical aperture (or diaphragm), as one of the most basic elements in optical imaging systems, is used to control the luminous flux reaching the focal plane and field of view, as well as the depth of focus of the system. It is apparent that a tunable optical aperture (also known as the iris diaphragm in modern cameras) is necessary for an imaging system to handle various scenes in order to balance the overall energy received by the imaging sensor. The most common technology used in present cameras is based on movable blades, simulating the iris of the human eye. This idea inspired engineers to develop the first kind of MEMS tunable apertures, which consist of a few blades controlled by individual MEMS actuators. By precisely controlling the positions of the blades, the size of the aperture formed by these blades is electrically adjustable. Depending on the motion of the blades, these MEMS solid tunable apertures can be further classified as sliding blades- and rotating blades-based apertures, as listed in Figure 8.2. Similar to liquid tunable lenses, the second kind of tunable apertures is based on optofluidic technology, which achieves a continuously adjustable aperture size by modulating the diameter of a circular opening formed by a dyed liquid with diverse mechanisms. There are no moving mechanical parts involved in such a liquid tunable aperture in contrast to its counterparts. To alter the size of such a liquid tunable aperture, one of the possible ways is to use electrostatic force or the electrowetting effect, while the other is based on deformable membranes driven by pressure.

In this chapter, we will start with the liquid tunable optics, and then move on to the recently emerged solid tunable optical devices. A review of these technologies together with references is provided. Furthermore, basic working principles, fabrication processes, and performances of some typical designs are discussed in detail to demonstrate their capabilities.

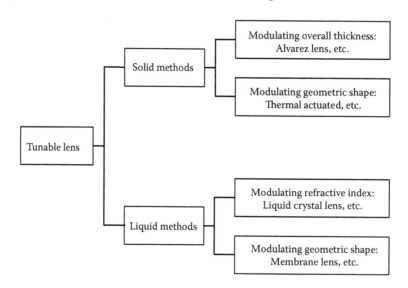

Figure 8.1 Summary of various tunable lenses.

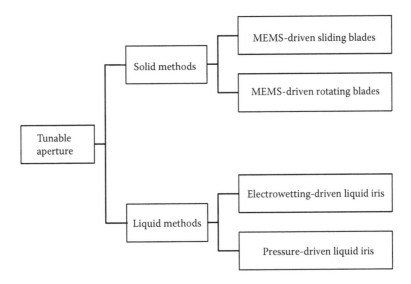

Figure 8.2 Summary of various tunable apertures.

8.2 Liquid tunable lenses

8.2.1 An overview of liquid tunable lenses

We have the experience of observing water droplets on windows or leaves, and it is not surprising to find the optical magnification effect of the water droplet. The reason is obvious, that is, the water droplet forms a natural lens as its shape is spherical due to the surface tension. It is easy to understand from basic optics that the focal length is changed once its radius or refractive index is varied due to temperature or pressure variations. Based on this idea, various liquid tunable lenses are proposed and demonstrated experimentally.

As mentioned above, to adjust the focal length of a lens, one needs to change the geometric profile of the lens or the refractive index of the lens material (Nguyen, 2010; Chiu et al., 2012). There are extensive reports to achieve a liquid tunable lens with deformable geometric shapes (Ahn and Kim, 1999; Chronis et al., 2003; Zhang et al., 2003, 2004; Agarwall et al., 2004; Chen et al., 2004; Jeong et al., 2004; Pang et al., 2005; Werber and Zappe, 2005; Cheng et al., 2006; Yu et al., 2008; Feng and Chou, 2009; Zhou et al., 2009; Leung et al., 2010). The first typical method is based on flexible membranes (Zhang et al., 2003; Werber and Zappe, 2005; Yu et al., 2009). More specifically, such a liquid tunable lens consists of a chamber holding the liquid and one deformable membrane to seal the chamber. One external or integrated pump is connected to the chamber through a narrow channel, and deforms the membrane with different pressures. Consequently, the focal lengths are tuned to various values due to the curvature variation of the lens surface profile. Miniature actuators, such as a piezoelectric actuator, can also be used to deform the membrane by pushing or pulling the side walls of the chamber (Oku et al., 2004). In addition, electric actuation mechanisms, including piezoelectric, electromagnetic, and electrochemical actuators, have also been developed to modulate the shape of the membrane (Lopez et al., 2005; Ren et al., 2006; Lee and Lee, 2007; Ren and Wu, 2007; Choi et al., 2009; Lee et al., 2009).

Besides the membrane-based liquid tunable lenses, there are also some other novel configurations. For example, temperature and pH value sensitive materials are used to configure various tunable lenses based on their capabilities of shape changing or refractive index variation as a function of the surrounding temperature or pH value. In 2006, Dong et al. proposed liquid tunable lenses using hydrogels which expand with decreasing temperature or increasing pH value (Dong et al., 2006). More specifically, a piece of ring-shaped hydrogel is used to contain water, and the whole configuration is covered by a layer of oil. The lens is formed by the interface between these two layers of different immiscible liquid materials. By varying the temperature or pH values, the hydrogel expands or contracts, leading to a pressure change and hence a curvature variation of the lens formed by the interface. The typical response time of such tunable lenses is in the range of tens of seconds. Zeng and Jiang also use a similar configuration to achieve tunable lenses, but the hydrogel is replaced by an IR light sensitive one (Zeng and Jiang, 2008).

Electrowetting, which can be used to modify the wetting properties of a surface with an applied electric field, is also employed to tune the shape of a liquid lens (Berge and Peseux, 2000; Berge, 2005; Lopez et al., 2005; Fan et al., 2009). A driving voltage can be applied across the conductive liquid and solid interface to modulate the contact angle, hence varying the lens shape for various focal lengths. In the following section, we will discuss this type of tunable lens in detail. In addition, dielectrophoresis is also used by the researchers in this field to achieve a tunable lens configuration. It is reported that dielectrophoresis forces on dielectric particles in a nonuniform electric field are able to change the curvature of the interface between two layers of different liquid materials (Cheng and Yeh, 2007). Different from electrowetting generating a field force, this effect generates a body force phenomenon, and hence consumes one-order-of-magnitude-lower power than the electrowetting one. Cheng et al. and Ren et al. experimentally demonstrated this idea with different dielectric materials (Cheng and Yeh, 2007; Ren and Wu, 2008). Another example is to use hydrodynamic force to tune the lens shape. Specifically, in a liquid-core-and-cladding configuration, the ratio between the widths of the core and cladding is related to the flow rate. Therefore, the flow rate can be used to modulate the ratio between the widths of the core and cladding and hence change the curvature of the interface formed by the core liquid and cladding liquid. Tang, and Seow et al. demonstrated tunable lenses using this configuration (Seow et al., 2008; Tang et al., 2008).

Rather than modulating the geometric shape of the lens, one can also achieve a liquid tunable lens by tuning the refractive index of the lens material. One of the typical examples is liquid crystal lenses (Lin et al., 2011a). Liquid crystal is a kind of single-axis optical material due to its rod-like molecule structure, which offers two distinct refractive indices for light with different polarizations. A typical liquid crystal lens is made of one or multiple liquid crystal cells, which consist of two indium tin oxide (ITO) glass substrates coated with alignment layers and sandwiching the liquid crystal material. The effective refractive index of the liquid crystal is modulated by the voltage applied to the cell as the orientations of the rod-like liquid crystal molecules are related to the electrical field. More detailed discussion is given in the following section.

Some other methods used to modify the refractive index of a lens include applying the liquid with an external electrical field, acoustic field, mechanical strain or temperature field as the optical properties of most of the materials are related to these external fields (Erickson et al., 2005). Another effective way to change the refractive index of a lens material is to use mixtures with variable concentrations (Erickson et al., 2005). In the section below, we will discuss some typical examples of liquid tunable lenses.

8.2.2 Typical examples: Membrane lenses, liquid crystal lenses, and electrowetting-driven lenses

Figure 8.3 schematically presents a general membrane lens, which consists of a circular chamber holding the liquid and sealed by a deformable membrane. The basic idea of such a liquid tunable lens is to modulate the geometric shape of the membrane which is, in most cases, made of soft materials such as PDMS, which changes the radius of the curvature of the lens. Usually, such a radius change is implemented through pressurizing the liquid inside the chamber. In addition to the planoconvex design as shown in the schematic, there are also planoconcave, biconvex, biconcave, and convex/concave designs, depending on the number of flexible membranes and driving pressures applied in the configuration (Chronis et al., 2003; Agarwall et al., 2004; Chen et al., 2004; Jeong et al., 2004; Zhang et al., 2004; Pang et al., 2005). In addition, interesting tunable double-focus lens designs are also reported using the pressure-driven membrane lens configuration (Leung et al., 2009; Yu et al., 2009). Instead of utilizing a uniform membrane in the configuration, a membrane with two different thicknesses is used in this case. When the lens liquid is subject to one driving pressure, the membrane is deformed to two different spherical profiles, which gives us two separate focal lengths.

As shown in Figure 8.3, one commonly used method to drive the deformable membrane is to apply the pressure to the lens liquid through a narrow channel. It is noted that such a design faces the problems of possible leakage and evaporation. One variation is to seal the chamber completely and drive the lens liquid through electric methods, including using integrated piezoelectric, electromagnetic, and electrothermal actuators for applying the pressure (Lopez et al., 2005; Ren et al., 2006; Lee and Lee, 2007; Ren and Wu, 2007; Choi et al., 2009; Lee et al., 2009).

Most of these pressure-driven membrane lenses can respond in milliseconds. For those driven by integrated actuators, the response speed of the tunable lens is mostly restricted by the actuators. The optical power of the liquid lens can be as large as 10^3 diopters with a tuning ratio of 10 (Werber and Zappe, 2005). The image quality of the membrane lenses reported varies greatly from device to device, and there is not much research work reported that focuses on the details of the aberration analysis and the technical methods for improvement. In addition, due to the liquid materials involved in the configuration, membrane lenses are confronted with a few practical challenges, including a complex fabrication process, vibration and thermal instability, and possible leakage and evaporation.

Nematic liquid crystal is one kind of excellent uniaxial optical material due to its rod-shaped molecule structures and roughly parallel orientations (Sato, 1979). As the basic component to form tunable lenses, one liquid crystal cell generally consists of two indium tin oxide (ITO) glass substrates coated with alignment layers sandwiching the liquid crystal material, as shown in Figure 8.4 (Wang et al., 2005, 2006; Ye et al., 2006). Once there is an electric field applied to the liquid crystal cell, the orientations of the rod-shaped molecules are changed, leading to a variation of the effective refractive index. For a linearly polarized incident light, the effective refractive index is determined by (Lin et al., 2011a)

$$n_{eff} = n_e n_o \sqrt{\frac{1}{n_o^2 \cos^2\theta + n_e^2 \sin^2\theta}} \qquad (8.1)$$

where θ is the angle between the polarization direction and the long axis of the liquid crystal molecules. n_e and n_o represent the extraordinary and ordinary refractive indices, respectively. Hence,

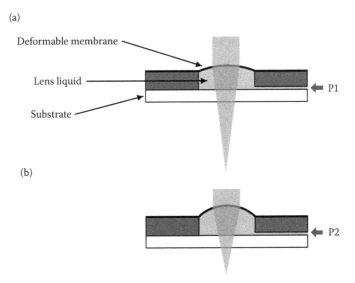

Figure 8.3 Schematic of the tunable membrane lenses driven by pressure: (a) with pressure P1 and (b) with pressure increased to P2.

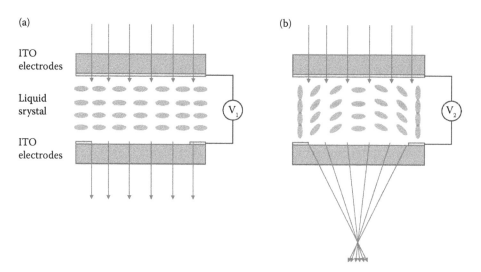

Figure 8.4 Schematic of the liquid crystal lenses. (a) and (b) show the reorientation of the liquid crystal molecules as the voltage changes.

the effective refractive index can be modulated from n_e to n_o by increasing the tilt angle of the liquid crystal molecules from 0° to 90° with a growing driving voltage.

There are two basic implementations of liquid crystal lenses. As shown in Figure 8.4, one of the methods is to use a homogeneous cell gap and hence two planar lens surfaces, but the surfaces are driven by inhomogeneous electric fields with discrete pads (Sato, 1999; Ye and Sato, 2002, 2003; Ye et al., 2004, 2006, 2008; Wang et al., 2006; Lin et al., 2011b; Chen and Lin, 2013). By driving the liquid crystal molecules with different tilt angles, a convergent or divergent phase-delay distribution can be achieved, resulting in a positive or negative lens, respectively. The other one is to implement the lens as a planococave/planocovex configuration with a curved indium tin oxide glass substrate (Choi et al., 2003; Ji et al., 2003; Jae-Hoon and Satyendra, 2004; Ren et al., 2004; Ren and Wu, 2005; Dai et al., 2009). Only one uniform electric field is required to simultaneously change the orientation of all the liquid crystal molecules. In this case, the driving mechanism is relatively simple, but the lens can only perform as a positive one or a negative one, depending on the initial design of the lens.

The response speed of liquid crystal lenses is related to the cell gap of the lens. Normally, such lenses can respond in seconds (Nguyen, 2010). It is easy to understand that a large optical power requires a large phase delay and hence a large cell gap. However, increasing the cell gap obviously leads to a longer response time. Therefore, one of the challenges of liquid crystal lens design is how to balance the conflict between the largest optical power (as well as the tuning range) and the response time of the lens. In addition, the polarization dependency of liquid crystal material poses another challenge for engineers. The complex optical configuration and low optical efficiency also limit the performance of such a lens (Lin et al., 2011a). To solve these problems, a few polarization-independent liquid crystal lenses have been reported in recent years (Ye et al., 2006; Lin and Chen, 2013; Hsu et al., 2014; Saito et al., 2015).

It is widely known that the shape of a liquid droplet is determined by the surface tension if there is no external electric field. Once a voltage is applied to the conductive droplet placed on an electrode surface, the contact angle, as shown in Figure 8.5a, is determined as follows (Mugele and Baret, 2005; Zeng and Jiang, 2013):

$$\cos\theta = \cos\theta_0 + \frac{\varepsilon_m}{2d\sigma_{lv}}(V - V_0) \tag{8.2}$$

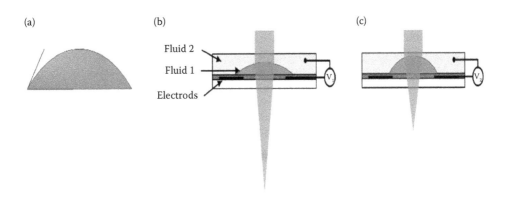

(a)   (b)   (c)

Fluid 2

Fluid 1

Electrods

Figure 8.5 Schematic of the liquid tunable lens driven by electrowetting. (a) shows the contact angle, (b) and (c) show the variation of the focal length as the voltage changes.

where θ_0 and θ are the contact angles before and after the external voltage is applied. V and V_0 are the applied voltage and initial potential difference; d is the thickness of the induced charge layer (electrical double layer), and ε_m is the permittivity of the material. σ_{lv} is the surface tension between the liquid and the electrode surface. It is apparent that the curvature of the liquid droplet is related to the contact angle, and hence can be modulated by the external voltage applied to the liquid.

Most of the electrowetting-driven lenses are implemented as shown in Figure 8.5b (Berge and Peseux, 2000; Krupenkin et al., 2003; Kuiper and Hendriks, 2004; Grilli et al., 2008; Li and Jiang, 2012). One droplet of lens liquid is positioned on the substrate and immersed in an electrolyte, as shown in the schematic. Thus, the lens is a combination of one positive lens formed by the lens liquid and a negative one formed by the electrolyte. Due to the refractive index difference between these two materials, an effective positive or negative lens is achieved. By applying an external voltage to the electrolyte, the contact angle of the interface between these two materials is changed, which leads to a focal length variation.

Electrowetting-driven lenses can respond in a few tens of milliseconds (Nguyen, 2010). The largest optical power of such a lens can reach a few hundreds of diopters, but the tuning range is relatively small due to the inherent limitations of the operation principle. The image quality of such lenses is similar to that of membrane lenses. Because of the liquid involved in the lens configuration, electrowetting-driven lenses also pose a series of practical challenges to engineers, including the complex fabrication process, possible leakage and evaporation, and performance instability.

8.3 Liquid tunable apertures

8.3.1 Methods: Pressure-driven and electrowetting

It is widely noticed that the irises in human eyes can automatically enlarge or shrink the pupil in response to light levels. Bright light causes a small pupil size, and on the contrary, darkness leads to a relative large pupil size. Inspired by the irises of human eyes, moveable blades-based iris diaphragms are widely used in modern cameras to control the light energy reaching the image sensors, as well as the field of view and depth of focus of the imaging system. Such an iris diaphragm consists of a few moveable blades linked by a rotary actuation mechanism. By simultaneously moving the blades, the polygonal aperture formed by these blades is variable in size. However, the relatively complicated mechanical configurations and driving mechanism as well as the possible friction encountered in operation greatly limit its applications in the emerging miniature imaging systems. To solve this problem, various miniature tunable aperture designs have been reported in recent years. In this section, we will briefly discuss some of the typical examples using optofluidics.

Similar to the aforementioned liquid tunable lenses, Yu et al. reported one variable aperture based on optofluidic technology (Hongbin et al., 2008). As shown in Figure 8.6a, a cavity filled with dyed liquid is sandwiched between two transparent substrates. In addition, there is an air pressure chamber designed below the cavity, where a flexible membrane is used to seal the liquid. Through a microchannel connected to the air pressure chamber, the flexible membrane will be deformed by pressurizing the air inside the chamber to form various aperture sizes. The results show that the aperture diameter can be varied from 0 to 6.35 mm with an excellent repeatability. Draheim et al. reported a similar membrane-based tunable aperture in 2011 (Draheim et al., 2011). Instead of using the pneumatic driving method, a piezo bending actuator is used to deform the flexible membrane for various aperture sizes in this case. In the membrane-based tunable apertures, the large amount of liquid may increase the net weight of the whole configuration, which may limit their applications in miniature imaging systems.

Various electric tuning methods are also employed to achieve tunable apertures, one typical example of which is the electrowetting-driven adaptive iris. As shown in Figure 8.6b, the basic configuration of such a tunable aperture consists of a chamber holding two different liquids, which are typically one kind of dyed liquid for light absorption and one kind of transparent oil. The bottom of the chamber is coated with conductive electrodes followed by a layer of low-friction material such as Teflon. By applying voltages to the liquid, the contact angle between the liquid droplet and the solid substrate is changed, leading to various aperture sizes. The examples are listed in the references (Murade et al., 2011; Muller et al., 2012; Li et al., 2013a,b). Furthermore, the dielectric force, which is generated by an external electric field on the dielectric liquid-liquid interface, is also used to realize tunable apertures (Tsai and Yeh, 2010; Xu et al., 2015). In addition, electromagnetic and electrostatic actuation mechanisms can also be used to achieve the tunable apertures (Chang et al., 2013; Seo et al., 2015).

Capillary force, which can lift a liquid in narrow tubes, is also reported for achieving tunable apertures. The configuration normally consists of narrow capillary channels filled with two non-mixing fluids, one opaque and one transparent. By controlling the position of the opaque liquid, various aperture sizes can be achieved (Muller et al., 2010; Kimmle et al., 2011). Due to the inherent property of the capillary force, such designs may have slow responses. In 2014, Schuhladen et al. reported a novel iris-like tunable aperture employing liquid crystal elastomers (Schuhladen

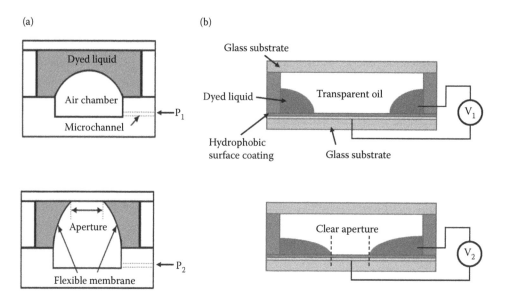

Figure 8.6 (a) Schematic of membrane-based tunable aperture driven by pressure and (b) electrowetting-driven tunable aperture.

et al., 2014). It is noted that the liquid crystal elastomers can deform reversibly and are expected to be applied in artificial muscles. Inspired by this actuation property, they implement a circularly symmetric actuation mechanism to drive a radially oriented liquid crystal elastomers plate. By driving the ring-shaped liquid crystal elastomers plate to expand or recover to the original shape, a tunable aperture is achieved. There is no liquid involved in this configuration and no extra optical interface at the open pupil, which promises a relative stable performance and a 100% transmission compared with its counterparts. Such a design provides a reported tuning range from 2.6 mm to 3.8 mm, with a relatively slow response time of a few tens of seconds.

8.4 MEMS solid tunable lenses

Because of the liquid involved, liquid tunable lenses face some challenges, including complex fabrication processes, vibration and thermal instability, and possible leakage and evaporation. In addition, limited by the driving mechanisms, the aperture sizes of the majority of the liquid tunable lenses are relatively small. Although pneumatic tunable lenses may have a relatively large aperture size, the required liquid volume may greatly increase the mass and thickness of the whole lens configuration as well as the gravitational effect, and hence limit their applications. In recent years, a novel type of tunable lenses involving pure solid-state materials has drawn increasing attention as such solid tunable lenses offer a few obvious advantages over the liquid ones. In the absence of the liquid materials, the solid tunable lenses perform more robustly and can be fabricated with relatively simple processes. In addition, there are no concerns about the liquid leakage, evaporation or freezing under low temperature operation. Lee et al. reported a thermal-actuated solid tunable lens in 2016 (Lee et al., 2006). The lens is made of PDMS surrounded by a silicon ring. A microthermal actuator is used to heat the lens and silicon ring. Due to the mismatch of the thermal expansion coefficients of these two materials, the lens shape is deformed during heating. The experimental results show that a relatively small tuning range of 0.834 mm is achieved. Besides the thermal effect, external strains can also be employed to deform the solid lens materials to achieve focal length tuning (Beadie et al., 2008; Santiago-Alvarado et al., 2011; Liebetraut et al., 2013). Limited by the maximum deformation that can be achieved by either thermal actuation or external strain, such a type of shape-deformation-based solid tunable lenses can only provide a quite limited optical power tuning range.

Alvarez and Lohmann proposed another type of solid tunable lenses a few decades ago (Alvarez, 1967; Lohmann, 1970). Such a tunable lens consists of two elements, each having an optical flat surface and a cubic surface (freeform). By shifting these two elements by a small lateral displacement with respect to each other, perpendicular to the optical axis, the whole configuration behaves like a lens with the focal length related to the lateral displacement. Thanks to the rapid development of precision machining technology during the past decades, the realization of the cubic freeform surfaces has become possible. During the past a few years, both refractive and diffractive solid tunable lenses based on the Alvarez-Lohmann principle have been experimentally demonstrated (Barton et al., 2000; Huang et al., 2008; Smilie et al., 2012). It is noticed that such a compact solid tunable lens can achieve a large focal length tuning range together with a promising imaging performance, and hence shows great potential in miniature imaging systems. In this section, we will discuss the state of the art of this novel kind of solid tunable lenses in detail.

8.4.1 Tunable Alvarez lenses

The schematic of the tunable Alvarez lens is shown in Figure 8.7. Such a lens basically consists of two lens elements, each having a flat surface and one cubic freeform surface (Alvarez, 1970). Their thicknesses are, respectively,

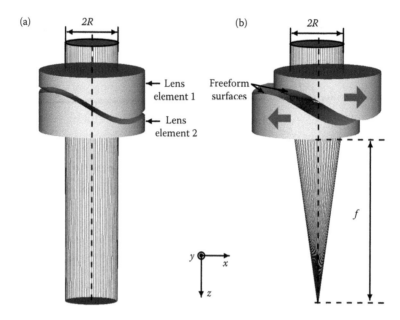

Figure 8.7 Schematic of the Alvarez lens. Lens elements are arranged (a) with lateral displacements and (b) without lateral displacements. (Copyright Optical Society of America. Reproduced with permission.)

$$t_1 = A\left(xy^2 + \frac{1}{3}x^3\right) + Dx + E_1$$

$$t_2 = -A\left(xy^2 + \frac{1}{3}x^3\right) - Dx + E_2$$

(8.3)

where A, D, E_1, and E_2 are the constants to be determined. Clearly, when these two elements are aligned perfectly as shown in Figure 8.7a, the whole configuration behaves like an optical plate with no optical power. However, when the two elements move a displacement of δ in the x and $-x$ directions, respectively, as shown in Figure 8.7b, the corresponding thicknesses of the two lens elements at position (x,y) change to

$$t_1 = A\left[(x-\delta)y^2 + \frac{1}{3}(x-\delta)^3\right] + D(x-\delta) + E_1$$

$$t_2 = -A\left[(x+\delta)y^2 + \frac{1}{3}(x+\delta)^3\right] - D(x+\delta) + E_2$$

(8.4)

which leads to an overall phase delay of $-2A(n-1)\delta(x^2 + y^2)$ by combining t_1 and t_2. The generated optical power makes the whole configuration equivalent to a lens with a focal length determined by

$$f = \frac{1}{4A\delta(n-1)}$$

(8.5)

where n is the refractive index of the material. It is further noticed that the focal length is modulated by the lateral displacement δ, and the value of A controls the "speed" of the tuning. E_1 and E_2 are constants that determine the central thickness of the element and contribute nothing to the optical performance. D is the coefficient of the tilt term, which can be used to reduce the overall thickness of the element, and hence improve the lens performance, as pointed out by Alvarez (1970).

One of the earliest materialized Alvarez lenses was reported by Barton et al., (2000). The two lens elements work in a diffractive way and are fabricated using standard multistep photolithography with 16 phase levels. The concept of the Alvarez lens principle is also used in the design of microscope objectives and accommodative intraocular lenses (Rege et al., 2004; Simonov et al., 2006). In 2008, a micro Alvarez lens array was reported (Huang et al., 2008). The Alvarez lens array is fabricated using a 5-axis ultra-precision diamond tuning followed by an injection molding process. The focal length of the micro Alvarez lens array is tuned from about 18 mm to 8 mm with a lateral displacement of about 0.11 mm.

To improve the Alvarez lens performance and guide the lens element design, Barbero analyzed the optical performance of the Alvarez-Lohmann lenses with different configurations, and proposed the optimum coefficient selection method (Barbero, 2009). For optimum lens performance, it was pointed out that the value of D should be chosen as the one that can minimize the overall lens element thickness. After a series of mathematical derivations, the equation for optimum D value selection is given by

$$D = -\frac{AR^2}{3} \tag{8.6}$$

where R is the radius of the lens element. In addition, it is shown that the configuration with inner cubic surfaces is superior to the one with the outer cubic surfaces.

In recent years, we have focused on the improvement of the design methods, fabrication technologies, driving mechanisms of miniature solid tunable lenses based on the Alvarez principle, and their applications in endoscopy and miniature cameras. We have spent much effort in integrating these lenses with various MEMS actuators and exploring their applications in miniature imaging systems. In 2013, the first MEMS-driven miniature adaptive Alvarez lens was reported (Zhou et al., 2013). The lens elements are fabricated by a single-point diamond turning technology followed by a molding replication process, as illustrated in Figure 8.8. The freeform surface is first achieved on top of a metallic mold using the single-point diamond turning technology, and then a standard PMDS replication process is employed to achieve a second mold with the inverse pattern, as shown in Figure 8.8b. After that, a droplet of UV adhesive is used to fill the concave chamber and it is sealed by another flat PDMS plate, as shown in Figure 8.8c. The whole configuration is then placed under a UV light source for hardening. Once hardened, the lens element can be easily separated from the PDMS mold, as shown in Figure 8.8d. The comb drive MEMS actuator is fabricated using a standard silicon-on-insulator (SOI) micromachining process. Photos of the fabricated MEMS

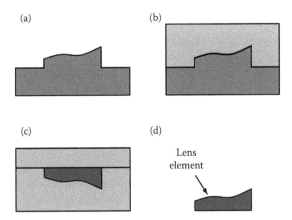

Figure 8.8 Fabrication process of the lens element. (a) Metallic mold from diamond turning, (b) PDMS replication, (c) lens element molding, and (d) final lens element. (Copyright Optical Society of America. Reproduced with permission.)

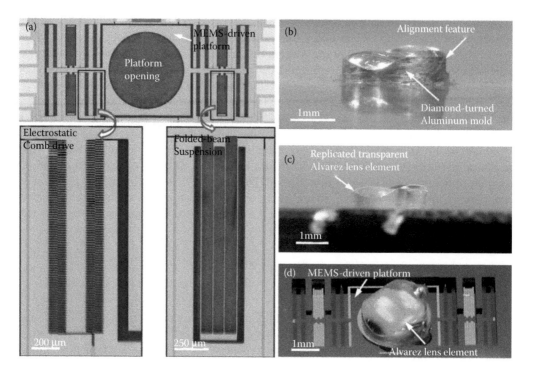

Figure 8.9 (a) MEMS actuator, (b) metallic mold, (c) lens element, and (d) final Alvarez lens after assembly. (Copyright Optical Society of America. Reproduced with permission.)

actuator, freeform surface mold, lens element, and MEMS-driven lens element after assembly are shown in Figure 8.9. The experimental results show that a dynamic tuning of the focal length of more than 1.5 times (from 3 mm to 4.65 mm) is achieved with a lateral displacement of 40 μm provided by the MEMS actuator. However, the image quality is found to be unsatisfactory due to the possible misalignments during the assembly.

To address this issue, numerical studies of the alignment tolerances of Alvarez lenses, imaging performances with various surface parameters, and development of the optimum lens design methods were subsequently conducted (Zou et al., 2013). The Alvarez lens with the surface profiles governed by Equation 8.3 is first established in Zemax® (13 EE). The effects of all lens parameters on the lens performance are then numerically analyzed. The results show that the gap between the two elements should be minimized for the best imaging performance, and at the same time, the lens performance also slightly degrades with the increase of the lateral displacements (namely, the increase of optical power). In addition, although it is obvious that parameter A and the lateral displacement δ mathematically play an equivalent role in Equation 8.5 for a particular focal length, reducing the value of A should be taken as a higher design priority than reducing the maximum lateral displacement. This is for better aberration control and lens performance.

The misalignment tolerances of the lens are also studied. It is shown that the lens performance is most sensitive to misalignment in the y direction. For those MEMS-driven Alvarez lenses, the misalignment in the y direction should be well controlled below 10 μm for a satisfactory lens performance. Among all the tilt errors, the tilt about the z axis causes the most serious performance degradation. It is shown that the tilt errors about any axis should not exceed 1° to maintain the lens performance at a satisfactory level. Furthermore, reducing the value of A can also help to slightly improve the alignment tolerance. Finally, the optimal coefficient selection method is analytically given and numerically investigated. Alvarez and Barbero pointed out that the minimum lens element thickness leads to the optimum lens performance, based on which the value of D is determined by Equation 8.6. However, both experimental and simulation results show that it is the air gap between the two elements rather than the lens thickness that dominantly affects the lens performance. Hence, based on this principle, the optimal value of D is given by (Zou et al., 2013)

$$D_{optimal} = \begin{cases} -\dfrac{A\delta_{max}^2}{3} - \dfrac{g_0}{2\delta_{max}} & \text{if} \quad \delta_{max} \leq \left(\dfrac{3g_0}{4A}\right)^{1/3} \\[3mm] \left(-\dfrac{9Ag_0^2}{16}\right)^{1/3} & \text{if} \quad \delta_{min} < \left(\dfrac{3g_0}{4A}\right)^{1/3} < \delta_{max} \\[3mm] -\dfrac{A\delta_{min}^2}{3} - \dfrac{g_0}{2\delta_{min}} & \text{if} \quad \delta_{min} \geq \left(\dfrac{3g_0}{4A}\right)^{1/3} \end{cases} \tag{8.7}$$

where g_0 is the gap distance between the two elements and δ_{max} is the maximum lateral displacement during the tuning. It is then compared with the conventional method given by Equation 8.6. The results show that the proposed method remarkably improves the lens performance. Furthermore, the optimal D value computed with Equation 8.7 can also help to reduce the lens performance degradation with the increasing lateral displacements, and at the same time, slightly increase the misalignment tolerances.

To experimentally test the validity of the proposed lens design method, a miniature tunable Alvarez lens is designed based on the aforementioned principle. The lens performance is shown in Figure 8.10. It is shown that the designed Alvarez lens provides a focal length tuning range from about 30.5 mm to about 64.4 mm with an overall lens element lateral movement of 0.1 mm. Furthermore, the lens behaves as a near-diffraction-limit one within the whole tuning range since

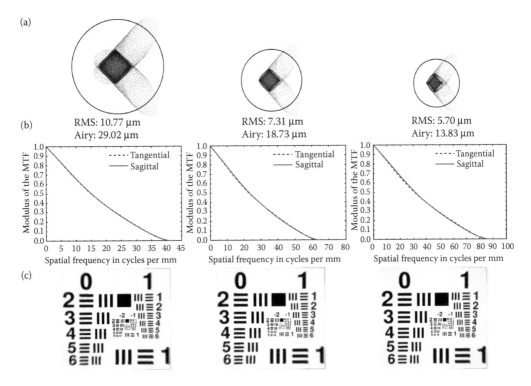

Figure 8.10 Lens performance of the designed Alvarez lens. Focal lengths are 64.4 mm (lateral displacement 0.1 mm), 41.4 mm (lateral displacement 0.15 mm), and 30.5 mm (lateral displacement 0.2 mm) from left to right. (a) Ray-tracing spot diagrams, (b) MTF curves, and (c) image simulation results of the lens with different lateral displacements.

the wavefront aberration PV values at all the tuning points are controlled below λ/4. To drive the lens elements, a commercially available piezo actuator (3 mm × 3 mm × 18 mm) is used to provide the initial displacements, which are then amplified by a compact mechanical displacement amplifier with V-shaped suspensions optimized by the finite element method. The lens elements are fabricated by the aforementioned diamond turning and replication process, while the mechanical actuator is fabricated by precision machining technology. The results show that the mechanical displacement amplifier is able to provide a maximum displacement of about 0.1 mm when the input voltage reaches 130 V. Figure 8.11 demonstrates the images formed by the MEMS-driven Alvarez lens (Zhou et al., 2013) and the ones formed by the new Alvarez lens (Zou et al., 2013). It is shown that the image quality is remarkably improved, which experimentally proves the validity of the proposed method.

Inspired by the Alvarez principle, Zou et al. also reported a novel solid tunable dual-focus lens in 2015 (Zou et al., 2016). As shown in Figure 8.12a, the tunable dual-focus lens consists of two

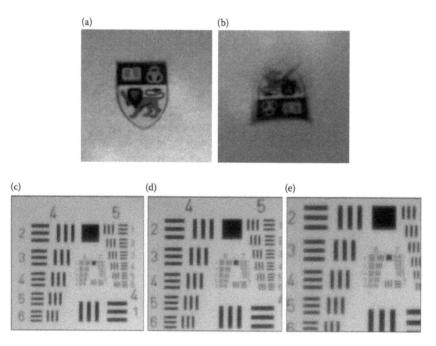

Figure 8.11 (a) and (b) Images formed by the MEMS-driven Alvarez lens designed through the conventional method and (c) to (e) images formed by the new one designed by the proposed method at different focal lengths.

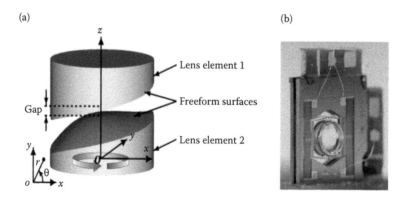

Figure 8.12 (a) Schematic of the dual-focus lens driven by rotatory actuators and (b) final dual-focus lens. (Copyright Optical Society of America. Reproduced with permission.)

lens elements, each having a freeform surface that faces the other. The freeform surface profile is given by

$$Z = \frac{cr^2}{1+\sqrt{1-(1+k)c^2r^2}} + \sum_{i=1}^{N} B_i Z_i(r,\theta)$$
$$+ \begin{cases} Ar^2(\theta+\phi) & \pi \geq \theta \geq 0 \\ Ar^2(2\pi-\theta+\phi) & 2\pi > \theta > \pi \end{cases} \tag{8.8}$$

where r and θ are the radius and azimuth angle, respectively, in the polar coordinate system. The first term is a standard conic term, where c and k present the curvature and conic constant, respectively. The second term is a linear combination of standard Zernike polynomials with the coefficient of each term given by B_i. The third term is the effective one to equip the lens with the tunable dual-focus capability, where A is a constant determining the tuning speed of the lens. Using the first-order approximation, the phase delay generated by such a configuration is given by

$$OPD = (n-1)(z_2 - z_1) + C$$
$$= \begin{cases} -Ar^2(n-1)(\theta_0+\alpha)+C & \pi \geq \theta \geq 0 \\ -Ar^2(n-1)(\theta_0-\alpha)+C & 2\pi > \theta > \pi \end{cases} \tag{8.9}$$

where n is the refractive index of the material and C is a constant determined by the central thickness of the elements. θ_0 is the initial azimuth angle and α is the rotated angle of lens element 2. It is noted that the optical phase delay in Equation 8.9 equips the whole configuration with two separate focal lengths, given by

$$f = \begin{cases} \left| \dfrac{1}{2A(n-1)(\theta_0+\alpha)} \right| & \pi \geq \theta \geq 0 \\ \left| \dfrac{1}{2A(n-1)(\theta_0-\alpha)} \right| & 2\pi > \theta > \pi \end{cases} \tag{8.10}$$

Such a configuration focuses the incident rays to two separate foci, depending on the region where they hit the lens. Furthermore, the two focal lengths are related to the relative rotation angle α. To achieve a tunable dual-focus lens, a MEMS rotary actuator is developed to drive one of the lens elements to rotate about its optical axis. The MEMS actuator is an electrothermal actuator, again fabricated using SOI micromachining. The fabrication of the freeform lens elements and assembly of the elements to the actuators are similar to those of the MEMS Alvarez lens shown in Figures 8.8 and 8.9. The final assembled tunable dual-focus lens is shown in Figure 8.12b. The experimental results show that the designed MEMS rotary actuator provides a maximum rotation angle of about 8.2° with an input DC voltage of 6.5 V. Driven by such an actuator, the dual-focus lens offers a large dynamic tuning range of both the two focal lengths, one changed from about 30 mm to 20 mm, while the other one is varied from about 30 mm to 60 mm. Such a solid miniature dual-focus lens may be useful in various optical systems, including laser cutting systems, microscopy objectives, and interferometer-based surface profilers.

8.4.2 Multi-element Alvarez lenses

In miniature Alvarez lens designs, the maximum optical power and the tuning range are limited by the design and the MEMS actuator, and hence may not meet the requirements of some practical

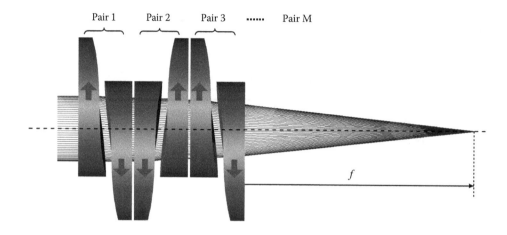

Figure 8.13 Schematic of a multi-element Alvarez lens.

imaging systems. From Equation 8.5, it is known that the optical power is proportional to the product of A and δ, where A is a constant controlling the tuning speed and δ is the lateral displacement provided by the miniature actuator. However, the maximum displacement from most of the MEMS actuators is limited to a few hundreds of microns. It is also proved that the lens performance degrades with the increase of A. Therefore, it is challenging to achieve a miniature Alvarez lens with a large optical power tuning range and satisfactory lens performance using the conventional two-element configuration.

To solve this problem, a multi-element Alvarez lens configuration is proposed and experimentally demonstrated (Zou et al., 2015b). As shown in Figure 8.13, M pairs of Alvarez lens elements are stacked in the design and synchronously driven with the same lateral displacement δ. Compared with the conventional one, the optical phase delay generated by this configuration is now given by

$$OPD = -M[2A\delta(n-1)(x^2 + y^2)] \tag{8.11}$$

if M pairs of lens elements are all the same as the one in Section 8.4.1. It is clear that the optical phase delay is increased by M times, thus the optical power generated by the lens is given by

$$\phi = -M[2A\delta(n-1)] \tag{8.12}$$

A multi-element Alvarez lens also helps to improve the imaging performance when compared with the standard Alvarez lens for a fixed focal length tuneable range. To guide the design, the performances of multi-element Alvarez lenses and conventional two-element Alvarez lenses are first compared numerically. As shown in Table 8.1, the parameters of all the Alvarez lenses with different configurations are listed. The target focal length tuning range is kept the same as for the configurations, which is from 10 mm to 30 mm. The elements' diameters are set at 1.5 mm. The lateral displacement is also set as the same, which is expected to drive the lens element to move from position $x = 0.1$ mm to $x = 0.3$ mm. The values of A and D are then easily determined by Equations 8.5, 8.7, and 8.12.

To indicate the lens performance, Strehl ratios are used in the comparison. As shown in Figure 8.14, it is shown that the Strehl ratios within the whole tuning range are increased with the increasing number of lens elements, which means that the lens performance is improved by using the multi-element configurations. The reason is that the value of A can be remarkably reduced with a multi-element configuration for a particular target focal length tuning range and a known maximum lateral displacement. As mentioned above, reduced A values lead to "flatter" freeform surfaces and

Table 8.1 Lens parameters for simulation

	Two-element lens	**Four-element lens**	**Six-element lens**
A (mm^{-2})	0.15	0.075	0.05
D	−0.15	−0.15	−0.15
E (mm)	0.3	0.3	0.3
δ (mm)	[0.1, 0.3]	[0.1, 0.3]	[0.1, 0.3]
f (mm)	≈[10, 30]	≈[10, 30]	≈[10, 30]
D (mm)	1.5	1.5	1.5
g_1 (mm)	0.3	0.3	0.3
g_2 (mm)	None	0.05	0.05

Note: A, D, and E are surface coefficients; δ is the lateral displacement; f is the expected focal length; D is the diameter of the optical stop; g_1 and g_2 are the center-to center gaps between two adjacent elements. All parameters are kept consistent in all simulations.

hence improved lens performances as well as larger alignment tolerances. However, the increased number of lens elements would obviously lead to more serious scattering and stray light issues, which should be taken into account in practical designs. Furthermore, for a multi-element Alvarez lens, it is noted that the centrosymmetric arrangement of lens elements leads to better lens performance than the non-centrosymmetric one, as also shown in Figure 8.14.

To deliver the basic idea of a multi-element Alvarez lens design, a four-element Alvarez lens is experimentally demonstrated in comparison with a conventional two-element lens. Two miniature piezo actuators are employed to drive the lens elements, which offer a maximum lateral displacement of about 125 μm for the lens elements when the input voltage reaches 130 V. When there are two elements mounted, the conventional two-element Alvarez lens is found to have a dynamic focal length tuning range from 39.48 mm to 21.55 mm. When the number of lens elements is increased to four, the optical power of the lens is increased with the same actuation mechanism, leading to a dynamic focal length tuning range from 19.64 mm to 10.63 mm. Compared with the ones formed by the two-element lens, there is no obvious image quality degradation except the reduced contrast. The main reason for this contrast reduction is the larger amount of stray light caused by the scattering and reflection from the increased number of optical surfaces. By applying anti-reflecting

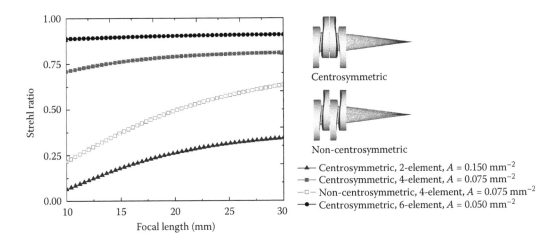

Figure 8.14 Comparison between the Alvarez lenses with different configurations. The focal length tuning ranges and the actuator stokes for different configurations are kept the same.

coatings on all the optical surfaces, this problem may be solved. The results suggest that the multi-element configuration provides an efficient way to achieve large optical power and desirable tuning ranges with small-stroke micro-actuators.

8.4.3 Applications of solid tunable lenses in endoscopes

Endoscopic systems are widely used in clinical practice for virtualizing the organs and tissues inside human bodies. To alleviate patients' discomfort, it is of great significance to minimize the physical movement of the endoscope probe during the endoscopy procedure. Hence, much effort is spent on the development of an optical autofocus or zooming function in endoscopes, which can help to freely switch the field of view among multiple areas of interest with minimum endoscope probe movement. There are a few reports in both academia and industry for such endoscopes, including the one based on traditional movable rod lenses to achieve the adjustable focus function (Daweke et al., 2014). However, due to the inevitable large stroke linear actuators and support mechanisms used to drive the lens elements to move along the optical axis for various focal lengths, the endoscope probe is always bulky, and this greatly limits its applications. Another example is based on the aforementioned liquid tunable lenses (Seo et al., 2009; Kuiper, 2011). However, such lenses are confronted with various challenges, including vibration and thermal instability, leakage and evaporation, and complicated packaging processes. In addition, liquid crystal lenses are also used to achieve adjustable-focus endoscopes (Chen and Lin, 2013; Chen et al., 2014), but it is noticed that the light efficiency is relatively low due to the necessary polarizer required by the liquid crystal, and the maximum optical power tuning range is limited.

Zou et al. reported a novel miniature adjustable-focus endoscope using the solid tunable lenses driven by slim piezo benders in 2015 (Zou et al., 2015a). As presented in Figure 8.15, such an

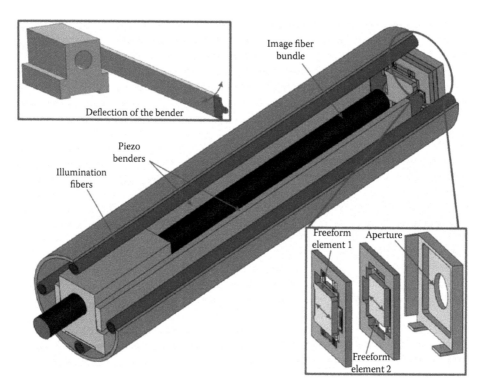

Figure 8.15 Schematic of the adjustable-focus endoscope using solid tunable lenses driven by two slim piezo actuators. (Copyright Optical Society of America. Reproduced with permission.)

adjustable focus endoscope consists of a solid tunable lens supported by two micromachined SOI chips and driven by two slim piezo benders aligned along the optical axis, where one aperture is placed at the tip of the probe to minimize the stray light, and one image fiber bundle is used to transfer the images formed by the lens to an external camera. Rather than the conventional cubic expression, in this case, a more general type of extended polynomial is used to describe the lens surface, as shown below

$$z = \sum_{i=1}^{N} A_i E_i(x, y) \tag{8.13}$$

where A_i is the coefficient of the *ith* extended polynomial term $E_i(x, y)$. The polynomials are a power series of x and y, which is arranged in order from the 1st degree terms x and y, followed by the 2nd degree terms x^2, xy, y^2, and so on. Two 6-order polynomials are used to separately govern the two freeform surfaces in Zemax for optimization. The results show that the back focal length of the designed solid tunable lens can be varied from 7.5 mm to 5.5 mm with the two elements moved from 1 mm to 1.4 mm along the x and $-x$ directions (overall 400 µm displacement for each), respectively. Furthermore, it is shown that the Strehl ratios are maintained to be greater than 0.9 throughout the whole tuning range and the sizes of the ray-tracing spot diagrams are always smaller than that of the Airy spots, which indicate that the designed lens performs as a near-diffraction-limit one. Compared with the conventional Alvarez principle using the cubic equation, it is noted that the new method, which employs two independent extended polynomials to describe the two freeform surfaces for lens element surface optimization, can greatly improve the lens performance.

The lens elements are fabricated using the same method as shown in Figure 8.8, while the SOI chips are fabricated using the standard silicon micromachining process. The SOI chip contains a platform for mounting the lens element and micrometer-sized silicon springs suspending the platform. The piezo benders and image fiber bundle are commercially available models. The assembly process starts with the insertion and fixing of the lens element into a window opening of the suspended silicon platform on the SOI chips, and is followed by stacking the two lens element integrated SOI chips with the freeform surfaces facing each other, as shown in Figure 8.16a After that,

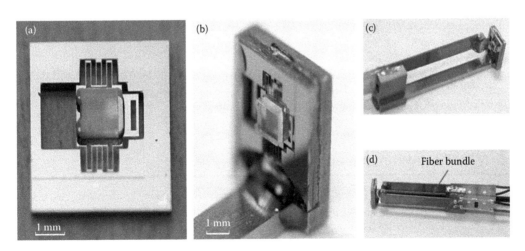

Figure 8.16 Device pictures. (a) Suspended freeform lens element integrated with the SOI chip; (b) top of the endoscope integrated with the tunable lens having two stacked SOI chips; and (c) endoscope with piezoelectric bender assembled and (d) final endoscope with fiber bundle inserted. (Copyright Optical Society of America. Reproduced with permission.)

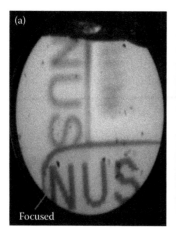
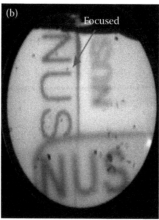

Figure 8.17 Adjustable focus of the endoscope. (a) The driving voltages of the two piezoelectric benders are set at −35 V and −37 V, respectively, and the target at an object distance of about 20 mm is focused; (b) the driving voltages of the two piezoelectric benders are set at 61 V and 62 V, respectively, and the target at an object distance of about 50 mm is focused; (c) the driving voltages of the two piezoelectric benders are set at 85 V and 86 V, respectively and the target at an object distance of about 150 mm is focused. (Copyright Optical Society of America. Reproduced with permission.)

the two stacked SIO chips are placed onto an aluminum plate with the center of the lens aligned to the center of the aperture as shown in Figure 8.16b. The piezo benders and image fiber bundle are then assembled with the assistance of two 3-axis stages and an optical microscope, forming the final adjustable focus endoscope probe, as shown in Figure 8.16c and d.

The experimental results show that the two piezo benders provide a maximum deflection of about ±225 μm with a driving DC voltage of ±90V, and subsequently, the focal length of the solid tunable lens is varied from 7.4 mm to 4.9 mm. The dynamic response speed of the solid tunable lens is determined to be about 0.11s (about 9 Hz). Finally, the output of the designed endoscope is placed under a microscope with the output port of the image fiber bundle clearly focused. One high-resolution CCD camera is connected to the microscope to record the images of the output port of the fiber bundle. To test the adjustable focus capability of the endoscope, three targets are placed under the endoscope probe with different object distances, roughly at 20 mm, 50 mm, and 150 mm, respectively. As shown in Figure 8.17, the three targets are individually focused by varying the input voltages without moving the endoscope, which clearly demonstrates the adjustable focus functionality of the designed endoscope. Compared with endoscopes based on other types of tunable lenses, such a configuration offers a few advantages, including mechanical and thermal robustness, stability against external environments, superior image quality, and compact size. It may be useful in various medical and industrial applications.

8.5 MEMS solid tunable apertures

Similar to the liquid tunable lenses, the liquid tunable apertures are also confronted with a series of challenges involving the liquid materials, including the complicated fabrication and assembly processes, instability rising from the external environment disturbances, and possible leakage and evaporation, which greatly limit their practical applications. Recently, a few solid tunable apertures based on the MEMS technology have been reported. Such solid tunable apertures integrated with the MEMS actuators demonstrate a greater potential in miniaturization and possess the capability to solve the challenges encountered by the liquid ones.

8.5.1 MEMS-driven sliding blades

Inspired by the iris diaphragms used in modern still and video cameras, researchers have developed various sliding blades-based tunable apertures. The basic principle of such a tunable aperture is to simultaneously move a few linked sliding blades, and hence change the size of a window opening formed by these blades, as shown in Figure 8.18. The simplest example is given by one single moveable blade integrated with a fixed aperture, as shown in Figure 8.18a. By moving the blade along the direction indicated by the arrow, the size of the aperture is tunable. Such a type of single blade-based tunable apertures is mostly used as optical attenuators and choppers, and various MEMS actuation mechanisms are developed to drive the single moveable blade (Ching et al., 1994; Toshiyoshi et al., 1995; Barber et al., 1998; Marxer et al., 1999). Similarly, two moveable blades can be arranged as shown in Figure 8.18b, to generate a symmetric tunable aperture for fiber optic applications (Li and Uttamchandani, 2004). To guide the design of sliding blade-based tunable apertures, Syms et al. established the optical model for such a configuration with various numbers of blades to theoretically analyze the diffraction efficiency of the aperture and the effect of the inevitable clearances between blades, and then demonstrated a four-sliding-blade-based MEMS iris in 2004 (Syms et al., 2004), as shown schematically in Figure 8.18c. The four blades are supported by paired beams and driven by four electrothermal actuators, which are connected together electrically and can be driven by a single voltage. All the blades are simultaneously driven to move along the directions of the arrows indicated in the figure once an external voltage is applied, leading to a physical aperture with tunable size. The device is fabricated by deep reactive ion etching of SOI wafers, and the experimental results show that the aperture size can be tuned from 0 to about 40 μm with a maximum input electrical power of 1.1W.

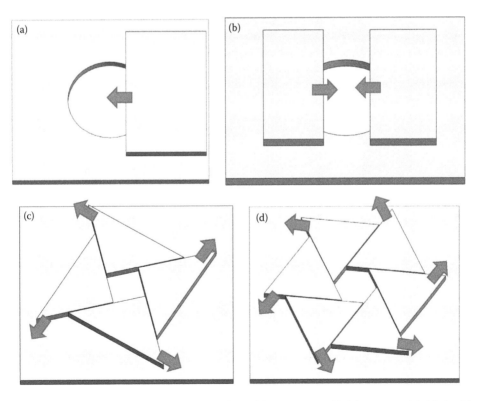

Figure 8.18 Schematic of the sliding-blade based tunable apertures with (a) one movable blade, (b) two movable blades, (c) four movable blades, and (d) six movable blades.

With the MEMS technology, the driving and supporting mechanisms of the sliding blade-based apertures can be easily mass produced with high precision and yield at the wafer level. Therefore, to form a polygon with an increased number of sides to approximate a circular aperture, one can choose to stack the aforementioned one-blade or two-blade-based tunable apertures layer by layer with the necessary relative tilt angles. Such a method is relatively simple to design, but may increase the overall thickness of the whole configuration. One can also choose to increase the number of blades in a single chip design, as shown in Figure 8.18d, to form a more near-circular aperture, but the driving mechanism as well as the supporting structures become more and more complicated, which may lead to a relatively larger device footprint and more design challenges.

8.5.2 MEMS-driven rotating blades

Limited by the fabrication process and maximum displacements provided by the integrated MEMS actuators, sliding blade-based tunable apertures face a few drawbacks, including the inevitable clearances between the moveable blades and the limited aperture size tuning range. Zhou et al. reported a novel rotary blade-based tunable aperture driven by MEMS electrostatic rotary actuators in 2012 (Yu et al., 2012; Zhou et al., 2012), which shows a greater potential to solve these problems. Similar to a conventional iris diaphragm used in modern still or video cameras which achieves tunable aperture size by simultaneously rotating a series of overlapped shutter blades, the demonstrated device consists of two noncontacting layers, as shown in Figure 8.19, each having a set of rotary blades supported by flexural springs and driven by separated rotary actuators. As shown schematically in Figure 8.19a and b, when each of the eight blades rotates clockwise, the size of the octagon-shaped aperture increases. On the contrary, to reduce the size of the aperture, one needs to rotate the eight blades counterclockwise. Due to a small air gap between these two layers, there is no contacting surface between any of the movable blades. Unlike those in conventional iris diaphragms, no friction force is generated, hence leading to the ease of driving using MEMS, reduced wear and tear, and longer operation life time. To achieve a more near-circular-shaped aperture, one can choose to increase the number of rotary blades in each layer to increase the number of sides of the polygon.

To demonstrate the proposed principle, a prototype of an eight-blade MEMS-driven tunable aperture is developed. Such a device consists of two MEMS chips, each designed to have four rotary blades and their driven actuators. They are fabricated using the SOI micromachining process. Each blade is

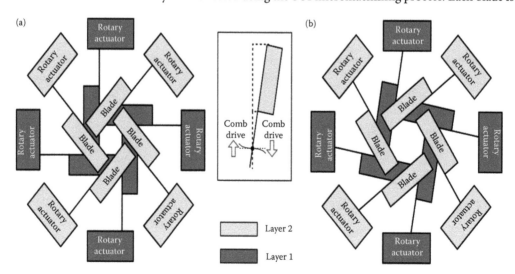

Figure 8.19 Schematic of the MEMS-driven rotary tunable aperture. (a) Original shape and (b) all blades rotate clockwise and the aperture diameter increases. (Copyright Optical Society of America. Reproduced with permission.)

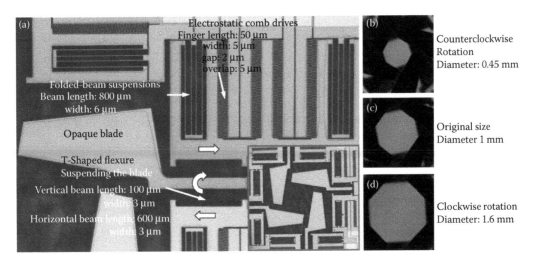

Figure 8.20 (a) An image showing a part of a MEMS rotatory blade with the inset showing a single layer of four MEMS blades. (b) Minimum aperture diameter size when all blades rotate counterclockwise. (c) Original aperture size without blades' rotation. (d) Maximum aperture size when all blades rotate clockwise. (Copyright Optical Society of America. Reproduced with permission.)

supported by flexural silicon beams and driven by two symmetric electrostatic comb drive actuators to rotate about a pivot point, as shown in the inset in Figure 8.19. Detailed device images are shown in Figure 8.20. The experimental results show that the blade is able to rotate by an angle of 10 degrees clockwise and 11 degrees counterclockwise when the input voltage reaches 100 V for those two directions, respectively. Two identical SOI chips are then stacked together with a relative tilt angle of 45 degrees with respect to each other to form the octagon-shaped aperture as shown in Figure 8.20. The testing results show that the diameter of the aperture can be tuned from 1.0 mm to 1.6 mm when the blades are rotated clockwise, and from 1.0 mm to 0.45 mm when the blades are rotated counterclockwise. It is noted that such a miniature tunable aperture offers a relative larger adjustable range compared with its counterparts, for example, those with sliding blades. It may be useful in miniature cameras integrated in current portable electronic devices and medical imaging systems.

8.6 Summary

Miniature tunable optical components, including tunable lenses and tunable apertures, are important building blocks for modern miniaturized imaging systems, for example, endoscopes and ultracompact camera modules. Recent years have witnessed increasing research and development work in this area, which can be categorized into two approaches involving either liquid or solid materials as the key functional elements to achieve tunability. Liquid tunable optical components can be implemented using pressure-driven deformable membranes, an electrowetting effect, and varying refractive index liquid crystal materials. Solid tunable lenses can be implemented using transparent deformable elastomer lenses and freeform optical elements with the Alvarez principle. The latter hold the greater potential for practical use, because they are compact, robust, provide ease of large optical power variation, long operation life time without wear and tear, and require only small-stroke micro actuators. The freeform surface elements also help to provide more degrees of freedom in the design for optical aberration correction for high quality optical imaging, and they can be mass produced by precision surface forming and inject molding technologies. Solid tunable apertures or irises can be implemented using MEMS-driven sliding and rotating blades. The research and development in miniature tunable optics will continue for both liquid and solid approaches. Further applications utilizing these ultraminiature dynamically adjustable optical components are expected to emerge in the near future.

References

Agarwall M, Gunasekaran RA, Coane P, Varahramyan K, 2004. Polymer-based variable focal length microlens system. *Journal of Micromechanics and Microengineering* 14:1665–1673.

Ahn SH, Kim YK, 1999. Proposal of human eye's crystalline lens-like variable focusing lens. *Sensor Actuat a-Phys* 78:48–53.

Alvarez LW, 1967. *Two-Element Variable-Power Spherical Lens*. In. US Patent 3305294 USA.

Alvarez LW, 1970. *Variable-Power Lens and System*. In. US Patent 3507565A USA.

Barber B, Giles CR, Askyuk V, Ruel R, Stulz L, Bishop D, 1998. A fiber connectorized MEMS variable optical attenuator. *Ieee Photonics Technology Letters* 10:1262–1264.

Barbero S, 2009. The Alvarez and Lohmann refractive lenses revisited. *Optics Express* 17:9376–9390.

Barton IM, Dixit SN, Summers LJ, Avicola K, Wilhelmsen J, 2000. Diffractive Alvarez lens. *Opt Lett* 25:1–3.

Beadie G, Sandrock ML, Wiggins MJ, Lepkowicz RS, Shirk JS, Ponting M, Yang Y, Kazmierczak T, Hiltner A, Baer E, 2008. Tunable polymer lens. *Optics Express* 16:11847.

Berge B, 2005. Liquid lens technology: Principle of electrowetting based lenses and applications to imaging. In: *MEMS 2005*, Miami: Technical Digest, pp. 227–230.

Berge B, Peseux J, 2000. Variable focal lens controlled by an external voltage: An application of electrowetting. *The European Physical Journal E* 3:159–163.

Chang J.-H, Jung K.-D, Lee E, Choi M, Lee S, Kim W, 2013. Variable aperture controlled by micro-electrofluidic iris. *Optics Letters* 38:2919–2922.

Chen HS, Lin YH, 2013. An endoscopic system adopting a liquid crystal lens with an electrically tunable depth-of-field. *Opt Express* 21:18079–18088.

Chen J, Wang WS, Fang J, Varahramyan K, 2004. Variable-focusing microlens with microfluidic chip. *Journal of Micromechanics and Microengineering* 14:675–680.

Chen M.-S, Chen P.-J, Chen M, Lin Y.-H, 2014. An electrically tunable imaging system with separable focus and zoom functions using composite liquid crystal lenses. *Optics Express* 22:11427.

Cheng C.-C, Yeh JA, 2007. Dielectrically actuated liquid lens. *Optics Express* 15:7140–7145.

Cheng CC, Chang CA, Yeh JA, 2006. Variable focus dielectric liquid droplet lens. *Optics Express* 14:4101–4106.

Ching MT, Brennen RA, White RM, 1994. Microfabricated optical chopper. *Opt Eng* 33:3634–3642.

Chiu CP, Chiang TJ, Chen JK, Chang FC, Ko FH, Chu CW, Kuo SW, Fan SK, 2012. Liquid lenses and driving mechanisms: A review. *Journal of Adhesion Science and Technology* 26:1773–1788.

Choi ST, Lee JY, Kwon JO, Lee S, Kim W, 2009. Liquid-filled varifocal lens on a chip. In: Proc. SPIE Vol. 7208, p. 72080P, San Jose, California.

Choi Y, Park J.-H, Kim J.-H, Lee S.-D, 2003. Fabrication of a focal length variable microlens array based on a nematic liquid crystal. *Optical Materials* 21:643–646.

Chronis N, Liu GL, Jeong KH, Lee LP, 2003. Tunable liquid-filled microlens array integrated with microfluidic network. *Optics Express* 11:2370–2378.

Dai HT, Liu YJ, Sun XW, Luo D, 2009. A negative-positive tunable liquid-crystal microlens array by printing. *Optics Express* 17:4317–4323.

Daweke RDG, Kelp M, Lehr H, Moennich O, Osiak P, IEEE, 2014. Electromagnetic direct linear drives for medical endoscopes. *2014 International Conference on Optimization of Electrical and Electronic Equipment*, Bran, Romania (Optim):245–251.

Dong L, Agarwal AK, Beebe DJ, Jiang H, 2006. Adaptive liquid microlenses activated by stimuli-responsive hydrogels. *Nature* 442:551–554.

Draheim J, Burger T, Korvink JG, Wallrabe U, 2011. Variable aperture stop based on the design of a single chamber silicone membrane lens with integrated actuation. *Optics Letters* 36:2032–2034.

Erickson D, Heng X, Li Z, Rockwood T, Emery T, Zhang Z, Scherer A, Yang C, Psaltis D, 2005. Optofluidics. In: SPIE Optics and Photonics – Special Session on Optofluidics. San Diego, California, pp. 59080S.

Fan S.-K, Chiu C.-P, Lin J.-W, 2009. Electrowetting on polymer dispersed liquid crystal. *Applied Physics Letters* 94:164109.

Feng G.-H, Chou Y.-C, 2009. Fabrication and characterization of optofluidic flexible meniscus-biconvex lens system. *Sensor Actuat a-Phys* 156:342–349.

Grilli S, Miccio L, Vespini V, Finizio A, De Nicola S, Ferraro P, 2008. Liquid micro-lens array activated by selective electrowetting on lithium niobate substrates. *Optics Express* 16:8084–8093.

Hongbin Y, Guangya Z, Fook Siong C, Feiwen L, 2008. Optofluidic variable aperture. *Optics Letters* 33:548.

Hsu CJ, Liao CH, Chen BL, Chih SY, Huang CY, 2014. Polarization-insensitive liquid crystal micro-lens array with dual focal modes. *Optics Express* 22:25925–25930.

Huang C, Li L, Yi AY, 2008. Design and fabrication of a micro Alvarez lens array with a variable focal length. *Microsystem Technologies* 15:559–563.

Jae-Hoon K, Satyendra K, 2004. Fast switchable and bistable microlens array using ferroelectric liquid crystals. *Japanese Journal of Applied Physics* 43:7050.

Jeong KH, Liu GL, Chronis N, Lee LP, 2004. Tunable microdoublet lens array. *Optics Express* 12:2494–2500.

Ji H.-S, Kim J.-H, Kumar S, 2003. Electrically controllable microlens array fabricated by anisotropic phase separation from liquid-crystal and polymer composite materials. *Optics Letters* 28:1147–1149.

Kimmle C, Schmittat U, Doering C, Fouckhardt H, 2011. Compact dynamic microfluidic iris for active optics. *Microelectronic Engineering* 88:1772–1774.

Krupenkin T, Yang S, Mach P, 2003. Tunable liquid microlens. *Applied Physics Letters* 82:316–318.

Kuiper S, 2011. Electrowetting-based liquid lenses for endoscopy. In: *Moems and Miniaturized Systems X* (Schenk H, Piyawattanametha W, eds). *Proc.* SPIE Vol. 7930, pp. 793008. San Francisco, California.

Kuiper S, Hendriks BHW, 2004. Variable-focus liquid lens for miniature cameras. *Applied Physics Letters* 85:1128–1130.

Lee J.-Y, Choi S.-T, Lee S.-W, Kim W, 2009. Microfluidic design and fabrication of wafer-scale varifocal liquid lens. In, 742603–742611.

Lee S-y, Tung H-w, Chen W-c, Fang W, 2006. Thermal actuated solid tunable lens. *IEEE Photonics Technology Letters* 18:2191–2193.

Lee SW, Lee SS, 2007. Focal tunable liquid lens integrated with an electromagnetic actuator. *Applied Physics Letters* 90:121129.

Leung HM, Zhou G, Yu H, Chau FS, Kumar AS, 2009. Liquid tunable double-focus lens fabricated with diamond cutting and soft lithography. *Applied Optics* 48:5733–5740.

Leung HM, Zhou G, Yu H, Chau FS, Kumar AS, 2010. Diamond turning and soft lithography processes for liquid tunable lenses. *Journal of Micromechanics and Microengineering* 20:025021.

Li CH, Jiang HR, 2012. Electrowetting-driven variable-focus microlens on flexible surfaces. *Applied Physics Letters* 100:231105.

Li L, Liu C, Ren H, Wang Q.-H, 2013b. Adaptive liquid iris based on electrowetting. *Optics Letters* 38:2336–2338.

Li L, Liu C, Wang Q.-H, 2013a. Electrowetting-based liquid Iris. *IEEE Photonics Technology Letters* 25:989–991.

Li L, Uttamchandani D, 2004. Design and evaluation of a MEMS optical chopper for fibre optic applications. *IEE Proceedings - Science, Measurement and Technology* 151:77–84.

Liebetraut P, Petsch S, Liebeskind J, Zappe H, 2013. Elastomeric lenses with tunable astigmatism. *Light: Science & Applications* 2:e98.

Lin H.-C, Chen M.-S, Lin Y.-H, 2011a. A review of electrically tunable focusing liquid crystal lenses. *Transactions on Electrical and Electronic Materials* 12:234–240.

Lin YH, Chen HS, 2013. Electrically tunable-focusing and polarizer-free liquid crystal lenses for ophthalmic applications. *Optics Express* 21:9428–9436.

Lin YH, Chen MS, Lin HC, 2011b. An electrically tunable optical zoom system using two composite liquid crystal lenses with a large zoom ratio. *Optics Express* 19:4714–4721.

Lohmann AW, 1970. A new class of varifocal lenses. *Applied Optics* 9:1669–1671.

Lopez CA, Lee CC, Hirsa AH, 2005. Electrochemically activated adaptive liquid lens. *Applied Physics Letters* 87:134102.

Marxer C, Griss P, de Rooij NF, 1999. A variable optical attenuator based on silicon micromechanics. *Ieee Photonics Technology Letters* 11:233–235.

Mugele F, Baret JC, 2005. Electrowetting: From basics to applications. *Journal of Physics: Condensed Matter* 17:R705–R774.

Muller P, Feuerstein R, Zappe H, 2012. Integrated optofluidic Iris. *Journal of Microelectromechanical Systems* 21:1156–1164.

Muller P, Spengler N, Zappe H, Monch W, 2010. An optofluidic concept for a tunable micro-iris. *Journal of Microelectromechanical Systems* 19:1477–1484.

Murade CU, Oh JM, van den Ende D, Mugele F, 2011. Electrowetting driven optical switch and tunable aperture. *Optics Express* 19:15525–15531.

Nguyen NT, 2010. Micro-optofluidic lenses: A review. *Biomicrofluidics* 4:031501–031515.

Oku H, Hashimoto K, Ishikawa M, 2004. Variable-focus lens with 1-kHz bandwidth. *Optics Express* 12:2138–2149.

Pang L, Levy U, Campbell K, Groisman A, Fainman Y, 2005. Set of two orthogonal adaptive cylindrical lenses in a monolith elastomer device. *Optics Express* 13:9003–9013.

Rege S, Tkaczyk T, Descour M, 2004. Application of the alvarez-humphrey concept to the design of a miniaturized scanning microscope. *Optics Express* 12:2574–2588.

Ren H, Fan Y.-H, Wu S.-T, 2004. Liquid-crystal microlens arrays using patterned polymer networks. *Optics Letters* 29:1608–1610.

Ren H, Fox D, Anderson PA, Wu B, Wu S.-T, 2006. Tunable-focus liquid lens controlled using a servo motor. *Optics Express* 14:8031–8036.

Ren H, Wu S.-T, 2005. Polymer-based flexible microlens arrays with hermaphroditic focusing properties. *Applied Optics* 44:7730–7734.

Ren H, Wu S.-T, 2007. Variable-focus liquid lens. *Optics Express* 15:5931–5936.

Ren H, Wu S.-T, 2008. Tunable-focus liquid microlens array using dielectrophoretic effect. *Optics Express* 16:2646–2652.

Saito M, Maruyama A, Fujiwara J, 2015. Polarization-independent refractive-index change of a cholesteric liquid crystal. *Optical Materials Express* 5:1588–1597.

Santiago-Alvarado A, Vazquez-Montiel S, Munoz-López J, Cruz-Martínez VM, Díaz-González G, Campos-García M, 2011. Comparison between liquid and solid tunable focus lenses. *Journal of Physics: Conference Series* 274:012101.

Sato S, 1979. Liquid-crystal lens-cells with variable focal length. *Jpn J Appl Phys* 18:1679–1684.

Sato S, 1999. Applications of liquid crystals to variable-focusing lenses. *Opt Rev* 6:471–485.

Schuhladen S, Preller F, Rix R, Petsch S, Zentel R, Zappe H, 2014. Iris-like tunable aperture employing liquid-crystal elastomers. *Journal of Advanced Materials* 26:7247–7251.

Seo HW, Chae JB, Hong SJ, Rhee K, Chang J-h, Chung SK, 2015. Electromagnetically driven liquid iris. *Sensors and Actuators A: Physical* 231:52–58.

Seo SW, Han S, Seo JH, Kim YM, Kang MS, Min NK, Choi WB, Sung MY, 2009. Microelectromechanical-system-based variable-focus liquid lens for capsule endoscopes. *Japanese Journal of Applied Physics* 48:052404.

Seow YC, Liu AQ, Chin LK, Li XC, Huang HJ, Cheng TH, Zhou XQ, 2008. Different curvatures of tunable liquid microlens via the control of laminar flow rate. *Applied Physics Letters* 93:084101.

Simonov AN, Vdovin G, Rombach MC, 2006. Cubic optical elements for an accommodative intra-ocular lens. *Optics Express* 14:7757–7775.

Smilie PJ, Dutterer BS, Lineberger JL, Davies MA, Suleski TJ, 2012. Design and characterization of an infrared Alvarez lens. *Optical Engineering* 51:013006.

Syms RRA, Zou H, Stagg J, Veladi H, 2004. Sliding-blade MEMS iris and variable optical attenuator. *Journal of Micromechanics and Microengineering* 14:1700–1710.

Tang SKY, Stan CA, Whitesides GM, 2008. Dynamically reconfigurable liquid-core liquid-cladding lens in a microfluidic channel. *Lab on a Chip* 8:395–401.

Toshiyoshi H, Fujita H, Ueda T, 1995. A piezoelectrically operated optical chopper by quartz micromachining. *Journal of Microelectromechanical Systems* 4:3–9.

Tsai CG, Yeh JA, 2010. Circular dielectric liquid iris. *Optics Letters* 35:2484–2486.

Wang B, Ye M, Sato S, 2005. Liquid crystal lens with stacked structure of liquid-crystal layers. *Optics Communications* 250:266–273.

Wang B, Ye MO, Sato S, 2006. Liquid crystal lens with focal length variable from negative to positive values. *Ieee Photonics Technology Letters* 18:79–81.

Werber A, Zappe H, 2005. Tunable microfluidic microlenses. *Applied Optics* 44:3238–3245.

Xu M, Ren H, Lin YH, 2015. Electrically actuated liquid iris. *Optics Letters* 40:831–834.

Ye M, Sato S, 2002. Optical properties of liquid crystal lens of any size. *Japanese Journal of Applied Physics Part 2-Letters* 41:L571–L573.

Ye M, Sato S, 2003. Liquid crystal lens with focus movable along and off axis. *Optics Communications* 225:277–280.

Ye M, Wang B, Sato S, 2004. Liquid-crystal lens with a focal length that is variable in a wide range. *Applied Optics* 43:6407–6412.

Ye M, Wang B, Sato S, 2006. Polarization-independent liquid crystal lens with four liquid crystal layers. *Ieee Photonics Technology Letters* 18:505–507.

Ye M, Wang B, Sato S, 2008. Realization of liquid crystal lens of large aperture and low driving voltages using thin layer of weakly conductive material. *Optics Express* 16:4302–4308.

Yu H, Zhou G, Du Y, Mu X, Chau FS, 2012. MEMS-based tunable iris diaphragm. *J Microelectromech S* 21:1136–1145.

Yu H, Zhou G, Siong CF, Lee F, Wang S, 2008. A tunable Shack-Hartmann wavefront sensor based on a liquid-filled microlens array. *Journal of Micromechanics and Microengineering* 18:105017.

Yu HB, Zhou GY, Chau FK, Lee FW, Wang SH, Leung HM, 2009. A liquid-filled tunable double-focus microlens. *Optics Express* 17:4782–4790.

Zeng X, Jiang H, 2008. Tunable liquid microlens actuated by infrared light-responsive hydrogel. *Applied Physics Letters* 93:151101.

Zeng X, Jiang H, 2013. Liquid tunable microlenses based on MEMS techniques. *Journal of Physics D (Applied Physics)* 46:323001.

Zhang DY, Justis N, Lo YH, 2004. Fluidic adaptive lens of transformable lens type. *Applied Physics Letters* 84:4194–4196.

Zhang DY, Lien V, Berdichevsky Y, Choi J, Lo YH, 2003. Fluidic adaptive lens with high focal length tunability. *Applied Physics Letters* 82:3171–3172.

Zhou G, Leung HM, Yu H, Kumar AS, Chau FS, 2009. Liquid tunable diffractive/refractive hybrid lens. *Optics Letters* 34:2793–2795.

Zhou G, Yu H, Chau FS, 2013. Microelectromechanically-driven miniature adaptive Alvarez lens. *Optics Express* 21:1226–1233.

Zhou G, Yu H, Du Y, Chau FS, 2012. Microelectromechanical-systems-driven two-layer rotary-blade-based adjustable iris diaphragm. *Optics Letters* 37:1745–1747.

Zou Y, Zhang W, Chau FS, Zhou G, 2015a. Miniature adjustable-focus endoscope with a solid electrically tunable lens. *Optics Express* 23:20582–20592.

Zou Y, Zhang W, Chau FS, Zhou G, 2016. Solid electrically tunable dual-focus lens using freeform surfaces and microelectro-mechanical-systems actuator. *Optics Letters* 41:1.

Zou Y, Zhou G, Du Y, Chau FS, 2013. Alignment tolerances and optimal design of MEMS-driven Alvarez lenses. *Journal of Optics* 15:125711.

Zou YC, Zhang W, Tian F, Chau FS, Zhou GY, 2015b. Development of miniature tunable multi-element Alvarez lenses. *IEEE J Sel Top Quant* 21:2700408.

SECTION II

Nanophotonics for communication, imaging, and sensing applications

9

Physical sensors based on photonic crystals

Bo Li and Chengkuo Lee

Contents

Optical sensors have been considered the most promising technology that have various applications including quantum study, healthcare, environmental monitoring, motion sensors, and pharmaceuticals. They have received noticeable research attention in the past decades and will dramatically change human life. In practical applications, the demands from the points of global interest and human need lead to the various optical sensors becoming smaller, faster, more sensitive, and cheaper. Si photonics resonator-based sensors have the potential to fulfill all these characteristics.

Optical sensors are powerful tools for detection and analysis of many kinds of physical and chemical disturbances in and from the environment. A vast number of optical resonance devices have been developed for optical sensing, such as optical waveguides, ring resonators, surface plasmon resonators, optical interferometers, optical fibers, and photonics crystals. They are typically robust to electrical and magnetic interference, capable of fast and multiplexing sensing within a single chip, and perform sensing without damage or contact with the samples. Generally, any one or a combination of the above-mentioned optical devices can give rise to a high-quality factor resonance and greatly enhance the light-matter interactions, which results in an increment of the sensitivity and detection limit of optical sensors. In addition, the advantages of Si photonics include the very high refractive index and almost lossless media of near infrared light. Meanwhile, the key benefits of silicon photonics also include the ability to produce integrated photonics with electronics, as well as the sharing of the mature CMOS fabrication technique. Optical resonators play a key role in the design of optical sensors. The energy stored in the resonator acts with the surroundings and responds in the resonance condition.

Optical microgeometry resonators, such as rings and disks, are the most commonly used optical sensor geometries and have been studied over the past decades [1–10]. This is mainly due to the unique features of the circulating mode inside the rings and the disks, such as ultra-high Q-factor and ultra-compact footprint. In a ring resonator, the effective light-matter interaction length is no longer determined by the physical dimension of the structure itself, but by the Q-factor of the resonator, which indicates the revolution of the light supported by the resonators. The circulating propagated optical mode within the ring and disk are usually known as the whispering gallery mode (WGM). Light is confined by total internal reflection by the high index contrast at the local boundary in a WGM resonator.

On the other hand, a photonics crystal (PhC), owing to its fundamental significance and promising optical performance, is attracting more and more research efforts around the world. Since the pioneering work of photonic band gaps (PBG) and PhCs done by E. Yablonovitch [11] and S. John [12], it has received considerable attention for fundamental physics study as well as various potential applications. Photonic crystals have periodic structures that are in the order of the wavelength of the light. Within each period, there is a high refractive index contrast that scatters the input light. The scattered energy will either sum up or subtract each other. Thus, PhC is able to control the propagation of light within its structure. Due to the fact that band gaps exist in the structure, the PhC can exist either as a lossless mirror or a transparent material and prohibit the propagation of light in certain frequency ranges [13]. The concept of photonic crystals was first introduced by Eli Yablonovitch in 1987 as a method of controlling spontaneous emission in semiconductors. He combined the tools of classical electromagnetism and solid-state physics, and this led to concepts of photonic band gaps in two and three dimensions. The general concept is that photonic crystals do to photons what semiconductors crystals do to electrons. There is a range of energies in which electrons are blocked from traveling through the semiconductor. Similarly, by structuring a material in carefully designed patterns at the nanoscopic size scale, a range of wavelengths of light is blocked by the material. This constitutes a photonic crystal and it functions as a "semiconductor of light."

A PhC cavity is another well-known approach to generate a high Q-factor resonance. The PBG effect helps with the light confinement and various optical devices can be realized by creating a lattice defect from a complete PhC lattice. In addition, some engineering works have been done to slow down the speed of light in order to enhance the performance of the optical sensors. The general purposes of the resonance are to enhance the light intensity in certain areas, and hence enhance the light-matter interaction as well as the electro-optics and thermo-optics. By extending the cavity defect in the PhC to a line defect of missing air holes, a PhC waveguide is created and works in a slow light regime [14]. It allows the propagation of a certain wavelength of light, which is confined by total internal reflection and Bragg reflection in a lateral direction. A PhC waveguide also slows down the light travelling speed and results in a strong localized electrical field. Despite optical dispersion engineering addressing the loss limitation feature of slow light, it plays a key role in all-optical circuit related applications such as all-optical storage, switching, and data regeneration in telecommunications.

9.1 Backgrounds

The ultra-confinement of light in a PhC nanocavity structure in a subwavelength scale, typically in $(\lambda/n)^3$ order, enables the existence of a very high quality factor optical localized resonance [15]. Such a high confinement of energy in a small volume enables a variety of scientific and engineering applications, such as high sensitivity sensors [16], ultra-compact filters [17,18], fast modulators [19], and low power all-optical switches [20].The photons can be arbitrarily controlled by introducing defect modes within the PBG, for example, point and line defects. The PBG effect of PhC reveals highly confined and localized light. Many optical communication devices based on PhC, such as channel drop filters, power splitters, PhC couplers, and light sources are reported [21–25]. Particularly, a channel drop filter plays a key role in photonic integrated circuits (ICs) [25] because it integrates various functional elements, for example, a multiplexer [26,27], switch [28], and directional coupler [29], together. The ultimate sizes of such PhC devices are suggested to be less than 1/10,000 of those of conventional optical devices. Compared with those conventional optical waveguide devices, PhCs devices can achieve the same function with their nanometric sizes, hence raising channel densities within a small chip of photonic ICs.

Combining photonic resonators with micro-/nanoelectromechanical systems (M/NEMS) structure such as a comb drive, cantilever or diaphragm as the sensing element, has been proposed as an intriguing solution for nanobiosensors of the next generation [30–32]. Since they share the same material and fabrication techniques, the future of the integration of such devices on a single chip is quite attractive. Suspended silicon cantilevers integrated with a PhC nanocavity-based optical resonator have been investigated as a NEMS sensor [30]. The output resonant wavelength is sensitive to the shape of air holes and defect length of the nanocavity resonator. Therefore, integrating PhC and MEMS together as a micro-/nanosensor leverages the advantage of those two structures, which are the ultra-fine light confinement and ultra-small size from the PhC side, and the platform for mechanical/biochemical sensing and easy packaging and handling, which comes from the MEMS side.

In summary, because of the fundamental and intrinsic properties of nanophotonics devices, the resonance condition responds to many internal and external disturbances. Hence, nanophotonics resonators play important roles as wonderful candidates for various kinds of sensors. The sensing elements are numerous, including biomolecules, chemicals, strain, stress, forces, displacement, and so on. Compared with the conventional sensors, nanophotonics sensors show many advantages, including ultra-compact size, ultra-sensitivity, low power consumption, and low cost. Those extraordinary properties come from the intrinsic material properties and fundamental resonance behaviors. As the nanophotonics resonants in a volume of wavelength scale, the overall dimension of the optical sensor can be scaled down to a few micrometers. The ultra-compact size enables the possibility of multiple devices embedded in a single chip. With proper design of the optical sensors, such as the operating wavelength, many sensing signals can be detected in a single testing. Moreover, the expensive fabrication cost of the nanostructure can be averaged by the large output and can lower the cost of each device. Furthermore, the sensitivity and detection limit are important characterizations with which to evaluate the sensor's performance. The ultra-confinement of light and high light-matter interaction bring the advantage of ultra-sensitivity to the optical sensors. Finally, the power required for testing is very small because of the high Q-factor and compact size of the optical resonators.

9.1.1 Design principle

Conventional optical resonance in the early stage utilized two or more optical mirrors, which are also called Bragg mirrors. The Bragg mirrors are constructed by several dielectric layers with different refractive indices stacked together. The reflectance of such structures can be close to 100% in a wide frequency range. If the stacking layer number is high enough, the index difference need not be very huge to achieve high reflectance. Once two mirrors are placed side by side across a certain range, the reflected light can achieve the resonance condition. This resonance, known as the Fabry–Pérot (FP) resonance, is highly used in laser generation and oscillation to increase the effective pathlength in spectroscopic applications or resolution in interferometer measurements. However, the multi-stacks optical mirror design is fairly complex in the fabrication process and results in expensive and large devices. Modern techniques require the structure to be low cost and easily fabricated, but the performance should be at the same level. PhC cavity mode resonators will be demonstrated as the new generation candidates for the optical resonator designs in this chapter.

In the design of a PhC cavity resonator, a high Q-factor is always the first and most important characteristic that needs to be optimized; it is also the key component in the photonic sensor design. PhC-based resonators provide ultra-low bending loss due to very good optical confinement. In contrast to the Si slab waveguide which leverages the total internal reflection principle as the confinement method for the light, PhCs use the PBG created by its unique structure. The alternative variance of the dielectric properties help create the bandgap, which forbids light from passing through. Hence, the energy lost when light propagates through the waveguide is minimized. As the PhC structure typically has a comparable size with the wavelength, it offers a good alternative for ultra-compact resonators with a high quality factor and high wavelength selectivity in contrast to the current microring resonators. Therefore, as we introduce some imperfections in the PhC lattice, for example, a defect, the defect mode will be presented in the band gap region. Since the mode is trapped inside the cavity, that is, the PBG is still present in the other lattice structures, the light can only resonant in the defect area. The resonance volume can be as small as a single point defect and results in an extremely high Q-factor. Various mode shapes of the defect can be designed according to the different applications and wavelength of interest. We can have one, two, and multiple missing lattice points in order to create the resonance volume. PhC waveguides can also be formed if we remove a line of air holes. Hence, we can guide the propagation light in the PhC slabs. However, a point defect in the PhC slab faces difficulties in resonance excitation and signal collection. Hence, some of the easily coupled designs are established, for example, a hexagonal PhC ring together with two PhC waveguides. The coupling between the waveguide and the PhC rings can be manipulated by the spacing between the two PhC devices. The increase in the coupling length of a PhC ring can greatly enhance the coupling efficiency of the PhC cavity resonator. On the other hand, the resonance condition of the PhC cavity is highly dependent on the PBG location, and the PBG is determined by the air holes' radii and their lattice properties. Therefore, the cavity resonance will respond to any modification of the lattice shape of the PhC structures.

9.1.2 PhC bandgap design methods

Utilizing the periodic pattern of the high and low index dielectric structure, a PBG structure is generated in the transmission spectrum. The formation of these band gaps relies on coherent Bragg reflection of the electromagnetic modes of the periodic lattice. Two microscopic scattering resonances are presented here, the first from the different dielectric material in a unit cell, and the second from the lattice structure constructed by the unit cells. The spacing between the unit cells will be in the order of the wavelength. Two options need to be considered; we can either have low index holes or high index pillars to construct the periodic pattern of PhC. The PBG acts quite similarly to the electronic band gap which prohibits the existence of electron states in certain energy levels. The PBG block all the propagation modes of the light in any direction. PhC can be classified by the degree of periodicity as one-, two- and three-dimensional PhC. A multilayer dielectric mirror is actually a 1D PhC, as shown in Figure 9.1a. Figure 9.1b illustrate the 2D PhC with high index pillars structure and low index holes structure, respectively. The 2D PhC is the most commonly used structure since it has two-direction PBG light confinement and requires a relatively easier fabrication process. The 3D PhC is illustrated in Figure 9.1c. It has shown some extraordinary properties,

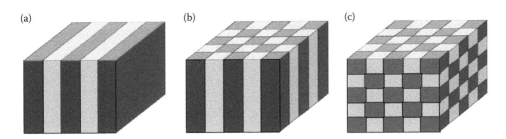

Figure 9.1 Examples of a) one dimensional b) two dimensional c) three dimensional photonic crystal structures.

because of the all-directional PGB effect. However, it is not well developed because of the complexity of the fabrication process.

In order to simulate the PhC behavior, in the 1970s, a direct time domain solution for Maxwell's differential curl equation was explored, in particular, the finite difference time domain (FDTD) method. The innovations and design keys of the optical engineering technologies that benefit from this method cover all kinds of optical passive and active devices. The FDTD method was first introduced by Yee in 1966 [33]. It is used as a primary means to computationally model many scientific and engineering problems which involve electromagnetic wave interactions with material structures. Since it is a time domain method, solutions can cover a wide frequency range with a single simulation run. Based on the Yee's mesh, the materials used in the simulation can be determined for every node in the entire computational domain. According to Maxwell's differential equations, the change in the electric field in the time domain is dependent on the change in the magnetic field across space and vice versa. This results in the basic FDTD relation by which, at any point in space, the updated value of the electric field in time is dependent on the stored value of the electric field and the numerical curl of the local distribution of the magnetic field in space. Hence, with a proper source defined and a boundary condition, all points of the EW field can be studied.

The reasons for the FDTD method being widely used are: first of all, the FDTD is accurate and robust; the sources of error in FDTD calculation are well understood and can be bounded to permit accurate models. Second, FDTD is a fully explicit computation and uses no linear algebra. Finally, FDTD treats impulsive and nonlinear behavior naturally. In this work, we mainly use the FDTD method to design and optimize the photonics resonators. In addition, some of other well-known optical simulation methods such as the plane wave expansion (PWE) method and beam propagation method (BPM) are used for photonics bandgap calculation and propagation mode estimation, and will be discussed in the following chapters.

An example of a 2D PhC band structure plot is shown in Figure 9.2. By using the PWE method, the band structure of a silicon PhC slab with a hexagonal lattice of air holes is derived. The example consists of a silicon PhC slab of 220 nm thickness, which is released from the silicon-on-insulator (SOI) substrate, where the refractive index of air/220 nm-Si/air is derived as 1, 3.46, and 1, respectively. According to the derived band gap map in Figure 9.2, the normalized frequency range of the first photonic band gap extends from 0.26 to 0.33 in a TM polarization electromagnetic wave, that is, the magnetic field parallel to the surface of the silicon slab. The ratio between the radius of air holes (r) and lattice constant (a) is selected as 0.292, the center of the band gap range. The corresponding directions of ΓM and ΓK with respect to the hexagonal lattice are indicated in the inset of Figure 9.2. The corresponding band gap wavelength range extends from 1.242 to 1.577 μm. A combinational approach of the 2D FDTD method and the effective refractive index (ERI) approximation was deployed to calculate and predict the performance of a PhC resonator and the field distribution of the resonant mode. M. Qiu reported a good agreement between the data derived by this combination approach and the full vector 3D FDTD method [34]. Furthermore, experimental data have been predicted well by this combination approach by other groups as well [35,36]. As shown in the inset of Figure 9.2, a hexagonal nanoring, which is the basic PhC structure we use in this

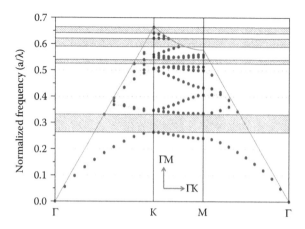

Figure 9.2 Band structure of the PhCs structure of hexagonal lattice.

work, is formed by removing air holes of the hexagonal trace in the 220 nm thick silicon PhC slab. Some line defect, which is not shown in the plot, can be treated as a waveguide that can guide the input and output light beams. The hexagonal lattice gives the nano-ring resonator a better photon confinement, such that a higher Q-factor is achieved [37–39].

9.2 Sensing principle

9.2.1 PhC sensing with defect resonance

A PhC-based cavity resonator is one of the perfect sensing elements to simultaneously achieve high sensitivity, high accuracy, and low cost. As a typical structure type, a PhC cavity resonator is formed by defects in the orderly arranged lattices. It exhibits strong spatial and temporal light confinement and a high Q-factor, thus greatly enhancing the light-matter interaction strength in the defect region. As for sensing applications, the enhanced optical field and materials interaction effect give rise to an optical mode of PhC with a resonant wavelength highly sensitive to the local variations in its surrounding medium. Another attractive characteristic of the PhC is that we have full control of the PBG. We can change the materials we are using, and modify the ratio of the lattice constant and hole radius to change the location and width of one particular PhC bandgap. When we have a defect in the PhC slab, the appeared resonance peak is always associated with the PBG, hence, we usually use the resonance peak as the signature label to perform the sensing activities.

In addition, the effective sensing area of PhC, on the order of a micrometer or less, has great potential as the array-based sensing platform is attractive in very near future applications, thus the sensing accuracy and sensing capability can be greatly enhanced. Leveraging with other techniques, such as MEMS, forming up the LOC design of the PhC sensor array provides an advanced sensing platform for in situ monitoring with smart design. Figure 9.3 shows some typical PhC cavity structures in 2D PhC slab examples, which have more variations than the 1D PhC cavity. One or a few point defects in the same line create the L type PhC cavity, which is the most popular PhC cavity resonator type, as shown in Figure 9.3a. The H type PhC cavity keeps all the PhC holes, but shifts their lattice position to create the defect, as shown in Figure 9.3b. The third type of PhC resonance cavity is the mode gap cavity, which is formed by two designs of PhC structures with different PBGs (Figure 9.3c). In certain wavelengths, the middle PhC structure with the pass band is sandwiched between two PhC structures and has their bandgap, hence this creates the resonance structure. The PhC defect can also form more complex structures. Figure 9.3d is an example of a PhC defect ring cavity. Based on the hexagonal lattice structure, the ring cavity has a hexagonal shape. In Figure 9.3e,

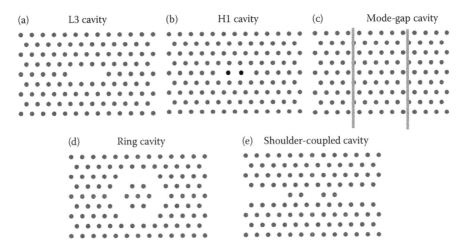

Figure 9.3 Typical PhC cavity structure design examples in 2D PhC slab: (a) L3 cavity; (b) H1 cavity; (c) mode-gap cavity; (d) ring cavity; and (e) shoulder-coupled cavity.

the PhC cavity is assisted by a PhC waveguide and the coupling between the waveguide mode and resonance mode can be manipulated with the period number of the shoulder lattices.

9.2.2 PhC sensing without resonance

Sometimes, a PhC cavity resonance is not necessarily needed in optical sensing. A line defect in PhC creates a waveguide which operates in the slow light regime. Slow light means that the light travels with a very low group velocity. In general, the proposed PhC waveguide can support the slow light because of the standing wave pattern generated by the PhC lattice. In addition, some engineering works have been done to slow down the light in order to enhance the performance of the optical sensors. The general purpose of the slow light is to enhance the light intensity in certain areas, and hence enhance the light-matter interaction as well as the electro-optics and thermo-optics. By extending the cavity defect in the PhC to a line defect of missing air holes, a waveguide is created and works in a slow light regime. The PhC waveguide allows the propagation of certain wavelengths of light, which is confined by total internal reflection and Bragg reflection in the lateral direction. A PhC waveguide also slows down the light travelling speed and results in a strong localized electrical field. Hence, by leveraging this property and the strong light-matter coupling properties in the slow light region, the sensitivity of the optical sensor can be enhanced and can be comparable to the resonance-based sensing mechanism.

To illustrate the slow light effect in an optical readout, the slow light device is always associated with the Mach–Zender interferometer (MZI), which shows the phase differences between the slow light and the reference light in the output. The change of the MZI ripple interval provides us the information of the slow light performance, and hence becomes the sensing element. The change of the slow light position, the bandwidth, and the group index number illustrate the change of the local variation of the surroundings in the slow light region.

9.2.3 PhC sensing with local physical variations

There are many local variations which can cause the interaction with the optical mode in PhC and change the optical properties in the PhC structure. In this chapter, we will only discuss the physical sensing, which means the disturbance to the PhC occurs in physical forms. Integrating with

MEMS structures is the most popular and common way to transfer the physical disturbance to the PhC structure.

Deformations: The first thing to be discussed regarding physical disturbances is deformations. The deformations could be produced by a bended cantilever or stretched thin films. Either bending or stretching results in the same consequence, during which the shape of the lattice will be affected. From our discussion in the last section, the shapes of the lattice define the PBG performance and cause the cavity resonance wavelength shifts. By finding out the relationship between the shifts and the deformation and the cause of the deformation, either by a force or a pressure, we can sense these physical parameters by reading the wavelength shifts.

Distance: Light travels between two optical waveguides and we call this evanescent wave coupling. It is a near field interaction between two waves, hence it is highly sensitive to the distance between the two waveguides. By putting two resonators together, we can find the resonance energy and the resonance amplitudes that change as the distances getting closer to each other. The one with the energy before coupling will lose energy when they are close enough; and the other one will gain the energy until some critical point is reached. We can find out the distance between the two objects by comparing their optical signal strengths. This way has a much better resolution since the critical coupling distance is usually in the range of 100 nm and is much smaller than the optical diffraction limits.

Stress: Some crystal materials' indices will change as a result of the influence of external variables, such as static or mechanical stresses. This is known as the effect of photoelasticity. The index change can be calculated using the photoelastic tensor and this provides us with a good platform to monitor the stress under the crystal of one material by measuring its refractive index changes. Applied on the PhC cavity resonator, the resonance peak shifts with the change in the material's index changes. However, if the stresses come from the deformation of the PhC slab, the shift caused by the stresses will be much less than the deformation caused. Hence, the stress sensing elements are mainly applied on bulky materials.

Temperature: The thermo-optical effect is well known and most of the materials have either a positive or negative effect, which means their refractive indices will increase as temperature increases and vice-versa. The best property of this effect as a platform for sensing applications is that the change of refractive index and temperature has a linear relationship. This prevents further transformation of the optical signal to the real external physical signal and enhances the accuracy of the sensor performance.

9.3 One-dimensional PhC sensor

The first PhC structure is measured in one-dimensional form. The smaller footprint and easy preparation are the advantages of the 1D PhC resonator. However, because the light confinement is supported by the total internal reflection in all directions except the direction with the PhC structure, the photon lifetime inside the PhC cavity is much lower than that of the other forms of the PhC cavity. Hence, the coupling distance sensing is the main streaming method for 1D PhC cavity structures.

9.3.1 Optomechanical accelerometer using PhC beams

The monitoring of acceleration is essential for a variety of applications. Because of the rapid development of MEMS technology, accelerometers have become exceedingly popular. With a canonical sample of an accelerometer, the mass will experience a force when it is under a constant acceleration. Krause et al. presented their work on the accelerometer sensing in a microchip in 2012 [40]. As shown in Figure 9.4d, if the force is balanced with the spring, a constant displacement of the testing mass m will be presented. From here, we can bring in the optical coupling concept with two one-dimensional photonic crystal nanobeams, which attached to the testing mass and the anchor frame separately, as illustrated in Figure 9.4 a and b. The SEM image shows that the PhC cavity is located at the center of the beam and will be coupled when the two objects get close to each other. The testing mass, which is made of a big bulk of SiN, with dimensions of 150 μm × 60 μm × 400 nm, is suspended on nanotether springs. The nanotether is made of 150 nm wide and 560 μm long highly stressed SiN and allows high oscillator frequencies and a high mechanical Q-factor. Figure 9.4 c demonstrates an array of devices with different test mass sizes, which could extend the dynamic testing range of photonics accelerometers. The PhC cavity is designed to operate in the telecommunication band, with a measured optical resonance at 1537 nm and an optical Q-factor of 9600. When the two beams get closer, the optical cavity field is largely confined to the slot between the nanobeams. Figure 9.4e is the transmission curve of this PhC cavity. A displacement of the test mass caused by in-plane acceleration can be readout optically since the resonance frequency is sensitively coupled to the relative motion of the PhC nanobeams. A pumping and reading setup is shown in Figure 9.14f. A tapered fiber is moving closely to the coupled optical cavity region, then we scan the optical wavelength pumping into the fiber and read the transmissions power from the other end of the fiber. The device is attached to a vibration shaker of a shear piezo actuator to give an AC acceleration to the device chip.

Figure 9.4g demonstrates the linear dynamic range of the devices. While varying the amplitude of the acceleration applied with the calibrated shake table at 9.92 kHz, optical signal transduced via

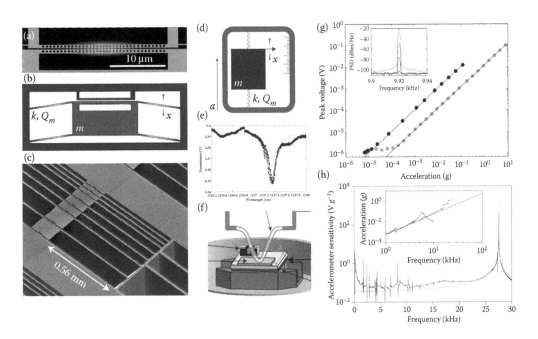

Figure 9.4 (a) Zoom-in of the optical cavity region showing the magnitude of the electric field |E(r)| for the fundamental bonded mode of the zipper cavity. The top beam is mechanically anchored to the bulk SiN and the bottom beam is attached to the test mass. (b) Schematic displacement profile (not to scale) of the fundamental in-plane mechanical mode used for acceleration sensing. (c) SEM image of an array of devices with different test mass sizes. (d) Canonical example of an accelerometer. When the device experiences a constant acceleration a, a test mass m undergoes a displacement x = ma/k. (e) Example transmission curve of the PhC cavity. The curve is obtained by scanning an external cavity diode laser across the cavity resonance at 1537.36 nm while monitoring the fiber taper transmission. (f) Illustrations of experimental system. The light is coupled to the optical cavity via a closed taper fiber. (g) Demonstration of larger linear dynamic range of this device. (h) Sensitivity curve as a function of frequency. The dashed line corresponds to the theoretical expectation for the sensitivity without fit parameters. Inset image illustrates the data from commercial accelerometers also attached to the shake table, which were used for calibrating the applied acceleration.

the mechanical mode is measured. The black lines are linear fits to the data and it shows highly linear response over 40 dBs. The inset shows the corresponding power spectral density (PSD) spectra. Figure 9.4h shows the demodulated photodiode signal normalized to the applied acceleration as a function of drive frequency, corresponding to the frequency-dependent acceleration sensitivity of the PhC cavity. The highest sensitivity appears at the nature frequency of the test mass which gives rise to the highest displacement to the vibrating mass.

In summary, the presented optical accelerometer structure leverages the different resonance wavelength under different coupling distances to show their high sensitivity and linear relationship over a large bandwidth. By having arrays of different mass size and nanobeam spring dimension, the sensing range can be further extended. The whole structure is defined in a 400 nm thick SiN layer formed on top of a 500 μm thick Si wafer. The structures are defined in a single electron beam lithography step and transferred into the SiN layer using inductively coupled plasma (ICP). The structures are undercut by antisotropic wet-etching in 70° hot KOH solutions.

9.3.2 PhC distance sensor using coupled 1D PhC cavities

For the in-plane coupled PhC cavity, Tian et al. demonstrated a succinct and compact out-of-plane displacement sensing mechanism using double-coupled structure [41]. The SEM image of the structure is shown in Figure 9.5a. The two cavities are coupled with a gap of 144 nm. From the cross-section

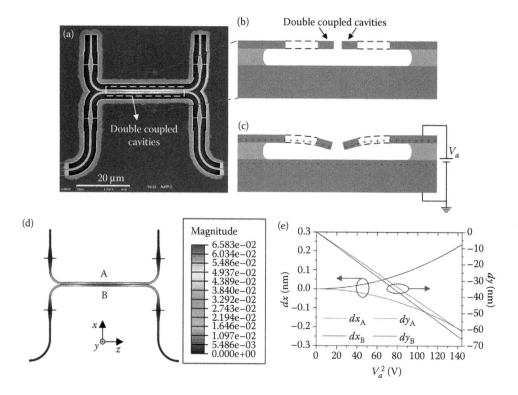

Figure 9.5 (a) SEM image of the double coupled one-dimensional PhC cavities supported by the bended waveguides. (b) Schematic of the cross-section along the center of coupled cavities. (c) Using voltage to simulate when there is an out-of-plane displacement. (d) Deformation of the suspended structures simulated by the finite element method (FEM) under an applied voltage of 12V. The A and B denote the highest magnitudes of deformation of 65.83 nm. (e) Displacement components along the x and y axes versus the square applied voltage. (*Continued*)

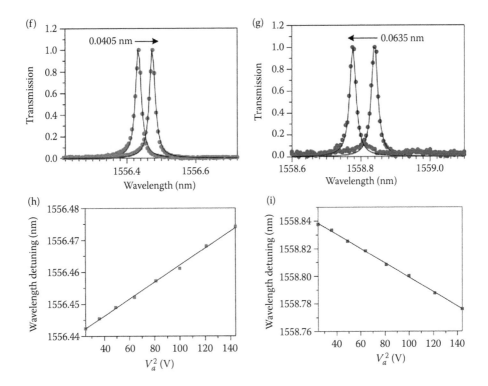

Figure 9.5 (Continued) (f) Measured wavelength shifts of the resonance peaks of the second-order odd modes driven by a voltage switching from 0 V to 12 V. (g) Wavelength shifts of the resonance peaks of the second-order even modes. (h) The wavelength detunings of the second-order odd resonance modes under various applied voltage. (i) The wavelength detunings of the second-order even resonance modes.

view of such structure (Figure 9.5b), we find that the residual stress in the device layer needs to be prevented to avoid influence of the suspended structure. To achieve this, the author designed a bended shape supporting waveguide to release the stress in plane. To simulate the displacement in vertical direction, a voltage is applied between the device layer and the handle layer. A FEM simulation is used to find the transfer function of the voltage to displacement, as shown in Figure 9.5d, e. The optical characterization of the device is conducted with the tunable laser source and the optical spectrum analyzer. The two coupled 1D PhC cavities used here are identical and designed following the principles of high Q-factors and high transmissions. The author chose the second-order odd (TE_{o2}) and even (TE_{e2}) resonance modes to characterize the cavities' resonance detuning driven by the vertical electrostatic force. Figure 9.5f, g show the tuning process of the TE_{o2} and TE_{e2} modes when the voltage is applied from 0 V to 12 V. There is a red shift of 0.0405 nm for the $TE_{o,2}$ mode and a blue shift of 0.0635 nm for the $TE_{e,2}$ mode. Figure 9.5h, i summarize the linear relationships between the detuning of two second-order modes and the square applied voltages, respectively. The two linear relations of the voltage and displacement and resonance wavelength shift imply that the transfer function of the displacement to the optical readout is also in linear relationships.

9.4 Cantilever-based PhC sensor

Cantilever is one of the most commonly used MEMS structure. It has wide freedom to design the proper mechanical characteristics, such as the mechanical resonance frequency and mode, stress and strain distributions and mechanical tunabilities. Hence, integration of the

PhC structure together with the MEMS cantilever can easily transfer the mechanical variation of the surroundings to the PhC optical mode region and generate high sensitivity signals.
Moreover, since the cantilever usually contains a bigger active area, two-dimensional PhC slab
is usually implied to obtain better optical confinement and higher mechanical variation to the
PhC structure.

9.4.1 Operation mechanism of PhC cantilever sensor

Most of the PhCs resonators proposed here have channel drop effect. The drop mechanism obeys
the field distribution symmetry theory discussed by S. Fan et al. [31]. More specifically, the resonator is only active in the single channel in some particular wavelength range [42]. A dual nanoring
(DNR) resonator is depicted in the inset of Figure 9.6a. Two nanorings are sandwiched in between
two bus waveguides. The input port is indicated by an arrow, while the corresponding spectra
is observed a strong resonant peak in the backward drop (BD), port and the forward drop (FD),
and transmission (TR), ports are in low energy state. Figure 9.6b shows frames of the resonator
magnetic field distribution in the steady state condition which is varying with time. We define
the right-hand side ring as R1 and the left-hand side ring as R2, respectively. The input light beam
comes from the bottom right side and travels along the input waveguide and excites the nanorings
region. The nanorings then couple the light energy of certain wavelength to the output waveguide.
The field distribution graphs show that the two rings are always in the same phase, and light
energy strength changes periodically and simultaneously. Strong field is observed from the input
waveguide to both nanorings as well as the middle 4 holes between the two nanorings. A strong
coupling effect between rings and the drop waveguide can be obtained too. But only the BD signal
is active. High level of energy exchange in the middle 4 holes between the DNR indicates that the
two rings are not working independently of each other. Light coming from the input waveguide
will travel within the two rings first before coupling to the drop port at the output waveguide.
Figure 9.6c shows a proposed light drop mechanism in a DNR resonator-based channel drop filter.
When the light of R1B and R2B is in-phase and the light of R1F and R2F is out of phase, constructive and deconstructive interference occurs between R1B, R2B and R1F, R2F, respectively.
Therefore, 100% of dropped light goes to the BD port, that is, the total backward drop condition,
due to the conservation of energy.

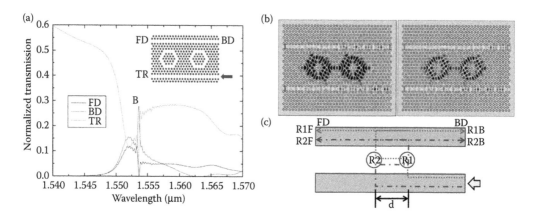

Figure 9.6 (a) Spectra of ports FD, BD, and TR for DNR channel drop filter; (b) no-load case: out-of-
plane magnetic field distribution plot at resonant steady state; (c) conceptual drawing of light paths
dropped through DNR resonators to the FD and BD ports. (Adapted from Li B et al. *IEEE Transactions on
Nanotechnology*, 10(4), 789–796, 2011.)

Based on the drop mechanism of DNR resonator, the size of the ring and the separation distance of DNR have great effect on the resonant wavelength. These properties of DNR make it suitable for use as a sensing element for force and strain detection, since the resonant wavelength measured at BD port is strongly dependent on the ring size and separation distance. When the DNR is deformed, the resonant wavelength will change accordingly. Hence, the DNR resonator can be placed at the junction of a free-standing silicon cantilever and an edge of substrate as shown in Figure 9.7a. When an external force is applied on the cantilever's free-end, strain at the junction is created. Such strain is in linear proportion to the applied force. More precisely, the vertical displacement (Z-displacement) at different positions along the longitudinal direction of the cantilever gradually increases as the position moves to the cantilever edge under an applied force. By leveraging this unique feature, the input and output ports of the DNR resonator are arranged on the same side of the cantilever sensor as shown in Figure 9.7a. It renders a simple configuration of cantilever sensor with advantage of easy setup of resonator wavelength measurement with respect to the optical fiber alignment task. To optimize the performance of DNR in terms of a sensing element for cantilever sensor, three types of cantilever sensors are proposed in Figure 9.7b. The junction of cantilever and edge of substrate is placed at the right corner of the first ring (R1), that is, type 1 cantilever. The junction of cantilever and edge of substrate is placed at the left corner of the R1 for type 2 cantilever, and is arranged at the right corner of the second ring (R2) for type 3 cantilever, respectively.

In the applications, either external force or surface stress will introduce a vertical displacement at the cantilever end and strain at the DNR resonator sensing element region. FEA modeling is deployed to obtain the deformation data of holes among the PhC air hole array under various loading force applied at the center point of the cantilever free end, as shown in Figure 9.7c. The FEA simulation results show that the location of DNR resonators can lead to different strain and deformation for all three types of cantilevers. We further applied numerical 2D FDTD method to simulate the propagation of the electromagnetic waves in the deformed PhC DNR resonator structure in the presence of various applied force loads. We expect the resonant peak at the BD port will vary according to the strain of DNR resonator. Due to the different location of DNR resonator in the three cantilevers, we can investigate the resonant behavior with respect to ring deformation and strain in the 4-hole separation. Figure 9.8a through c show

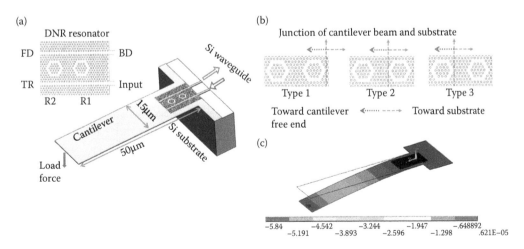

Figure 9.7 (a) 3D illustration of the PhC cantilever sensor using hexagonal dual nano-ring (DNR) resonator. (b) three types of cantilever sensors with different locations of DNR resonator around the junction of cantilever and substrate. (c) Side view of FEA simulation result of vertical displacement along the cantilever sensor using DNR when there is 0.2 μN force applied at the end of cantilever. (Adapted from Li B et al. *IEEE Transactions on Nanotechnology*, 10(4), 789–796, 2011.)

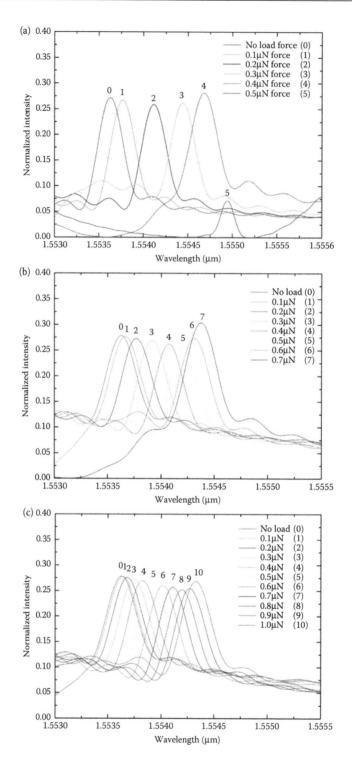

Figure 9.8 Resonant wavelength peak at the backward drop (BD) port under various force loads: (a) Type 1 cantilever; (b) Type 2 cantilever; (c) Type 3 cantilever. (Adapted from Li B et al. *IEEE Transactions on Nanotechnology*, 10(4), 789–796, 2011.)

the resonant wavelength peak derived at the BD port under various force loads with a step of 0.1 μN for all three cantilevers. The resonant wavelength will shift to the higher wavelength as the load force increases, that is, a red shift, in all three cases. For the type 1 cantilever (Figure 9.8a), the resonant peak shifts from 1553.6 nm to 1554.9 nm with the same Q-factor when the load force changes from 0 to 0.4 uN. However, as the loading force approaches 0.5 μN, the output signal intensity at the BD port becomes relatively low in comparison to the other cases of smaller force loads, while the FD signal shows comparable intensity as the BD signal. Major portion of input energy does not couple into the BD and FD ports. Thus, we observed strong output intensity at the TR port. It refers to the resonant behavior for BD is degraded due to the deformed DNR resonator.

On the other hand, for the type 2 cantilever (Figure 9.8b), the left nanoring (R2) and 4-hole separation will experience deformation under force loads, while the equivalent deformation within R2 region in type 2 cantilever is expected to be smaller than such deformation of R2 in the case of type 1 cantilever. The resonant behavior maintains the channel drop mechanism up to load force of 0.7 μN. The BD behavior is degraded in the case of higher force load. Figure 9.8c shows the output resonant peaks at the BD port for various force loads in the type 3 cantilevers. In this case, only the R2 kept within the deformation range. Since the shape of the right nanoring (R1) and the 4-hole separation are maintained as the same as the no load case, the deformation of R2 provides the largest detectable force range of 1 μN. In other words, the BD behavior is degraded in the case of force load higher than 1 μN. However, it also implies that type 3 cantilever becomes less sensitive due to a smaller portion of DNR resonator maintained within the deformation region, that is, less strain experienced.

Figure 9.9a illustrates the relationship between the resonant peak wavelength versus Z-displacement and applied force at the cantilever end, and the relevant data is depicted in Figure 9.9b. We observed good linear relationship of resonant peak wavelength versus applied force as well as Z-displacement for type 1 cantilever in Figure 9.9a. The slope of linear data of type 1 cantilever is derived as 370 nN/nm (applied force wavelength shift) and 10.30 μm/nm (Z-displacement wavelength shift) in Figure 9.9b. It is suggested that the minimum detectable force and minimum detectable displacement will be 37 nN and 1.03 μm, respectively, if the wavelength detection resolution is 0.1 nm for the testing setups [39]. Second, the data points of type 2 and type 3 cantilevers follows linear behavior approximately in both figures. The slope of data fitting lines in Figure 9.9b is derived as 830 nN/nm or 23.26 μm/nm for type 2 cantilever, and 1250 nN/nm or 35.57 μm/nm for type 2 cantilever, respectively. Therefore, the minimum detectable force and the minimum detectable displacement are derived as 83 nN and 2.33 μm for type 2 cantilever, and 125 nN and 3.56 μm for type 3 cantilever, respectively. It implies that type 1 cantilever is more sensitive than the other two types but having the smallest detectable force range, that is, less than 0.5 μN. In the case of type 3 cantilever, the detectable force range is up to 1 μN, although the minimum detectable force is 125 nN. Compared with the nanocavity resonator cantilever and the single nanoring resonator cantilever, the proposed type 1 DNR cantilever achieves better data of the minimum detectable force [43,44].

For type 3 cantilever, the linearity looks less perfect than the other two cases. This can perhaps be attributable to the presence of only a nanoring located within the deformation range. To further clarify the origin of dependence of resonant wavelength versus force loads in the SNR resonator, we conducted the 2D FDTD modeling for a cantilever with SNR at the position the same as the case of the type 3 cantilever. In contrast to this 2nd-order polynomial data, relatively linear fitting line is given for data derived for type the 3 cantilever. We conclude the linear relationship between the resonant wavelength and applied force or Z-displacement is obtained for cantilevers using a DNR resonator, while the same linear data is derived for resonant wavelength shift as well. Slight degraded linearity is observed for type 3 cantilever because only R2 experienced the deformation. The resonant behavior is still dominant by DNR configuration. Thus, data of type 3 cantilever keeps the linearity.

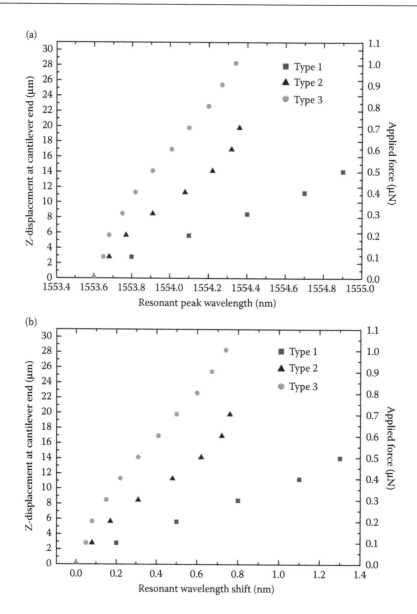

Figure 9.9 (a) Resonant wavelength as functions of Z-displacement and applied force for three types of cantilevers; (b) resonant wavelength shift as functions of the Z-displacement and the applied force for three types of cantilevers. (Adapted from Li B et al. *IEEE Transactions on Nanotechnology*, 10(4), 789–796, 2011.)

9.4.2 Configuration analysis of DNR resonator as sensing element

Various positions adjacent to the junction of the microcantilever and the substrate can also show great differences in cantilever-based PhC sensor. Two types of DNR PhC resonator are shown in Figure 9.10, the hexagonal nanorings and the waveguide are formed by removing the air holes in the PhC trace. For the spectra intensity plot, TR, FD, and BD ports are shown as green, red, and blue curves, respectively. For DNR resonator with 1-hole separation between the rings, shown in the top of Figure 9.10, resonant peak at 1551.41 nm is observed at the FD port with Q-factors about 3800. The intensity of the BD port is only 6% of the FD port. For DNR resonator with 4-holes separation between the rings, shown in the bottom of Figure 9.10, resonant peak at 1553.59 nm is

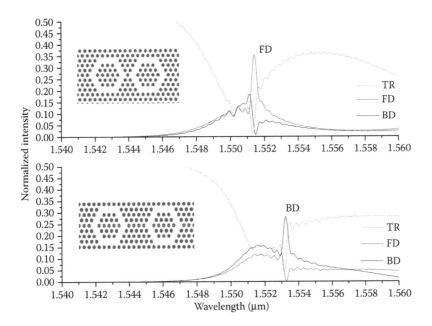

Figure 9.10 Spectra of ports FD, BD, and TR for 1 hole and 4 holes separation DNR channel drop filter. (Adapted from Li B et al. *Sensors and Actuators A: Physical*, 169(2), 352–361, 2011.)

observed at the BD port with Q-factors about 3800. The intensity of the FD port is only 8% of the BD port. Hence, these two designs of DNR resonators are good channel drop filters.

The schematic of force sensors is shown in Figure 9.11, while the DNR resonator is aligned along the longitudinal direction (e.g., V4 type) or the lateral direction of the cantilever (e.g., H4 type), and is placed at the junction of a free-standing silicon cantilever and an edge of substrate. The cantilever with length and width of 50 μm and 15 μm can be patterned and released from the 220 nm thick silicon device layer of an SOI substrate. The lattice constant (a) of 410 nm and air hole radius (r) of 180 nm of the PhC slab of hexagonal lattice are deployed in modeling. The two hexagonal nanorings are placed along the longitudinal direction of the cantilever, and are separated by 4 air holes along their symmetric axis. The whole layout configuration including the lattice constant, a, air hole radius, r, and the air-hole separation between waveguides and rings has been optimized in order to achieve resonant peak with a higher Q-factor. Four types of microcantilever sensors are characterized in this study, that is, where the DNR resonator is placed longitudinally (Y direction) and transversely (X direction) against the cantilever beam with regard to two types of hole separation, for example, 1-hole and 4-holes cases, as shown in Figure 9.11. The microcantilever sensors

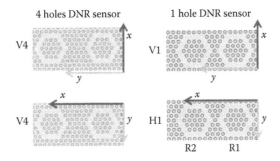

Figure 9.11 Define the 4 types of DNR cantilever sensors by varying the orientation of the DNR resonator. (Adapted from Li B et al. *Sensors and Actuators A: Physical*, 169(2), 352–361, 2011.)

with a DNR resonator of 4-holes separation for longitudinal and transverse layout arrangements are named V4 and H4, respectively, while the equivalent layout arrangements for microcantilever sensors of 1-hole separation are named V1 and H1, respectively.

The well-defined linear relationship of the z direction displacement at the cantilever end and the applied force of the H1 sensor is illustrated in this session. In the following discussion, we investigate the resonant wavelength and resonant wavelength shift with respect to the applied force. We can foresee the same trend of curves could be derived for curves of resonant wavelength versus Z-direction displacement, or resonant wavelength shift versus Z-direction displacement. We further applied the numerical 2D FDTD method with the ERI method to simulate the propagation of the electromagnetic waves in the deformed PhC DNR resonator structure in the presence of various applied force loads. We expect the resonant peak at the BD port to vary according to the strain of the DNR resonator. As mentioned above, the shape change of the air holes among the deformed PhC nanocavity resonator structure does not affect the output resonant behavior. More importantly, the relative position shifts of these air holes among the deformed PhC nanocavity resonator structure play a major role in contributing to the resonant behavior. Thus, we record the position of the holes in the deformed region of the whole sensing element of the air hole array of the microcantilever under a force load according to the FEA results, and conduct the 2D FDTD modeling based on the layout of such deformed DNR resonators.

Figure 9.12 shows the resonant wavelength peak for V4 and H4 sensors derived at the BD port under different force loads with a step of 0.1 μN. The resonant peak, which is located at 1553.6 nm with a Q-factor of about 3800, is referred to as the no-load case. As the force increases, both sensors show a red shift of the resonant wavelength and demonstrate a well-defined resonant peak within the force range of 0.4 μN. However, in the case of the V4 sensor, the intensity of the resonant peak at the BD port becomes relatively weak when the force reaches 0.5 μN. It is suggested that the channel drop effect becomes weak such that a limited amount of light is coupled to the BD terminal via the DNR, and most of the pumped energy goes to the TR port. In other words, the resonance of the DNR is slightly deviated while the DNR resonator experiences a large deformation. Thus, the coupling becomes worse. In the H4 sensor case (Figure 9.12b), both rings of the DNR experience the same amount of deformation due to the cantilever bending, since the two rings are placed in the transverse direction of the cantilever. A minor peak of the shorter wavelength in addition to the major peak is observed for the H4 sensor under force loads of 0.5 μN and 0.6 μN, because the deformed ring shape brings extra coupled energy which matches another resonance at the BD port as well.

Figure 9.13 shows the resonant wavelength peak for V1 and H1 sensors derived at the FD port under different force loads with a step of 0.1 μN. Different from the 4-holes separation case, the resonant wavelength of the no-load case locates at 1551.41 nm. The resonant peaks shift to the higher wavelength range when the loaded force increases for both the V1 and H1 sensors. However, the Q-factor of the resonant peak for the V1 sensor drops dramatically as force increases, from over 3800 for the no-load case to 873 for the 0.5 μN applied force case. The output intensity also drops as the force increases. Thus, these imply that the deformation of the V1 DNR resonator changes the resonant properties as well as the light confinement of the resonator. Hence, the sensible range for the V1 would be the smallest among all the proposed sensors. On the other hand, the H1 resonator shows good Q-factors and output intensities for the large force sensing range. The resonant peaks lose their comparable intensities until the applied force reaches 0.7 μN, which is the best result among all the four types of microcantilever sensors.

9.4.3 Strain sensing effect of PhC cantilever sensor

A theoretical and experimental work of the strain sensing effect of a PhC cantilever sensor was investigated in 2011 [46]. A 2D PhC cavity resonator is integrated with a MEMS cantilever structure and the strain sensitivity of a high Q-factor PhC resonance is calculated by FDTD and FEM simulations. In this work, a cavity structure is created by removing three air holes, which means it is an

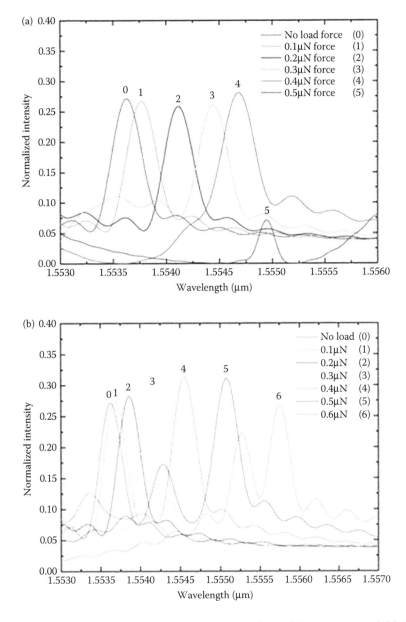

Figure 9.12 Resonant wavelength peak with various load force: (a) V4 sensor and (b) H4 sensors. (Adapted from Li B et al. *Sensors and Actuators A: Physical*, 169(2), 352–361, 2011.)

L3 cavity, as shown in Figure 9.14a. The in-plane confinement of light was guaranteed by the PhC structure, and the out-of-plane confinement was due to the total internal reflection at the Si–SiO_2 and Si–air interfaces. Figure 9.14b shows the output transmission spectrum in the free stress/strain status. The shifts in these resonant wavelengths were recorded during the application of mechanical strain. To analyze the strain effect of the PhC structure, changes in the geometry of the PhC when subjected to stress/strain were simulated using FEM. Then, these changes in geometry, such as the hole's position and shape, were used as input parameters for the FDTD simulation in order to obtain the transmission spectrum. The FDTD simulation is shown in Figure 9.14d and e. Under a longitudinal strain, the wavelength tends to shift to a longer wavelength, whereas it tended to shift to a shorter wavelength under a transverse tensile strain. The relationship between the applied stress/strain and the resonant shift is linear in the tested stress/strain range. Figure 9.14e shows the simulation results

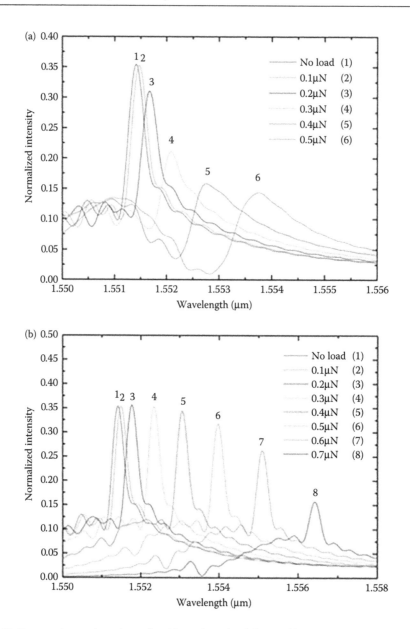

Figure 9.13 Resonant wavelength peak with various load force: (a) V1 sensor and (b) H1 sensors. (Adapted from Li B et al. *Sensors and Actuators A: Physical*, 169(2), 352–361, 2011.)

for the shift in the resonant wavelength due to longitudinal and transverse strains. The difference in the shifts in resonant wavelength because of the longitudinal and transverse strains highlights the anisotropic property of the strain sensing effect of the PhC cavity.

Figure 9.14c shows the SEM image of a fabricated device. An EBL and a fast atom beam etching process are applied to create the PhC structure. The characteristics of the fabricated PhC cavity were evaluated using the transmission measurement system. Figure 9.14f illustrates the cantilever strain generating system. The cantilever is mounted on a fixed base and the PhC cavity is placed near the fixed end of the cantilever in order to realize a large strain. To avoid the alignment issue during the strain testing, the coupling facets are designed on the fixed position of the cantilever to ensure the maximum coupling during the entire testing processes. Figure 9.14g and h show the measurement results of the cantilever sensor. The inset of Figure 9.14g is the near field image from

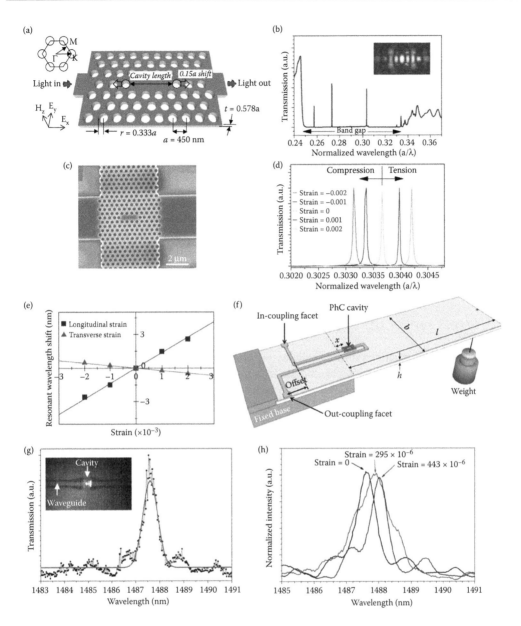

Figure 9.14 (a) Configuration of photonic crystal cavity to investigate strain sensing effect where r/a = 0.333 and lattice constant a = 0.45 nm. The arrows indicate the launching direction of light used for exciting the resonator. (b) Transmission spectrum of resonant modes showing cavity peaks at 0.2573(a/λ), 0.2733(a/λ), and 0.3037(a/λ). The inset shows the intensity profile of a degenerate mode obtained using FDTD simulations. (c) SEM images of photonic crystal structure after fabrication. (d) Transmission spectrum showing resonant wavelength versus different strain values. (e) Resonant wavelength shift versus strain. (f) Schematic of experimental setup of mechanical strain loading model. (g) Transmission spectrum of fabricated cavity. The inset shows the near field image observed using an infrared camera from the top of the cavity. (h) Shift in resonant wavelength of cavity due to application of strain.

the top of the cavity. The strong resonance field is observed in the cavity region and the measured transmission spectrum has a good match with the simulation data. Figure 9.14h shows the resonance shift under the strain effect. Because of the setups, only the longitude tension strain is plotted in the image. A shift of resonance to the higher wavelength is observed and matches with the simulation data very well.

Microcantilevers using integrated PhC are proven effective for NEMS mechanical sensor applications. Different structures of PhC cavity resonators, as well as the location arrangements, for example, longitudinal direction and transverse direction, show various responses in terms of resonance wavelength shifting directions and sensitivities. The combining of FEM and FDTD simulation methods for this PhC-based mechanical sensor is working very accurately. The various potential designs of the MEMS cantilever call for further research to apply these potential PhC-based mechanical sensing devices.

9.5 Diaphragm-based PhC sensors

A diaphragm is another common MEMS structure that can be integrated with a PhC structure. In the MEMS process, releasing holes are needed to define a diaphragm on a substrate. This creates an advantage in that the PhC hole is the perfect releasing hole for the diaphragm, and this makes this process become a mask-free process. Second, compared with the cantilever-based structure, the diaphragm experiences more uniform strain and stress change along the PhC slab. Hence, the linear sensing performance shows a better result than that of the cantilever case.

9.5.1 Stress sensor using PhC cavities film

The geometrics and positions of the air holes are modified by applying mechanical stress to the PhC slab, hence the stress sensing can be realized from the measurement of changes in the optical property. To achieve a high Q-factor and hence a high sensitivity for the PhC stress sensor, the author, Yang et al. [47] proposed an aslant nanocavity which is 60° from the Γ-K direction so that the applied stress in both directions generated almost the same geometry variations on the cavity, as shown in Figure 9.15. Figure 9.15a is the 3D FDTD simulation result of the electrical

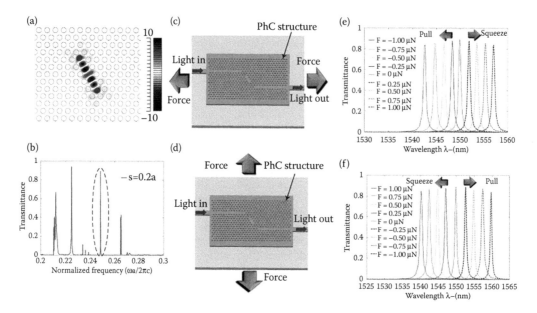

Figure 9.15 (a) Electric field distribution in the x–y plane of the cavity by using 3D FDTD simulations, where lattice constant a = 385 nm, diameter of air holes is 0.6a, thickness t = 0.56a. (b) Transmittance spectra of original PhC cavity which shows four peaks. (c) Schematic design of the stress sensor under stress in the horizontal direction. (d) Stress in the vertical direction. (e) Transmittance spectra showing the shift of resonant wavelength under different force in horizontal directions. (f) Transmittance spectra showing the shift of resonant wavelength under different force in the vertical direction.

field distribution in the PhC plane. This result proves that the aslant cavity has great light confinement ability. The transmittance spectrum of the PhC cavity is shown in Figure 9.15b. The highest Q-factor peaks, labeled with the dashed line circle, is chosen to observe the shift under the stress applied. Figure 9. 15c and d illustrate the schematic of the stress sensor design. The direction of the stress can be differentiated by detecting the resonant wavelength shift directions. The value of the stress can be known by measuring the shift wavelength of the resonant peak.

The simulation is done by utilizing FEM and 3D FDTD simulations together. In the next step, all the different shapes of a PhC structure under different applied stresses are modeled in 3D FDTD to obtain the transmittance spectra. Figure 9.15e and f show the transmittance spectra under different forces in the horizontal and vertical directions. The shifts of the resonant wavelength in both directions are evenly distributed with the force variations. The resonant wavelength has a blue shift when the structure is pulled and a red shift when the structure is squeezed in the horizontal direction. On the contrary, opposite results are obtained in the vertical direction, when a blue shift occurs when squeezed and a red shift when pulled. Therefore, we can differentiate the stress directions by monitoring the resonance wavelength shift direction.

9.5.2 PhC diaphragm sensor using triple-nano-ring resonator

The triple-nano-ring (TNR) PhC resonator is formed by three triangularly arranged hexagonal nanorings with a size of 2.87 μm for each ring, as shown in Figure 9.16 [45]. A diaphragm of 10 μm in diameter (d) is patterned and released from the 220 nm thick silicon device layer of a silicon-on-insulator (SOI) wafer. Many methods are available to fabricate such a structure in the literature, such as deep UV lithography [44]. The two waveguides are placed at two sides of

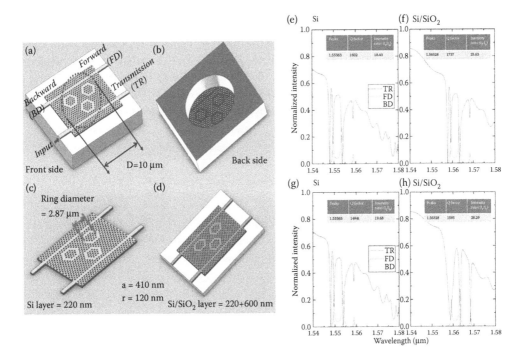

Figure 9.16 (a,b) 3D illustration of the PhC diaphragm sensor using hexagonal triple-nano-ring (TNR) resonator; (c) Si layer diaphragm; (d) Si/SiO₂ layer diaphragm. Spectra of ports FD, BD, and TR for TNR channel drop filter; (e) Si layer diaphragm; (f) Si/SiO₂ layer diaphragm. (g) Si layer diaphragm when input port was swapped with BD port and (h) Si/SiO₂ layer diaphragm when input port was swapped with BD port. (Adapted from Li B, Lee C. *Sensors and Actuators A: Physical*, 172(1), 61–68, 2011.)

the resonator and have two lines of air holes' separation regarding the edge of the nanorings. The input light port located at the bottom left is denoted by a red arrow in Figure 9.16a, and the other three output ports are called transmission (TR), forward drop (FD), and backward drop (BD), respectively. The TNR resonators give strong resonant peaks at the FD port, as shown in Figure 9.16. The spectra are shown in similar fashion, but with different locations of the resonant peak wavelength for the Si layer and Si/SiO$_2$ layer cases in Figure 9.16e and f, respectively. Two major peaks are shown in the transmission spectra; we are more interested in the peak of the longer wavelength. Figure 9.16g and h are the spectra plots of the TNR resonator where we swapped the input port and BD port.

The mechanical deformation or displacement of the diaphragm can be introduced by an externally applied force. Hence, force sensing can be explored by using a diaphragm with the TNR resonator on the Si layer Due to the symmetric nature of a circular diaphragm, the linear relationship of the center displacement along the normal direction of the diaphragm and the applied force is obtained in the FEM modeling results. Figure 9.17 shows the resonant peak plot obtained at the FD port of the TRN resonator. The resonant peak of the no-load case falls at 1553.63 nm with a Q-factor of 1602. As we increase the constant loading force at the center of the Si-diaphragm, the FD resonant peak shifts accordingly. It shows that the FD resonant peak shifts to the higher wavelength region as the applied force increases. Furthermore, the resonant peaks maintain similar Q-factor values as the load force increases and reaches about 20 μN. This implies a wide force-sensing range for the Si-diaphragm as a NEMS sensor.

When the SiO$_2$ layer is retained underneath the Si layer, an Si/SiO$_2$ bilayer diaphragm is created. The enclosed area is sealed by the SiO$_2$ layer and enables the pressure sensing. The inset of Figure 9.18 illustrates the FEA modeling of the Si/SiO$_2$ bilayer diaphragm under various pressure loads. The stiffness of the bilayer structure is higher than that of the Si layer case, hence the deformation of the Si/SiO$_2$ bilayer diaphragm is much less than that of the Si-diaphragm case. However, a similar linear relationship between the loaded pressure and the center normal displacement of the diaphragm is obtained. The maximum deformation occurs at the center part of the diagram, hence we put our TNR resonator at the center of the diagram. Figure 9.18 shows the resonant peak plot obtained at the FD port of the TRN resonator in the case of the Si/SiO$_2$ bilayer diaphragm. The resonant peak of the no-load case falls at 1563.28 nm with a Q-factor of 1737. The first two pressure data correspond to the force loads of 10 μN and 20 μN, that is, the force loads investigated in the previous Si-diaphragm case, but no wavelength shift is observed since the deformation is

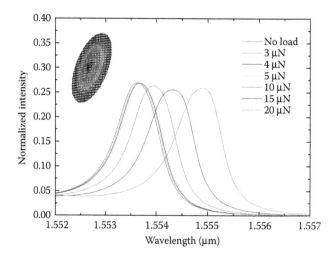

Figure 9.17 Resonant wavelength peak at the backward drop (BD) port under various force loads of Si layer diaphragm force sensor. (Adapted from Li B, Lee C. *Sensors and Actuators A: Physical*, 172(1), 61–68, 2011.)

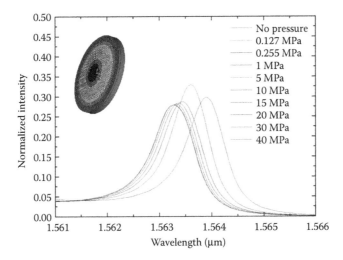

Figure 9.18 Resonant wavelength peak at the backward drop (BD) port under various pressure loads of Si/SiO$_2$ layer diaphragm pressure sensor. The inset picture illustrates the side view of FEA simulation result of Si/SiO$_2$ layer diaphragm under 30 MPa applied pressure. (Adapted from Li B, Lee C. *Sensors and Actuators A: Physical*, 172(1), 61–68, 2011.)

negligible. As the load pressure increases, the same red shift behavior of the resonant wavelength as before is observed and the obtained Q-factor values do not change, which shows that good light confinement is still valid.

Figure 9.19a illustrates the relationship of the resonant peak wavelength with the applied pressure in the case of the Si/SiO$_2$ diaphragm. A 2nd-order polynomial curve (red line) fits the derived data well. In order to evaluate the pressure-sensing sensitivity in the linear region, we divide the data into groups of 5 to 20 MPa and groups of 20 to 40 MPa. By taking the blue dotted fitting lines in these two regions, we derived the slope of 0.0094 nm/MPa in the pressure range of 5 to 20 MPa, and 0.024 nm/MPa for the range of 20 to 40 MPa. Therefore, the corresponding minimum detectable pressures are calculated as 10.64 MPa and 4.17 MPa, respectively. Figure 9.19b illustrates the Q-factor change with respect to applied external pressure. Different from the force sensor, the Q-factor has a significant increase at the 30 Mpa pressure load. Hence, we can conclude that the bottom layer of SiO$_2$ in the pressure sensor provides better light confinement than the single Si layer force sensor.

So far, the resonant wavelength has been characterized and plotted as a function of the applied force and pressure. In fact, deformation at the center of the diaphragm along the normal direction, that is, the Z-direction, is introduced by such force and pressure loads. To compare the sensing sensitivity of these two TNR resonator diaphragm sensors, the resonant wavelength shift was derived according to the Z-displacement measured at the center of the diaphragm, created by either the force load or the pressure load. As shown in Figure 9.20a, in the range of a small Z-displacement, for example, smaller than 0.03 μm, both diaphragm sensors show the same relationship in the derived curves. From the slope of the curves in this region, the derived minimum detectable change in the Z-displacement is 49 nm when the resonant peak wavelength shifts 0.1 nm. In the range of a large Z-displacement, the Si diaphragm shows better sensitivity in terms of resonant wavelength shift under the same incremental Z-displacement. Briefly speaking, the derived data of 49 nm displacement at the center of a 10 μm diaphragm refers to a good sensing capability in the NEMS field. The resonant wavelength shift versus the diaphragm strains are plotted as shown in Figure 9.20b. A linear relationship is found for both the Si and Si/SiO$_2$ diaphragm sensors. The minimum detectable strain is calculated as 0.00427% and 0.00608% under 0.1 nm of wavelength shift for the Si-layer diaphragm and Si/SiO$_2$ bilayer diaphragm, respectively.

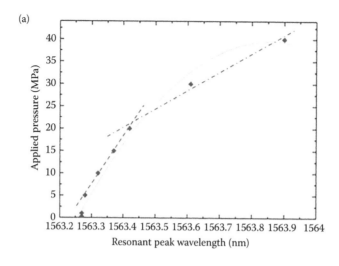

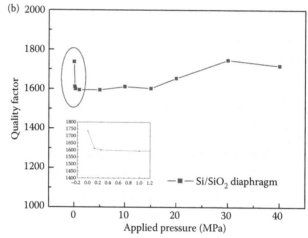

Figure 9.19 (a) Resonant wavelength as functions applied pressure. (b) Q-factor versus applied pressure of Si/SiO₂ layer diaphragm pressure sensor. (Adapted from B. Li and C. Lee, *Sensors and Actuators A: Physical*, 172(1), 61–68, 2011.)

9.5.3 Nonresonance PhC pressure sensor

A similar work was done in 2015 which used PhC and a suspended diaphragm to work as a pressure sensor [49]. The line defect suspended PhC waveguide is used as a sensing element and demonstrated an improved sensor response. In order to design a more robust and stable sensor, a thinner than conventional silicon slab photonic crystal was chosen to increase the dynamic range of the sensor. Figure 9.21a and b show the schematic drawing of the fabricated device. The PhC is etched from an SOI wafer and released from the BOX layer. The diaphragm will deform toward the substrate under uniform pressure, as shown in Figure 9.21b. The principle of operation of the devices is based on the evanescent field coupling between the PhC waveguide and the substrate while the suspended diaphragm is bending to the substrate under the uniform pressure. In order to optimize the sensitivity, the Si film is chosen to enhance the evanescent field coupling. Figure 9.21c shows the transmission spectrum from a 3D FDTD simulation. The transmission power drops while the distance between the PhC waveguide and substrate decreases. The optical power is coupled into the substrate as predicted. In Figure 9.21d, the transmission at a target wavelength of 1550 nm and

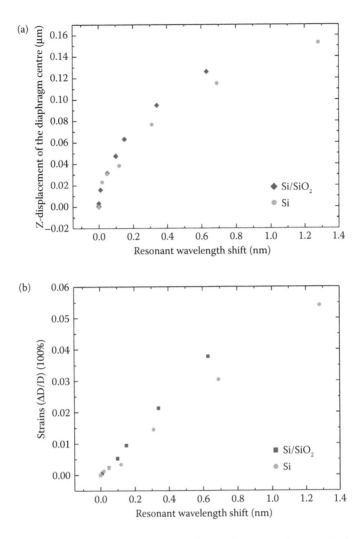

Figure 9.20 (a) Resonant wavelength as functions of Z-displacement for two diaphragm sensors. (b) Resonant wavelength as functions of strains for two diaphragm sensors. (Adapted from Li B, Lee C. *Sensors and Actuators A: Physical*, 172(1), 61–68, 2011.)

the transmission sensitivity versus bridge height are plotted. The best optical sensitivity is found to be 0.4%/nm, and it is centered on the bridge height of 280 nm.

The second parameter to consider is the mechanical design of the suspended bridge, which will be important for the sensitivity as described previously, and the dynamic response of the sensor to applied pressure. To improve the mechanical sensitivity, one could increase the length of the supporting input/output waveguide, which would effectively reduce the stiffness of the central platform. However, a better approach is to increase the applied force by increasing the area exposed to pressure while keeping the sensor footage constant.

PhC-based resonator and waveguide optical devices embedded on a diaphragm show the potential of microforce and pressure sensing applications. The release of the diaphragm using PhC air holes is an easy and straightforward step in MEMS design. Moreover, except for the strain and stress modification of the PhC structure in the diaphragm causing the resonance wavelength shift, the nonresonance concept is demonstrated using PhC waveguides that enhance the evanescent field coupling to monitor the distance between the PhC waveguide and the substrate.

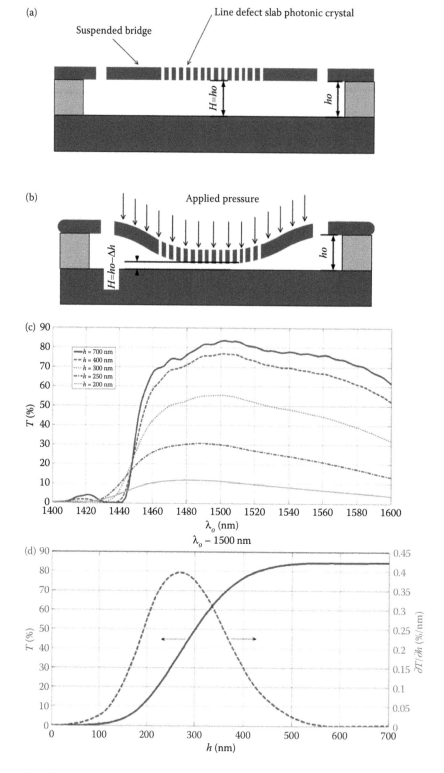

Figure 9.21 Schematic drawing of PhC sensor under (a) no pressure and (b) under uniform pressure over the top surface. (c) Transmission spectrum of the PC line defect waveguide at different bridge heights. (d) Transmission at wavelength of 1500 nm and sensitivity versus bridge substrate gap height.

9.6 2D FP-based PhC physical sensors

In the PhC bandgap, the PhC prohibits the propagation of light and acts as a lossless mirror. As we place two PhC slabs close together, for example, a short distance apart in the wavelength range, the two-PhC slab system acts as a Fabry–Pérot interferometer and creates a strong resonance at the output port. Since the reflection of the PhC slab is near 100%, the performance of the FP resonator is enhanced. In this case, the PhC device becomes a perfect candidate for a displacement sensor.

9.6.1 In-plane displacement sensor using PhC cavities

The in-plane displacement of a PhC cavity shows remarkable sensitivity to the coupling distance, which has been demonstrated in many applications, such as a microdisplacement sensor based on line-defect resonant cavity, as shown in Figure 9.22a,b [50]. A quasi-linear measurement of microdisplacement is realized with high sensitivity and a low Q-factor. The layout of the microdisplacement sensor based on a line-defect PhC resonant cavity is shown in Figure 9.22a. The system includes two planar PhC segments, where one is fixed and the other is movable. The travelling optical wave is guided in the PhC waveguide and monitored at the end of the moving PhC segment. Figure 9.22b illustrates the electric field distributions when the line cavity is perpendicular to the PhC waveguide and the displacement is 0.7a. There is a strong electric field resonance in the gap

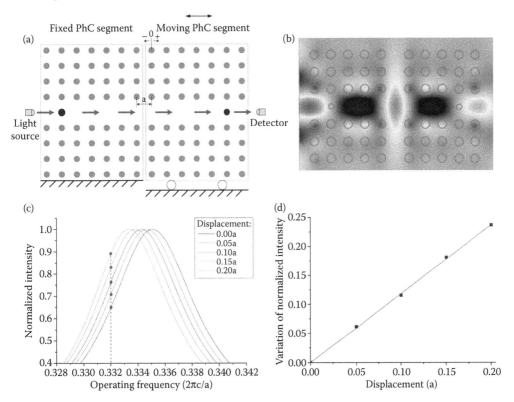

Figure 9.22 (a) Layout of the linear displacement sensor based on 2D PhC with square lattice of dielectric rods. The structure uses the gap between the two segments of PhC slab to generate the resonance peaks. The resonance wavelength will shift as the moving PhC segment changes its position. (b) Electric-field distribution when a PhC waveguide perpendicular to the line-defect cavity is formed. (c) Lorentzian curves of normalized intensity for different displacement of moving PhC segment. (d) Variation of normalized intensity as a linear function of displacement, the square dots present the simulation values, and the solid line is the linear regression result.

between the two PhC segments. The resonant wavelength is determined by the distance between the two PhC segments. Figure 9.22c shows the resonance wavelength shift as the distance between the two PhC segment increases. The resonance wavelength shifts to a higher wavelength as the gap increases between the two PhC segments. However, if we consider the fixed wavelength laser input, the sensing element becomes the nonresonance sensing mechanism. The power decreases as the gap becomes smaller at a 0.332a wavelength optical wave input, as shown in Figure 9.22c. Figure 9.22d shows the device is working in a linear region with a sensitivity of $1.15a^{-1}$. Furthermore, the sensing range can be broadened by using multiple operating frequencies. However, by adding an actuator, such as the comb drive, the test mass always requires a large space and enlarges the footprint of the whole device very much.

9.6.2 Out-of-plane displacement sensor using PhC cavities

An out-of-plane mechanism is favorable for large-scale integration for mass product fabrications. One of the applications is that we have a multilayer structure, as illustrated in Figure 9.23a. The structure consists of two PhC slabs. Because of the existence of the bandgap in a PhC slab, it works

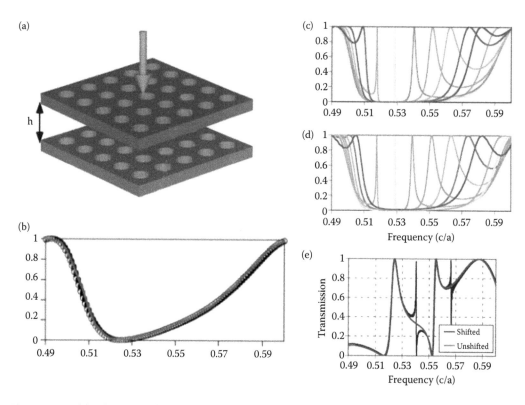

Figure 9.23 (a) Schematic of displacement sensitive PhC slab. The arrow represents the direction of the incident light. The PhC slab is design to have high reflectivity in its bandgap. (b) Transmission spectrum through a single photonic crystal slab for normally incident light. The crystal structure consists of a square lattice of air holes of radius 0.4a, where a is the lattice constant, introduced into a dielectric slab. The slab has a dielectric constant of 12 and a thickness of 0.55a. The open circles are the numerical results from an FDTD simulation. (c) Theoretical transmission spectra for a two-slab structure. (d) FDTD simulation result of the transmission spectra for the two-slab structure. The slabs are the same and the solid lines represents different spacing h between the slabs decreasing from left to right. (e) Transmission spectra through two-slab structures. The spacing between the slabs is 0.1a. The shifted lattice (0.05a) of holes introduces two additional peaks in the spectrum near 0.54 and 0. 57.

as a perfect reflector inside its bandgap. Hence, when we place two slabs together, it behaves like a Fabry–Pérot resonator. Hence, the transmission spectra is strongly influenced by the spacing h between the PhC slabs, as shown in Figure 9.23c through e [51,52]. When the size of the air holes is large, the guided resonance is not a narrow band phenomenon and strong reflection can occur over a fairly wide range of frequencies. When the reflectivity for each slab is high, the transmission becomes highly sensitive to the longitudinal displacement. Figure 9.23c and d show the transmission spectra of two slabs using the theoretical approach and the FDTD simulation, respectively. A 100% transmission peak appears within the high reflection range of a single PhC slab reflector. As the displacement of the two slabs is reduced, the transmission peak moves to a shorter wavelength range. Moreover, the width of the peak critically depends on the reflectivity of a single slab, and can be made arbitrarily small at certain frequencies when the single slab shows complete reflection. Therefore, by changing the distance between the slabs, the system can reconfigure the sensitivity to the displacement. Figure 9.23e shows the properties of the two PhC slabs, providing capabilities for lateral displacement sensing. To achieve this purpose, the system needs to ensure evanescent coupling between the slabs, which means the two slabs need to be close enough. Because of that, the evanescent tunneling and near-field coupling between the slabs become important at smaller spacings. In the transmission spectra shown in Figure 9.23e, the near-field coupling between the slabs splits the resonance, leading to strong reflection. Such resonances correspond to the single degenerate states that are uncoupled when the two slabs are aligned. The calculation thus demonstrates that the near-field coupling regime provides the additional possibility of lateral displacement sensing.

9.7 Slow light-based PhC sensors

Slow light means that the light travels with a very low group velocity in a waveguide. Hence, the optical field is accumulated in the guided media and the light interaction is highly enhanced. In general, the PhC waveguide can support the slow light because of the standing wave pattern generated by the PhC lattice. Hence, by leveraging the shape change of the PhC lattice design, the slow light structure can provide potential for sensing the stress and strain presented in the PhC slab.

In most of the PhC waveguide designs, both the size and shape of the PhC air holes and their lattice position are good sensing elements which can modify the slow light properties. Once the circular air holes from the PhC waveguide in a triangular lattice are changed to elliptical air holes under the effect of stress and strain, the slow-down factor of the propagation mode is enhanced, and the slow light wavelength is also shifted [53].

9.7.1 Stress sensor using PhC waveguide slow light

The basic PhC structure containing circular air holes in a triangular lattice is shown in Figure 9.24a. The radius of each air hole R is designed to be 0.286**a**, where **a** is the lattice constant of the PhC lattice, in order to obtain the large photonic bandgap in transverse electric polarization. The corresponding propagation mode of the electric field distribution is also shown in Figure 9.24a. The main propagation mode of light inside the PhC waveguide is the TE mode in simulation. A highly concentrated field energy in the waveguide region is observed, utilizing the high confinement by the PBG in the triangular PhC lattice. Stress induced the line defect width change [54,55], and the selected air hole position change can be monitored by the change of the slow light effect [56,57]. Briefly speaking, the group index and the bandwidth can be modified by shifting the lattice line and modifying the air hole shape. The propagation mode of the PhC waveguide will change as the circular air holes are changed into elliptical air holes, as shown in Figure 9.24b. The modeling suggests a slightly modified mode profile in comparison with the data shown in Figure 9.24a, that is, a result derived from the PhC waveguide of circular air holes. Because of the elliptical air holes placed along the lateral direction, the longer radius is parallel to the light propagation direction.

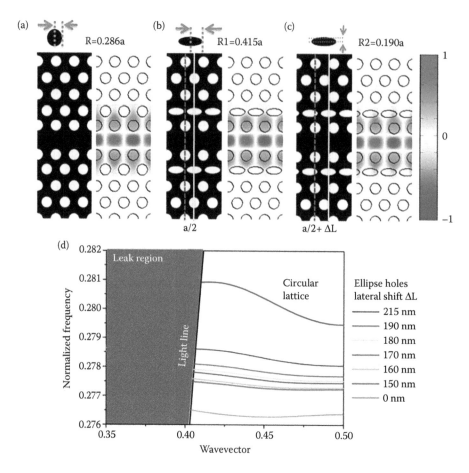

Figure 9.24 Schematic of proposed structure and simulated electric field distribution. (a) Normal photonics crystal lattice; (b) one line of circular air holes is replaced by ellipse-shaped air holes; (c) the ellipse air holes shift to half of the lattice period; the red dashed line indicates the original lattice position. (d) Band structure plot of the ellipse PhC waveguide shows that the slow light frequency region changes as the lateral shift of PhC ellipse holes. (Adapted from Li B et al. *Journal of Nanophotonics*, 8(1), 084090–084090, 2014.)

The longer and shorter radii of the ellipses are denoted as R1 and R2 in Figure 9.24b and c, and correspond to the stress magnitude. The PhC waveguide propagation mode can be further reproduced by laterally shifting the elliptical shape air holes to off-lattice positions [57]; we define ΔL to represent the offset from the lattice position, as shown in Figure 9.24c. The red dashed line represents the original position of the triangular lattice, while the solid yellow line indicates the original elliptical hole lattice position. We shifted the elliptical air holes along the light propagation direction by ΔL, meaning the yellow solid line shifted away from the red dashed line by ΔL. The difference of the index profile of the electric field introduced by the shifted elliptical air holes modifies the standing wave pattern, which is attributed to the light scattering in the waveguide. Hence, the propagation mode within the PhC waveguide is modified and also affects the slow light profile. Figure 9.24d illustrates the band structure plot of the PhC waveguide as the lattice shifts in the longitudinal direction, which causes the slow light region to also shift to a lower wavelength.

As shown in Figure 9.25a, to enable testing of the PhC slow light sensor performance, a PhC waveguide embedded in one of the MZI arms with a total length of 17 μm, is fabricated to demonstrate the change of the slow light effect. The reference arm is a 440 nm wide Si strip waveguide. The total length of each arm of the MZI is designed to be 300 μm between the two arms of the Y-shaped

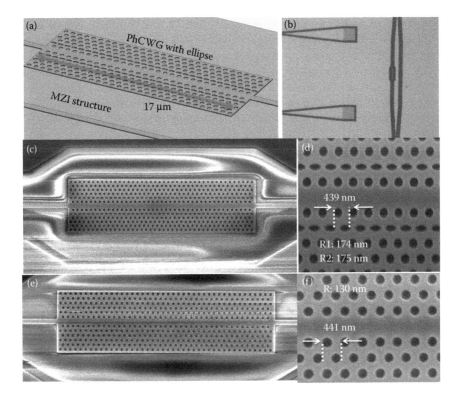

Figure 9.25 (a) Schematic drawing of Mach–Zehnder interferometer (MZI) embedded with PhC wave-guide. (b) Optical microscopy image of fabricated MZI device. (c) SEM image of fabricated PhC wave-guide. (d) Zoom-in image of ellipse-shaped holes. (Adapted from Li B et al. *Journal of Nanophotonics*, 8(1), 084090–084090, 2014.)

beam splitter. The 440 nm wide Si bus waveguide splits the light into two beams at the 15° Y-shaped junction. An optical microscopy image of the fabricated device is shown in Figure 9.25b. The light is coupled into and out of the bus waveguide via two adjacent grating couplers, which are located at the bottom of Figure 9.25b. The spacing between the two grating couplers is 127 μm, which is designed to fit the pm-fiber array for easy alignment. The entire structure is patterned using electron beam lithography and then etched using plasma etching (DRIE, Oxford 100 Plus) and processed on a 220 nm Si–on-insulator (SOI) wafer with a 2 μm buried oxide layer.

A scanning electron microscopy (SEM) image of the fabricated device is shown in Figure 9.25c. Figure 9.25d is the magnified image of the fabricated PhC lattice. The lattice constant is designed to be 430 nm in order to center the slow light region at 1550 nm. Because of the imperfection of the fabrication process, the lattice constant and the radius of the air holes are measured as 438.8 and 118.35 nm, respectively. For the elliptical design, the longer and shorter radii are measured as 173.65 and 75 nm, respectively. PhC waveguides with a circular lattice are also fabricated as a reference and are shown in Figure 9.25(e, f). The radii and the lattice constant are measured as 130 and 440.7 nm, respectively.

9.8 Conclusions

This chapter covers various examples of the design and development of various photonic crystal-based physical sensors. The focus of this chapter is to explore and investigate novel optical resonance structures with high Q-factors and to optimize the sensitivity of the proposed resonators working as physical and chemical sensors.

Optical sensors are highly promising because of their high sensitivity, compact size, and low cost. By leveraging the total internal reflection and PBG structure design, a high Q-factor PhC cavity mode resonator can be designed, optimized, and realized by advanced fabrication technology. The modeling methods are mainly focused on the 3D and 2D FDTD simulations, together with some FEM simulation methods. The defect constructed in a 2D PhC slab creates different types of PhC cavity resonators and has different advantages in sensing applications. Since the PBG is highly dependent on the lattice structure of the PhC, the deformation of the lattice period will be sensed by the optical resonance inside the PhC resonator. Hence, utilizing these unique properties and integrating with a microcantilever MEMS structure, micromechanical sensors such as force sensors and pressure sensors are proposed and developed. Engineering work has been done on the organization of the laterally coupled 1D PhC cavity waveguides, which can be used as displacement sensors with their resonance wavelength shift. An FP cavity structure is another promising sensing candidate for large-scale measurement, leveraging the high reflectivity of a PhC slab. In summary, the PhC physical sensor, with its high Q-factor and high sensitivity, is a promising candidate for physical sensing.

References

1. S. Lacey, I.M. White, Y. Sun, S.I. Shopova, J. M. Cupps, P. Zhang and X. Fan, Versatile optofluidic ring resonator lasers with ultra-low threshold, *Opt. Express*, 15, 15523, 2007.
2. A. Melloni, R. Costa, P. Monguzzi and M. Martinelli, Ring-resonator filters in silicon oxynitride technology for dense wavelength-division multiplexing systems, *Opt. Lett.*, 28, 1567–1569, 2003.
3. M. Soltani, S. Yegnanarayanan and A. Adibi, Ultra-high Q plannar silicon microdisk resonators for chip-scale silicon photonics, *Opt. Express*, 15, 4694–4704, 2007.
4. C. Baker, S. Stapfner, D. Parrain, S. Ducci, G. Leo, E. M. Weig and I. Favero, Optical instability and self-pilsing in silicon notride whispering gllery resonators, *Opt. Express*, 20, 29076–29089, 2012.
5. S. Ghosh and G. Piazza, Photonic microdisk resonators in aluminum nitide, *Appl. Phys. Lett.*, 102, 016101, 2013.
6. M. Bürger, M. Ruth, S. Declair, J. Förstner and C. Meier, Whispering gallery modes in zinc-blende ALN microdisks containing non-polar GaN quantum dots, *Appl. Phys. Lett.*, 102, 081105, 2013.
7. K. Volyanskiy, P. Salzenstein, H. Tavernier, M. Pogurmirskiy, Y. K. Chemb and L. Larger, Compact optoelectronic microwave oscollators using ultra-high Q whispering gllery mode disk-resonators and phse modulation, *Opt. Express*, 18, 22358–22363, 2012.
8. T. Baehr-Jones, M. Hochberg, C. Walker and A. Scherer, High-Q ring resonators in thin silicon-on-insulator, *Appl. Phys. Lett.*, 85, 3346, 2004.
9. W. S. Fegadolli, G. Vargas, X. Wang, F. Valini, L. A. M. Barea, J. E. B. Oliveira, N. Frateschi, A. Scherer, V. R. Almeida and R. R. Panepucci, Reconfigurable silicon thermo-optical ring resonator switch based on Vernier effect control, *Opt. Express*, 20, 14722–14733, 2012.
10. A. Einat and U. Levy, Analysis of the optical force in the Micro Ring Resonator, *Opt. Express*, 19, 20405–20419, 2011.
11. E. Yablonovitch, Inhibited spontaneous emission in solid-state physics and electronics, *Phys. Rev. Lett.*, 58, 2059–2062, 1987.
12. S. John, Strong localization of photons in certain disordered dielectric superlattices, *Phys. Rev. Lett.*, 58, 2486–2489, 1987.
13. S. G. Johnson and J. D. Joannopoulos, Designing synthetic optical media: Photonic crystals, *ActaMaterialia*, 51, 5823, 2003.
14. T. Baba, Slow light in photonics crystal, *Nature Photo*, 2, 465–473, 2008.
15. Y. Akahane, T. Asano, B.-S. Song and S. Noda, Fine-tuned high-Q photonic-crystal nanocavity, *Opt. Express*, 13, 1202, 2005.

16. M. Loncar, A. Scherer and Y. Qiu, Photonic crystal laser sources for chemical detection, *Appl. Phys. Lett.*, 82, 4648–4650, 2003.
17. S. Noda, A. Chutinan and M. Imada, Tapping and emission of photons by a single defect in a photonic bandgap structure, *Nature*, 407, 608–610, 2000.
18. B. S. Song, S. Noda and T. Asano, Photonics devices based on in-plane hetreo photonic crystals, *Science*, 300, 1537, 2003.
19. T. Baba, S. Akiyama, M. Imai, N. Hirayama, H. Takahashi, Y. Noguchi, T. Horikawa and T. Usuki, 50-Gb-s ring-resonator-based silicon modulator, *Opt. Express*, 21, 11869–11876, 2013.
20. K. Nozaki, A. Shinya, S. Matsuo, T. Sato, E. Kuramochi and M. Notomi, Ultralow-energy and high-contrast all-optical switch involving Fano resonance based on coupled photonic crystal nanocavities, *Opt. Express*, 21, 11877, 2013.
21. S. Fan, P. R. Villeneuve, J. D. Joannopoulos and H. A. Haus, Channel drop filters in photonic crystals, *Opt. Express*, 3, 4–11, 1998.
22. C. C. Chen, H. D. Chien and P. G. Luan, Photonic crystal beam splitters, *Appl. Opt.*, 43, 6188–6190, 2004.
23. H.-T. Chien, C. Lee, H.-K. Chiu, K.-C. Hsu, C.-C. Chen, J. A. Ho and C. Chou, The comparison between the graded photonic crystal coupler and various couplers, *IEEE J. Lightwave Technol.*, 27(7), 2570–2574, 2009.
24. L. M. Chang, C. H. Hou, Y. C. Ting, C. C. Chen, C. L. Hsu, J. Y. Chang, C. C. Lee, G. T. Chen and J. I. Chyi, Laser emission from GaN photonic crystals, *Appl. Phys. Lett.*, 89, 1116–1118, 2006.
25. S. Kim, I. Park and H. Lim, Highly efficient photonic crystal-based multi-channel drop filters of three-port system with reflection feedback, *Opt. Express*, 12, 5518–5525, 2004.
26. K. H. Hwang and G. H. Song, Design of a high-Q channel-drop multiplexer based on the two-dimensional photonic-crystal membrane structure, *Opt. Express*, 13, 1948–1957, 2005.
27. D. S. Park, B. H. O., S. G. Park, E. H. Lee and S. G. Lee, Photonic crystal-based GE-PON triplexer using point defects, *Proc. SPIE*, 6897, Feb, 2008.
28. E. A. Camargo, H. M. H. Chong, and R. M. De La Rue, 2D photonic crystal thermo-optic switch based on AlGaAs/GaAs epitaxial structure, *Opt. Express*, 12, 588–592, 2004.
29. Y. D. Wu, M. L. Huang and T. T. Shih, Optical interleavers based on two-dimensional photonic crystals, *Appl. Opt.*, 46, 7212–7217, 2007.
30. C. Lee, J. Thillaigovindan, C. Chen, X. T. Chen, Y. Chao, S. Tao, W. Xiang, A. Yu, H. Feng and G. Q. Lo, Si nanophotonics based cantilever sensor, *Appl. Phys. Lett.*, 93 113113, 2008.
31. C. Lee and J. Thillaigovindan, Optical nanomechanical sensor using a silicon photonic crystals cantilever embedded with a nanocavity resonator, *Appl. Optics*, 48, 1797–1803, 2009.
32. W. Xiang and C. Lee, Nanophotonics sensor based on microcantilever for chemical analysis, *IEEE J. Sel. Top. Quantum Electron*, 15, 1323–1326, 2009.
33. K. S. Yee, Numerical solution of initial boundary value problems involving Maxwell's equations in isotropic media, *IEEE Trans. Antennas Propag.*, 14, 302–307, 1966.
34. M. Qiu, Effective index method for the heterostructure-slab-waveguides-based two-dimensional photonic crystals, *Appl. Phys. Lett.*, 81, 1163, 2002.
35. W. Zheng, M. Xing, G. Ren, S. G. Johnson, W. Zhou, W. Chen and L. Chen, Integration of a photonic crystal polarization beam splitter and waveguide bend, *Opt. Express*, 17, 8657–8668, 2009.
36. Y. Chassagneux, R. Colombelli, W. Maineults, S. Barbieri, S. P. Khanna, E. H. Linfield and A. G. Davies, Predictable surface emission patterns in terahertz photonic-crystal quantum cascade lasers, *Opt. Express*, 17, 9492–9502, 2009.
37. A. D'Orazio, M. De Sario, V. Marrocco, V. Petruzzelli and F. Prudenzano, Photonic crystal drop filter exploiting resonant cavity conFiguration, *IEEE Trans Nanotechnol.*, 7, 10–13, 2008.
38. F.-L. Hsiao and C. Lee, Computational study of photonic crystals nano-ring resonator for biochemical sensing, *IEEE Sensors J.*, 10(7), 1185–1191, 2010.
39. C. Lee, R. Radhakrishnan, C.-C. Chen, J. Li, J. Thillaigovindan and N. Balasubramanian, Design and modeling of a nanomechanical sensor using silicon photonic crystal, *IEEE J. Lightwave Technol.*, 26, 839–846, 2008.

40. G. Krause, M. Winger, T. D. Blasius, Q. Lin and O. Painter, A high-resolution microchip opto-mechanical accelerometer. *Nature Photonics*, 6(11), 768–772, 2012.
41. F. Tian, G. Zhou, Y. Du, F. S. Chau, J. Deng and R. Akkipeddi, Out-of-plane nanomechanical tuning of double-coupled one-dimensional photonic crystal cavities, *Optics Letters*, 38(12), 2005-2007, 2013.
42. B. Li, F. L. Hsiao, C. Lee, Computational characterization of a photonic crystal cantilever sensor using a hexagonal dual-nanoring-based channel drop filter. *IEEE Trans. Nanotechnol.*, 10(4), 789–796, 2011.
43. T. T. Mai, F.-L. Hsiao, C. Lee, W. Xiang, C.-C. Chen and W. K. Choi, Optimization and comparison of photonic crystal resonators for silicon microcantilever sensors, *Sens. Actuators A*, 165(1), 16-25, 2011.
44. C. Stampfer, T. Helbling, D. Obergfell, B. Scholberle, M. K. Tripp, A. Jungen, S. Roth, V. M. Bright and C. Hierold, Fabrication of single-walled carbon-nanotube-based pressure sensors, *Nano Letters*, 6(2), 233–237, 2006.
45. B. Li, F. L. Hsiao, and C. Lee, Configuration analysis of sensing element for photonic crystal based NEMS cantilever using dual nano-ring resonator. *Sensors and Actuators A: Physical*, 169(2), 352–361, 2011.
46. B. T. Tung et al., Investigation of strain sensing effect in modified single-defect photonic crystal nanocavity, *Optics Express*, 19.9, 8821–8882, 2011.
47. Y. Yang, D. Yang, H. Tian and Y. Ji, Photonic crystal stress sensor with high sensitivity in double directions based on shoulder-coupled aslant nanocavity. *Sensors and Actuators A: Physical*, 193, 149–154, 2013.
48. B. Li and C. Lee, NEMS diaphragm sensors integrated with triple-nano-ring resonator. *Sensors and Actuators A: Physical*, 172(1), 61–68, 2011.
49. J. Sabarinathan et al. Photonic crystal thin-film micro-pressure sensors. *J. Phys., Conf. Ser.* 619, (1): IOP Publishing, 2015.
50. Z. Xu, L. Cao, C. Gu, Q. He and G. Jin, Micro displacement sensor based on line-defect resonant cavity in photonic crystal. *Optics Express*, 14(1), 298–305, 2006.
51. W. Suh, M. F., Yanik, O. Solgaard and S. Fan, Displacement-sensitive photonic crystal structures based on guided resonance in photonic crystal slabs. *Appl. Phys. Lett.*, 82(13), 1999–2001, 2003.
52. W. Suh, O. Solgaard and S. Fan, Displacement sensing using evanescent tunneling between guided resonances in photonic crystal slabs. *J. Appl. Phys.*, 98(3), 033102, 2005.
53. B. Li et al., Lateral lattice shift engineered slow light in elliptical photonics crystal waveguides. *Journal of Nanophotonics*, 8(1), 084090–084090, 2014.
54. E. Kuramochi, M. Notomi, S. Hughes, A. Shinya, T. Watanabe and L. Ramunno, Disorder-induced scattering loss of line-defect waveguides in photonic crystal slabs, *Phys. Rev. B.*, 72, 161318, 2005.
55. M. Notomi, K. Yamada, A. Shinya, J. Takahashi, C. Takahashi and I. Yokohama, Extrmely large group-velocity dispersion of line-defect waveguides in photonic crytal slabs, *Phys. Rev. Lett.*, 87, 253902, 2001.
56. S. Rahimi, A. Hosseini, X. Xu, H. Subbaraman and R. T. Chen, Group-index independent coupling to band engineered SOI photonic crystal waveguide with large slow-down factor, *Opt. Express*, 19, 21832–21841, 2011.
57. Y. Hamachi, S. Kubo and T. Baba, slow light with low dispersion and nonlinear enhancement in a lattice-shifted photonic crystal waveguide. *Opt. Lett.*, 34, 1072–1074, 2009.

Silicon photonic variable waveguide coupler devices

Kazuhiro Hane

Contents

10.1 Introduction

Silicon waveguide circuits are promising for optical telecommunication and optical interconnec-tions (Jalali and Fathpour, 2006). Monolithic fabrication of silicon waveguides and silicon elec-tronics is attractive for future integration in opt-electronic circuits. Due to the high refractive index of silicon (~3.5 at 1.5 μm wavelength), the silicon waveguide circuits can be miniaturized in area by a few orders of magnitude smaller than silica waveguide circuits. A silicon submicron-wide waveguide works as a single-mode waveguide with a bending radius as small as a few micro-meters. Several silicon submicron-wide waveguide devices, such as a waveguide splitter/coupler and microring resonator, have been studied (Janz et al., 2006; Koonath et al., 2006; Yamada et al.,

2006). Combining waveguide components, complex circuits such as arrayed waveguide devices have also been studied (Sasaki et al., 2005).

One of the key components of waveguide circuits is the waveguide optical switch. Waveguide optical switches are often used for routing light signals from input waveguides to respective output waveguides through different optical paths. An $m \times n$ switch has m input ports and n output ports to connect the respective ports. A 2×2 switch is a fundamental element for composing $m \times n$ switches by connecting them in series and in parallel.

A few different principles have been studied for waveguide switches. In silicon submicron-wide waveguide switches, the refractive index change by heat in a waveguide interferometer has often been employed (Koonath et al., 2006) because of the high temperature coefficient of the silicon refractive index ($5.7 \times 10\%$ -3%/deg.). However, the power consumption of the switch is generally large (>10 mW per switch) and the switching time is usually slow (~ 1 msec). Recently, an ultra-fast silicon waveguide light modulator using the refractive index change induced by a carrier injection (Green et al., 2007) has attracted much attention for telecommunication and interconnections. The switch using the silicon refractive index change by carrier injection also consumes a large amount of power if it is used as a matrix switch.

Recently, submicron-wide waveguide switches for optical path change using nanomechanical motion by an electrostatic actuator have been reported (Yao et al., 2007; Bulgan et al., 2008). Due to the very small power consumption by electrostatic capacitive operation, this technology suits the large-scale integration and reduction of energy consumption in telecommunication. Moreover, the small mass of motion makes the switching time as small as a few microseconds. Since the silicon submicron-wide waveguide switches based on microelectromechanical systems (MEMS) technology are scalable with silicon photonics (Han et al., 2015; Seok et al., 2016), further developments are expected for large-scale integration. Several functional devices of silicon submicron-wide waveguides using tunable mechanisms with actuators are studied (Chew et al., 2010, 2011). In this chapter, variable gap coupler switches and tunable microring filters for routing lightwaves at telecommunication wavelengths are mainly described.

10.2 Freestanding silicon photonic waveguides and support structures

Freestanding waveguides are utilized for movable waveguide devices combined with MEMS actuators. In the case of silica fiber optics, several types of optical MEMS switches moving input optical fibers to connect to output optical fibers were previously studied (Kanamori et al., 2005; Wu et al., 2006). The optical fibers are bulky and stiff for small actuators, even if the fibers are thinned. The actuators generating large forces, such as thermal actuators, are often used for the movement of the optical fibers. On the other hand, because silicon submicron waveguides are thinner and lighter than the silica waveguides, the waveguide can be easily moved by an electrostatic actuator driving at a few volts. Therefore, freestanding silicon waveguides can be moved mechanically for switching optical routes. There are a few freestanding waveguide structures that can be moved without disturbing the lightwave transmitting in the movable waveguide. Figure 10.1a and b show the schematics of freestanding waveguides supported, respectively, by an elliptical bridge (Fukazawa et al., 2004; Martines and Lipson, 2006) and a rib waveguide supported by thin plate as its cladding (Seok et al., 2016).

It is important to minimize the loss of the supporting bridge for the lightwave propagating inside the waveguide. In Figure 10.1a, the elliptical-shaped bridge supporting a rectangular waveguide with air cladding is schematically shown. An elliptical mirror efficiently focalizes light from one focus to another. The losses of the elliptical bridges are studied and minimized to be less than 0.1 dB for rectangular waveguides (Fukazawa et al., 2004). A simple rectangular supporting arm without an elliptical part may be used for a waveguide having a wide width with a small loss. However, in the case of the narrow waveguide such as a single-mode waveguide, the loss introduced by the simple rectangular supporting arm is significant. The waveguide supported by the elliptical bridge can be moved horizontally (x-axis) and vertically (y-axis) by an actuator.

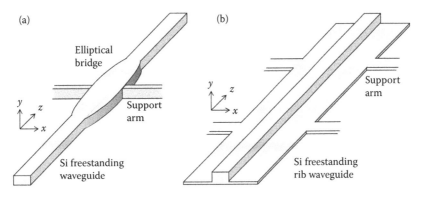

Figure 10.1 (a) Silicon freestanding waveguide supported by elliptical bridge, (b) silicon freestanding rib waveguide supported by plate.

The freestanding waveguide can also be supported by a thin plate as shown in Figure 10.1b (Seok et al., 2016). In this case, the waveguide consists of a low-loss rib waveguide. The core of the waveguide is the rib part, where most of the propagating lightwave energy is concentrated. Therefore, the plate part beside the rib part can be connected to a MEMS actuator without disturbing the lightwave if the width of the plate part is wide enough. In this case, the motion of the waveguide is generally restricted to the vertical direction (y-axis) due to the stiffness of bending.

10.3 Waveguide modes of the freestanding silicon waveguide

A rectangular cross-sectional silicon waveguide works as a high-index waveguide if it is supported in air, because the refractive index of silicon is much higher than that of the surrounding air. The surrounding air works as a cladding material for the high-index contrast waveguide. Here, we consider freestanding silicon waveguides of 260 nm in thickness. The propagation constant is calculated as a function of the waveguide width in the transverse electric (TE) mode using the effective index method for the 260 nm thick waveguide, as shown in Figure 10.2. For a width of 400 nm, a single TE mode wave exists. By increasing the waveguide width wider than about 500 nm, the second mode wave is included in the allowable modes. A single-mode waveguide is preferable for telecommunication because of its precise manipulation of the propagating lightwave. A similar tendency is also obtained for the transverse magnetic (TM) mode. For calculating the exact modes held in the waveguide, more precise simulations are needed.

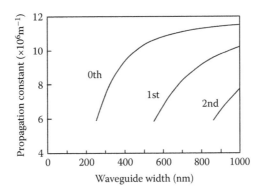

Figure 10.2 Propagation constant calculated as a function of waveguide width for 260-nm-thick silicon waveguide.

10.4 Silicon submicron waveguide switch using variable directional coupler

10.4.1 Principle of switch using variable directional coupler

Figure 10.3a shows the basic structure of a silicon waveguide coupler switch (Akihama et al., 2011; Akihama and Hane, 2012). The coupler consists of two freestanding silicon waveguides. The cross-sectional schematic of the coupled waveguides is also shown in Figure 10.3b. The two waveguides are located close together in their interaction region. While keeping the coupled waveguides parallel, the gap between the waveguides is varied in the horizontal plane. One of the waveguides of the coupler is connected to an electrostatic MEMS comb drive actuator with suspension arms. The movable waveguide is displaced by the actuator. The other waveguide of the coupler is also a freestanding waveguide fixed to the substrate with elliptical bridges. By applying a voltage to the actuator, the movable waveguide is displaced to decrease the gap, thus the coupling between the waveguides becomes stronger.

In a general expression for two parallel coupled waveguides (Marcatili, 1969; Okamoto, 1992), the coupling coefficient is defined by

$$\chi_{pq} = \frac{\omega\varepsilon_0 \int_{-\infty}^{\infty}\int_{-\infty}^{\infty}\left(N^2 - N_2^2\right)\mathbf{E}_{1p}^* \cdot \mathbf{E}_{2q}\,dx\,dy}{\int_{-\infty}^{\infty}\int_{-\infty}^{\infty}\mathbf{u}_z \cdot \left(\mathbf{E}_{1p}^* \times \mathbf{H}_{1p} + \mathbf{E}_{1p} \times \mathbf{H}_{1p}^*\right)dx\,dy}, \tag{10.1}$$

where it is assumed that a wave in p-mode is propagating in the first waveguide of the coupler and a wave in q-mode is in the second waveguide. $\mathbf{E}_{1p}$, $\mathbf{E}_{2q}$, $\mathbf{H}_{1p}$, and $\mathbf{H}_{2q}$ are the electric and magnetic fields in the p and q modes of waveguides 1 and 2, respectively. $N(x,y)$ expresses the cross-sectional refractive index distribution in the case where the two waveguides exist, and $N_2(x,y)$ is that in the case where only the second waveguide exists. The symbols ω, ε_0, and $\mathbf{u}_z$ are, respectively, the angular frequency of the lightwave, the permittivity of vacuum, and the unit vector along the z-axis. The waves are propagating along the z-axis.

We consider here that the silicon waveguides are rectangular and are located in air. The two waveguides of the coupler are identical in cross-sectional dimensions with a width of $2a$ and a thickness

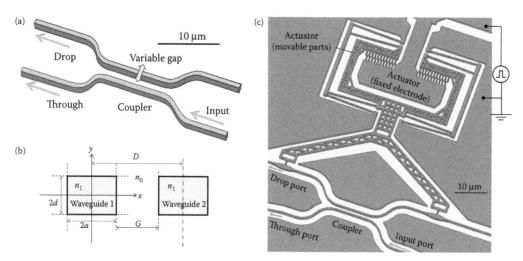

Figure 10.3 (a) Schematic diagram of a variable gap silicon waveguide coupler. (b) Cross-section of coupler. (c) Design of variable-gap silicon waveguide coupler switch with a comb actuator.

of $2d$. The distance between the centers of the two waveguides is D and the gap between waveguide walls is G. The refractive index of the silicon waveguide is n_1 and that of the surrounding air is n_0 ($n_0 = 1$). The numerator in Equation 10.1 is an overlap integral between the electric fields of the modes p and q in the region of waveguide 1, since $N^2(x, y) - N_2^2(x, y)$ has a non-zero value only in the region of waveguide 1. In order to analytically obtain the electromagnetic fields in the waveguide, a simple approximation is usually introduced which assumes that the fields are in transverse electromagnetic (TEM) modes (Marcatili, 1969; Okamoto, 1992). This assumption is valid only when the refractive index difference between the core and cladding is much smaller than their indices. Since the freestanding waveguide is a high-index contrast waveguide, the assumption does not hold. However, for simplicity and qualitative explanation, we dare to use this assumption for analytical calculation. In TE mode, it is assumed that an electric field E_x and a magnetic field H_y exist in the x and y directions, respectively, and fields in the other directions are negligible. The coupling coefficient is obtained as follows, after some mathematical manipulations (Okamoto, 1992, Equation 4.91)

$$\chi = \frac{\sqrt{2\Delta}}{a} \frac{(k_x a)^2 (\gamma_x a)^2}{(1 + \gamma_x a) v^3} \exp(-\gamma_x G). \tag{10.2}$$

Here, $\Delta = (n_1^2 - n_0^2)/2n_1^2$ is the relative refractive index difference and $v = kn_1 a \sqrt{2\Delta}$ is a normalized frequency. k_x is the wave number and γ_x is the decay constant of the field outside the waveguide along the x-axis, respectively. The symbol k represents the wave number in vacuum. The propagation constant β is given in the waveguide by $\beta^2 = k^2 n_1^2 - k_x^2 - k_y^2$ with the wave number k_y along the y-axis.

Using the coupling coefficient, the output intensity I_1 of transmitted light in waveguide 1 at the through port and the output intensity I_2 of transmitted light in waveguide 2 at the drop port are given by

$$I_1 = I_0 \cos^2(\chi L_z), \quad I_2 = I_0 \sin^2(\chi L_z), \tag{10.3}$$

where I_0 is input light intensity and L_z is coupler length. With the increase of L_z, the intensities change sinusoidally. The intensities are also oscillatory with the decrease of gap G.

The mode profile is assumed to be that of the lowest TE mode. The refractive index of silicon is 3.47 at a wavelength of 1.55 μm. The width and the thickness of the waveguide are designed to be 400 and 260 nm, respectively. The values of the parameters in the equation at a wavelength of 1.55 μm are as follows: $\Delta = 0.46$, $a = 200$ nm, $k_x = 7.6 \times 10^6$ m^{-1}, $\gamma_x = 1.1 \times 10^7$ m^{-1}, $\beta = 8.7 \times 10^6$ m^{-1}, and $v = 2.7$. The calculated coupling coefficient is shown in Figure 10.4 as an analytical calculation. The coupling coefficient decreases nearly exponentially.

The numerical simulation for obtaining the precise properties of the coupler switch was also carried out using a finite difference time domain (FDTD) method of rigorous electromagnetism. In the simulation, the model, consisting of two straight parallel waveguides of 10 μm in length, was used to calculate the coupling coefficient. Figure 10.4 shows the calculated coupling coefficients as a function of gap. The coupling coefficient by the FDTD method is approximately four times smaller at a gap of 100 nm compared with the analytical coupling coefficient for the TE mode. The model using the actual shape of the coupler was also simulated, and it was found that the value of the coupling coefficient was larger by a factor of 1.3 than that of the straight coupled waveguides due to the bent parts of the coupler waveguides.

Figure 10.5a and b show the square of the electric field distributions in the waveguide coupler at gaps of 95 and 50 nm, respectively. In Figure 10.5a, the input light in the lower waveguide is totally transmitted to the upper waveguide and no light is transmitted to the through port. In Figure 10.5b, the coupling between the waveguides becomes stronger and the light in the waveguide is transmitted back and forth in the coupler region and is split into the two ports.

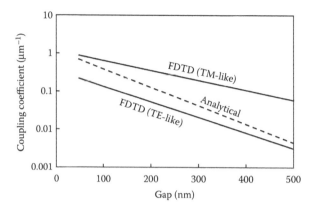

Figure 10.4 Coupling coefficients of 10 μm straight waveguide coupler calculated as a function of gap.

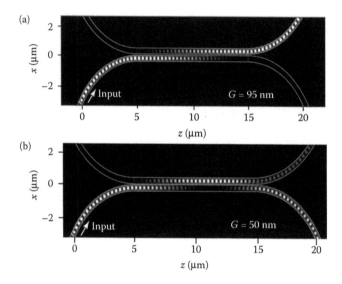

Figure 10.5 Intensity distribution in the waveguide coupler calculated by FDTD at the coupler gap G of (a) 95 nm and (b) 50 nm.

Figure 10.6a shows the output intensities at the through and drop ports, calculated as a function of gap for the coupler with the bent waveguides. The coupling coefficient calculated by FDTD in the TE-like mode is used for the calculation. Decreasing the gap from a relatively large value, the intensity at the through port decreases and that at the drop port increases. Those intensities reach a maximum and a minimum at a gap of 98 nm, respectively. Under these conditions, the intensity distribution in the coupler is nearly equal to the result shown in Figure 10.5a. Further decreasing the gap from 98 nm, the intensities oscillate, as shown in Figure 10.6a. Therefore, before reaching a zero gap, a few switch points appear. The switch point at the gap of 98 nm is useful because of the largest tolerance in positioning. Figure 10.6b shows the wavelength dependences of the logarithmic output intensities around a wavelength of 1.55 μm. The wavelength region where the extinction ratio is larger than 20 dB is about 25 nm, and that larger than 30 dB is approximately 8 nm.

10.4.2 Fabrication and measurement setup

Figure 10.3c shows the design of the single 2 × 2 switch. The length of the waveguides in the coupler region is 10 μm. The gap can be decreased from 1000 to 20 nm by the actuator. The movable

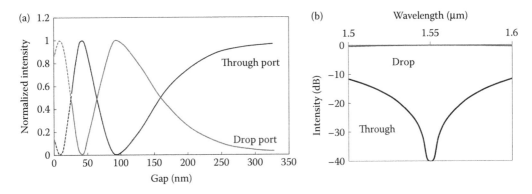

Figure 10.6 (a) Normalized output intensities of coupler switch calculated as a function of gap. (b) The wavelength dependence at the gap of 98 nm.

freestanding silicon waveguide is suspended by 1.3-μm-wide and 8.0-μm-long elliptical bridges. The bridges connect the movable waveguide to the suspension arms of the actuator with 0.2-μm-wide 1.6-μm-long silicon beams. The coupling part of the coupler is connected to two bent waveguides with a radius of 6 μm.

As the waveguide cross-sectional dimensions are smaller than 1 μm, a very small actuator is needed to be embedded in the silicon waveguide circuits. An ultra-small electrostatic comb drive actuator is connected to the movable waveguide with the suspension arms (Takahashi et al., 2009; Chew et al., 2010). Compared with conventional electrostatic comb drive actuators used for several applications, the actuator used here is smaller in area by two orders of magnitude. The actuator is as thin as the waveguide thickness, because both the waveguide and the actuator are formed from the same top silicon device layer of an SOI wafer.

Figure 10.7 shows the fabrication steps. In the fabrication, an SOI wafer with a 260-nm-thick top silicon device layer and 2-μm-thick buried oxide layer on a 625-μm-thick silicon substrate was used. First, a positive resist polymer was coated on the SOI wafer, and it was exposed using an electron beam patterning machine. After developing the resist polymer, the top silicon layer was etched by a fast atom beam. The resist polymer was removed by an H_2SO_4/H_2O_2 solution. The SOI wafer was cleaved after partially dicing the silicon substrate to obtain a facet of input waveguide for coupling light. Finally, the SiO_2 layer was etched by hydrofluoric acid vapor to obtain the freestanding structure.

The switch operations were measured using an infrared (IR) optical microscope. A tunable laser was used as a light source at a wavelength around 1.55 μm. For butt-coupling the laser light into the input port of the switch, a lensed single-mode fiber was used. The light intensities at the output ports were measured from the spot images of scattered light at the ends of the waveguides using an IR camera. In addition to the spot intensity measurement, another lensed fiber and a photodetector were also used for the measurement of the output light intensities.

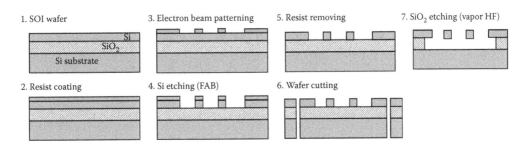

Figure 10.7 Fabrication steps.

10.4.3 Ultra-small comb drive electrostatic actuator embedded in the coupler switch

The movable waveguide is connected to the actuator as shown in Figure 10.3c, and is also supported by the neighboring elliptical bridges (not shown in the figure) at a distance of 100 μm from the elliptical bridges connected to the actuator arms. The comb finger of the actuator is 1.73 μm in length, 225 nm in width, and 260 nm in thickness, and the gap between the comb fingers is 225 nm. The number of finger pairs is 20. The movable part of the actuator is supported by four double-folded springs. Each spring beam is 15 μm in length, 250 nm in width, and 260 mm in thickness. The spring constant is calculated by the equation

$$k = \frac{2Etw^3}{h^3}.$$
(10.4)

Here, $E(=1.5 \times 1011 \text{ N/m}_2)$ is Young's modulus of silicon. The symbols t, w, and h are thickness, width, and length of the spring beam, respectively. Using the design sizes, the value of k is calculated to be 0.36 N/m. The spring constant of the freestanding waveguide connected to the actuator is 1/40 of the value of k.

Figure 10.8a shows the displacement of the actuator as a function of the applied voltage. The maximum displacement of the actuator is limited by an integrated mechanical stopper at a displacement of 980 nm. Therefore, the minimum gap of the coupler is designed to be 20 nm. A maximum displacement of about 1 μm is obtained at a voltage of about 28 V. The dotted curve shows the displacement calculated on the basis of a simple electrostatic comb actuator model, where the electrostatic force between parallel comb fingers is considered. The force F generated by the comb actuator is obtained by (Jaecklin et al., 1992)

$$F = \varepsilon_0 tN \left(\frac{1}{g} + \frac{w_c}{(d_0 - x)^2} \right) V^2,$$
(10.5)

where ε_0, t, N, and g are the permittivity of vacuum, comb thickness, number of comb fingers, and gap between the fingers, respectively. The symbols w_c and d_0 are the width of each finger and initial distance between the top and the bottom of the facing combs. The displacement x is obtained by equalizing the generated force F to the restoring force k_x of the spring at a voltage V. For simplicity, the second term of Equation 10.5 is often neglected when d_0 is large.

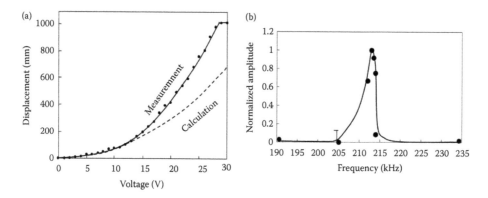

Figure 10.8 (a) Displacement of the actuator as a function of applied voltage. (b) Normalized oscillation amplitude measured as a function of frequency of the applied voltage.

Mechanical natural frequency is an essential property of a mass-spring system. Figure 10.8b shows the normalized oscillation amplitude as a function of frequency around the mechanical natural frequency measured in vacuum. The oscillation has a peak at the frequency of 213 kHz, while the calculated natural frequency using the mass and the spring constant is about 200 kHz, which agrees well with the experiment. Because of the higher natural frequency, a fast switch response time of less than 10 μs is expected.

10.4.4 Fabricated (2 × 2) switch and the gap dependence of output intensities

Figure 10.9a shows an electron micrograph of the fabricated waveguide coupler switch (Akihama et al., 2011; Munemasa and Hane, 2013). The silicon submicron waveguides are freestanding, as seen in the micrograph. The movable waveguide is connected to an electrostatic comb drive actuator and the movable part of the actuator is also freely suspended by silicon springs. Figure 10.9b shows the output intensities measured at the through and drop ports as functions of the coupler gap. The measured output intensities are normalized by the intensity at the through port under the initial gap condition of 1 μm. The light intensity at the drop port starts increasing at $G \sim 400$ nm, and reaches a maximum at $G \sim 110$ nm. Further increasing the voltage, the output intensity at the drop port decreases and reaches a minimum, where the transmitted light returns to the original waveguide. Although the switch point in the theoretical calculation shown in Figure 10.6a is shifted about 15 nm from that measured in the experiment, the gap dependences of the measured intensities at the output ports are explained well with the theory. The loss of the switch from the input port to the drop port was typically less than −1 dB, and the loss from the input to the through port is smaller than that to the drop port. The loss of the elliptical bridges connecting the waveguide to the actuator was about −0.2 dB in our experiment.

The coupling coefficient of the fabricated coupler is also obtained from the measured intensity, assuming the sinusoidal dependence of output intensity given by Equation 10.3, and is shown in Figure 10.10. The coupling length is assumed to be 10 μm. The measured values are shown with the solid circles. The calculated coupling coefficient by FDTD for 10 μm straight parallel waveguides is shown again in the Figure for the TE-like mode. In the experiment, a TE mode selector is added to the input waveguide to eliminate the lightwave in the TM mode. The measured coupling coefficient is close to the calculated coefficient in the gap ranging from 70 to 250 nm. The difference is considered to be caused by the influence of the bent waveguides connected to the parallel waveguides of the coupler and the errors in the actuator displacement measurement.

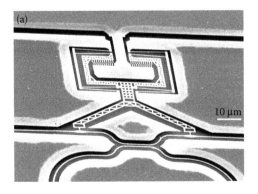
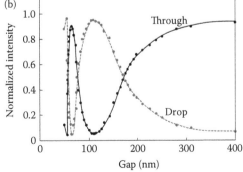

Figure 10.9 (a) Fabricated coupler switch. (b) Output intensities measured as a function of the gap of the coupler.

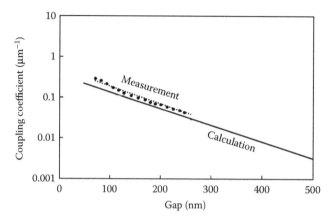

Figure 10.10 Coupling coefficients as a function of coupler gap.

10.4.5 Time response of switch

An important property of a switch is the response time. A minimum response time is obtained by optimizing the driving force and damping coefficient. The critical damping condition is obtained by adjusting both mechanical and electrical responses. The dynamic response of a mechanical system driven by an electric circuit can be expressed by the coupled equations (Abe et al., 2014)

$$m\ddot{x} + \zeta\dot{x} + kx = \frac{V}{\ell}q \tag{10.6}$$

$$R\dot{q} + \frac{q}{C} - \frac{V}{\ell}x = V_0 u_{-1}(t), \tag{10.7}$$

where m is the mass of the movable part and x is the small displacement from the equilibrium position where a constant voltage V is applied to the comb electrode in the initial switch-off condition. The symbol ζ is the coefficient of the mechanical damping caused mainly by air drag. A stepped voltage is superposed on the initial static voltage V for switching. The stepped voltage is expressed by $V_0 u_{-1}(t)$, where V_0 and $u_{-1}(t)$ are the amplitude of the stepped voltage and the unit step function at $t = 0$. The symbol ℓ is the overlapping length of the comb fingers. The incremental charge from the equilibrium charge Q is expressed by q (=dQ). The right side of Equation 10.6 and the third term of Equation 10.7 are the interaction terms between the two equations. The former is the incremental external force generated by the actuator. Electrostatic force is obtained by differentiating stored energy with respect to displacement. The incremental force $(V/\ell)q$ is derived for linearization by differentiating the force with respect to the interacting charge Q. The latter is the incremental voltage $(V/\ell)x$ generated by an incremental displacement, which is obtained by differentiating the capacitor voltage with respect to displacement. The electrical resistance of the wiring and the capacitance of the comb actuator are expressed by R and C, respectively. The value of C is usually very small ($\sim 10^{-16}$ F) because of the small area and air gap. The general response of x is nonlinear and complex, since the values of C and the electric field V/ℓ are dependent on x. In order to obtain a basic property of the system, it is assumed that x and q are very small, that is, the values of C and V/ℓ are nearly constant.

Laplace transforming Equations 10.5 and 10.6, and eliminating the variable q, we obtain the expression for Laplace transformed displacement $L\{x\} = X(s)$

$$X(s) = \frac{VV_0}{\ell s\{(Rs + (1/C))(ms^2 + \zeta s + k) - (V^2/\ell^2)\}}. \tag{10.8}$$

The system is expressed by the fourth degree function of s. The damping coefficient ζ of the mechanical system is estimated by a Couette flow between the movable plate of the actuator and the silicon substrate with an air gap of 2 μm (Cho et al., 1994). The calculated value of ζ is 1.8×10^{-8} kg/s at most, although some hole structures are formed on the movable plate for sacrificial etching. The step motion governed by Equation 10.6 is oscillatory, that is, $\zeta_2 < 4mk$ when $q = 0$. The third-order term in the denominator of Equation (10.8) has three solutions. Two of the solutions are complex conjugate pairs, $\alpha_1 = -a + ib$ and $\alpha_2 = -a - ib$ ($a > 0$, $b > 0$), and the third solution is a negative real number α_3. Inversely transforming Equation 10.8, the time-dependent displacement x is given by

$$x(t) = \frac{VV_0}{m\ell R} \left[\frac{1}{-\alpha_3} \left(\frac{u_{-1}(t)}{|\alpha_1|^2} - \frac{\exp(\alpha_3 t)}{|\alpha_3 - \alpha_1|^2} \right) - \frac{\exp(-at)\sqrt{(b^2 - a^2 - a\alpha_3)^2 + b^2(2a + \alpha_3)^2} \, \sin(bt - \phi)}{|\alpha_1|^2 |\alpha_3 - \alpha_1|^2 b} \right],$$

(10.9)

with the initial phase ϕ of the oscillatory motion expressed by $\phi = \tan^{-1}(b(2a + \alpha^3)/(b^2 - a^2 - a\alpha^3)$ $(-\pi/2 < \phi < \pi/2)$. The value $-\alpha^3$ is equal to the inverse of the time constant of the first exponential term in Equation 10.9. The value of b is equal to the angular frequency of the oscillatory motion in the last term in Equation 10.9, with the decay time equal to the inverse of a. The former term, including $u_{-1}(t)$ and $\exp(\alpha^3 t)$, determines the response to the stepped increase of the applied voltage, and the latter term, including the sinusoidal dependence, expresses the decaying oscillation at a frequency close to the mechanical natural frequency $f_0 = (1/2\pi)\sqrt{k/m}$.

Figure 10.11a shows the displacement calculated as a function of time by using Equation 10.9 with the R setting as a variable. The calculated value of mass m from the designed dimensions is 2.3×10^{-13} kg. The calculated capacitance for the comb finger pairs is 1.13×10^{-16} F, assuming an overlap length ℓ of 550 nm. In Figure 10.11a, the calculated displacements are shown when R is equal to 1, 10, and 100 GΩ at a driving condition of $V = 20$ V and $V_0 = 1$ V. Although the oscillation component always exists in Equation 10.8, the component becomes smaller when R is larger than 10 GΩ. Therefore, the motion of the switch is oscillatory in the condition of $R < 10$ GΩ. The switch response is over-damped in the condition of $R > 10$ GΩ. When $R = 100$ GΩ, the oscillation component is smaller than 5% of the total displacement. The oscillation frequency varies from 200 to 152 kHz for an R from 1 to 100 GΩ. The calculated mechanical natural frequency is 200 kHz.

The optical intensity at the drop port is expressed from Equations 10.2 and 10.3 by

$$I_2 = I_0 \sin^2(L_z \chi_0 \exp(-\gamma_x(G - x)))$$

(10.10)

The symbol χ_0 is the effective coupling coefficient at zero gap. The gap G is an initial gap, which corresponds to a switch-off state. The gap of about 100 nm corresponds to the switch-on state. Therefore, using the calculated x, the optical output I_2 may be calculated as a function of time by combining Equations 10.9 and 10.10. The normalized optical intensity I/I_0 is calculated as a function of time with the R setting as a variable, as shown in Figure 10.11b. The intensity is calculated in the non-oscillation conditions of $R > 10$ GΩ, assuming that the values of χ_0 and γ_x are 0.473 and 11.0 μm^{-1}, respectively. Figure 10.11b shows the calculated results, for which the rise time of the switch, which is defined by the time between 10% and 90% of the settled intensity, ranges from 4.8 to 36 μsec for R from 10 to 100 GΩ. The delay time of the switch, defined by the time at 50% intensity position, increases from 3.4 to 46 μsec with the increase of R. Therefore, the minimum rise time in optical intensity is roughly estimated to be around 10 μsec.

Figure 10.12a and b show the optical output intensity measured as a function of time for switching-on and -off, respectively. The applied stepped voltages are also shown for referencing. The rise time is about 18 μsec, and the fall time is about 14 μsec. The measured resistivity of the top silicon layer of the SOI wafer was approximately 10 kΩ-cm, and the wiring resistance of the spring beam

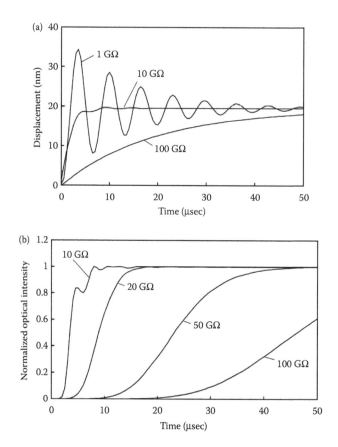

Figure 10.11 (a) Displacements calculated as a function of time with the resistance R as a parameter. (b) Normalized optical intensity calculated as a function of time.

was calculated to be about 40 GΩ. From the calculations shown in Figure 10.11b, the rise time is from 9 to 23 μsec for R from 20 to 50 GΩ, which roughly explains the measured value. The energy consumption is caused by the losses of mechanical friction and wiring resistance. It is estimated by fitting the theoretical curves shown in Figure 10.11b to that shown in Figure 10.12a. In the case of a minimum displacement of 300 nm for switching, the consumed energy is estimated to be 0.0041 pJ.

10.4.6 A (1 × 3) silicon submicron waveguide coupler switch

Using the variable gap waveguide coupler, a 1 × 3 switch was designed and fabricated (Munemasa and Hane, 2013). Figure 10.13 shows the scanning electron micrograph of the fabricated 1 × 3 switch and the spot images at the three output ports. Two variable gap couplers are located on both sides of the center waveguide which is connected to the input port. The upper movable waveguide can approach the center waveguide to transmit the lightwave to drop port 1. The lower movable waveguide can also approach the center waveguide to transmit the lightwave to drop port 2.

The spot images of the three output ports were taken at a wavelength of 1.55 μm by an infrared microscope. In the spot image (a), the input lightwave is equally divided into the three waveguides when the two movable waveguides are in the switch-on states. In the spot image (b), the upper switch is in the on-state and the lower switch is in the off-state, thus the spot at drop port 1 is bright. When both switches are in the off-states, the spot at the through port is bright as shown in the spot image (c). In the spot image (d), the upper switch is in the off-state and the lower switch is in the on-state, thus the spot at drop port 2 is bright. The two couplers in the switches work well to generate

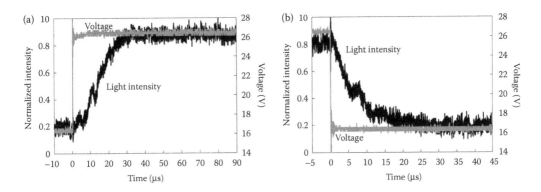

Figure 10.12 (a) Optical output intensity measured as a function of time for switching-on, and (b) for switching-off.

a 1 × 3 switch. The port isolation between drop port 1 and the through port was 12.0 dB, and that between drop port 2 and the through port was 11.3 dB, derived from the intensities of the measured images. The loss of the 1 × 3 switch at the drop ports was less than 0.3 dB.

10.4.7 A multiple (1 × 6) switch consisting of silicon submicron waveguide coupler switches

A multiple switch is designed and fabricated by connecting five 2 × 2 waveguide coupler switches in series and in parallel (Akihama and Hane, 2012). Five 2 × 2 switches are connected with a 200-μm-long straight waveguide. As one of the input waveguides of the 2 × 2 switch is not used for connections, a 1 × 6 multiple switch is constructed by the connections of 1 × 2 switches, as shown

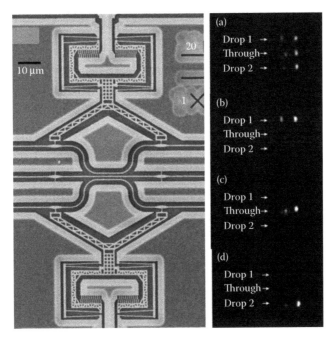

Figure 10.13 Electron micrograph of 1 × 3 silicon waveguide coupler switch with the spot images at the output ports: (a) equal outputs at the three ports, (b) switch-on at drop port 1, (c) switch-off, and (d) switch-on at drop port 2.

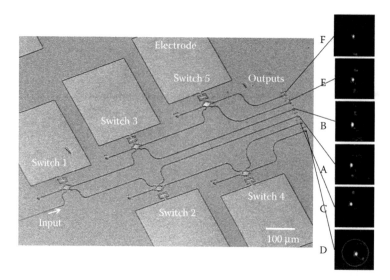

Figure 10.14 1 x 6 switch consisting of 1 x 2 silicon submicron waveguide coupler switches with the spot images at each output port when the output port is at the on-state.

in Figure 10.14. In the switch test, the input light at the 1.55 μm wavelength was incident at the input of the first switch (Switch 1), and spot images at the respective output ports from A to F were observed by the IR camera. The switches from 1 to 5 were selectively turned on to construct an optical path to one of the output ports. Therefore, one of the six ports was in the switch-on state (bright spot) and the others were in the switch-off states (no spot). In the right column of Figure 10.14, the spot images are shown for the respective ports. In the case of the spot image at port A, a bright spot is observed after the light passes two through ports of switches 1 and 2. In the case of port B, the lightwave passes a drop port of switch 1 and a through port of switch 3. When port B is compared with port A, similar intensities of the spots are observed. A spot intensity similar to that at port B is also observed at port C, where the lightwave passes three switches, including one drop port and two through ports. In the cases of ports D and E, the lightwave passes three switches, including two drop ports and one through port, and spots having slightly weaker intensities are obtained. In the case of port F, the intensity is the weakest after passing the three drop ports. The output differences among the respective ports at switch-on states are within 2.3 dB. The port isolation ranges from 6.2 to 27.3 dB at the adjacent ports and at the ports apart from the output port, respectively. The crosstalk ranges from 6.9 to 28.1 dB. The insertion loss of each 1×2 switch is evaluated to be less than 1 dB. The measured port isolations and the crosstalks are dependent on the precision of the voltages applied to the respective switches and the background intensity in the spot intensity measurement.

10.4.8 Silicon submicron waveguide coupler switch with mechanical latch

A mechanical latch for the silicon submicron waveguide coupler switch is studied (Abe and Hane, 2013). Using a latch mechanism, the switch-on state is held without applying voltage. Although several studies of microelectromechanical switches with latch mechanisms have been reported in micrometer scale optical switches (Ko et al., 2002), the effective latch mechanism using a buckled arm is not always applicable to the silicon subwavelength waveguide switches because of the very thin silicon layer. A variable gap silicon submicron waveguide coupler switch with a nanolatch mechanism is designed and fabricated.

Figure 10.15a shows the coupler switch with a mechanical latch. The switch consists of a similar directional coupler of two freestanding 400-nm-wide 260-nm-thick silicon waveguides, with an

(a)

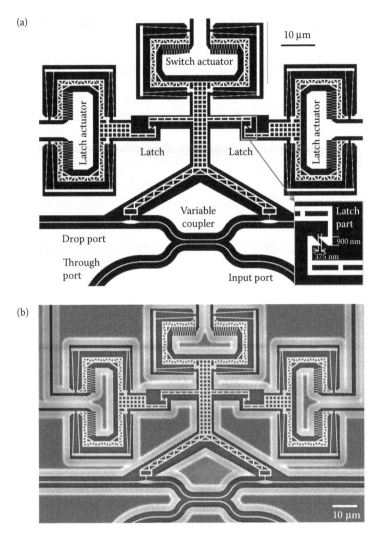

Figure 10.15 (a) Design of the coupler switch with a nano-latch mechanism, (b) electron micrograph of the fabricated switch.

actuator (switch-actuator) for varying the air gap of the coupler and a small latch mechanism with twin actuators (latch-actuators) for engaging and disengaging the latches, as shown in Figure 10.15a. The coupler is composed of 10-μm-long straight waveguides and 6-μm-radius bent waveguides. The nanolatch mechanism consists of four hook-shaped latches for holding a switch-on state. The designed overlap length of the hooks is 375 nm, as shown in the inset of Figure 10.15a. When the switch actuator pushes the latches to the engaged position, the waveguide is held by the latches in the switch-on state. The latches are released by applying a voltage to the twin latch actuators.

The switch was fabricated as shown in Figure 10.15b. The whole area of the switch is about 110 μm square. Figure 10.16 shows the displacement of the switch actuator measured as a function of the voltage. The displacement increases quadratically when increasing the voltage below 22 V. At 22 V, the hooks of the latches come into contact and the displacement increases slowly due to the load by the springs of the latch actuators. At 35 V, the hook is pushed through the maximum point and the waveguide is latched. The close-up views of the coupler and the latch are shown in Figure 10.16a–d, with the device in the switch-on and switch-off states. The overlap length of the latches is 375 nm and the position of the hook is well fixed, as shown in Figure 10.16c.

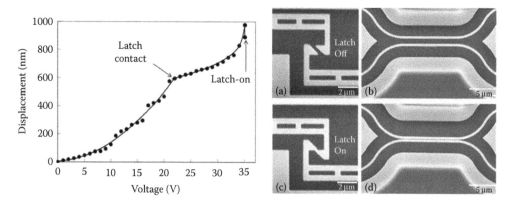

Figure 10.16 The displacement of switch-actuator measured as a function of the applied voltage. The latch part and the waveguide coupler are shown in images (a) and (b) in the conditions of latch-off and (c) and (d) latch-on, respectively.

Figure 10.17a shows the output intensities at the drop and through ports measured as a function of gap. The output intensity at the drop port increases around the gap of 400 nm and jumps at the gap smaller than 200 nm. The switch actuator pushes the hooks of latches, and the movable waveguide is latched at the voltage of 35 V. Since the spring constant of the latch actuator is 0.315 N/m, the force needed to latch is 236 nN, assuming the friction at the hook is negligibly small. From the images obtained in the latch-on state (Figure 10.16d), the gap between the coupler waveguides was about 109 nm, which was 9 nm larger than the designed value. The port isolation after latching is 17 dB.

Figure 10.17b shows the infrared spot images at the through and the drop ports and the normalized output intensities at the respective voltages of the switch actuator and latch actuators in a sequential off-on-off switching operation. The switch is turned on at a switch actuator voltage of 44 V in this experiment. After removing the switch actuator voltage, the switch-on state is held, with a little decrease of the output intensity (−0.18 dB) due to a small return of the waveguide to the latched position. The loss at the drop port under the latch-on condition is −0.30 dB. Without the actuator voltages, the output intensities of the switch are held by the latch mechanism. The latches are released by the application of a latch actuator voltage of 27 V, and the actuators return to the switch-off state.

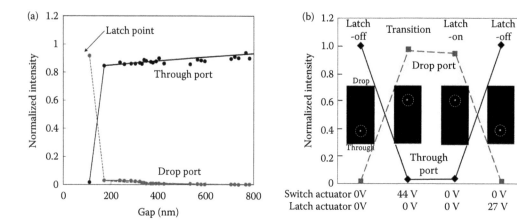

Figure 10.17 (a) Normalized output intensities measured as a function of gap. (b) Infrared images of the waveguide ends at output ports and the output intensities in the sequence of latching-on and -off.

10.5 Tunable silicon photonic microring filter for wavelength selective switch

10.5.1 Background and principle of microring resonator

A microring resonator is a key component for a narrow band filter and nonlinear functional devices in waveguide optics. Microring resonators have been intensively studied for wavelength selective filters and switches (Kato et al., 2002; Vorckel et al., 2003; Ito and Kokubun, 2004; Bogaerts et al., 2006). Silicon submicron waveguide microring resonators are also compact and can be densely integrated in silicon photonic circuits. The tunability of a microring resonator also opens a new technology for reconfigurable photonic circuits on the basis of the wavelength division multiplex system, which is now the main network system for fiber optic telecommunication (Green et al., 2005; Levy et al., 2006; Moselund et al., 2006; Wang et al., 2007). Lightwaves at different wavelengths are transmitted in the same optical fiber to efficiently utilize the capacity of a single fiber. The superposition and separation of the lightwaves at different wavelengths are carried out by multiplex and demultiplex systems. Single wavelength selection from densely multiplexed lightwaves (a few tens of lightwaves within a wavelength interval less than 10 nm) and the routing of the lightwaves are indispensable. Therefore, the microring resonator as a narrow wavelength filter is attractive to be used to drop and/or add a lightwave at a specific wavelength. A wavelength selective switch extracts a lightwave having a specific wavelength from the multiplexed lightwaves propagating in a busline waveguide, and then routes the lightwave to one of the output ports. For wavelength selection, tunability of the resonant condition of a microring is needed. Thermo-optic tuning by a heater is a conventional method using the temperature dependence of the material refractive index. However, heating is power consuming, and the device response time is relatively slow. Tuning by an electrostatic actuator is an attractive method due to its inherent advantages including fast response and low power consumption. In this chapter, silicon submicron waveguide microring resonators are studied for wavelength selective switches using integrated MEMS electrostatic actuators.

The basic optical configuration of a microring filter is shown in Figure 10.18 with a schematic filter response. A lightwave traveling from the input port couples with the microring waveguide and generates a standing wave along the circumference of the microring under the resonant condition. In the case where the wavelength of lightwave does not match with the resonant wavelength of the ring, the lightwave passes over the microring to the through port. Another waveguide for the drop and add ports is also coupled with the microring to extract the lightwave to the drop port and to add a new lightwave to the busline. At the resonant wavelength, the output intensity at the through port is minimized and that at the drop port is maximized, as shown in Figure 10.18b, which means that the lightwave is transmitted to the drop port. Using the propagation constant β,

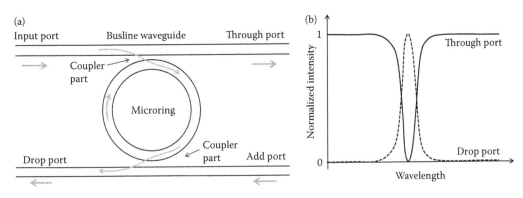

Figure 10.18 (a) Basic optical configuration of microring filter. (b) Schematic wavelength dependence of the microring filter.

the resonant condition of the microring having circumference length L is given by $\beta L = 2m\pi$ (m is an integer), which is rewritten as follows:

$$L = \frac{\lambda}{n} m, \tag{10.11}$$

where λ and n are the wavelength and effective index, respectively. The resonance spacing (free spectral range: FSR) is given by

$$\Delta\lambda = \frac{\lambda^2}{NL}, \tag{10.12}$$

where N is the group index defined as: $N \equiv n - \lambda(dn/d\lambda)$.

The intensity transmittance at the through port of the busline waveguide is given by (Okamoto, 1992)

$$T(\phi) = 1 - \frac{(1-x^2)(1-y^2)}{(1-xy)^2 + 4xy\sin^2(\phi/2)}, \tag{10.13}$$

where the round trip amplitude transmittance of the microring is given by $x = A \exp(-\rho L/2)$, the busline amplitude transmittance is expressed by $y = \cos(\chi_1 l_1)$, and $\phi = \beta L$. The symbol ρ is the intensity attenuation coefficient of the microring waveguide. The symbol A expresses the crossport amplitude transmittance by the coupler. The coupling coefficient and the effective length of the busline coupler are expressed by χ_1 and l_1, respectively. The intensity transmittance T as a function of wavelength at the through port is schematically shown in Figure 10.18b.

10.5.2 Wavelength selective switch with variable busline coupler

In this section, a wavelength selective add-drop switch using a silicon microring resonator is described, where the coupling between a busline and the microring is varied by an electrostatic actuator (Takahashi et al., 2008). The microring resonator is suspended in air by connecting to the actuator. Figure 10.19a shows an application of the wavelength selective switch in the wavelength division multiplex system. The input lightwaves at the wavelengths of λ_1, λ_2 and λ_n are selectively dropped by the respective switches using the microrings having the same resonant wavelengths by applying voltages to the actuators. Figure 10.19b shows the structure of the microring switch consisting of three silicon submicron waveguides, that is, a busline waveguide, an add/drop waveguide, and a microring waveguide. The input lightwave is selectively transmitted to the drop port when the microactuator moves the microring close to the busline waveguide in the condition that the input wavelength is equal to the resonant wavelength of the microring. The lightwaves at non-resonant wavelengths pass the microring to the through port. On the other hand, when the actuator is not operated, the microring is away from the busline waveguide and does not couple to it. The input light passes to the through port without interacting with the microring resonator.

The microring resonator has a rounded square shape with four 5 μm radius corners. Those waveguides are 260 nm in thickness and 500 nm in width. The circumferential length of the microring resonator is 63.4 μm. The initial gap between the ring and the busline is 775 nm. All the waveguides are suspended in air by elliptical bridges. The elliptical bridge is 8 μm long and 1.5 μm wide, and the suspension arm is 200 nm wide and 1 μm long. The input lightwave is in the lowest TE-like mode after passing through a TE mode selector. The comb finger is 250 nm in width, 260 nm in

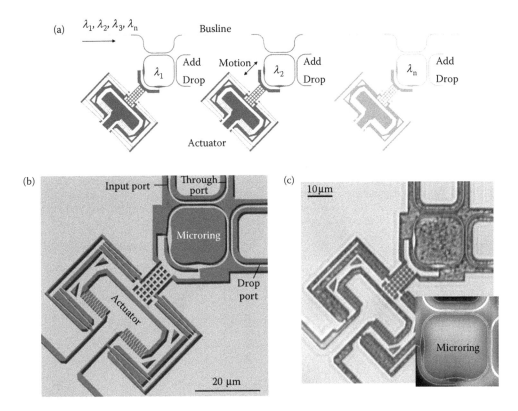

Figure 10.19 (a) An example of wavelength selective switch. (b) Schematic diagram of a microring wavelength selective add-drop switch. (c) Fabricated switch.

thickness, and 2 μm in length, and the gap between each finger pair is 350 nm. The spring constant of the actuator spring is 0.23 N/m. Figure 10.19c shows the optical micrograph of the fabricated switch with the electron micrograph of the microring part.

Figure 10.20a shows an infrared optical image of the switch. A bright light spot is observed at the drop port when laser light is incident on the input port at the resonant wavelength of the microring (1557.94 nm), and the actuator moves the microring close to the busline at a voltage of 28.2 V (switch-on state). A weak light spot at the through port is also observed, since the coupling of microing to busline waveguide does not fulfill the critical coupling condition (Okamoto, 1992, p. 199). Figure 10.20b shows the image obtained without applying the voltage (switch-off state) in the same wavelength condition that was used in Figure 10.2a. Because of a large distance between the microring and the busline, the lightwave is not transmitted to the drop port even if the wavelength is equal to the resonant wavelength of the microring.

Figure 10.21a shows the drop efficiency measured as a function of the input laser wavelength. The drop efficiency is obtained from the infrared spot intensity at the drop port in the switch-on state. The FSR is 8.45 nm. From the full width at half maximum (FWHM) of approximately 0.5 nm, the Q-value is determined to be about 3150.

Figure 10.21b shows the drop efficiency measured as a function of the distance between the microring and the busline at a resonant wavelength of 1576.04 nm. The actuator voltage is also shown in the Figure. Decreasing the distance from the initial position (775 nm), the light intensity at the drop port starts to increase at around 300 nm with the increase of the actuator voltage. It steeply increases between 300 and 120 nm. The critical coupling occurs at a gap of 120 nm when the maximum intensity is obtained. In the experiment, the maximum drop efficiency was −2.6 dB at a resonant wavelength of 1557.94 nm, and the through efficiency was −7.8 dB at the same time.

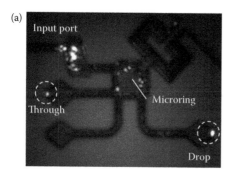
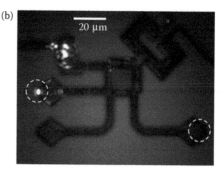

Figure 10.20 (a) Infrared camera image in the switch-on state at 1557.94 nm resonant wavelength and the actuator voltage of 28.2 V. (b) The switch-off state at the same resonant wavelength and voltage of 0 V.

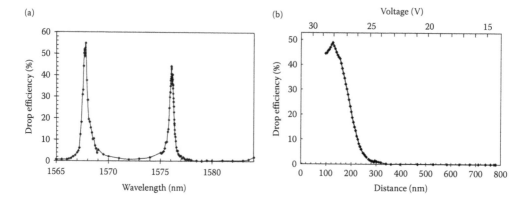

Figure 10.21 (a) Drop efficiency measured as a function of the wavelength at the voltage of 28.2 V, (b) Drop efficiency measured as a function of the applied voltage at the resonant wavelength (1576.04 nm).

10.5.3 Tunable microring composed of directional couplers and a hitless wavelength selective switch

By tuning the resonant wavelength of a microring, a lightwave at a specific wavelength is dropped and routed from a busline waveguide. The wavelength tuning of a microring is often based on the thermo-optic effect by heating the microring waveguide (Ng et al., 2007; Geng et al., 2009; Atabaki et al., 2010). The power consumption for the tuning is on the order of 1 mW/nm. In this section, a wavelength tunable microring consisting of two semicircular waveguides is described. The resonant wavelength of the microring is tuned by mechanically varying the ring round trip length with an actuator. In addition, the busline waveguide is disconnected by another actuator in order to not affect the lightwave propagating in the busline during the wavelength tuning of the microring. Therefore, a hitless wavelength selective switch can be constructed by the proposed mechanism. The variable mechanisms using the MEMS actuators have the advantages of a wide tuning range and low power consumption.

Figure 10.22a shows that the schematic diagram of an element of the proposed wavelength selective switch and the elements is composed of a wavelength-tunable microring resonator with a busline switch mechanism (Ikeda et al., 2013a). The microring consists of two freestanding silicon submicron waveguides. One is a semicircular waveguide coupling with a baseline waveguide. The second semicircular waveguide of the microring is movable, and is connected to an electrostatic comb drive actuator (tuning actuator). The two semicircular waveguides are optically coupled by two air gaps. Each overlapping region of the two semicircular waveguides operates as a directional waveguide coupler. While keeping the coupling rate close to unity, the movable waveguide

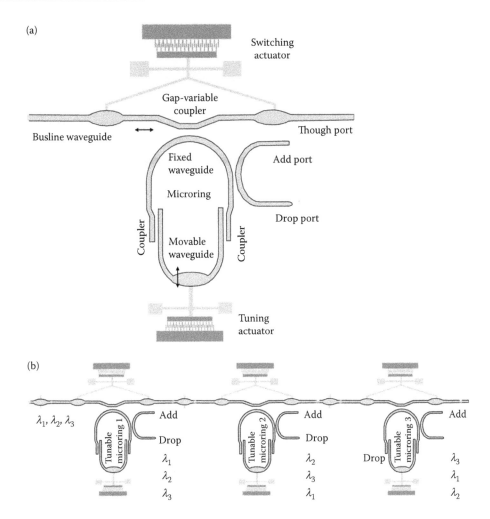

Figure 10.22 (a) Schematic diagram of wavelength-tunable microring resonator with busline switch mechanism and (b) schematic diagram of a wavelength selective switch using three tunable microrings with busline switches.

is translated to vary the circumferential length of the microring (Ikeda et al., 2013b). The coupling length of the coupler is unchanged. In order to construct a hitless wavelength selective switch using the tunable microrings, a busline switch is integrated to disconnect the microring. The resonant wavelength of the microring is varied after disconnecting the busline so as not to disturb the light-waves in the busline. In Figure 10.22a, the busline waveguide is coupled/decoupled with the fixed waveguide of the microring using another actuator (switching actuator).

Figure 10.22b shows an example connection of the wavelength-tunable microrings for a hitless wavelength selective switch. Assuming that lightwaves at different wavelengths ($\lambda 1$, $\lambda 2$, $\lambda 3$) are traveling simultaneously in the busline waveguide, the respective lightwaves are dropped at the respective ports if the resonant wavelengths of the microrings are tuned to the wavelengths ($\lambda 1$, $\lambda 2$, $\lambda 3$) of the lightwaves, as shown in the first set of wavelengths ($\lambda 1$, $\lambda 2$, $\lambda 3$) in Figure 10.22b. In order to change the set of wavelengths dropped at the respective ports, as shown in the second set of wavelengths ($\lambda 2$, $\lambda 3$, $\lambda 1$) in the Figure, the busline waveguide is disconnected from the respective microrings and the resonant wavelengths of the microrings are changed to match the new set of wavelengths. After tuning the wavelengths of the microrings, the busline waveguide is again con-nected to the microrings to drop the lightwaves at the set wavelengths. Therefore, the lightwaves in any set of wavelengths can be dropped to the designated drop ports. Without the busline switch

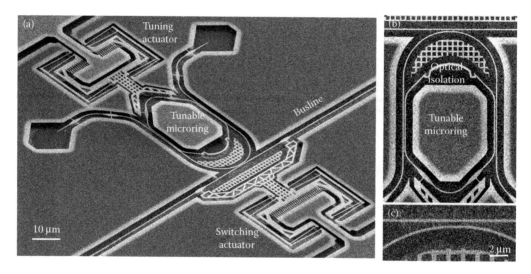

Figure 10.23 (a) Scanning electron micrograph of the fabricated tunable microring resonator with the busline coupling mechanism, (b) magnified microring, and (c) magnified busline coupler.

mechanism, the lightwaves at different wavelengths are blocked when tuning the resonant wavelengths of microrings.

Figure 10.23a shows an electron micrograph of the fabricated microring. The waveguides of the couplers in the microring are formed and aligned parallel to each other with an air gap of 248 nm. The waveguides used for the device are 320 nm in width and 340 nm in thickness. The coupling lengths are 14 μm. These gaps and lengths were determined to obtain the maximum coupling coefficient nearly equal to unity at the wavelength of 1.55 μm. The two ends of the movable waveguide are physically connected by an optical isolation structure for reinforcing the freestanding waveguide. The optical isolation structure with strong bends prevents light from propagating through. The total round trip length of the microring is approximately 122 μm under the initial condition. Figure 10.23b and c show magnified views of the microring and busline coupler, respectively.

Figure 10.24a and b show the output light intensities at the through port measured as a function of wavelength with tuning actuator displacements of 330 nm (15 V) and 875 nm (24 V), respectively. The busline coupling is close to the critical coupling, where the gap of the coupler is about 383 nm. The periodic dips are observed as a function of wavelength. The value of FSR is 3.3 nm and the Q-factor is about 4000. The depth of the resonant dip is approximately −10 dB. Using Equation 10.12, the transmittances x and y can be evaluated from the depth and width of the resonant dip. The estimated values of x and y are 0.86 and 0.76, respectively. The round trip loss of the resonator is evaluated to be approximately −1.28 dB. The transmittance $T(\phi)$, calculated by Equation 10.12, is also shown by the dotted curve in Figure 10.24a. When increasing the displacement from 330 nm in Figure 10.24a to 785 nm in Figure 10.24b, the resonant dip denoted by dip A shifts to the lower wavelength region. The wavelength shift is 6.6 nm for a voltage increase of 9 V.

The measured wavelength shift of the resonant dip is shown as a function of the displacement and the voltage in Figure 10.25. The wavelength shift is almost linearly dependent on the displacement. The maximum wavelength shift is approximately 10 nm at a voltage of 26 V.

Applying a voltage to the switching actuator, the switch function of the busline is examined. The minimum output intensity of a resonant dip at a wavelength of 1478.24 nm at the through port is monitored and measured as a function of the gap of the busline coupler, as shown in Figure 10.26. At the initial gap of 380 nm, the coupling condition is slightly overcoupled. The critical coupling condition is obtained at a gap of 383 nm and 2.0 V. Further increasing the voltage of the switching actuator pulls the busline apart from the microring, and the busline is gradually decoupled from the microring, hence the resonant dips become shallower. The dips disappear at a voltage of 22 V, then the busline is almost disconnected from the microring resonator.

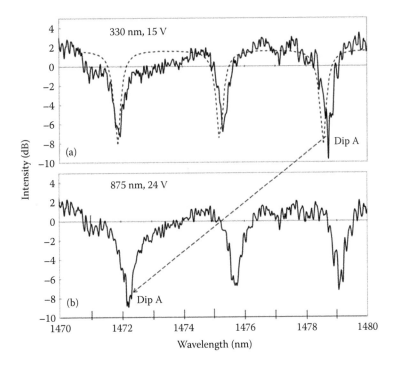

Figure 10.24 Output intensity at the through port as a function of wavelength, (a) the displacement of tuning actuator: 330 nm and the applied voltage: 15 V, (b) 875 nm and 24 V. The dip A in (a) moves to the position shown in (b).

In order to construct the wavelength selective switch, the tunable microrings are connected as shown in Figure 10.27. The microrings are arrayed with a period of about 100 μm along the busline waveguide, and the waveguides for the respective drop ports are also arranged. In order to cover a large wavelength band in the wavelength division multiplexing optical telecommunication system and to select a lightwave at a single wavelength in the region, a wide-range wavelength-tunable silicon microring resonator is also designed and fabricated (Chu and Hane, 2014). The initial round trip length of the microring is 19.1 μm and the free spectral range is 21 nm. The wavelength dependence of the couplers used in the microring is compensated by adjusting the coupling length while

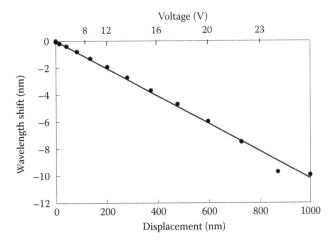

Figure 10.25 Wavelength shift of microring resonator measured as a function of the displacement and the voltage of tuning actuator.

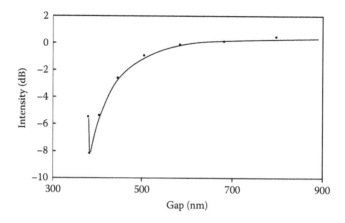

Figure 10.26 Output intensity at a dip (1478.24 nm) measured as a function of the gap of busline coupler.

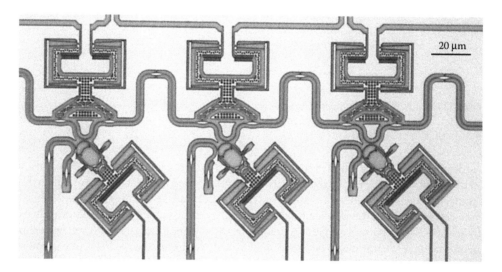

Figure 10.27 Optical micrograph of an arrangement of tunable microring for wavelength selective switch.

varying the round trip length of the ring. The resonant wavelength of the microring is varied by 27 nm at an applied voltage of 25 V.

10.6 Conclusions

MEMS technology is promising for the future development of silicon photonic/electronic circuit integration in optical telecommunication and interconnections, as well as the integration of photonics and electronics. Variable functional devices using MEMS/NEMS actuators are attractive for reconfigurable densely integrated optical circuits. In this chapter, silicon photonic waveguide devices on the basis of variable waveguide couplers were described. The variable couplers were operated by ultra-small electrostatic comb drive actuators. Using the variable gap silicon photonic waveguide couplers, single and multiple waveguide switches were studied. The fabricated 2 × 2 coupler switch was as small as 100 μm by 100 μm. The fundamental properties of the variable-gap coupler and the optical and mechanical characteristics of the switch were investigated theoretically and experimentally. In addition, devices using the silicon photonic variable couplers, including a microring switch and a tunable microring resonator for a wavelength selective filter in the wavelength-division multiplexing system, were also discussed. The microring was connected and disconnected to a busline waveguide, thus preventing undesirable disturbances during tuning. The resonant wavelength of the microring for filtering was widely tuned by varying the round trip length of the microring by the ultra-small MEMS actuator. The proposed mechanisms were viable to realize hitless wavelength selective switches. From those proposed devices, dense reconfigurable photonic circuits are feasible on a small silicon chip.

Acknowledgments

The author thanks Y. Kanamori, T. Sasaki, K. Takahashi, Y. Akihama, T. Ikeda, S. Abe, H.M. Chu, and Y. Munemasa for their experiments and useful advice. This work was supported by JSPS, SCOPE, and µSIC. The device fabrications were carried out in MNC in Tohoku University.

References

Abe S, Chu M H, Sasaki T, Hane K 2014. Time response of a microelectromechanical silicon photonic waveguide coupler switch. *IEEE Photon Technol Lett*. 26: 1553–1556.

Abe S, Hane K 2013. Variable-gap silicon photonic waveguide coupler switch with a nanolatch mechanism. *IEEE Photon Technol Lett*. 25: 675–677.

Akihama Y, Hane K 2012. Single and multiple optical switches that use freestanding silicon nanowire waveguide couplers. *Light: Sci Appl*. 1: e16(8pp).

Akihama Y, Kanamori Y, Hane K 2011. Ultra-small silicon waveguide coupler switch using gap-variable mechanism. *Opt Express*. 19: 23658–23663.

Atabaki A H, Shah Hosseini E, Eftekhar A A, Yegnanarayanan S, Adibi A 2010. Optimization of metallic microheaters for high-speed reconfigurable silicon photonics. *Opt Express*. 18: 18312–18323.

Bogaerts W, Dumon P, Thourhout D V, Taillaert D, Jaenen P, Wouters J et al. 2006. Compact wavelength-selective functions in silicon-on-insulator photonics wires. *IEEE J SelTop Quantum Electron*. 12: 1394–1401.

Bulgan E, Kanamori Y, Hane K 2008. Submicron silicon waveguide optical switch driven by microelectromechanical actuator. *Appl Phys Lett*. 92: 101110.

Chew X, Zhou G, Chau F S, Deng J 2011. Nanomechanically tunable photonic crystal resonators utilizing triple-beam coupled nanocavities. *IEEE Photon Technol Let*. 23: 1310–1312.

Chew X, Zhou G, Chau F S, Deng J, Tang X, Loke Y C 2010. Dynamic tuning of an optical resonator through MEMS-driven coupled photonic crystal nanocavities. *Opt Lett*. 35: 2517–2519.

Cho Y-H, Kawk B M, Pisano A P, Howe R T 1994. Slide film damping in laterally driven microstructures. *Sens Act*. A40: 31–39.

Chu H M, Hane K 2014. A wide-tuning silicon ring-resonator composed of coupled freestanding waveguides. *IEEE Photo Technol Lett*. 26: 1411–1413.

Fukazawa T, Hirano T, Ohno F, Baba T 2004. Low loss intersection of Si photonic wire waveguides. *Jpn J Appl Phys*. 43: 646–647.

Geng M, Jia L, Zhang L, Yang L, Chen P, Wang T, Liu Y 2009. Four-channel reconfigurable optical add-drop multiplexer based on photonic wire waveguide. *Opt. Express*. 17: 5502–5516.

Green W J, Lee R L, DeRose G A, Scherer A, Yariv A 2005. A Hybrid InGaAsP-InP Mach-Zehnder racetrack resonator for thermooptic switching and coupling control. *Opt Express*. 13: 1651–1659.

Green W M J, Rooks M J, Sekaric L, Vlasov Y A 2007. Ultra-compact, low RF power 10 Gb/s silicon Mach-Zehnder modulator. *Opt Express*. 15: 17106–17113.

Han S, Seok T J, Quack N, Yoo B-W, Wu M C 2015. Large-scale silicon photonic switches with movable directional couplers. *Optica*. 2: 370–375.

Ikeda T, Hane K 2013a. A tunable notch filter using microelectromechanical microring with gap-variable busline coupler. *Opt Express*. 21: 22034–22042.

Ikeda T, Hane K 2013b. A microelectromechanically tunable microring resonator composed of freestanding silicon photonic waveguide couplers. *Appl Phys Lett*. 102: 221113(4pp).

Ito T, Kokubun Y 2004. Nondestructive measurement of propagation loss and coupling efficiency in microring resonator filters using filter responses. *Jpn J Appl Phys*. 43: 1002–1005.

Jaecklin V P, Linder C, de Rooij N F, Moret J M 1992. Micromechanical comb actuators with low driving voltage. *J. Micromech. Microeng*. 2: 250–255.

Jalali B, Fathpour S 2006. Silicon photonics. *J Lightwave Technol*. 24: 4600–4615.

Janz S, Cheben P, Dalage D, Densmore A, Lamontague B, Picard MJ et al. 2006. Microphotonic elements for integration on the silicon-on-insulator waveguide platform. *IEEE J Sel Top Quantum Electron*. 12: 1402–1415.

Kanamori Y, Aoki Y, Sasaki M, Hosoya H, Wada A, Hane K 2005. Fiber-optical switch using cam-micromotor driven by scratch drive actuators. *J Micromech Microeng*. 15: 118–123.

Kato T, Suzuki S, Kokubun Y, Chu S T 2002. Coupling-loss reduction of a vertically coupled microring resonator filter by spot-size matched busline waveguide. *Appl Opt*. 41: 4394–4399.

Ko J S, Lee M L, Lee D-S, Choi C A, Kim Y T 2002. Development and application of a laterally driven electromagnetic microactuator. *Appl Phys Lett*. 81: 547–549.

Koonath P, Indukuri T, Jalali B 2006. Monolithic 3-D silicon photonics. *J Lightwave Technol*. 24: 1796–1804.

Levy U, Campbell K, Groisman A, Mookherjea S, Fainman Y 2006. On-chip microfluidic tuning of an optical microring resonator. *Appl Phys Lett*. 88: 11107.

Marcatili E A J 1969. Dielectric rectangular waveguide and dielectric coupler for integrated optics. *Bell Syst Tech J*. 47: 2071–2102.

Martines L, Lipson M 2006. High confinement suspended micro-ring resonators in silicon-on-insulator. *Opt Express*. 14: 6259–6263.

Moselund K E, Dainesi P, Declercq M, Bopp M, Coronel P, Skotnicki T, Ionescu A M 2006. Compact gate-all-around silicon light modulator for ultra high speed operation. *Sens Act A*. 130–131: 220–227.

Munemasa Y, Hane K 2013. A compact 1 × 3 silicon photonic waveguide switch based on precise investigation of coupling charcteristics of variable-gap coupler. *Jpn Appl Phys*. 52: 06GL15(6pp).

Ng H-Y, Wang M R, Li D, Wang X, Martinez J, Panepucci R R, Pathak K 2007. 1 × 4 wavelength reconfigurable photonic switch using thermally tuned microring resonators fabricated on silicon substrate. *IEEE Photon Technol Lett*. 19: 704–706.

Okamoto K 1992. *Fundamentals of Optical Waveguides*. Burlington FL: Academic Press/Elsevier.

Sasaki K, Ohno F, Motegi A, Baba T 2005. Arrayed waveguide grating of 70 × 60 μm² size based on Si photonic wire waveguides. *Electron Lett*. 41: 801–802.

Seok T J, Quack N, Han S, Muller R S, Wu M C 2016. Large-scale broadband digital silicon photonic switches with vertical adiabatic couplers. *Optica*. 3: 64–70.

Takahashi K, Bulgan E, Kanamori Y, Hane K 2009. Submicron comb-drive actuators fabricated on thin single crystalline silicon layer. *IEEE Trans Ind Electron*. 56: 991–995.

Takahashi K, Kanamori Y, Kokubun Y, Hane K 2008. A wavelength-selective add-drop switch using silicon microring resonator with a submicron-comb electrostatic actuator. *Opt Express*. 16: 14421–14428.

Vorckel A, Monster M, Henschel W, Bolivar P H, Kurz H 2003. Asymmetrically coupled silicon-on-insulator microring resonators for compact add-drop multiplexers. *IEEE Photon Technol Lett*. 15: 921–923.

Wang T J, Chu C H, Lin C Y 2007. Electro-optically tunable microring resonators on lithium niobate. *Opt Lett*. 32: 2777–2779.

Wu M C, Solgaard O, Ford J E 2006. Optical MEMS for lightwave communication. *J Lightwave Technol*. 24: 4433–4454.

Yamada H, Chu T, Ishida S, Arakawa Y 2006. Si photonic wire waveguide devices. *IEEE J Sel Top Quantum Electron*. 12: 1371–1379.

Yao J, Leuenberger D, Lee M C M, Wu M C 2007. Silicon microtoroidal resonators with integrated MEMS tunable coupler. *IEEE J Sel Top Quantum Electron*. 13: 202–208.

11

Metasurface and ultrathin optical devices

Xianzhong Chen, Dandan Wen, and Fuyong Yue

Contents

11.1 Introduction

Traditional optical devices can modify the wave front of light by altering its phase, amplitude, and polarization. For example, a lens or prism is usually realized by reshaping the wave front of the light that relies on gradual phase changes along the optical paths, which is accomplished by either controlling the surface topography or varying the spatial profile of the refractive index. A wave plate utilizes bulk birefringent crystals with optical anisotropy to change the polarization of light, and a hologram is based on the interference of diffracted waves from different parts of the components in the far-field to produce the desired optical pattern. All of these components shape optical wave fronts using the propagation effect. Thus, it is hard to accumulate sufficient phase change once the device size is further reduced to micro- and even nano-scale due to the finite permittivity and permeability of natural materials. While there has been great progress in the miniaturization of optical elements, such photonic integration largely depends on the technical advancement. On the other hand, for optical systems to continue to be established as economically viable in a range of emerging application areas, it is necessary to continue the trend of miniaturization and integration. Therefore, to meet the growing requirement of device miniaturization and system integration, a new design methodology is urgently needed to develop ultrathin optical devices.

Optical metamaterials, the artificially engineered subwavelength structures whose optical properties are determined by their geometrical structures instead of their constituent material composition, have been used to control the propagation of electromagnetic waves to an unprecedented level, leading to highly unconventional and versatile functionalities compared with their natural counterparts. Broad applications of metamaterials that have been reported to date include invisibility cloaks (Schurig, 2006, Edwards et al., 2009, Gabrielli et al., 2009, Liu, 2009, Valentine et al., 2009, Ergin et al., 2010), negative refraction (Shelby et al., 2001, Parazzoli et al., 2003, Shalaev, 2007, Valentine, 2008), and super imaging (Fang et al., 2005). Recently, optical metasurfaces, a two-dimensional counterpart to optical metamaterials, have attracted much attention due to their distinguished features and simplicity of fabrication. Metasurfaces consisting of a monolayer of artificial atoms are capable of manipulating light in a desirable manner by imparting local and space-variant abrupt phase change on an incident electromagnetic wave, breaking our dependence on phase accumulation due to the propagation effect. At the interface of a metasurface, wave front shaping is accomplished within a distance much smaller than the wavelength of the light beam, thus providing a new opportunity to develop ultrathin devices that are easy to integrate into compact platforms. Metasurfaces enable us to engineer the spatial distribution of amplitude, phase, and polarization response with subwavelength resolution, and have been used to develop a plethora of ultrathin devices, such as wave plates for generating vortex beams (D'Aguanno et al., 2008, Genevet et al., 2012), ultrathin metalenses (Chen et al., 2012, 2013), aberration-free quarter-wave plates (Yu et al., 2012), spin-hall effect of light and spin-controlled photonics (Shitrit et al., 2013, Yin et al., 2013), unidirectional surface plasmon polariton excitation (Huang et al., 2013b, Lin et al., 2013), and three-dimensional optical holography (Huang et al., 2013a). In this chapter, we are going to focus on phase discontinuities at the interface of metasurface, conversion efficiency, and some ultrathin optical devices such as vortex beam generators, metalenses, metasurface holograms, and multifunction optical devices.

11.2 Interfacial phase discontinuities

The phase of the emitted light from an optical resonator can be changed appreciably through the resonance (Yu et al., 2011). By placing an array of nanoresonators on the surface of a substrate, the phase of the refracted or reflected light can be changed locally, which induces phase discontinuities. As an example, V-shaped nano-antennas (Yu et al., 2011) support both the symmetric and anti-symmetric modes, which are both excited when a light beam with arbitrary polarization is incident onto the antenna. Since both modes have different resonant conditions, the refracted or reflected light contains a substantial component that has the polarization orthogonal to that of the incident light. The phase and amplitude of the cross-polarized light can be controlled by modifying the geometry of the V-shaped antennas. If each supercell of the metasurface consists of eight antennas, which have equal scattering amplitudes and constant phase difference $\Delta\Phi = \pi/4$ between neighbors, the phase gradient along the x direction is then formed, and the anomalous refraction and reflection are realized for the cross-polarized light. Besides the V-shaped antennas, the choice of nanoresonators is wide-ranging, such as the silicon posts (Arbabi et al., 2015b), H-shaped (Sun et al., 2012), and C-shaped antennas (Zhang et al., 2013, Liu et al., 2014).

Different from the metasurfaces presented above that depend on the dispersion of antenna resonance, the geometric metasurface utilizing the Pancharatnam–Berry phase provides an alternative to realize the phase discontinuity. The Pancharatnam–Berry phase is a geometric phase associated with the polarization of light. When the polarization of a beam traverses a closed loop on the Poincaré sphere, the final state differs from the initial state by a phase factor equal to half of the area encompassed by the loop upon the sphere. It is well known that each state of polarization of a monochromatic light beam can be uniquely represented by a point on the Poincaré sphere. By convention, the north and south poles represent the states of circular polarization, and points on the equator correspond to linear polarization of different orientations with diametrically opposite points being orthogonal to each other. It is important to note that we can vary the phase by changing the state of polarization without changing the optical path length. With the help of optical elements (e.g., polarizers and retarders), the state of polarization is made to trace out a closed contour on the sphere and a geometric phase for the light beam can be acquired. This phase change can be considered as the Pancharatnam–Berry phase that is acquired when the polarization state of light travels around a contour on the Poincaré sphere. Specifically, for an incident beam with circular polarization σ, the phase difference between the scattered waves of opposite circular polarization σ^- scattered by two dipole antennas of different orientations, φ_1 and φ_2, can be viewed as half of the solid angle enclosed between two paths on the Poincaré sphere, $\sigma \to L(\varphi_1) \to \sigma^-$, and $\sigma \to L(\varphi_2) \to \sigma^-$, where σ and σ^- are represented by the north and south poles of the Poincaré sphere, respectively, and $L(\varphi_i)$ are the linear polarization states residing on the equator of the Poincaré sphere (Berry, 1987).

One example of such a metasurface using Pancharatnam–Berry phase is shown in Figure 11.1a, which consists of gold nanorods with identical geometry but varying orientations (Huang et al., 2012). Each nanorod can be regarded as an anisotropic scatter which is able to change part of the incident circularly polarized light into its opposite handedness and impart a Pancharatnam–Berry phase $\pm 2\psi$ to it (ψ is the orientation of the nanorod, the sign "+" for the conversion from left circular polarization (LCP) to right circular polarization (RCP) and the sign "−" for RCP to LCP). As a

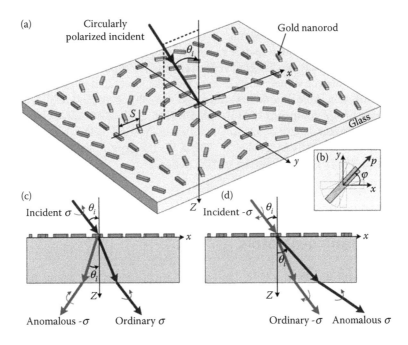

Figure 11.1 Images and schematic illustrations of a refractive dipole array. (a) Schematic illustration of a representative dipole array. A rectangular Cartesian coordinate system $Oxyz$ is established so that the z-axis is parallel to the normal of the surface and the structure is periodically modulated in the x-direction and repeated in y-direction (with one unit cell highlighted). θ_i and θ_t are the incident and transmission angles, respectively. (b) Excitation of dipole moment when illuminating one nanorod with the azimuth angle φ with the x-axis. (c–d) Schematic illustration of normal and anomalous refraction by dipole arrays when illuminated with σ and $-\sigma$ polarized circularly polarized light, respectively. $\pm\sigma$ correspond to the RCP/LCP. (Figures reproduced with permission from ACS.)

result, the phase response of the metasurface is controlled solely by the distribution of the nanorods with designed orientations. When the circularly polarized light is incident onto the metasurface, the ordinary refracted light with the same handedness of the incident light follows Snell's law, while the anomalously refracted light with the opposite handedness has to obey the general laws for refraction under which the anomalous refraction is observed (Figure 11.1c, d). The geometric metasurface may have various shapes of nanoresonators, such as silicon nanobeams (Lin et al., 2014), the U-shaped (Kang et al., 2012), and cross-shaped antennas (Luo et al., 2015).

11.3 Conversion efficiency

One of most important parameters to evaluate the performance of the designed metasurface is conversion efficiency, which is defined as the ratio of the power of the modulated refracted or reflected light to that of the incident light. The conversion efficiency for the cross-polarized light is low for the transmissive-type metasurface that consists of a single layer of metallic antennas (Ding et al., 2015). It is reported that the maximum conversion efficiency is 25% for such a kind of metasurface due to the fundamental limit (Arbabi, 2014).

The conversion efficiency can be greatly improved using the reflective-type metasurface, which usually consists of three layers: a metallic ground layer, a dielectric spacer, and a nano-antenna layer on the top (Zheng et al., 2015) (Figure 11.2a, b). By optimizing the parameters of the three-layer structure, the reflection efficiencies for the light polarized along the long and short axes of the nanorod can both reach over 80%, and the phase difference between them approaches π within 600–1000 nm (Figure 11.2d, e). As a result, the three-layer structure functions as a reflective-type half-wave plate that is able to change the incident circularly polarized light into its cross-polarization with high efficiency (Figure 11.2f). The simulated conversion efficiency is over 80% in a broad wavelength range between 550 and 1000 nm.

Although the reflection-type metasurfaces can achieve high efficiency (Grady et al., 2013, Luo et al., 2015), they are generally less popular than transmissive-type devices for practical applications. Fortunately, the dielectric metasurfaces have been demonstrated to possess high conversion efficiency in transmission since they do not suffer from the Ohmic loss in metal. It is reported that the dielectric metasurface composed of Si nanobeam (Lin et al., 2014) wave plates can reach an efficiency of 75% at the wavelength of 500 nm. Another interesting dielectric metasurface consists of dielectric elliptical nanoposts (Arbabi et al., 2015a, b). The experimentally measured efficiency ranges from 72% to 97%, depending on the specific design.

11.4 Ultrathin optical devices

11.4.1 Vortex beam generators

As a special light beam that carries orbital angular momentums, vortex beams have found wide applications (Yi et al., 2014) such as optical tweezers, microscopy, and optical communication. Generally, the vortex beam is generated using the computer-generated hologram, spatial light modulator or Q-plate. In this section, a method to generate an optical vortex beam over a broad wavelength range is proposed by using the metasurface (Huang et al., 2012).

11.4.1.1 Basic principles of the transmissive-type geometric metasurface

The metasurface used here consists of an array of metallic nanorods with the same geometry but spatially varying orientations (Huang et al., 2012) (Figure 11.3a). Within a certain range of incident angles around the surface normal, a circularly polarized beam is primarily scattered into waves of the same polarization as that of the incident beam without phase change, and waves of the opposite circular polarization with a phase change of twice the angle formed between the dipole and

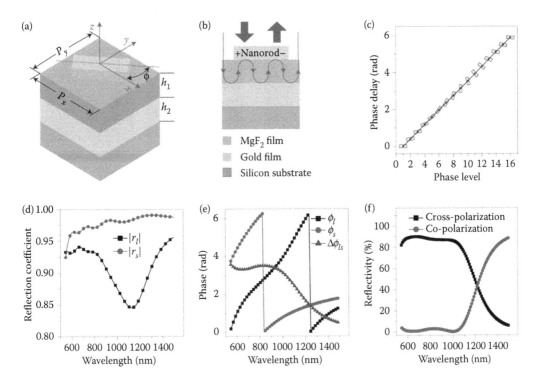

Figure 11.2 Illustration of the unit-cell structure and its polarization conversion efficiency by numerical simulations. (a) One-pixel cell structure of the nanorod-based hologram. The nanorod can rotate in the x–y plane with an orientation angle ϕ to create a different phase delay. The pixels are arranged with periods $P_x = 300$ nm and $P_y = 300$ nm. The nanorods have length $L = 200$ nm, width $W = 80$ nm, and height $H = 30$ nm. The MgF$_2$ and gold films have thicknesses of $h_1 = 90$ nm and $h_2 = 130$ nm, respectively. (b) Cross-section of the pixel cell. The MgF$_2$ film acts as a Fabry–Pérot cavity, which can allow the reflected beam to continue to excite the nanorod and generate the output beam with a phase delay. The gold film acts as a mirror to reflect the incident light. (c) Phase delay for the different phase levels. On each knot, the orientation of the nanorod is annotated. (d–e) Simulated amplitude $|r_l|$, $|r_s|$ (d), phase ϕ_l, ϕ_s and phase difference $\Delta\phi_{ls}$ (e) of the reflection coefficients r_l and r_s, where l and s denote the long and short axis directions of the nanorods, respectively. (f) Simulated cross-polarization and co-polarization reflectivity with normal light incidence. (Figures reproduced with permission from NPG.)

the x-axis ($\Phi = 2\sigma\varphi$), where $\sigma = \pm 1$ correspond to the helicity of left- (LCP) and right-circularly polarized (RCP) incident light.

The conversion efficiency to the opposite polarization is only determined by the geometry of the dipole antenna, instead of its orientation. That is, the scattering amplitudes of light of the opposite helicity automatically satisfy the equal-amplitude condition, with the highest efficiency occurring at the peak of the resonance, and they decrease when the wavelength of light is away from the resonance frequency. As the phase discontinuity does not depend on the dispersion property of the dipole antenna, the proposed structure can operate for a wide range of wavelengths with dispersion-less phase discontinuity. Hence, the proposed phase control scheme represents a robust and facile approach for controlling the phase and wave front.

11.4.1.2 Sample design and experimental results

The vortex beam is characterized by an $\exp(il\varphi)$ azimuthal phase dependence, that is, the orbital angular momentum in the propagation direction has the discrete value of $l\hbar$ per photon, where l is the topological charge and can be an integer or a non-integer. As shown by Figure 11.3a, the plasmonic interface is created by arranging the dipoles with spatially varying orientation angles that rotate

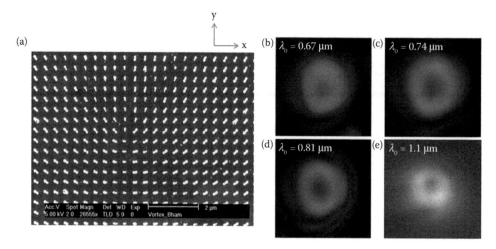

Figure 11.3 (a) Scanning electron microscopy image of the dipole array which was designed for generating an optical vortex beam. (b–e) Measured intensity distribution of the vortex beam patterns for different wavelengths from 670 to 1100 nm. (Figures reproduced with permission from ACS.)

180° when the locations of the dipoles rotate by 360° around the singular point. Mathematically, the rotation angle φ is given by tan $2\varphi = (x/y)$ for $y > 0$ and tan $2(\varphi-(\pi/2)) = (x/y)$ for $y < 0$. The interface introduces a spiral-like phase shift with respect to the planar wave front of the incident light, creating a vortex beam with topological charge $l = 1$. Because the azimuthal phase profile is abruptly introduced through the interaction with the dipole plane, the screw-like phase profile characteristic of the vortex beam can be created over a negligible propagation distance, in contrast to conventional methods. In the experiment, we measured the evolution of the vortex beam for different incident wavelengths and clearly observed the characteristic beam profile for all wavelengths (Figure 11.3b–e): an annular intensity distribution in the cross-section, and a characteristic dark spot with zero intensity in the center.

11.4.2 Ultrathin metalenses

Traditional lenses change the wave front of light by accumulating the phase difference through the propagation. As a result, the thickness of the lenses is usually much larger than the incident wavelength. Benefiting from the geometry phase introduced by the metasurface, it is possible to realize ultrathin metalenses whose thickness is deep subwavelength. In the following, the cylindrical lens and spherical lens based on metasurfaces are proposed and experimentally verified (Chen et al., 2012, 2013).

11.4.2.1 The cylindrical lens

11.4.2.1.1 Design principle

The transmissive metasurface consists of gold nanorods, whose mechanism is illustrated above. In order to focus an incident circularly polarized plane wave, the flat lensing surface must undergo a spatially varying phase shift. To achieve the phase profile equivalent to a conventional cylindrical lens, the following expression governs the relationship between the rotation angle φ and the location x of the dipole antenna

$$\phi(x) = \pm 0.5 k_0 \left(\sqrt{f^2 + x^2} - |f| \right)$$

(11.1)

where $k_0 = 2\pi/\lambda$ is the free-space wave vector and f is the focal length of the lens. Note that the + and − sign in Equation 11.1 correspond to a positive (convex) and negative (concave) lens, respectively, for an RCP incident wave, and the opposite holds for an LCP incident wave.

To implement the cylindrical lens, the rotation angle of the dipole antennas should vary according to Equation 11.1. Figure 11.4 shows a schematic of the designed lens that consists of dipole nano-antennas with the directional orientation corresponding to a + sign in Equation 11.1. The dipoles are arranged in a two-dimensional array with a subwavelength period of S in both x and y directions.

11.4.2.1.2 Sample fabrication

Based on the interfacial phase discontinuity, we designed and fabricated two plasmonic metalenses, Lens A and Lens B, with a negative and a positive polarity for an incident beam with RCP polarization, respectively. The plasmonic dipole antennas are fabricated by electron beam lithography on an indium-tin-oxide (ITO)-coated glass substrate. The dipole antennas are made from gold with a thickness of 40 nm. Each lens has an aperture of $80 \times 80\ \mu m^2$ and a focal length $f_A = -60\ \mu m$ and $f_B = 60\ \mu m$ for an incident wave with RCP polarization, respectively. The dipole antennas are 200 nm long and 50 nm wide, exhibiting a longitudinal resonance around 970 nm and a transverse resonance around 730 nm. The rotation angles for the dipoles far from the lens center change more rapidly than those near the center.

11.4.2.1.3 Characterization of the lens

We experimentally demonstrate the performance of the focusing of the plasmonic lens by using a circularly polarized laser beam at the wavelength of 740 nm. A positive lens causes the incident laser beam to converge at a focal plane on the transmission side of the lens forming a real focus line, while a negative lens causes the incident laser beam to emerge from the lens as though it is emanated from a virtual focal plane on the incident side of the lens. In the measurement, by gradually adjusting the distance between the objective and the plasmonic lens, we are able to examine the optical intensity distribution at different z locations along the propagation direction to determine the focal plane.

Figure 11.5 shows the optical microscopy images for two different incident/transmission polarization combinations: RCP/LCP (Figure 11.5a) and LCP/RCP (Figure 11.5b). The experimental setup is shown in Figure 11.6. As shown by Figure 11.5a (left) for the RCP incident beam, we observe a bright focused line along the y direction for Lens A at $z = -61$ mm, which agrees well with the designed focal length. This is a virtual focal point as it lies on the incident side of the plasmonic lens. Hence, it verifies that Lens A is a negative (concave) lens for RCP incident light. On the

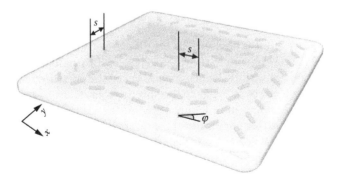

Figure 11.4 Diagram of the bipolar plasmonic lens. The lens consists of an array of plasmonic dipole antennas on a glass substrate with orientations varied along the focusing direction (*x*). The distance between neighboring dipoles, $S = 400$ nm, is the same along the two in-plane directions. φ is the rotation angle of the dipole relative to the *x*-axis. The abrupt phase shift is solely determined by the orientation of the dipoles. (Figure reproduced with permission from NPG.)

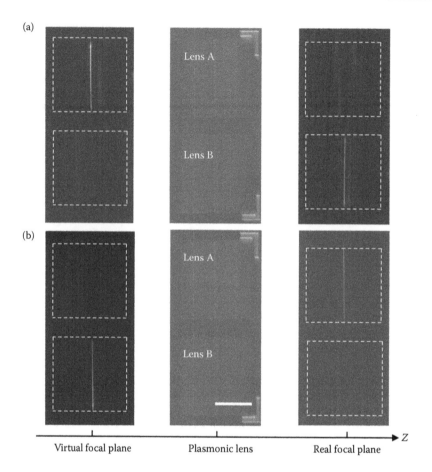

Figure 11.5 Lens polarity is reversed by changing the handedness of the incident light. Optical microscope images at virtual focal plane (left), lens surface (middle), and real focal plane (right) for the incident light with (a) RCP and (b) LCP. CP laser beam is incident on the plasmonic lens from the left along the z direction, and the lens is located at $z = 0$. Positions of lenses are marked by the white dashed squares. The scale bar is 50 mm. The polarity of Lens A is different from that of B for the same CP light. The distance between the real focal plane and the lens is measured to be 60 mm, which is the real focal length. The distance between the real focal plane and the virtual one is 121 mm, which corresponds to 2f. The wavelength of the incident laser beam used in the above images is 740 nm. (Figures reproduced with permission from NPG.)

other hand, a bright focused line is observed for Lens B on the transmission side of the plasmonic lens at $z = 60$ mm, which corresponds to the real focal plane. This confirms that Lens B is positive (convex) for the incident light with RCP polarization. When the polarizations of the incident and transmitted beams are switched to LCP and RCP, respectively, the focusing behaviors for both Lens A and Lens B are reversed, as shown in Figure 11.5b. At the virtual focal plane $z = -61$ mm, a virtual focused line is observed for Lens B and at the real focal plane $z = 60$ mm, a real focal line is observed for Lens A. The conversion in the focusing properties from positive (negative) to negative (positive) is solely attributed to the handedness change of the CP for the incident light, which agrees perfectly with the theoretical prediction.

11.4.2.2 The spherical lens

11.4.2.2.1 Design and fabrication

As shown by Figure 11.7a, the spherical metalens is composed of dipole antennas arranged in a number of evenly spaced concentric rings, with the radius of the rings increasing by a step size

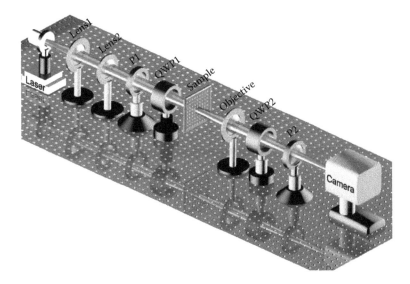

Figure 11.6 Schematic of the optical measurement setup. The polarization directions of the two polarizers are parallel to each other. The incident CP light is generated by the linear polarizer P1 and the QWP1. The opposite circular polarization in transmission is detected by a QWP2 and the linear polarizer P2. The microscope objective is mounted on a three-dimensional stage. To image an object, light from a white laser source is incident on the backside of the lens. The transmission through the sample (object and plasmonic lens) is collected with an original magnification ×20/0.40 objective and imaged on a charge-coupled device camera. (Figure reproduced with permission from NPG.)

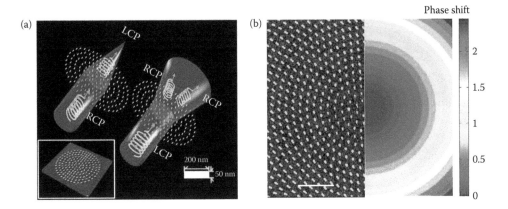

Figure 11.7 Schematic of a metalens with interchangeable polarity and an SEM image of the fabricated plasmonic lens. (a) The focusing properties of the same metalens can be switched between a convex lens and a concave lens by controlling the helicity of the incident light. Each dipole antenna is 200 nm long and 50 nm wide. (b) SEM image of the fabricated two-dimensional dual-polarity plasmonic lens (top view) with a focal length of 80 μm (left). The corresponding phase shift profile is displayed on the right. The dipoles that are arranged on the same annulus have the same orientation. The distance between two neighboring annuluses is 400 nm along both the radial and azimuth directions. The scale bar is 2 μm. (Figures reproduced with permission from Wiley-VCH.)

of 400 nm. Within each ring, the dipole antennas have the same orientation and the separation between two neighboring dipoles along the ring is 400 nm. The antenna structures are defined in a resist film by using standard electron beam lithography. Thereafter, a 40 nm gold film is deposited via thermal evaporation. The lens structures are obtained by a lift-off procedure. A scanning electron microscopy (SEM) image of the resulting patterns for the lens (left) designed at a wavelength of 740 nm and the corresponding phase shift profile (right) are shown in Figure 11.7b. The lenses have a diameter of 180 μm with a focal length $f = 80$ μm. Each dipole antenna is 200 nm long and 50 nm wide (shown in Figure 11.7a).

11.4.2.2.2 Characterization of the lens

We first characterize the focusing performance of the metalens with a laser beam at $\lambda = 740$ nm. The experimental setup is similar to that of the cylindrical lens. By gradually tuning the distance between the microscope objective lens and the plasmonic metalens, the optical intensity distribution is examined along the propagation direction to determine the focal point. Figure 11.8 shows the microscope image of the metalens illuminated by white light (Figure 11.8a) and the focal point at the focal plane (Figure 11.8b) by using the laser beam. From the intensity distribution of the laser beam at the focal plane (Figure 11.8c), a focused spot with a diameter of 7.2 μm at full width of half maximum is experimentally obtained.

11.4.2.2.3 Imaging property

We further study the imaging functionality of such spherical metalenses. The lens used for imaging has a diameter of 80 μm and a focal length of 80 μm, and its polarity is positive for RCP incident

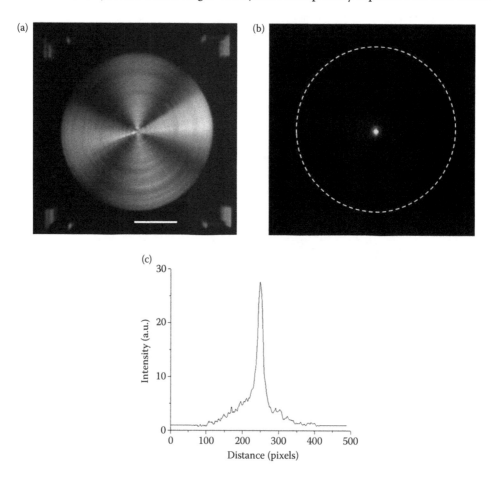

Figure 11.8 Optical microscope images of (a) the metalens illuminated with white light and (b) the focal point at real focal plane illuminated with an RCP incident laser beam at 740 nm wavelength. The scale bar is 50 μm. The position of the lens is marked by the white dashed circles. (c) Experimental measurement of the intensity distribution at the focal plane along one direction. Each pixel is equal to 0.37 μm. The spot diameter at full width of half maximum is 7.2 μm. (Figures reproduced with permission from Wiley-VCH.)

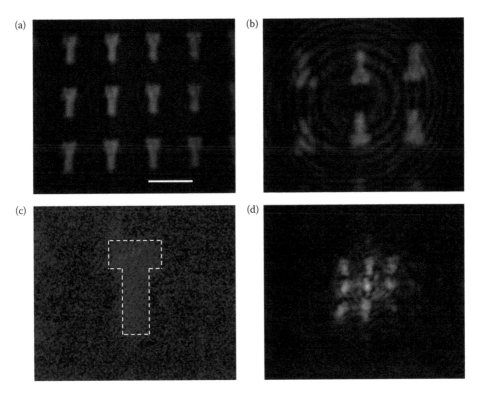

Figure 11.9 Experimental demonstration of imaging. (a) A "T" pattern array with a pitch of 20 μm along both in-plane directions is used as an object. The scale bar is 20 μm. The object and the lens are separated by an air gap whose thickness is controlled by spacers at the edge of the samples. (b) A real and inverted image with a magnification of 1.3 is achieved for RCP incident light, in which the polarity of the lens is positive (convex lens). The air gap is 150 μm. (c) A virtual and upright image with a magnification of 3.1 on a positive lens with an air gap of 55 μm. The edge of the image is marked by the white dashed lines. (d) For the same air gap as in (c), a virtual and upright image with a magnification of 0.58 is obtained when switching the incident light from RCP to LCP, with the polarity of the same lens being negative (concave lens). (Figures reproduced with permission from Wiley-VCH.)

light. The object to be imaged consists of an array of "T" apertures with a pitch of 20 μm along both in-plane directions (shown in Figure 11.9a), which are fabricated on a chromium film with a thickness of 50 nm. Each "T" aperture is designed with a height of 11 μm and a width of 6 μm. The object and metalens are separated by an air gap whose thickness is controlled by the spacers at the edge of the samples. For an object distance greater than the focal length of the metalens, it is expected that an inverted real image is formed for a positive (convex) lens. This is confirmed by the measurement with an object distance of 150 μm with RCP incident light, where the metalens functions as a positive lens (Figure 11.9b). By comparing with the image of the "T" patterns without the plasmonic lens, the magnification of the image formed by the metalens can be obtained. Experimentally, we measured the magnification of the image to be 1.3, which shows reasonable agreement with a simple ray calculation of 1.14. The slight difference in magnification is due to the substrate flatness and air gap accuracy. Due to the resolution limitation of the measurement system, the upright virtual image for LCP incident light with a predicted magnification of 0.34 cannot be resolved.

In order to observe the upright virtual image, we decrease the air gap to 55 μm. As the air gap is within the focal length of the plasmonic metalens, upright virtual images are observed for both circular polarizations of incident light, as shown in Figure 11.9c and d. The magnifications in Figure 11.9c and d are 3.1 and 0.58, respectively, which coincide with the magnification of 3.2 for a positive lens (convex) with RCP incident light, and magnification of 0.59 for a negative lens (concave) with LCP incident light. Thus, the reconfigurable imaging functionality of the same plasmonic metalens as a convex lens or a concave lens is clearly demonstrated by controlling the helicity of the incident CP light.

11.4.3 Metasurface hologram

Here we propose and experimentally demonstrate a helicity multiplexed metasurface hologram with high efficiency and good image fidelity over a broad range of frequencies (Wen et al., 2015). The metasurface hologram features the combination of two sets of hologram patterns operating with opposite incident helicities. Two symmetrically distributed off-axis images are interchangeable by controlling the helicity of the input light. The demonstrated helicity multiplexed metasurface hologram with its high performance opens avenues for future applications with functionality-switchable optical devices.

11.4.3.1 Design of the helicity multiplexed metasurface hologram

To realize the polarization-dependent phase profile while maintaining constant amplitude, the reflective-type metasurface adopted here consists of three layers: a two-dimensional (2D) array of elongated silver nanorods on the top, a SiO_2 spacer, and a silver background layer on the silicon substrate. For the circularly polarized light at normal incidence, the reflected light by a metallic mirror is defined to have the opposite helicity compared to that of the incident light. The reflected light from the nanowaveplate consists of two circular polarization states: one has the same helicity as the incident circularly polarized light, but with an additional phase delay, and the other has the opposite helicity without the additional phase delay. This is equivalent to flipping the circular polarization in transmission or maintaining the same circular polarization in reflection (Zheng et al., 2015). The additional phase delay is known as Pancharatnam–Berry phase with a value of $\pm 2\varphi$, where φ is the orientation angle of each nanorod. Specifically, "+" and "−" represent the sign of the phase shift for the incident RCP light and that for the incident LCP light, respectively. Hence the reversion of the phase profile can be achieved by either changing the helicity of the circularly polarized light or flipping the sign of the orientation angle φ of the nanorod for the same circular polarization, offering a new design methodology to achieve helicity multiplexed functionalities.

The combination of two sets of hologram patterns operating with opposite incident helicities on the same metasurface is the key step for the realization of the helicity multiplexed functionality. First, the Gerchberg–Saxton algorithm (Gerchberg and Saxton 1972) is used to generate two phase profiles, which can reconstruct two off-axis images on the different sides of the incident light. Both images ("bee" and "flower") have a projection angle of 22° × 22°, which are the opening angles of the image in both x and y directions. The off-axis angle is designed to be 10.35° in the imaging area. Then, the two phase profiles are encoded onto the metasurfaces (see Figure 11.10a), where the nth phase pixel φ_n of the hologram is represented by a nanorod with the orientation angle $\varphi_n/2$ defined in the metasurface. After that, two sets of data are merged together with a displacement vector of $(d/2, d/2)$, as shown in Figure 11.10a. d is the distance between neighboring antennas with a value of 424 nm, which is the pixel size before merging. Therefore, although the new metasurface contains two sets of hologram data, the size of the sample is still the same and the equivalent pixel size is 300 nm × 300 nm, leading to the increase of the nano-antenna density. Upon the illumination of LCP light, the merged metasurface can reconstruct the "flower" on the left and the "bee" on the right side of the metasurface, viewing from the incident beam, respectively (Figure 11.10a). Since the sign of the phase profile can be flipped by controlling the helicity of the incident light, the positions of reconstructed images in Figure 11.10b are swapped in contrast to those in Figure 11.10a when the helicity of the incident light is changed from LCP to RCP. As a result, the reconstructed images on the same position are switchable, that is, either "bee" or "flower," depending on the helicity of the incident light.

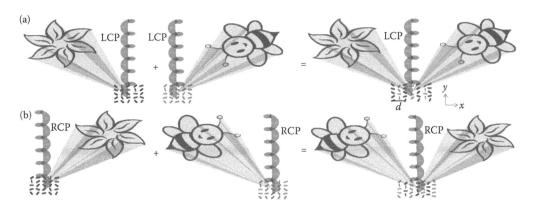

Figure 11.10 Generation of the helicity multiplexed metasurface hologram. Two sets of hologram patterns are designed to operate with opposite incident helicities and merged together with a displacement vector of $(d/2, d/2)$. d is the distance between neighboring antennas with a value of 424 nm along x and y directions. (a) Upon the illumination of LCP light at normal incidence, the reconstructed "flower" and the "bee" are on the left side and on the right side of the incident beam, respectively. The two off-axis images are symmetrically distributed. Two kinds of nanorods represent the metasurface holograms for "flower" and "bee," respectively. (b) The position for the "flower" and that for the "bee" are swapped upon the illumination of RCP light. (Figures reproduced with permission from NGP.)

11.4.3.2 Characterization of helicity multiplexed metasurface hologram

Figure 11.11a shows the experimental setup to characterize the performance of the fabricated metasurface. A laser beam with circular polarization from a tunable laser source (Fianium-SC400-PP) is generated by a polarizer and a quarter-wave plate (QWP) in front of the sample (see Figure 11.11a). Then the light is incident onto the fabricated sample, which is mounted on a 3D translation stage, allowing adjustment of the sample position. The incident beam has a radius of ~1 mm to ensure that the entire metasurface is illuminated by a plane-wave-like wave front. The

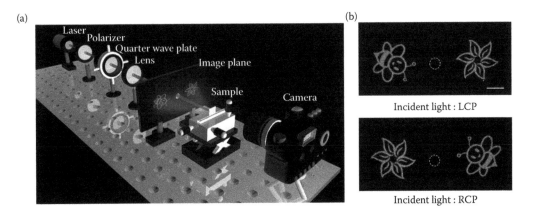

Figure 11.11 Illustration of the experiment setup and the dependence of experimental results on the helicity of the incident light. (a) The incident light with various polarization states is obtained by controlling the angle θ between the polarization axis of the polarizer and the fast axis of the quarter-wave plate. The incident light impinges normally onto the metasurface and the reconstructed images are projected onto the image plane. The screen is a white paper board with an opening (diameter 6 mm) in the middle, which allows the incident light and zero order reflected light passing through. (b) The experimentally obtained images for the incident light with LCP (top) and RCP (bottom). The wavelength of the incident light is 633 nm. The dashed circles mark the edge of the opening. The scale bar is 1 cm. (Figures reproduced with permission from NGP.)

reflected holographic image is either projected to a screen in the far field for inspection of the image quality, or focused by a condenser lens with high numerical aperture to measure the conversion efficiency. The screen is a white paper board with an opening (diameter 6 mm) in the middle, which allows the incident light and zero-order reflected light to pass through. A normal camera instead of a CCD camera is used to capture images on the image screen.

Figure 11.11b shows the two reconstructed images on the image plane with a distance of 50 mm between the screen and the metasurface at the wavelength of 633 nm. It should be noted that the size of the reconstructed images without distortion increases linearly with the reconstruction distance. Two clear images ("bee" on the left and "flower" on the right) with very high fidelity and no distortion (Figure 11.11b top) are observed when the metasurface is illuminated with the LCP light. When the light polarization is switched from LCP to RCP, the positions of the two images ("bee" and "flower") are swapped (Figure 11.11b bottom). In addition, the positions of the two identical images are centrosymmetric, agreeing very well with the theoretical prediction.

11.4.3.3 Discussion

The unique nature of the Gerchberg–Saxton algorithm and symmetrical distribution of the two patterns enables the optimization of the metasurface design, where each individual nanorod simultaneously contributes to both images. In this design, a target image includes two separated images, each one projected to a different direction. To theoretically explain the performance enhancement of the optimized design, the angular spectrum representation is adopted here to analyze the spatial frequency distribution of the scattered light from these holograms. From simulation results, we can clearly see that a significant part of the energy is transferred to high spatial frequencies after the two separate holograms are merged together, while most of the energy for the optimized design is still allocated to the low spatial frequencies. The propagating wave contributes to the target image while the evanescent wave is absorbed by the metal. Simulation also shows that, in our optimized design, if we neglect the optical losses, the window efficiency, which is defined as the ratio between the optical power projected into the two image regions and the input power, reaches 90.5%. Figure 11.12 shows the experimentally measured conversion efficiency versus wavelength, which is higher than 40% over a broad wavelength band ranging from 620 to 1020 nm. The maximum conversion efficiency in our experiment is 59.2% at $\lambda_0 = 860$ nm in comparison with the theoretical value of 86% (without the titanium layer). The metasurface hologram proposed here

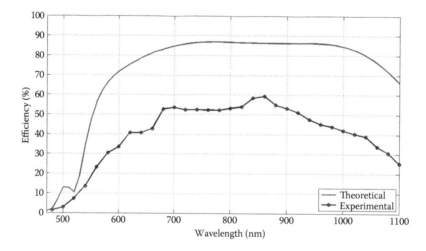

Figure 11.12 Conversion efficiency of the metasurface hologram versus wavelength of the incident light. The conversion efficiency is defined as the power of "bee" and "flower" divided by the power of the incident light. The blue circles represent the experimentally measured efficiencies over a broad range of wavelengths. The titanium layer is not considered in the simulation. (Figure reproduced with permission from NGP.)

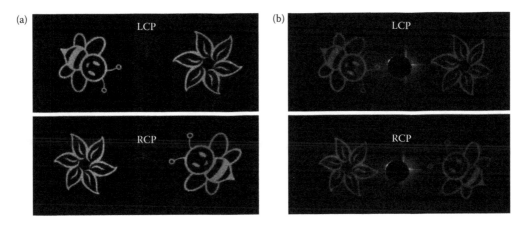

Figure 11.13 Experimentally obtained images at other visible wavelengths. The wavelengths of the incident light are (a) 524 nm and (b) 475 nm, respectively. (Figures reproduced with permission from NGP.)

shows the capability to reconstruct clear images in the visible spectrum. Figure 11.13 shows the experimentally captured images at the wavelengths of 524 and 475 nm for (a) and (b), respectively.

11.4.4 Multifunction optical devices

Great effort has been made to incorporate multifunctions into a single device, such as the optical Janus device (Zentgraf et al., 2010). As one of the important multifunction optical elements, polarization selective optical elements (PSOE) can achieve multiple functionalities according to the polarization states of the incident beam. Traditional polarization-selective optical elements are mainly based on birefringence, which is realized by using the well designed structure of each phase pixel. However, further reduction of the pixel size and improvement of the phase levels are hindered by the complicated fabrication process. In this section, an approach is proposed to realize a metasurface device that possesses two distinct functionalities (Wen et al., 2016). The designed metasurface device, consisting of gold nanorods with spatially varying orientation, has been experimentally demonstrated to function as either a lens or a hologram, depending on the helicity of the incident light.

11.4.4.1 Method

The multifunctional optical devices are based on the transmissive-type metasurface, as proposed in Sections 11.4.1 and 11.4.2. First, a phase-only hologram for the target image of "cat" is generated. The target image is discretized and regarded as a collection of point sources. After each point source is added with a random phase, the phase-only hologram is obtained by superimposing the light emitting from all the point sources. Then the phase distribution of the hologram is sampled and encoded onto the metasurface. Each nanorod defined in the metasurface represents a sampled phase value in the hologram. Under the illumination of LCP light, the desired continuous local phase profile is generated for the transmitted RCP light and the image "cat" is reconstructed (Figure 11.14a). Similarly, by encoding the phase function of $\Phi(x, y) = -(2\pi/\lambda)(\sqrt{x^2 + y^2 + f^2} - f)$ onto the metasurface, a convex lens with the focal length f is created for the incident/detected combination of RCP/LCP. However, the metasurface lens turns into concave when the combination changes to LCP/RCP, and the refracted light diverges as though it is emanated from a virtual image plane on the incident side of the metasurface (Figure 11.14b). Finally, the two sets of data are merged together with a displacement vector of $(d/2, d/2)$ (Figure 11.14c), where d is the distance between neighboring nanorods. Although the merged metasurface contains nanorods of two categories, the size of the sample is still the same and the equivalent pixel size changes into $\sqrt{2}d/2$ (viewing from the 45° direction), leading to the increase of the nanorod density.

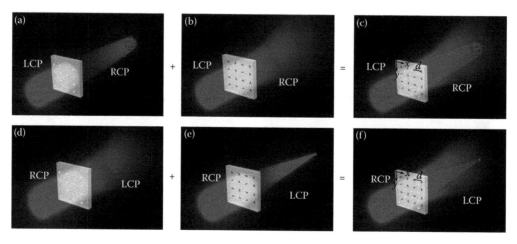

Figure 11.14 Design methodology for obtaining two distinct functionalities on a single metasurface device. Two different metasurfaces consisting of (a) a hologram that can reconstruct a real image of "cat" and (b) a concave lens that diverges the refracted beam for the incident/detected combination of LCP/RCP. (c) The two metasurfaces are merged together with a displacement vector of (d/2, d/2). Only a reconstructed image of "cat" is observed upon the illumination of LCP. As the polarity of the hologram and the lens depend on the polarization of the incident light, no real image is observed in (d) but a real focal point appears in (e) when the incident/detected combination is changed from LCP/RCP to RCP/LCP. (f) Similarly, a focal point will be observed in the real focal plane of the designed metasurface. (Figures reproduced with permission from Wiley-VCH.)

The two types of nanorods contributing to the hologram and the lens are named type A and type B for convenience, respectively. As the dipolar coupling between the neighboring nanorods is negligible, the nanorods of the two categories work independently. For the incident/detected combination of LCP/RCP, the RCP light emitting from the nanorods A reconstructs the image of "cat," and the RCP light from the nanorods B diverges and forms a subtle background (Figure 11.14c). In contrast, if the combination is switched to RCP/LCP, the phase profiles of the metasurface in Figure 11.14c are all reversed and the merged metasurface changes from a hologram to a convex lens (Figure 11.14d–f). By comparing Figure 11.14c with Figure 11.14f, it clearly shows that the functionality of the merged metasurface is switchable, depending on the incident/detected combination.

11.4.4.2 Experimental verification

First, we calculate the phase profiles for the hologram and the lens separately, then merge them together and encode the result onto the metasurface (Figure 11.15a). The metasurface is fabricated on the ITO-coated glass substrate with standard electron beam lithography, followed by the gold film deposition and lift-off process. The scanning electron microscopy (SEM) image of the fabricated metasurface is shown in Figure 11.15b. The incident CP light is generated by a polarizer and a quarter-wave plate, which are placed in front of the spectrally tunable laser source (NKT, SuperK Extreme). The fabricated sample is mounted on the 2D translational stage with the metasurface side facing the incident light. To ensure the uniform illumination of the metasurface, the incident beam has a diameter (about 2 mm) much larger than the size of the metasurface (400 μm). The scattered light in the transmission side is collected by an objective (10×/0.3) and a lens ($f = 50$ mm) that are used for the far-field microscopy detection. As the objective is fixed on the 3D translational stage whose resolution along the optical path is 1 μm, the distance between the objective and the metasurface can be finely adjusted. Another pair of quarter-wave plate and polarizer is placed in the transmission side to select the CP light with opposite helicity to the incident light. Finally, the amplified images are captured by a CCD camera.

The polarization selectivity of the metasurface is experimentally verified. The metasurface functions as a hologram when the incident/detected polarization combination has the state

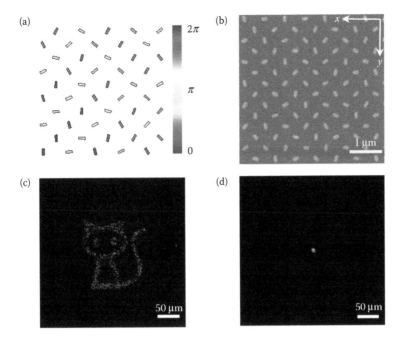

Figure 11.15 Metasurface encoding process and the experimental results. (a) Nanorod distribution for part of the designed multifunction metasurface device. The phases of the nanorods are denoted by different gray scales. The metasurface has the size of 400 μm × 400 μm in total and the sampling number is 666 × 666 × 2 (666 × 666 for the hologram and the same for the lens). The target image of "cat" is designed to be 150 μm × 150 μm in size and the reconstruction distance is 200 μm. The focal length of the lens is 200 μm. (b) Part of the scanning electron microscope image of the fabricated metasurface. The gold nanorods are fabricated on the ITO coated glass substrate with a standard electron-beam lithography and lift-off process. Each nanorod is 50 nm wide, 150 nm long, and 40 nm high. The nanorods in odd and even rows contribute to the lens and hologram, respectively. The distance between the neighboring nanorods is 600 nm both in the x and y directions. (c) When the incident/detected light is LCP/RCP, the merged metasurface functions as a hologram and the image of "cat" is reconstructed experimentally. (d) When the incident/detected light (633 nm) is RCP/LCP, the metasurface turns into a lens that converges the refracted light to the focal point. (Figures reproduced with permission from Wiley-VCH.)

of LCP/RCP (Figure 11.15c). By gradually tuning the distance between the objective and the metasurface, the optical intensity distribution is examined along the propagation direction to determine the position of the image "cat." When the polarizations of the incident and detected beams are switched to RCP and LCP, respectively, the metasurface functions as a convex lens that causes the incident laser beam to converge at a focal plane on the transmission side (Figure 11.15d). In comparison with Figure 11.15c, it clearly shows that the selectivity of the metasurface from the hologram to the convex lens is solely attributed to the helicity change of the CP light for the incident/detected combination, which is in good agreement with the theoretical prediction.

11.5 Conclusion

Traditional optical elements are based on refraction, reflection or diffraction of light, and the wave front shaping relies on light propagation over distances much longer than the wavelength to shape wave fronts. At the interface of a metasurface, wave front shaping is accomplished within a distance much smaller than the wavelength of the light beam, thus providing a new opportunity to develop ultrathin devices that are easy to integrate into compact platforms. Using phase discontinuities at the interface of metasurfaces as the design methodology not only provides the possibility to develop ultrathin devices with unusual functionalities, but also allows the integration of many different functions into a single optical element. These types of exciting new ultrathin optical devices open a new avenue to achieve a high density of functionality, effectively scaling down the size of photonic systems. Ultrathin optical devices are of great interest, since flat optical elements based on metasurfaces are comparable in size to conventional electronic components and compatible with standard semiconductor fabrication processes.

Acknowledgments

X.C. acknowledges the Engineering and Physical Sciences Research Council of the United Kingdom (Grant Ref: EP/M003175/1), EPSRC IAA grant, and Renishaw-Heriot Watt Strategic Alliance. We are grateful to Prof. Xiaofei Zang (University of Shanghai for Science and Technology) and Prof. Huigang Liu (Nankai University) for the proofreading.

References

Arbabi A 2014. Fundamental Limits of Ultrathin Metasurfaces. arXiv:1411.2537 [physics.optics] (2014).

Arbabi A, Horie Y, Bagheri M, Faraon A 2015a. Dielectric metasurfaces for complete control of phase and polarization with subwavelength spatial resolution and high transmission. *Nature Nanotechnology* 10:937–943.

Arbabi A, Horie Y, Ball AJ, Bagheri M, Faraon A 2015b. Subwavelength-thick lenses with high numerical apertures and large efficiency based on high-contrast transmitarrays. *Nature Communications* 6.

Berry M. V. 1987. The adiabatic phase and Pancharatnam's phase for polarized light. *Journal of Modern Optics* 34:1401–1407.

Chen X, Huang L, Muehlenbernd H, Li G, Bai B, Tan Q, Jin G, Qiu C-W, Zentgraf T, Zhang S 2013. Reversible three-dimensional focusing of visible light with ultrathin plasmonic flat lens. *Advanced Optical Materials* 1:517–521.

Chen XZ, Huang LL, Muhlenbernd H, Li GX, Bai BF, Tan QF, Jin GF, Qiu CW, Zhang S, Zentgraf T 2012. Dual-polarity plasmonic metalens for visible light. *Nature Communications* 3.

D'Aguanno G, Mattiucci N, Bloemer M, Desyatnikov A 2008. Optical vortices during a superresolution process in a metamaterial. *Physical Review A* 77:043825.

Ding XM, Monticone F, Zhang K, Zhang L, Gao DL, Burokur SN, de Lustrac A, Wu Q, Qiu CW, Alu A 2015. Ultrathin Pancharatnam-Berry metasurface with maximal cross-polarization efficiency. *Advanced Materials* 27:1195–1200.

Edwards B, Alu A, Silveirinha MG, Engheta N 2009. Experimental verification of plasmonic cloaking at microwave frequencies with metamaterials. *Physical Review Letters* 103:153901.

Ergin T, Stenger N, Brenner P, Pendry JB, Wegener M 2010. Three-dimensional invisibility cloak at optical wavelengths. *Science* 328:337–339.

Fang N, Lee H, Sun C, Zhang X 2005. Sub-diffraction-limited optical imaging with a silver superlens. *Science* 308:534–537.

Gabrielli LH, Cardenas J, Poitras CB, Lipson M 2009. Silicon nanostructure cloak operating at optical frequencies. *Nature Photonics* 3:461–463.

Genevet P, Yu N, Aieta F, Lin J, Kats MA, Blanchard R, Scully MO, Gaburro Z, Capasso F 2012. Ultra-thin plasmonic optical vortex plate based on phase discontinuities. *Applied Physics Letters* 100:013101.

Gerchberg RW, Saxton WO 1972. A practical algorithm for the determination of the phase from image and diffraction plane pictures, *Optik* 35, 237.

Grady NK, Heyes JE, Chowdhury DR, Zeng Y, Reiten MT, Azad AK, Taylor AJ, Dalvit DAR, Chen H-T 2013. Terahertz metamaterials for linear polarization conversion and anomalous refraction. *Science* 340:1304–1307.

Huang L, Chen X, Muehlenbernd H, Zhang H, Chen S, Bai B, Tan Q et al. 2013a. Three-dimensional optical holography using a plasmonic metasurface. *Nature Communications* 4.

Huang LL, Chen XZ, Bai BF, Tan QF, Jin GF, Zentgraf T, Zhang S 2013b. Helicity dependent directional surface plasmon polariton excitation using a metasurface with interfacial phase discontinuity. *Light, Science & Applications* 2.

Huang LL, Chen XZ, Muhlenbernd H, Li GX, Bai BF, Tan QF, Jin GF, Zentgraf T, Zhang S 2012. Dispersionless phase discontinuities for controlling light propagation. *Nano Letters* 12:5750–5755.

Kang M, Feng TH, Wang HT, Li JS 2012. Wave front engineering from an array of thin aperture antennas. *Optics Express* 20:15882–15890.

Lin D, Fan P, Hasman E, Brongersma ML 2014. Dielectric gradient metasurface optical elements. *Science* 345:298–302.

Lin J, Mueller JPB, Wang Q, Yuan G, Antoniou N, Yuan X-C, Capasso F 2013. Polarization-controlled tunable directional coupling of surface plasmon polaritons. *Science* 340:331–334.

Liu LX, Zhang XQ, Kenney M, Su XQ, Xu NN, Ouyang CM, Shi YL, Han JG, Zhang WL, Zhang S 2014. Broadband metasurfaces with simultaneous control of phase and amplitude. *Advanced Materials* 26:5031–5036.

Liu R 2009. Broadband ground-plane cloak. *Science* 323:366–369.

Luo WJ, Xiao SY, He Q, Sun SL, Zhou L 2015. Photonic spin hall effect with nearly 100% efficiency. *Advanced Optical Materials* 3:1102–1108.

Parazzoli CG, Greegor K, Li K, Koltenbah BEC, Tanielian M 2003. Experimental verification of negative index of refraction using Snell's law. *Physical Review Letters* 90:107401.

Schurig D 2006. Metamaterial electromagnetic cloak at microwave frequencies. *Science* 314:977–980.

Shalaev VM 2007. Optical negative-index metamaterials. *Nature Photonics* 1:41–48.

Shelby RA, Smith DR, Schultz S 2001. Experimental verification of a negative index of refraction. *Science* 292:77–79.

Shitrit N, Yulevich I, Maguid E, Ozeri D, Veksler D, Kleiner V, Hasman E 2013. Spin-optical metamaterial route to spin-controlled photonics. *Science* 340:724–726.

Sun SL, He Q, Xiao SY, Xu Q, Li X, Zhou L 2012. Gradient-index meta-surfaces as a bridge linking propagating waves and surface waves. *Nature Materials* 11:426–431.

Valentine J 2008. Three-dimensional optical metamaterial with a negative refractive index. *Nature* 455:376–380.

Valentine J, Li J, Zentgraf T, Bartal G, Zhang X 2009. An optical cloak made of dielectrics. *Nature Materials* 8:568–571.

Wen D, Chen S, Yue F, Chan K, Chen M, Ardron M, Li KF, Wong PWH, Cheah KW, Pun EYB, Li G, Zhang S, Chen X 2016. Metasurface device with helicity-dependent functionality. *Advanced Optical Materials* 4:321–327.

Wen D, Yue F, Li G, Zheng G, Chan K, Chen S, Chen M et al. 2015. Helicity multiplexed broadband metasurface holograms. *Nature Communications* 6.

Yi X, Ling X, Zhang Z, Li Y, Zhou X, Liu Y, Chen S, Luo H, Wen S 2014. Generation of cylindrical vector vortex beams by two cascaded metasurfaces. *Optics Express* 22:17207–17215.

Yin X, Ye Z, Rho J, Wang Y, Zhang X 2013. Photonic spin hall effect at metasurfaces. *Science* 339:1405–1407.

Yu N, Aieta F, Genevet P, Kats MA, Gaburro Z, Capasso F 2012. A broadband, background-free quarter-wave plate based on plasmonic metasurfaces. *Nano Letters* 12:6328–6333.

Yu N, Genevet P, Kats MA, Aieta F, Tetienne J-P, Capasso F, Gaburro Z 2011. Light propagation with phase discontinuities: Generalized laws of reflection and refraction. *Science* 334:333–337.

Zentgraf T, Valentine J, Tapia N, Li J, Zhang X 2010. An optical "Janus" device for integrated photonics. *Advanced Materials* 22:2561–2564.

Zhang XQ, Tian Z, Yue WS, Gu JQ, Zhang S, Han JG, Zhang WL 2013. Broadband terahertz wave deflection based on C-shape complex metamaterials with phase discontinuities. *Advanced Materials* 25:4567–4572.

Zheng G, Mühlenbernd H, Kenney M, Li G, Zentgraf T, Zhang S 2015. Metasurface holograms reaching 80% efficiency. *Nature Nanotechnology* 10:308–312.

12

Optical micro- and nanoresonators for biochemical sensing

Xingwang Zhang, Liying Liu, Lei Xu,
Xudong Fan, and Guangya Zhou

Contents

Biochemical sensors are indispensable to the biochemical detection system. According to their sensing principles, biochemical sensors can be divided into electronic sensors, optical sensors, thermal sensors, and so on, among which optical sensors have the unique advantages of fast response, high sensitivity, low detection limit, and immunity to electromagnetic disturbances. In recent years, optical biochemical sensors have been intensively investigated for their potential applications in biomedical diagnosis and environmental monitoring. Some of them have already been commercialized, such as enzyme-linked immunosorbent assay (ELISA), surface plasmon resonance (SPR) sensors, and optical fiber sensors. In the meantime, more and more novel optical biochemical sensors are emerging at a fast pace. There is especially a great deal of interest in biochemical sensors based on optical micro-/nanoresonators, due to their advantages of small size, small sample consumption, real time detection, fast response, high sensitivity, and low detection limit. Optical micro-/nanoresonators have micro-/nanoscale dimensions, which can confine light in small volumes, and have high Q/V ratios (i.e., Q-factor/mode volume ratio). In optical micro-/nanoresonator-based sensors, the light-analyte interaction is enhanced so that high sensitivity and low detection limit can be achieved. In this chapter, we focus on the optical micro-/nanoresonator-based biochemical sensors. We will discuss three kinds of optical micro-/nanoresonators, that is, Fabry–Pérot microresonators, photonic crystal nanoresonators, and whispering gallery mode microresonators, and their applications in biochemical detection.

12.1 Optical micro-/nanoresonators

Optical micro-/nanoresonators are optical resonators of micro-/nanometers in size, and can find a wide range of applications in microlasers, optical communications, optical sensing, cavity quantum electrodynamics, and so on (Vahala, 2003; Kavokin et al., 2007). In optical micro-/nanoresonators light is confined in a small volume and the light-matter interaction is enhanced. According to their light confinement schemes, optical micro-/nanoresonators are divided into three types: Fabry–Pérot microresonators, photonic crystal nanoresonators, and whispering gallery mode microresonators (see Figure 12.1).

12.1.1 Fabry–Pérot microresonators

As shown in Figure 12.1a, a Fabry–Pérot microresonator consists of two parallel reflecting mirrors spaced by a distance L with the reflective surfaces facing each other. The optical field inside the resonator exhibits strong resonance behavior when $2nL/\lambda_c = m$ (m is an integer), and the mode spacing is (Siegman, 1986)

$$\Delta\lambda = \frac{\lambda_c^2}{2nL},$$

(12.1)

where λ_c is the resonance wavelength and n is the refractive index of the material inside the resonator. The smaller the resonator size (L) is, the larger the mode spacing will be.

Usually, the quality factor (Q-factor) is used to characterize the energy decay rate inside a resonator. The Q-factor can be related to the resonance wavelength and mode linewidth $\delta\lambda$ by (Siegman, 1986)

$$Q = \frac{\lambda_c}{\delta\lambda}.$$

(12.2)

The higher the Q-factor is, the longer the photon lifetime ($\tau = Q\lambda_c/(2\pi c)$) will be. Generally, the Q-factor of the Fabry–Pérot microresonator is smaller than that of the other two kinds of microresonators (Vahala, 2003).

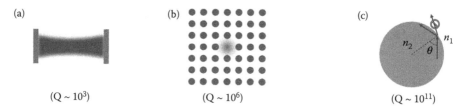

(a) $(Q \sim 10^3)$ (b) $(Q \sim 10^6)$ (c) $(Q \sim 10^{11})$

Figure 12.1 Schematics of Fabry–Pérot microresonator (a), photonic crystal nanoresonator (b), and whispering gallery mode microresonator (c).

12.1.2 Photonic crystal nanoresonators

The concept of "photonic crystal" originally emerged in 1987 when E. Yablonovitch and S. John, respectively, published two papers on periodic optical structures (i.e., photonic crystals) (Yablonovitch, 1987; Joannopoulos et al., 2008). Photonic crystals are periodic optical nanostructures that are capable of manipulating photons (Joannopoulos et al., 2008). Due to the photonic bandgap, which results from the periodic structure, only the light within a certain wavelength range can propagate in photonic crystals. The photonic crystal nanoresonator usually consists of a defect introduced in the photonic crystals as shown in Figure 12.1b, where some states with wavelengths within the photonic bandgap are created and form optical modes which are confined in the photonic crystal. Compared to Fabry–Pérot microresonators and whispering gallery mode microresonators, the photonic crystal nanoresonators have much smaller mode volumes, and much higher Q/V ratios (Vahala, 2003).

12.1.3 Whispering gallery mode microresonators

In 1910, Lord Rayleigh found and theoretically explained the whispering gallery mode sound wave resonators for the case of St Paul's Cathedral (Rayleigh, 1910). In 1969, E. A. J. Marcatili in Bell Lab expanded this concept into the optics regime and theoretically demonstrated whispering gallery modes in microring resonators (Marcatili, 1969). In whispering gallery mode resonators, light travels around the inner surface of a circular structure, which has a higher refractive index than the surroundings, via total internal reflection, as shown in Figure 12.1c. The optical field inside the resonator exhibits strong resonance behavior when the round trip phase shift is equal to an integer multiple of 2π. Compared to the Fabry–Pérot microresonators, the whispering gallery mode microresonators confine light via total internal reflection and usually have much higher Q-factors (up to 10^{11}) (Savchenkov et al., 2007). According to the geometric structures as shown in Figure 12.2, whispering gallery mode microresonators can be divided into microspheres (a), microdisks (b), microcylinders (c), microtoroids (d), hollow structural whispering gallery mode microresonators (e and f), and so on (Vahala, 2003; White et al., 2006; Pollinger et al., 2009; Sumetsky, 2010; Senthil Murugan et al., 2011).

12.2 Optical micro-/nanoresonator-based biochemical sensing

In recent decades, optical micro-/nanoresonators have been widely applied to biochemical sensing, such as bulk concentration detection, biomolecule surface density detection, single molecule

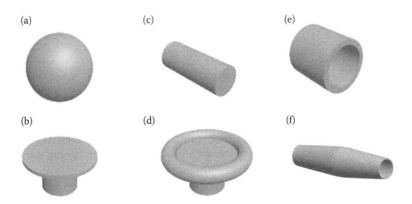

Figure 12.2 Schematics of microsphere (a), microdisk (b), microcylinder (c), microtoroid (d), micro-capillary (e), and hollow structural microbottle (f).

detection, gas detection, bio-imaging, and so on (Vollmer et al., 2002; Chao and Guo, 2003; Ozkumur et al., 2008; Jagerska et al., 2010; Li and Fan, 2010; Sun and Fan, 2011; Quan et al., 2013; Zhang et al., 2014a, 2015a,b) Due to the optical field enhancement of optical micro-/nanoresonators the interaction between the optical field and analyte is enhanced. Meanwhile, owing to the narrow mode linewidth of the optical micro-/nanoresonators high wavelength resolution can be achieved in biochemical sensing. Compared to the traditional optical biochemical sensing methods, the optical micro-/nanoresonator-based sensors have much higher sensitivity and much lower detection limits.

Sensitivity and detection limit are usually used to characterize the performances of the sensors. A higher sensitivity together with a lower detection limit means a better sensor performance. Generally, when a small change of the input signal (e.g., concentration) causes a change of the output signal (e.g., wavelength shift), the sensitivity (S) is defined by (Fassel, 1976)

$$S = \frac{\Delta[\text{Output}]}{\Delta[\text{Input}]}. \tag{12.3}$$

For example, if $\Delta[\text{Input}]$ is the refractive index change of the analyte, and $\Delta[\text{Output}]$ is the corresponding wavelength shift, the sensitivity is determined by $S = \Delta\lambda/\Delta n$ (nm/RIU). Additionally, the lowest quantity of the analyte that the sensor can detect (i.e., detection limit) is limited not only by the sensitivity, but also by the noise. The detection limit (DL) is defined by (Fassel, 1976)

$$\text{DL} = k \cdot \frac{\sigma}{S}, \tag{12.4}$$

where S is the sensitivity and σ is the standard deviation of the blank measures. The factor k is chosen according to the desired confidence level. Usually, when $k = 3$ is chosen, the corresponding DL can be considered as the lowest analyte amount that can be detected under the experimental conditions used. If $k = 1$ is chosen, the detection limit is at the same level as the noise, and is named the noise equivalent detection limit (NEDL) (Li and Fan, 2010). Clearly, in order to improve the detection limit, the sensitivity has to be increased and the noise has to be decreased at the same time.

Usually, the change of analyte can be transduced into the change of refractive index, photoluminescence, Raman spectroscopy, and so on. According to their sensing mechanisms, we divide the biochemical sensors that are based on optical micro-/nanoresonators into refractive index sensors and nonrefractive index sensors (Sun and Fan, 2011).

12.2.1 Refractive index sensing

12.2.1.1 Sensing based on resonance wavelength shift

As shown in Figure 12.3, the change of analyte introduces a perturbation of the effective refractive index of the resonator, and leads to a shift of the resonance wavelength (Lin, 1997). Therefore, the sensors can transduce the change of the analyte into a readable signal (i.e., wavelength shift). This method has been widely used for various optical micro-/nanoresonator sensors, with a commonly used setup shown in Figure 12.4a. Light launched from a tunable laser is coupled into the resonator, and the transmitted light is detected by a photodetector. The wavelength of the laser is swept and the transmission spectra of the resonator are monitored by a computer.

With such a sensing mechanism, rapid and high-throughput Fabry–Pérot microresonator sensor arrays are used to detect biochemical analytes (Moiseev et al., 2006; Ozkumur et al., 2008). Up to the present, the Fabry–Pérot microresonator sensors are able to reach a detection limit down to

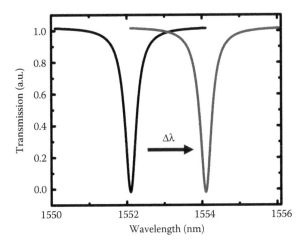

Figure 12.3 Resonance wavelength shift in response to the effective refractive index change, which results from the interaction between the analyte and the optical field of the micro-/nanoresonator.

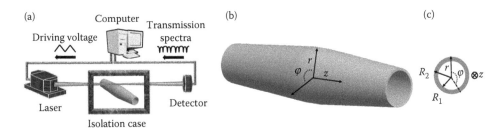

Figure 12.4 (a) Experimental setup of optical microresonator-based biochemical sensors: the tunable laser is driven by the data acquisition (DAQ) card embedded in the computer, and the laser wavelength is swept. The transmission spectra are monitored by the DAQ system; the geometry (b) and the cross-section (c) of the hollow structural microbottle resonator. (Reprinted from Zhang, X. et al. 2014. *Appl. Phys. Lett.*, 104:033703. with permission. DOI: http://dx.doi.org/10.1063/1.4861596.)

10^{-8} RIU for bulk refractive index sensing, and 60 pg/mL for biomolecule mass sensing (Lin, 1997; Liu et al., 2013).

The photonic crystal nanoresonator sensors can usually achieve a sensitivity of 10^2 nm/RIU and detection limit of 10^{-4}–10^{-5} RIU for bulk refractive index sensing (Jagerska et al., 2010). Due to having the smallest mode volume among the three kinds of optical micro-/nanoresonator the detection limit of the photonic crystal nanoresonator sensor for the biomolecule mass sensing is as low as 16 zeptomolar (zM) (Hachuda et al., 2013).

Owing to the ultrahigh Q-factor and easy fabrication process, the whispering gallery mode microresonator-based sensors are the most flourishing research topic among all of the three kinds of resonators (Sun and Fan, 2011; Foreman et al., 2015). Various cavity structures are employed to enhance the sensitivity, including hollow structural microresonators (microcapillary, microbottle), nanofiber ring resonators, and slotted microring resonators (Li and Fan, 2010; Zhang et al., 2011a, 2014b; Ren et al., 2012; Li et al., 2013; Ward et al., 2014; Liao et al., 2016; Tang et al., 2016). Here, we introduce a bottle-like hollow structural optical microresonator, which can be employed in biochemical refractive index-based detection with an ultralow detection limit.

The bottle-like hollow structural optical microresonator is schematically shown in Figure 12.4b. It consists of a microcapillary with a larger diameter in the central region (Sumetsky, 2010). The outer radius of the cross-section varies along the z-axis, and is given by (Zhang et al., 2014b)

$$R_2(z) = R_{20} \cdot \left[1 - \frac{1}{2}(\Delta k \cdot z)^2 \right]. \tag{12.5}$$

where R_{20} is the outer radius at $z = 0$, and Δk denotes the curvature. In the cylindrical coordinates, the electric field $E(r, \varphi, z)$ of the TM mode can be described by the Helmholtz function as below

$$\frac{1}{r}\frac{\partial}{\partial r}\left(r\frac{\partial E(r,\varphi,z)}{\partial r} \right) + \frac{1}{r^2}\frac{\partial^2 E(r,\varphi,z)}{\partial \varphi^2} + \frac{\partial^2 E(r,\varphi,z)}{\partial z^2} + n^2 k^2 E(r,\varphi,z) = 0. \tag{12.6}$$

By solving the Helmholtz function, the optical field $E(r,\varphi,z) = E_r(r)E_\varphi(\varphi)E_z(z)$ can be obtained (Zhang et al., 2014b)

$$E_\varphi(\varphi) = N \exp(jm\varphi), \tag{12.7}$$

$$E_r(r) = \begin{cases} AJ_m(n_{core}k_\varphi r), & r \le R_1(z) \\ BJ_m(n_{wall}k_\varphi r) + CH_m^{(1)}(n_{wall}k_\varphi r), & R_1(z) < r \le R_2(z), \\ DH_m^{(1)}(n_{out}k_\varphi r), & r > R_2(z) \end{cases} \tag{12.8}$$

$$E_z(z) = C_{mq}H_q\left(\sqrt{\frac{\Delta E_m}{2}} \cdot z \right) \cdot \exp\left(-\frac{\Delta E_m}{4} \cdot z^2 \right), \tag{12.9}$$

$$\Delta E_m = \frac{2m\Delta k}{R_{20}}, \tag{12.10}$$

$$C_{mq} = \left[\frac{\Delta E_m}{(\pi \cdot 2^{2q+1} \cdot (q!)^2)} \right]^{1/4}, \tag{12.11}$$

where A, B, C, and D are constants, and J_m and $H_m^{(1)}$ are the mth order Bessel function and the first kind of Hankel function, respectively. H_q is the qth order Hermite Polynomial. n_{core}, n_{wall}, and n_{out} are the refractive indices of the analyte inside the resonator, wall, and surroundings. As shown in Figure 12.5a, the analyte interacts with the evanescent field inside the capillary, and this leads to a change of the effective refractive index of the resonator, which in turn is reported by the sensor and generates an output signal (i.e., resonance wavelength shift).

However, the flowing of analyte inside the capillary can generate a breathing vibration of the bottle-like hollow structure, which introduces a vibration noise into the resonance wavelength shift. In the meantime, the laser light source wavelength drift can also introduce a drift noise (see Figure 12.5b, the ramping of the offset wavelength, in addition to the wavelength shift caused by the introduction of Bovine serum albumin (BSA)). All of these noises can degrade the detection limit. To eliminate these noises, a differential self-referenced detection method can be employed. In Figures 12.5a and 12.6a, Mode H has the largest mode field proportion inside the analyte, so Mode H has the highest sensitivity. Meanwhile, the detection limit of Mode H is limited by the vibration noise and drift noise. Compared to Mode H, the mode field of Mode A is mostly confined inside the wall, and the sensitivity is the lowest. Meanwhile Mode A and H suffer the same laser wavelength drift noise. Therefore, by subtracting the sensing signal of Mode A from Mode H, not only does the sensing signal remain almost unchanged, but the drift noise is nearly eliminated (see Figure 12.5c). Similarly, both modes H and E have the same mode profile along the z-axis direction, which means

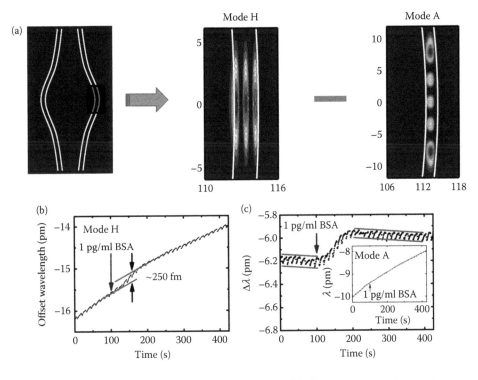

Figure 12.5 (a) Schematic view of the longitudinal section of hollow structural microbottle resonator, mode H, and mode A; (b) the sensorgram of mode H in response to 1 pg/mL bovine serum albumin (BSA); (c) the sensorgram of mode H after drift noise reduction. Inset shows the drift noise (the solid curve, red online) extracted from the polynomial fit of the sensorgram of mode A (the dotted curve). (Reprinted from Zhang, X. et al. 2014. *Appl. Phys. Lett.*, 104:033703, with permission. DOI: http://dx.doi.org/10.1063/1.4861596.)

their vibration noise is similar. Additionally, Mode E is also mostly confined in the wall. Therefore, by subtracting the sensing signal of Mode E from Mode H, the vibration noise of Mode H is greatly reduced (see Figure 12.6c). Finally, the vibration noise can be reduced by five times, down to 2.3 fm. The drift noise can be reduced by 23 times (see Figure 12.6b).

12.2.1.2 Sensing based on Vernier effect of coupled resonators

The sensitivity of the detection method based on the resonance wavelength shift is determined by the filling factor Γ, which is the overlapping of the optical field with the analyte (Zhang et al., 2011a). However, the filling factors of micro-/nanoresonators are usually quite small ($\Gamma \ll 1$), which limits the sensitivity to the range of 10–100 nm/RIU. To improve the sensitivity, the filling factor has to be increased. For example, by using a thinner microcapillary, the sensitivity can be increased up to 500–700 nm/RIU at 980 nm wavelength (Li and Fan, 2010). However, even if the filling factor reaches the maximum (i.e., $\Gamma = 1$), the sensitivity S is still limited by $S < \lambda/n$. For example, the maximum sensitivity is around 450 nm/RIU at $\lambda \sim 600$ nm and $n \sim 1.33$, where λ is the laser wavelength and n is the refractive index of the resonator. In this section, we use the Vernier effect of coupled microresonators to further enhance the sensitivity to break through the sensitivity theoretical limit mentioned above (Zhang et al., 2011a; Ren et al., 2012).

When two microresonators with mismatched cavity lengths couple to each other, due to the mismatch between their mode spacing (FSR), the overall transmission or reflection of the coupled cavity results in a modulated envelope in the transmission or reflection spectra, as shown in Figure 12.7b. When the modes of one microresonator shift as result of the introduction of analyte, the modulated spectra of the coupled cavity behaves in a way similar to the well-known Vernier effect,

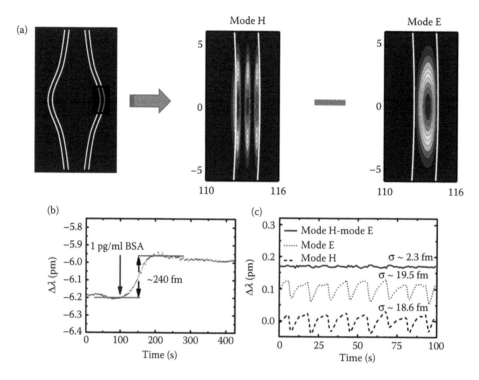

Figure 12.6 (a) Schematic view of the longitudinal section of hollow structural microbottle resonator, mode H, and mode E; (b) sensorgram of mode H in response to 1 pg/mL bovine serum albumin (BSA) after both drift noise and vibration noise reduction; (c) vibration noise comparison between mode H, mode E, and the difference between mode H and mode E. (Reprinted from Zhang, X. et al. 2014. *Appl. Phys. Lett.*, 104:033703, with permission. DOI: http://dx.doi.org/10.1063/1.4861596.)

which means the modulated envelope shifts more than that of the individual mode. Therefore, by taking the modulated envelope shift as the sensing signal, the sensitivity of the coupled cavity sensor is much larger than that of the single resonator sensor. Moreover, the smaller the difference between the two microresonators is, the larger the sensitivity will be.

To quantitatively characterize the sensitivity of coupled cavity sensors, a model as shown in Figure 12.7a, is used. Take cavity 1 as the master cavity, and cavity 2 as the slave cavity. Part of the light from cavity 1 (E_{inc}) is coupled into cavity 2, and the rest of the light is reflected back into cavity 1 (E_{ref}), so we have (Siegman, 1986; Zhang et al., 2011b)

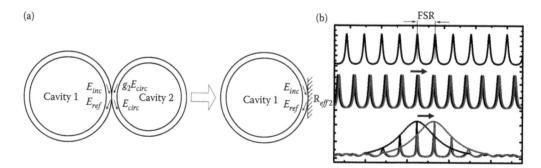

Figure 12.7 (a) Schematic drawing of the theoretical treatment of a coupled cavity; (b) Vernier effect of two coupled cavities with mismatched mode spacing (FSR). (Reprinted from Zhang, X. et al. 2011. *J. Opt. Soc. Am. B* 28:483–488, with permission (DOI: https://doi.org/10.1364/JOSAB.28.000483.))

$$E_{ref} = r \cdot E_{inc} + \kappa^* \cdot g_2 \cdot E_{circ}, \tag{12.12}$$

$$E_{circ} = \kappa \cdot E_{inc} + r^* \cdot g_2 \cdot E_{circ}, \tag{12.13}$$

$$r = \sqrt{1 - \kappa_0^2} \exp(j\varphi), \kappa = \kappa_0 \exp(j\varphi) \tag{12.14}$$

$$g_i = \exp(-\gamma L_i - j n_{effi} k L_i), \quad i = 1, 2, \tag{12.15}$$

where κ is the coupling coefficient, n_{eff2} is the effective refractive index of cavity 2, and $k = 2\pi/\lambda$. As shown in Figure 12.7a, the effect that cavity 2 has on cavity 1 can be treated as an effective reflection R_{eff2} (Zhang et al., 2011b)

$$R_{eff2} = \frac{E_{ref}}{E_{inc}} = \frac{r - 2|r|^2 g_2 + g_2}{1 - r^* g_2}. \tag{12.16}$$

In this way, when cavity 1 is initially excited, the electric field inside the coupled cavity is (Zhang et al., 2011b)

$$E = E_1 = \frac{1}{1 - R_{eff2} g_1}. \tag{12.17}$$

The spectra of the optical field intensity inside the coupled cavity can be calculated by Equation 12.17, and is shown in Figure 12.7b. The coupling between the two resonators with different cavity lengths results in the modulated spectra envelope, and a tiny shift of the resonance wavelength of one resonator can induce a giant shift of the modulated envelope. For the coupled-ring microresonator sensor shown in Figure 12.8a, one resonator is doped by dye to lase as the light source, while the other cavity is employed as a microfluidic channel for the flowing of the analyte. By measuring analytes with different concentrations, the sensitivity of the modulated spectrum envelope is determined as 5930 nm/RIU (see Figure 12.8c), which is one order of magnitude higher than that of the resonance wavelength shift (see Figure 12.8b).

12.2.1.3 Complex refractive index sensing

The complex refractive index is expressed as $\tilde{n} = n + i\kappa$. The traditional refractive index sensing method mentioned above detects the analyte by only using the real part of the refractive index. Therefore, only one transducing physical quantity can be obtained. However, if both the real and the imaginary parts of the complex refractive index are measured during the detection of the analyte, two independent transducing physical quantities can be obtained, which means two unknown parameters are able to be determined during the detection (Zhang et al., 2016). For example, in a ternary mixture system with unknown concentration composition (Cx, Cy, 1-Cx-Cy), it is impossible to use the traditional refractive index sensing method to obtain the concentration composition of this ternary mixture. However, the complex refractive index sensing method, which has two transducing physical quantities, is able to measure the concentration composition of such a ternary mixture system. The real part of the refractive index can be transduced by the resonance wavelength shift ($\Delta\lambda$), and the imaginary part of the refractive index can be transduced by the mode linewidth change ($\Delta[\delta\lambda]$).

Taking the $D_2O/H_2O/EtOH$ mixture as an example, compared with pure D_2O, the introduction of both H_2O (concentration: Cx) and EtOH (concentration: Cy) can result in both a resonance wavelength shift and a mode linewidth change, which can be illustrated by a matrix equation

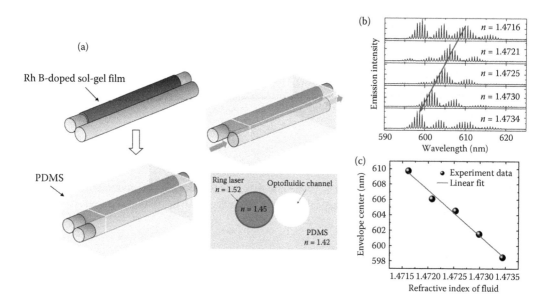

Figure 12.8 (a) Fabrication process and cross-section of an optofluidic coupled cavity; (b) emission spectra of a coupled optofluidic ring laser when analyte flows through the optofluidic channel. From top to bottom, the emission envelope moves to shorter wavelength when fluid refractive index increases. The arrow (red online) indicates the envelope shift; (c) fitted envelope center versus fluid refractive index. A linear fitting gives a sensitivity of 5930 ± 360 nm/RIU. (Reprinted from Zhang, X. et al. 2011. *Opt. Express* 19:22242–22247, with permission. DOI: https://doi.org/10.1364/OE.19.022242.)

$$AX = B, \tag{12.18}$$

$$A = \begin{bmatrix} \Delta\lambda/\Delta C_x & \Delta\lambda/\Delta C_y \\ \Delta[\delta\lambda]/\Delta C_x & \Delta[\delta\lambda]/\Delta C_y \end{bmatrix}, X = \begin{bmatrix} C_x \\ C_y \end{bmatrix}, \text{ and } B = \begin{bmatrix} \Delta\lambda \\ \Delta[\delta\lambda] \end{bmatrix}, \tag{12.19}$$

where $\Delta\lambda/\Delta C_x$ ($\Delta[\delta\lambda]/\Delta C_x$) and $\Delta\lambda/\Delta C_y$ ($\Delta[\delta\lambda]/\Delta C_y$) are the resonance wavelength (linewidth) change for every 1% concentration increase, respectively, for C_x and C_y. As for any vector B, Equation 12.18 has a single unique solution X only if the determinant of A is nonzero. We only need to prove that the determinant det(A) is not zero in order to demonstrate the one-to-one relationship between $[C_x, C_y]^T$ and $[\Delta\lambda, \Delta[\delta\lambda]]^T$, where the superscript "T" stands for the matrix transpose.

As an example, a complex refractive index sensor that is based on a photonic crystal nanobeam cavity is used to discriminate nine $D_2O/H_2O/EtOH$ mixtures with different concentration compositions with the complex refractive index sensing method (see Figure 12.9, and Table 12.1). The matrix A in Equation 12.19 can be experimentally determined with the results shown in Figures 12.10a and b

$$A = \begin{bmatrix} -1.2 & 65.9 \\ 0.14 & 0 \end{bmatrix}. \tag{12.20}$$

The determinant det(A) is clearly not zero, thus the concentration composition $[C_x, C_y]$ can be uniquely measured from the output signal set $[\Delta\lambda, \Delta[\delta\lambda]]$ in this sensor.

The transducing output signal sets ($\Delta\lambda, \Delta[\delta\lambda]$) of all the nine mixtures are shown in Figure 12.10c. Clearly, the detections of all the nine mixtures have totally different responses. Although A1, B2, and C3 have almost the same resonance wavelength shift, which means they cannot be distinguished

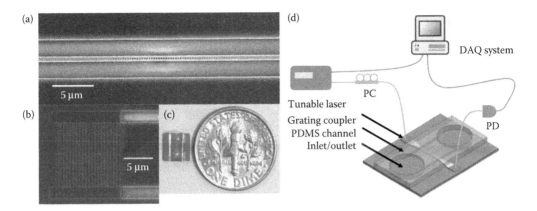

Figure 12.9 SEM images of the silicon photonic crystal nanobeam cavity (a) and grating coupler (b); (c) size comparison between the device (left) and a dime (right); (d) schematic view of the experiment setup. PC: polarization controller, DAQ system: data acquisition system, PD: photodetector. (Reprinted from Zhang, X. et al. 2016. *Opt. Lett.* 41:1197–1200, with permission. DOI: https://doi.org/10.1364/OL.41.001197.)

Table 12.1 Concentration compositions of $D_2O/H_2O/EtOH$ ternary mixtures used in the experiment

wt%	A1	A2	A3	B1	B2	B3	C1	C2	C3
D_2O	100	99.4	98.8	66.7	66.1	65.5	33.4	32.8	32.2
H_2O	0	0	0	33.3	33.3	33.3	66.6	66.6	66.6
EtOH	0	0.6	1.2	0	0.6	1.2	0	0.6	1.2

Source: Reprinted from Zhang, X. et al. 2016. *Opt. Lett.* 41:1197–1200, with permission. DOI: https://doi.org/10.1364/OL.41.001197.

by the traditional refractive index sensing method, they can still be distinguished by the complex refractive index sensing method. Similarly, A1, A2, and A3 cannot be distinguished by measuring the mode linewidth change, but they can be easily discriminated by measuring the resonance wavelength shift.

Moreover, the complex refractive index sensing method does not need any surface immobilization of functional groups for the multi-element detection. Therefore, compared to the traditional multi-element detection method, which usually relies on the sensor array and surface immobilization of different functional groups, the complex refractive index sensing method is much more versatile (Chakravarty et al., 2013). This method would have a wide range of applications in heavy metal ions detection, gas detection, biomolecule detection, and so on.

12.2.2 Nonrefractive index-based sensing

The nonrefractive index-based sensing methods include photoluminescence detection, Förster resonance energy transfer (FRET) detection, surface-enhanced Raman spectroscopy (SERS), and so on. (White et al., 2007; Sciacca et al., 2009; Anderson, 2010; Sun and Fan, 2012; Krismastuti et al., 2014). These methods have been widely used to detect DNA and protein concentration, DNA single base mismatch, DNA conformational changes, protein–protein interactions, and so on. (Sciacca et al., 2009; Sun and Fan, 2012; Zhang et al., 2012, 2013; Chen et al., 2013a,b; Wu et al., 2014). Based on optical micro-/nanoresonators photoluminescence intensity, FRET signal, and the Raman signal can all be improved by the cavity-enhanced electric field and optical feedback. Here, we focus on the FRET-based detection.

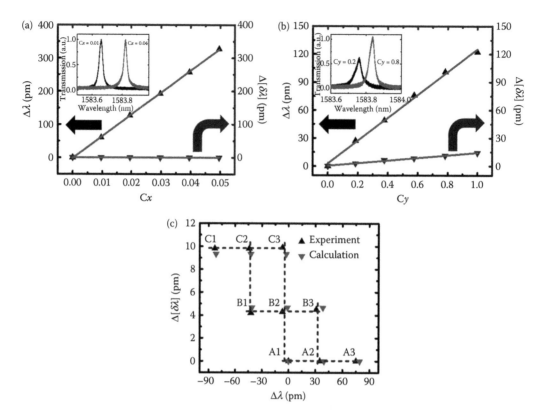

Figure 12.10 Resonance wavelength shift ($\Delta\lambda$) and linewidth change ($\Delta[\delta\lambda]$) of silicon photonic crystal nanobeam cavity in D_2O/EtOH mixtures with different EtOH concentrations (a), D_2O/H_2O mixture with different H_2O concentrations (b), and D_2O/H_2O/EtOH ternary mixtures with different concentration compositions as listed in Table 12.1 (c). The dots in (a) and (b) are experimental results and the solid lines in (a) and (b) indicate the linear fitting. The inset of (a) shows the transmission spectra when EtOH concentration C_x is around 0.01 and 0.04, respectively. The inset of (b) shows the transmission spectra when H_2O concentration C_y is around 0.2 and 0.8, respectively. (Reprinted from Zhang, X. et al. 2016. *Opt. Lett.* 41:1197–1200, with permission. DOI: https://doi.org/10.1364/OL.41.001197.)

When a fluorophore (e.g., dye, fluorescent protein) is excited, the excited fluorophore usually emits a photon and returns back to the ground state. However, the fluorophore can also transfer the energy to another molecule nearby instead of emitting a photon, and this process is called Förster resonance energy transfer (FRET) (Chirio-Lebrun and Prats, 1998). The fluorophore which provides the energy is the donor, and the molecule which accepts the energy is the acceptor. The energy transfer efficiency is determined by the distance between the donor and acceptor (R) (Chirio-Lebrun and Prats, 1998)

$$E = \frac{R_0^6}{R^6 + R_0^6},\qquad(12.21)$$

where R_0 is the Förster distance, which corresponds to the distance between the donor and acceptor when the energy transfer efficiency is 50%. R_0 is determined by the properties of the donor and acceptor, including the refractive index of the surroundings, quantum yield, molecule orientation, and spectral overlap between the donor and the acceptor (Chirio-Lebrun and Prats, 1998).

The traditional biomolecule conformational change detection method is based on the fluorescence FRET (see Figure 12.11). Usually, the biomolecule is labelled by fluorescent dyes or fluorescent

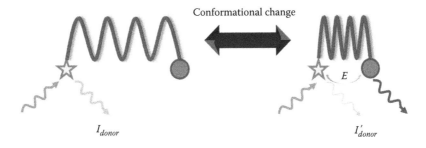

Conformational change

I_{donor} I'_{donor}

Figure 12.11 Schematic drawing of conformational change detection with the FRET sensing method.

proteins as donor and acceptor. When the biomolecule conformation changes, the distance between donor and acceptor also changes, as does the FRET efficiency. Therefore, by measuring the fluorescence emission spectra of the donor or acceptor, the biomolecule conformational changes can be analyzed (Bath and Turberfield, 2007; Sivaramakrishnan and Spudich, 2011; Douglas et al., 2012).

However, the changes of biomolecules are quite small, which leads to a very weak fluorescence FRET sensing signal (Fan and Yun, 2014). Here, a novel method, which is based on a FRET biolaser, is introduced to enhance the sensitivity of biomolecule conformational change detection.

In a FRET biolaser, the biomolecules labelled by the donor and acceptor are placed inside the resonator as the gain medium. When pumped above the lasing threshold, the emitted light is amplified by stimulated emission. Compared to the fluorescence FRET process, the light generated by the FRET biolaser has highly spatial and temporal coherence, which leads to a narrow emission angle and a narrow emission spectrum. In the meantime, due to the optical amplification by the stimulated emission, the light intensity is several orders higher than the fluorescence FRET, which will greatly enhance the sensing signal intensity. Moreover, the lasing behavior is highly sensitive to the gain medium. Even a tiny change of the donor–acceptor energy transfer efficiency which results from the conformational changes of biomolecules, will significantly affect the FRET laser output characteristics. In this section, we will discuss two examples of FRET biolaser applications on DNA conformational change and protein–protein interaction detection.

12.2.2.1 FRET biolaser for DNA conformational change detection

DNA is the essential constituent of genes. Due to the strict rule of base pairing in DNA assembly, DNA is used to construct DNA nanomachines, which are functional machines at the nanometer scale (Bath and Turberfield, 2007). Currently, various DNA nanomachines have been realized, including DNA logic gates, DNA nanorobots, DNA nanotweezers, and so on. (Beneson et al., 2004; Gu et al., 2010; Douglas et al., 2012). By monitoring the DNA conformational changes, the operation status of DNA nanomachines can be analyzed.

The DNA Holliday junction, a critical intermediate in homologous genetic recombination, is one of the simplest DNA nanomachines (Mount et al., 2006). It has four double-helical arms, as shown in Figure 12.12a. Due to the strong electrostatic repulsion among the backbone phosphate groups, the arms of the junction are fully extended into a planar square conformation (Duckett et al., 1990). However, in the presence of certain kinds of ions, such as magnesium, the repulsion will be largely suppressed, leading to a folded junction (Duckett et al., 1990). This conformational change can be detected by measuring the fluorescence intensity of the donor or acceptor labelled at the end of the arms. However, the fluorescence intensity change ratio is too small (see Figure 12.13). To enhance the sensitivity, the dye-labelled DNA Holliday junctions are put into an optofluidic ring resonator (see Figure 12.12b) (Zhang et al., 2012, 2013). Pumped by a pulsed laser, the lasing of the dyes on the DNA Holliday junctions is realized. When the Holliday junction DNA has a structural change, the laser spectra is greatly changed, which is 13 times more sensitive than the traditional fluorescence FRET method (see Figure 12.13).

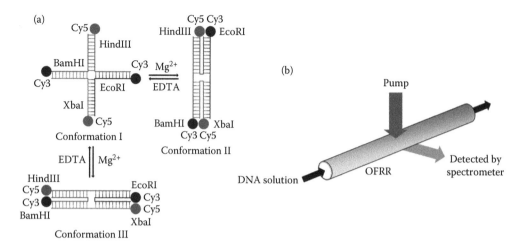

Figure 12.12 (a) Dye-labeled DNA Holliday junction and its two possible ion-dependent folding type. EDTA is used to remove Mg^{2+}; (b) schematic of the biolaser based on microcapillary resonator. (Reprinted from Zhang, X. et al. 2012. *Lab Chip* 12:3673–3675, with permission. DOI: https://doi.org/10.1039/C2LC40183E.)

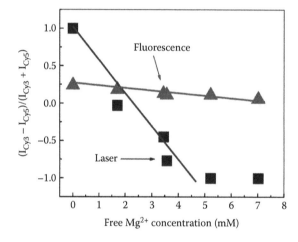

Figure 12.13 Comparison between laser and fluorescence FRET sensitivities to DNA conformational changes. Mg^{2+} is used to adjust the DNA conformation. Solid lines are the linear fit for the linear part of response. (Reprinted from Zhang, X. et al. 2012. *Lab Chip* 12:3673–3675, with permission. DOI: https://doi.org/10.1039/C2LC40183E.)

12.2.2.2 FRET biolaser for protein–protein interaction detection

Protein–protein interactions refer to the physical contacts between proteins which result from biochemical events. Many biological activities, such as signal transduction, transport across membranes, cell metabolism, and muscle contraction are all regulated by protein–protein interactions (Bhattacharyya et al., 2006; Bhardwaj et al., 2011). Protein–protein interactions can also be analyzed by monitoring the structural changes of proteins, the sensitivity of which can also be enhanced by using a FRET biolaser.

To investigate the FRET laser sensitivity to the protein–protein distance, two genetically encoded fluorescent proteins (donor: eGFP, acceptor: mCherry) with different distances between the donors and acceptors are used (see Figure 12.14a) (Chen et al., 2013a,b; Zhang et al., 2013). The experimental setup is the same as that shown in Figure 12.12b. The donor–acceptor energy transfer ratio of the

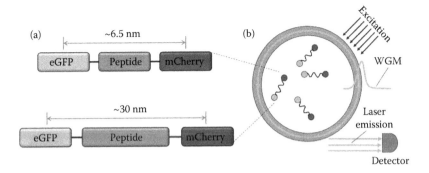

Figure 12.14 (a) Two types of genetically encoded fluorescent protein pairs (eGFP as donor and mCherry as acceptor) used in the experiment; (b) the cross-section of biolaser based on microcapillary resonator who supports a high-Q whispering gallery mode (WGM). When excited by a pulse laser, the genetically encoded fluorescent protein pairs placed inside the resonator achieve laser with the optical feedback provided by the resonator. (Reprinted from Chen, Q. et al. 2013. *Lab Chip* 13:2679–2681, with permission. DOI: https://doi.org/10.1039/C3LC50207D.)

short-linked pair is larger than that of the long-linked pair. Therefore, the donor lasing threshold of the long-linked pair is smaller than that of the short-linked pair. Meanwhile, under the same pump intensity, the laser intensity of the long-linked pair is nearly 25 times higher than that of the short-linked pair, which is 100 times more sensitive than the traditional fluorescence method (see Figure 12.15).

12.3 Optical micro-/nanoresonator-based single nanoparticle detection

The ultimate limit to biochemical analysis is single molecule (nanoparticle) detection (Tinnefeld, 2013). Single nanoparticle detections are uniquely poised to yield critical molecular information in biomedical applications (Dantham et al., 2013). In the meantime, single nanoparticle detection is a way to study the detailed physical properties of nanoparticle-surface interaction that allows for scrutiny of fundamental principles and mechanisms (Schein et al., 2015). Up until now, various single nanoparticle detection methods have been developed, such as fluorescence microscopy, localized surface plasmon resonance (LSPR), surface enhanced Raman spectroscopy (SERS), and

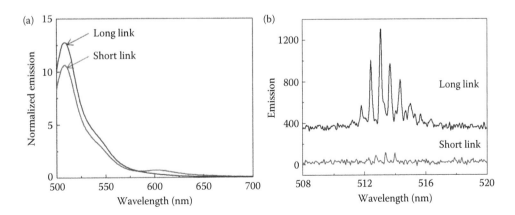

Figure 12.15 Comparison between fluorescence (a) and laser (b) FRET sensitivities to protein-protein distances. (Reprinted from Chen, Q. et al. 2013. *Lab Chip* 13:2679–2681, with permission. DOI: https://doi.org/10.1039/C3LC50207D.)

so on. (Kneipp et al., 1997; Chansin et al., 2007; Zijlstra et al., 2012). Alternatively, optical micro-/naoresonators also allow the detection of single nanoparticles due to the small mode volume and high optical field enhancement.

When a single nanoparticle enters the evanescent field of an optical micro-/nanoresonator, it will interact with the electric field and result in a resonance wavelength shift. Assuming that the resonator is in a single photon resonant state with a frequency of ω (Arnold et al., 2008), the nanoparticle will be polarized and generates an excess electric dipole moment $\Delta\vec{p}(\vec{r},t) = Re[\Delta\vec{p}_0(\vec{r})\exp(-j\omega t)]$, which in turn introduces a perturbation in the photon energy. The photon energy and its perturbation can be expressed by (Arnold et al., 2008)

$$\hbar\omega_r = -\frac{1}{2}\int \varepsilon(\vec{r})\vec{E}_0^*(\vec{r})\cdot\vec{E}_0(\vec{r})dV \tag{12.22}$$

$$\hbar\Delta\omega = -\frac{1}{2}\Delta\vec{p}(t)\cdot\vec{E}_0(\vec{r}_j,t)_t = -\frac{1}{4}Re\left[\vec{E}_0^*(\vec{r}_j)\cdot\Delta\ddot{\alpha}_j\cdot\vec{E}_0(\vec{r}_j)\right] \tag{12.23}$$

where $\Delta\ddot{\alpha}$ is the excess polarizability tensor. According to the equations above, the single nanoparticle detection sensitivity can be derived with the assumption that the nanoparticle is an isotropic sphere (Arnold et al., 2008)

$$\left(\frac{\Delta\omega}{\omega_r}\right)_j = -\frac{\Delta\alpha_j\cdot\vec{E}_0^*(\vec{r}_j)\cdot\vec{E}_0(\vec{r}_j)}{2\int \varepsilon(\vec{r})\vec{E}_0^*(\vec{r})\cdot\vec{E}_0(\vec{r})dV}. \tag{12.24}$$

Clearly, the sensitivity of the single nanoparticle detection is determined by the field intensity where the nanoparticle locates and the mode volume of the resonator.

Photonic crystal nanobeam cavities have the unique capability of single nanoparticle detection due to their ultra-small mode volumes and high Q-factor. Recently, a single nanoparticle 12.5 nm in radius has been experimentally detected (Quan et al., 2013). To push the single nanoparticle detection to the limit, the mode volume has to be further reduced and the field intensity in the sensing area has to be further enhanced. To achieve these goals, a slot structure in a photonic crystal nanobeam cavity is a good choice (Lin et al., 2015). As the optical mode is greatly localized inside the slot, both ultra-small mode volume and ultra-high localized field intensity will be realized.

As an example, a slotted photonic crystal nanobeam cavity is designed based on the silicon-on-insulator (SOI) wafers with a silicon thickness of 260 nm. As shown in Figure 12.16a, there is a slot in the central area of the nanobeam (width = 500 nm), and 40-airhole Bragg mirrors are placed on each side. The lattice period is 280 nm, and the radius of the mirror airholes changes quadratically from 100 nm to 54 nm after 39 periods. Additionally, the distance between the two mirrors in the

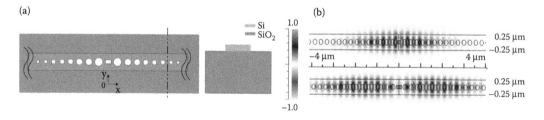

Figure 12.16 (a) Schematics of the slotted photonic crystal nanobeam cavity (left) and its cross-sectional view (right); (b) mode profiles of the first-order and second-order TE modes (the colorbar indicates the normalized electric field intensity ranging from −1 to 1). (Reprinted from Lin, T. et al. 2015. *J. Opt. Soc. Am. B* 32:1788–1791, with permission. DOI: https://doi.org/10.1364/JOSAB.32.001788.)

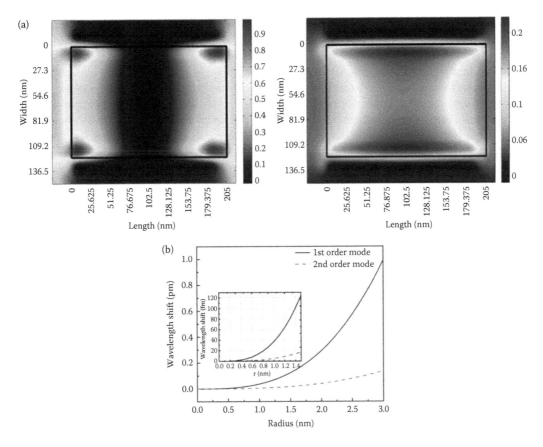

Figure 12.17 (a) The enlarged view of the mode profile inside the slot for the first-order TE mode (left) and the second-order TE mode (right); (b) the resonance wavelength shift induced by single nanoparticle with the radius ranging from 0 to 3 nm (inset shows the enlarged view when the nanoparticle radius is below 1.5 nm). (Reprinted from Lin, T. et al. 2015. *J. Opt. Soc. Am. B* 32:1788–1791, with permission. DOI: https://doi.org/10.1364/JOSAB.32.001788.)

central area is two times that of the lattice periods, and the slot is determined to be 205 nm long and 120 nm wide after optimization.

The mode profiles of such a slotted photonic crystal nanobeam cavity are shown in Figure 12.16b. Clearly, both the first-order mode and the second-order mode have hotspots inside the slot. In the meantime, the mode volume of the first and second modes are $0.07656(\lambda/n)^3$ and $0.1231(\lambda/n)^3$, respectively. Based on Equation 12.25, the wavelength shift induced by a single nanoparticle can be calculated and is shown in Figure 12.17b. By putting the nanoparticle in the hotspots of the first mode, the wavelength shift is around 40 fm for a nanoparticle with a radius of 1 nm, which is well above the noise level in the experimental condition (Quan et al., 2013; Zhang et al., 2014b). Moreover, the sensitivity is improved by 10-fold compared with that of the photonic crystal nanobeam cavity without an air slot (Quan et al., 2013).

12.4 Summary

In this chapter, we discussed three kinds of optical micro-/nanoresonators for biosensing, namely Fabry–Pérot microresonators, photonic crystal nanoresonators, and whispering gallery mode microresonators. Optical micro-/nanoresonators are able to confine light in small volumes at the micro-/nanometer scale, and significantly enhance the optical field intensity and light-matter interaction.

Therefore, these optical micro- and nanoresonators have the unique advantages in biochemical detection of high sensitivity and low detection limit. We discussed two sensing methods: refractive index sensing and non-refractive index sensing. For the refractive index sensing method that relies on the detection of the refractive index change, both the resonance wavelength shift and the Vernier effect can be utilized for the sensing of analyte concentrations. Additionally, a complex refractive index sensing scheme, which detects the resonance wavelength shift and mode linewidth change in response to the real and imaginary parts of the complex refractive index, can be used in the ternary mixture detection that has two unknown parameters. For the nonrefractive index-based sensing method that relies on the detection of other physical quantities apart from the refractive index, we focused on the FRET biolaser sensing. Compared with the traditional fluorescence FRET sensing that is based on spontaneous emission, the light emitted from the analyte inside the optical micro-/nanoresonator is amplified based on the stimulated emission. Therefore, the sensitivity of FRET biolaser-based biosensing is greatly enhanced. Finally, the optical micro-/nanoresonator-based single nanoparticle detection is also briefly discussed. Single nanoparticle detection has potential applications in biochemical analysis. The ultra-small mode volume coupled with an ultra-high optical field enhancement in optical micro-/nanoresonator can greatly boost the sensitivity and take the single nanoparticle detection to the limit.

References

Anderson MS 2010. Nonplasmonic surface enhanced Raman spectroscopy using silica microspheres. *Appl Phys Lett* 97:131116.

Arnold S, Ramjit R, Keng D, Kolchenko V, Teraoka I 2008. MicroParticle photophysics illuminates viral bio-sensing. *Faraday Discuss* 137:65–83.

Bath J, Turberfield AJ 2007. DNA nanomachines. *Nature Nanotech* 2:275–284.

Benenson Y, Gil B, Ben-Dor U, Adar R, Shapiro E 2004. An autonomous molecular computer for logical control of gene expression. *Nature* 429:423–429.

Bhardwaj N, Clarke D, Gerstein M 2011. Systematic control of protein interactions for systems biology. *Proc Natl Acad Sci U S A* 108:20279–20280.

Bhattacharyya RP, Remenyi A, Yeh BJ, Lim WA 2006. Domains, motifs, and scaffolds: The role of modular interactions in the evolution and wiring of cell signaling circuits. *Annu Rev Biochem* 75:655–680.

Chakravarty S, Lai WC, Zou Y, Drabkin HA, Gemmill RM, Simon GR, Chin SH, Chen RT 2013. Multiplexed specific label-free detection of NCI-H358 lung cancer cell line lysates with silicon based photonic crystal microcavity biosensors. *Biosens Bioelectron* 43:50–55.

Chansin GAT, Mulero R, Hong J, Kim MJ, Demello AJ, Edel JB 2007. Single-molecule Spectroscopy using nanoporous membranes. *Nano Lett* 7:2901–2906.

Chao C-Y, Guo LJ 2003. Biochemical sensors based on polymer microrings with sharp asymmetrical resonance. *Appl Phys Lett* 83:1527.

Chen Q, Zhang X, Sun Y, Ritt M, Sivaramakrishnan S, Fan X 2013a. Highly sensitive fluorescent protein FRET detection using optofluidic lasers. *Lab Chip* 13:2679–2681.

Chen QS, Zhang XW, Sun YZ, Ritt M, Sivaramakrishnan S, Fan XD 2013b. Highly sensitive Optofluidic FRET lasers with genetically encoded fluorescent protein pairs. In: *2013 Conference on Lasers and Electro-Optics (CLEO)*, p CM2H.2.

Chirio-Lebrun MC, Prats M 1998. Fluorescence resonance energy transfer (FRET): Theory and experiments. *Biochem Educ* 26:320–323.

Dantham VR, Holler S, Barbre C, Keng D, Kolchenko V, Arnold S 2013. Label-free detection of single protein using a nanoplasmonic-photonic hybrid microcavity. *Nano Lett* 13:3347–3351.

Douglas SM, Bachelet I, Church GM 2012. A logic-gated nanorobot for targeted transport of molecular payloads. *Science* 335:831–834.

Duckett DR, Murchie A, Lilley D 1990. The role of metal ions in the conformation of the four-way DNA junction. *EMBO J* 9:583–590.

Fan X, Yun SH 2014. The potential of optofluidic biolasers. *Nat Methods* 11:141–147.

Fassel VA 1976. Nomenclature, symbols, units and their usage in spectrochemical analysis-II. Data interpretation. *Pure & Appl Chem* 45:99–103.

Foreman MR, Swaim JD, Vollmer F 2015. Whispering gallery mode sensors. *Adv Opt Photonics* 7:168–240.

Gu H, Chao J, Xiao SJ, Seeman NC 2010. A proximity-based programmable DNA nanoscale assembly line. *Nature* 465:202–205.

Hachuda S, Otsuka S, Kita S, Isono T, Narimatsu M, Watanabe K, Goshima Y, Baba T 2013. Selective detection of sub-atto-molar Streptavidin in 10(13)-fold impure sample using photonic crystal nanolaser sensors. *Opt Express* 21:12815–12821.

Jagerska J, Zhang H, Diao Z, Le Thomas N, Houdre R 2010. Refractive index sensing with an air-slot photonic crystal nanocavity. *Opt Lett* 35:2523–2525.

Joannopoulos J, Johnson S, Winn J, Meade R 2008. *Photonic Crystals: Molding the Flow of Light*, 2nd Edition. Princeton Univ. Press.

Kavokin AV, Baumberg JJ, Malpuech G, Laussy FP 2007. *Microcavities*. Oxford University Press.

Kneipp YWK, Kneipp H, Perelman LT, Itzkan I, Dasari RR, Feld MS 1997. Single Molecule Detection Using Surface-Enhanced Raman Scattering (SERS). *Phys Rev Lett* 78:1667–1670.

Krismastuti FSH, Pace S, Voelcker NH 2014. Porous Silicon Resonant Microcavity Biosensor for Matrix Metalloproteinase Detection. *Adv Funct Mater* 24:3639–3650.

Li H, Fan X 2010. Characterization of sensing capability of optofluidic ring resonator biosensors. *Appl Phys Lett* 97:011105.

Li M, Wu X, Liu L, Fan X, Xu L 2013. Self-referencing optofluidic ring resonator sensor for highly sensitive biomolecular detection. *Anal Chem* 85:9328–9332.

Liao J, Wu X, Liu L, Xu L 2016. Fano resonance and improved sensing performance in a spectral-simplified optofluidic micro-bubble resonator by introducing selective modal losses. *Opt Express* 24:8574–8580.

Lin T, Zhang X, Zhou G, Siong CF, Deng J 2015. Design of an ultra-compact slotted photonic crystal nanobeam cavity for biosensing. *J Opt Soc AmB* 32:1788–1791.

Lin VS 1997. A Porous Silicon-Based Optical Interferometric Biosensor. *Science* 278:840–843.

Liu P, Huang H, Cao T, Liu X, Qi Z, Tang Z, Zhang J 2013. An ultra-low detection-limit optofluidic biosensor with integrated dual-channel Fabry-Pérot cavity. *Appl Phys Lett* 102:163701.

Marcatil.Ea 1969. Bends in Optical Dielectric Guides. *Bell Syst Tech J* 48:2103.

Moiseev L, Unlu MS, Swan AK, Goldberg BB, Cantor CR 2006. DNA conformation on surfaces measured by fluorescence self-interference. *Proc Natl Acad Sci U S A* 103:2623–2628.

Mount AR, Mountford CP, Evans SA, Su TJ, Buck AH, Dickinson P et al. 2006. The stability and characteristics of a DNA Holliday junction switch. *Biophys Chem* 124:214–221.

Ozkumur E, Needham JW, Bergstein DA, Gonzalez R, Cabodi M, Gershoni JM, Goldberg BB, Unlu MS 2008. Label-free and dynamic detection of biomolecular interactions for high-throughput microarray applications. *Proc Natl Acad Sci U S A* 105:7988–7992.

Pollinger M, O'Shea D, Warken F, Rauschenbeutel A 2009. Ultrahigh-Q tunable whispering-gallery-mode microresonator. *Phys Rev Lett* 103:053901.

Quan Q, Floyd DL, Burgess IB, Deotare PB, Frank IW, Tang SK, Ilic R, Loncar M 2013. Single particle detection in CMOS compatible photonic crystal nanobeam cavities. *Opt Express* 21:32225–32233.

Rayleigh L 1910. CXII.The problem of the whispering gallery. *Philos Mag Series 6* 20:1001–1004.

Ren L, Wu X, Li M, Zhang X, Liu L, Xu L 2012. Ultrasensitive label-free coupled optofluidic ring laser sensor. *Opt Lett* 37:3873–3875.

Savchenkov AA, Matsko AB, Ilchenko VS, Maleki L 2007. Optical resonators with ten million finesse. *Opt Express* 15:6768–6773.

Schein P, Ashcroft CK, O'Dell D, Adam IS, DiPaolo B, Sabharwal M, Shi C, Hart R, Earhart C, Erickson D 2015. Near-field Light Scattering Techniques for Measuring Nanoparticle-Surface Interaction Energies and Forces. *J Lightwave Technol* 33:3494–3502.

Sciacca B, Frascella F, Venturello A, Rivolo P, Descrovi E, Giorgis F, Geobaldo F 2009. Doubly resonant porous silicon microcavities for enhanced detection of fluorescent organic molecules.
Sens Actuator B-Chem 137:467–470.

Senthil Murugan G, Petrovich MN, Jung Y, Wilkinson JS, Zervas MN 2011. Hollow-bottle optical
microresonators. *Opt Express* 19:20773–20784.

Siegman AE 1986. *Lasers*. University Science Books.

Sivaramakrishnan S, Spudich JA 2011. Systematic control of protein interaction using a modular
ER/K alpha-helix linker. *Proc Natl Acad Sci U S A* 108:20467–20472.

Sumetsky M 2010. Mode localization and the Q-factor of a cylindrical microresonator. *Opt Lett*
35:2385–2387.

Sun Y, Fan X 2011. Optical ring resonators for biochemical and chemical sensing. *Anal Bioanal
Chem* 399:205–211.

Sun Y, Fan X 2012. Distinguishing DNA by analog-to-digital-like conversion by using optofluidic
lasers. *Angew Chem Int Ed Engl* 51:1236–1239.

Tang T, Wu X, Liu L, Xu L 2016. Packaged optofluidic microbubble resonators for optical sensing.
Appl Opt 55:395–399.

Tinnefeld P 2013. Single-molecule detection: Breaking the concentration barrier. *Nature Nanotech*
8:480–482.

Vahala KJ 2003. Optical microcavities. *Nature* 424:839–846.

Vollmer F, Braun D, Libchaber A, Khoshsima M, Teraoka I, Arnold S 2002. Protein detection by
optical shift of a resonant microcavity. *Appl Phys Lett* 80:4057–4059.

Ward JM, Dhasmana N, Nic Chormaic S 2014. Hollow core, whispering gallery resonator sensors.
Eur Phys J 223:1917–1935.

White IM, Gohring J, Fan X 2007. SERS-based detection in an optofluidic ring resonator platform.
Opt Express 15:17433–17442.

White IM, Oveys H, Fan X 2006. Liquid-core optical ring-resonator sensors. *Opt Lett* 31:1319–1321.

Wu X, Oo MK, Reddy K, Chen Q, Sun Y, Fan X 2014. Optofluidic laser for dual-mode sensitive biomolecular detection with a large dynamic range. *Nat Commun* 5:3779.

Yablonovitch E 1987. Inhibited spontaneous emission in solid-state physics and electronics. *Phys
Rev Lett* 58:2059–2062.

Zhang C, Chen SL, Ling T, Guo LJ 2015a. Imprinted Polymer Microrings as High-Performance
Ultrasound Detectors in Photoacoustic Imaging. *J Lightwave Technol* 33:4318–4328.

Zhang C, Chen SL, Ling T, Guo LJ 2015b. Review of imprinted polymer microrings as ultrasound
detectors: Design, fabrication, and characterization. *IEEE Sens J* 15:3241–3248.

Zhang C, Ling T, Chen S-L, Guo LJ 2014a. Ultrabroad bandwidth and highly sensitive optical ultra-
sonic detector for photoacoustic imaging. *ACS Photonics* 1:1093–1098.

Zhang X, Chen Q, Ritt M, Sivaramakrishnan S, Fan X, Reed GT 2013. Bioinspired optofluidic lasers
for DNA and protein detection. *Proc SPIE* 8629:862907.

Zhang X, Lee W, Fan X 2012. Bio-switchable optofluidic lasers based on DNA Holliday junctions.
Lab Chip 12:3673–3675.

Zhang X, Liu L, Xu L 2014b. Ultralow sensing limit in optofluidic micro-bottle resonator biosensor
by self-referenced differential-mode detection scheme. *Appl Phys Lett* 104:033703.

Zhang X, Ren L, Wu X, Li H, Liu L, Xu L 2011a. Coupled optofluidic ring laser for ultrahigh-sensitive
sensing. *Opt Express* 19:22242–22247.

Zhang X, Zhou G, Shi P, Du H, Lin T, Teng J, Chau FS 2016. On-chip integrated optofluidic complex
refractive index sensing using silicon photonic crystal nanobeam cavities. *Opt Lett* 41:1197.

Zhang XW, Li H, Tu X, Wu XA, Liu LY, Xu L 2011b. Suppression and hopping of whispering gallery
modes in multiple-ring-coupled microcavity lasers. *J Opt Soc AmB* 28:483–488.

Zijlstra P, Paulo PM, Orrit M 2012. Optical detection of single non-absorbing molecules using the
surface plasmon resonance of a gold nanorod. *Nature Nanotech* 7:379–382.

13

Terahertz MEMS metamaterials

Prakash Pitchappa and Chengkuo Lee

Contents

13.1 Introduction to THz metamaterials

Electromagnetic (EM) metamaterial is an array of subwavelength structures that can be engineered to achieve specific EM properties. The properties of the metamaterials strongly depend on the geometrical shape and size of the subwavelength unit cell, also termed as "meta-atoms." This has led to the demonstration of numerous interesting EM properties, such as artificial magnetism [1], negative refractive index [2], subwavelength focusing [3], perfect absorption [4], chirality [5], narrow band emission [6], electromagnetically induced transparency analogue [7–11], and many more, over a wide range of the EM spectrum. The disruptive feature of metamaterials lies in the scalable design and ultrathin size relative to the interacting wavelengths of EM waves. This scalability of the metamaterial design has been instrumental in enabling significant research progress in the field of terahertz (THz) wave interaction and manipulation. The THz part of the EM spectrum ($f \sim 0.1 - 10$ THz; $\lambda \sim 3000 - 30$ μm) lies in between the low energy microwaves and high energy infrared regions, and has minimal interaction with naturally occurring materials. Furthermore, the THz spectrum lies in between the electronics and photonics realms used for controlling the EM waves. The electronics components used for interacting with microwaves cannot be scaled up to operate at THz frequencies and the low energy of THz waves compared to IR waves limits the usage of photonics devices for THz wave interaction. This technological vacancy between the electronics and photonics realms of the electromagnetic spectrum is popularly termed the "THz gap" [12]. However, the versatility of the metamaterial resonator designs and the ease of fabrication of metamaterials have led to the realization of various THz devices. Metamaterial resonators are mostly designed to inter-act with either the magnetic field and/or the electric field of EM waves. The metamaterial acts as an effective medium, and its response to the incident magnetic and/or electric field will provide effective permeability (μ_r) and/or effective permittivity (ε_r) values, respectively. These two fundamental optical parameters can then be engineered to achieve other optical parameters, such as refractive index, surface impedance, and so on. This allows for the advanced interaction, control, and manipulation of EM waves using metamaterials. The geometrical shape and size of the meta-atom primarily determines the coupling mechanism of the incident field to the metamaterial and the operating spectral range. The most popular meta-atom design for enabling the magnetic response is the split ring resonator (SRR), which is a metal ring with a split gap in one side as shown in Figure 13.1a. The magnetic response of the SRR can be achieved either by having the incident magnetic field be perpendicular to the plane containing the SRR [1] or with the incident electric field parallel to the SRR gap [13]. The magnetic response is characterized by the circulating current induced in the metallic ring with a strong field confinement across the capacitive gap. The magnetic moment generated in response to the circulating currents in the planar SRR will always be in the plane perpendicular to the SRR. The SRR can be modelled as an electrical inductive-capacitive resonator (LC), as shown in Figure 13.1b. Thus, the resonant frequency can be given as $f_r = 1/\sqrt{(LC)}$, where "L" is the effective inductance of the metal ring and "C" is the effective capacitance across the split gap. The operational frequencies of the SRR can be scaled by varying the geometrical parameters of the SRR accordingly. Alternatively, a simple cut wire resonator (CWR), which is primarily a dipole resonator, is used as the meta-atom to achieve electrical response of the metamaterial by interacting directly with the electric field of the incident THz waves. The electric field response of the CWR is characterized by the induced surface current that is parallel to the incident electric field. Another popular meta-atom design for interacting with the incident electric field is the electrical split ring resonator (ESRR), which is formed by placing two SRRs facing each other [14]. When the incident electrical field is along the gap-bearing side of the ESRR, antiparallel currents are induced in the two SRRs with strong field confinement in the gap region. Due to causality, the response is electrical, and hence provides an effective permittivity value for the ESRR metamaterials. The resonance frequency of the ESRR, $f_r = 2/\sqrt{(LC)}$, and can be engineered based on the geometrical parameters of the ESRR. The metamaterial unit cell can also be made of multiple resonators that are coupled in either in-plane or out-of-plane directions. This enables more advanced THz manipulation. Hence, the versatility of the resonator design by itself offers a wide range of THz properties.

More interestingly, these THz properties realized through metamaterials can be actively controlled through external stimulus and this research field has recently been pursed with great vigor. Active control of THz metamaterial provides a new dimensionality for the efficient manipulation of THz waves and is critical for the realization of high performance THz devices. Conventionally, active materials whose properties can be varied through external control fields are integrated into

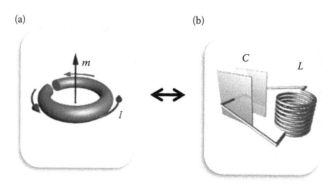

Figure 13.1 Schematic representation of (a) split ring resonator meta-atom and (b) its electrical equivalent circuit LC model, respectively.

the unit cell geometry or as surrounding media. The changes in material properties of the active materials to the external control field will cause the effective change in the unit cell inductance, capacitance or surrounding refractive index. This, in turn, varies the response of the metamaterial for the incident THz waves, based on the external control fields. Various studies have demonstrated the use of different types of active materials for dynamic control of the THz response. Each of these approaches offers complementary device level performance, which can be used based on the application needs. Some of the active material-based approaches that have been demonstrated for dynamic control of THz response are briefly elaborated next.

Initially, a photoconductive semiconductor was used as a substrate material on which SRR meta-atoms were fabricated [15]. When the substrate with SRR meta-atoms was pumped with an external optical beam, the conductivity of the overall substrate increased. This change in the surrounding medium of the metamaterial causes both the LC resonance and dipole resonance of the SRR to dynamically modulate with increasing pump power. Furthermore, the photoconductive material was selectively placed at the gap of ESRR meta-atoms and the electrical LC resonance of the ESRR metamaterial was modulated through optical pumping [16]. This selective placement of photoconductive materials in a meta-atom geometry allowed for the probing and manipulation of THz properties that are achievable through multi-resonator systems, such as electromagnetically induced transparency (EIT) analogue [17–20]. The most interesting feature of optical pump-based tunable metamaterials is the ultrafast response, which is in the order of picoseconds and is critical for the realization of high performance THz devices. Alternatively, thermally controlled phase changing vanadium oxide (VO_2) integrated into the meta-atom geometry was also demonstrated for ultrafast switching of THz resonances [21]. Thermal control was also used to actively modulate the conductivity of superconductors to realize active control of the THz metamaterial response, and is critical for realizing high Q resonator devices [19,22–24]. Electrical tuning of the THz metamaterial response is widely reported, owing to the ease of integration of the control signal lines with the metamaterial geometry. Some of the electrically tunable approaches involve the conductivity change of doped semiconductors [25], 2D materials such as graphene [26,27], molybdenum disulfide (MoS_2) [28], and tungsten disulfide (WS_2) [29], and the electrically tunable refractive index of liquid crystals [30–33]. Electrically tunable approaches provide a means of achieving highly miniaturized systems due to the ease of integration with ICs. Hence, the active materials-based tunable THz metamaterials provide a wide range of options with varying performance features to choose from, based on the application needs of the THz devices. However, they possess limitations that would seriously hinder their usage in specific applications. The frequency dependent properties of these active materials limit the scalability of tunable metamaterial over a wide spectral range. Additionally, the use of exotic materials demands for sophisticated fabrication processes and cannot be readily manufactured using batch processes. Furthermore, some of these approaches require bulky setups for providing the control signals, thereby making them not very attractive for miniaturized THz systems.

Alternatively, structurally reconfigurable metamaterials have been reported as ideal candidates for miniaturized THz devices with improved electro-optic performances. Owing to the strong dependence of metamaterial properties on the geometrical parameters, the structural reconfiguration approach is considered as a more straightforward and efficient way of realizing actively tunable metamaterials. The advancements in the design and fabrication techniques for the realization of microelectromechanical systems (MEMS) over the past 50 years, has greatly aided the direct integration of MEMS actuators into metamaterial unit cell geometry to achieve reconfiguration in a microscale. These metamaterials are popularly known as "microelectromechanically reconfigurable metamaterials (MRMs)." The geometrical size of microactuators and their electromechanical performance perfectly complements the THz metamaterial unit cell size and desired tunable EM properties. The MEMS actuators are realized using micromachining processes, and hence can be easily integrated with metamaterial unit cells and application-specific integrated circuits (ASIC) that can provide control signals on demand, thereby making the entire system highly miniaturized and realized with extremely low cost. The versatility of microactuators in the actuation direction and deformation range has enabled the demonstration of a wide range of MRMs. These MRMs complement each other to enable a wide range of tunable THz properties. Based on the direction of reconfiguration, the reported MRMs can be broadly classified into two categories; in-plane MRMs and out-of-plane MRMs. In the case of in-plane MRMs, the reconfiguration direction is along the plane containing the metamaterial resonators and is perpendicular to the direction of THz wave propagation. For out-of-plane MRMs, the reconfiguration direction is perpendicular to the plane containing the metamaterial resonator and is along the propagation direction of incident THz waves. The in-plane movable microactuators allow for dynamic reshaping of the unit cell or a continuously varying lattice constant. The out-of-plane deformable microactuators provide a varying response while preserving the unit cell geometry. Based on the application demands, the desired MEMS approach can be utilized. In this chapter, we will comprehensively describe the different classes of in-plane and out-of-plane MRMs, with major focus on the various MEMS actuator designs and control mechanisms used to achieve desired deformation of the metamaterial unit cell geometry. The active control of various THz properties based on each of the MEMS approaches reported so far will also be briefly described.

13.2 In-plane MEMS reconfigurable metamaterials

The in-plane MRMs provide reconfiguration in the same plane as the metamaterial unit cell patterns, and thus provide dynamic reshaping of the unit cell or laterally shifting of the resonators with respect to the fixed counterparts. The metamaterial unit cell can be continuously reconfigured along the incident electric or magnetic field directions in the case of normally incident THz waves. This allows for continuously tunable and completely switchable metamaterial resonances. There is a wide range of microactuators that can be adopted for realizing in-plane reconfiguration. The earlier proposed and reported in-plane THz MRMs include electrostatic comb drive actuators [34–39], electroactive π-conjugate polymer-based linear actuators [40], thermally actuated "V"-shaped [41] and "U" beam actuators, piezoelectric linear actuators, mechanically stretchable polymers [42,43], and so on. In the following subsections, various types of microactuators that are used for the realization of in-plane MRMs and their corresponding tunable THz responses will be elaborated.

13.2.1 Electrostatic comb drive-based in-plane MRMs

An electrostatic comb drive is one of the most popular MEMS actuators for achieving in-plane reconfiguration. A comb drive actuator consists of a pair of interdigitated comb finger sets, where one set is fixed to the substrate, and the other is suspended and made movable. Upon applying a voltage between these two sets of comb fingers, an electrostatic force is generated that will displace the suspended set of comb fingers in the lateral direction. When the voltage is removed, the movable comb finger set will return to the original position. Usually, for the electrostatically-driven in-plane MRMs, the movable part of the metamaterial is housed in a suspended frame which is attached to a comb drive actuator on either side. This allows for both positive and negative in-plane deformation from the initial rest position and this also doubles the displacement range.

The initial work on comb drive-based in-plane MRM at THz frequencies was reported for actively switchable magnetic resonance [34]. The metamaterial unit cell was made of two SRRs, with the gap sides facing each other forming a "[ ]" shaped unit cell as shown in Figure 13.2b, and is called an "open ring state." The presence of the in-plane gap between the SRRs allowed for the excitation of a magnetic resonance in the open ring state of the MRM, for the off-normal incidence of THz waves. One of the SRRs was hosted on the movable frame, while the other was placed in a silicon (Si) island fixed to the substrate. The movable frame housing one set of SRRs was connected to two comb drive actuators on either side, as shown in Figure 13.2a. Based on the direction of actuation and the displacement, the in-plane gap between the SRRs can be reduced or increased, as shown in Figure 13.2b–d. This allowed for the experimental demonstration of active tuning of the magnetic

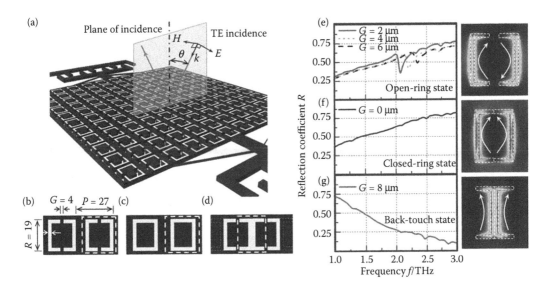

Figure 13.2 (a) Schematics of in-plane reconfigurable MRMs with electrostatic comb drive actuators. Schematic representation of the metamaterial unit cell in (b) open ring state, (c) closed ring state, and (d) back touch state. Measured magnetic response of the in-plane MRMs in (e) open ring state with gap varying from $g = 2$ μm to 6 μm, (f) closed ring state, and (g) back touch state, along with the simulated magnetic field coupled to the unit cell at the corresponding reconfiguration state. (Adapted from W. M. Zhu et al., *Advanced Materials*, 23, 1792–1796, 2011.)

resonance frequency in the open ring state with varying gaps, as shown in Figure 13.2e, and the effective permeability was altered from negative (-0.1) to positive (0.5) at 2.05 THz. With a large enough displacement along one direction, the in-plane gap between the SRRs can be completely closed to form a "[]" shape, also called the "closed ring state," as shown in Figure 13.2f. Due to the absence of a gap, the magnetic resonance is completely extinguished in this state as shown in Figure 13.2f. Similarly, along the opposite direction, the in-plane gap between the SRRs can be greatly increased, such that their backs touch each other to form the "I" shape, also called the "back touch state," as shown in Figure 13.2g, and the magnetic resonance ceases, as shown in Figure 13.2g. Hence, active switching of magnetic resonance through unit cell reshaping was experimentally demonstrated using a micromachined comb drive in the THz spectral region.

Similar works based on electrostatic comb drive driven in-plane MRMs were also reported for various polarization response controls in the THz spectral range. Linear polarization response switching was achieved by making and breaking of Maltese cross metamaterial unit cell symmetry, as shown in Figure 13.3a. This allowed for the realization of tunable THz anisotropy, which is defined as the relative response of the material for two orthogonally polarized incidences [37]. The Maltase cross unit cell structure has a four-fold rotation symmetry; hence the response of the metamaterial along orthogonally polarized THz waves will be identical. In order to achieve anisotropy, one-half of the cross along one direction is split and fabricated on the movable frame connected to comb drive actuators, as shown in Figure 13.3b. By changing the in-plane distance between the two halves of the cross structure along one direction, the symmetry of the entire unit cell can be varied as shown in Figure 13.3c–e. This subsequently provides a varying response of the metamaterial for the THz wave incident along the broken symmetry direction. While for the orthogonal polarization of the THz wave incidence, the two beams of cross structures are intact, thus the THz response does not vary. This symmetrical breaking in the Maltese cross metamaterial enables continuously tunable anisotropy in the THz spectral range. Alternatively, active switching of the polarization response of the metamaterial between a polarization independent state to a polarization dependent state was also reported by altering the lattice constant instead of the geometrical shape of the metamaterial unit cell [35]. The metamaterial consists of metallic closed microring resonators with

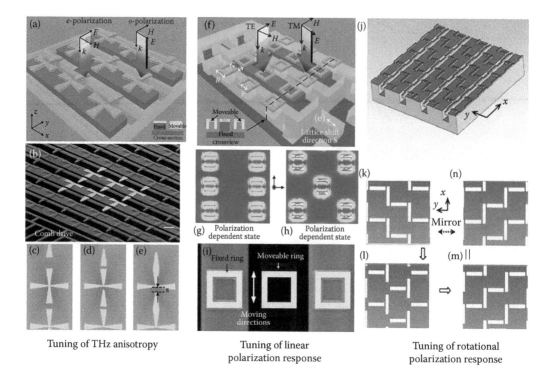

Tuning of THz anisotropy   Tuning of linear polarization response   Tuning of rotational polarization response

Figure 13.3 (a) Schematics and (b) SEM image of Maltase cross MRM for anisotropy switching, respectively. (Adapted from W. Zhu et al., *Nature Communications*, 3, 274, 2012.) Schematics of the different reconfiguration states with varying S (c) S = 0 μm, (d) S = 2.5 μm and (e) S = 5 μm, respectively. (f) Schematics of linear polarization response switching MRM, schematics of the (g) polarization-dependent state and (h) polarization-independent state of the in-plane MRM, respectively. (Adapted from W. Zhu et al., *Applied Physics Letters*, 99, 221102, 2011.) (i) Fabricated linear polarization response switching MRM showing the fixed and movable ring. (j) Schematics of the rotational polarization response switching MRM, (k) initial-state, (l) mid-state, and (m) final-state of the metamaterial; (n) mirror image of the initial-state. (Adapted from W. Zhang et al., THz polarizer using tunable metamaterials, in *Micro Electro Mechanical Systems (MEMS), 2013 IEEE 26th International Conference on*, 2013, Taipei, Taiwan, pp. 713–716.)

a periodicity of 28 μm along the x-direction and 56 μm along the y-direction, as shown in Figure 13.3f. Alternate lines of microrings are made movable by fabricating them over a movable frame, and are actuated by comb drives as shown in Figure 13.3i. In the rest position, when the x-polarized THz wave is incident on the metamaterial, the close spacing between the individual resonators causes coupling between their dipole resonances, thereby shifting the resonance frequency of the metamaterial. However, for y-polarized incidence, the closed ring resonators are placed further away, hence no coupling occurs. In this current state, the metamaterial response is polarization dependent due to the asymmetry in the metamaterial periodicity along the x- and y- directions, as shown in Figure 13.3g. By laterally displacing the alternate lines of microring resonators along the y-direction, the interpixelated geometry with $\pi/2$ rotational symmetry is achieved, as shown in Figure 13.3h. In this state, the periodicity of the unit cell along both x- and y-directions is 56 μm, hence the metamaterial provides a polarization independent response. Hence, the metamaterial can be switched between polarization dependent and independent states by varying the periodicity of the unit cell. Following these works, active control of the polarization rotation of the incident linear THz waves was also demonstrated by actively changing the geometry of the staircase metamaterial to its mirror image, as shown in Figure 13.3j [39]. A staircase structure was used as the metamaterial unit cell, and part of the unit cell was made movable by placing it on top of the comb drive driven-frame. Through lateral reconfiguration, the metamaterial unit cell could be in one of three states—up, mid-, and down states, as shown in Figure 13.3k, l and m, respectively. The up state and down state are nonsuperimposable mirror images of each other. Hence, when linearly x-polarized light is incident on the metamaterial, the polarization of the light is rotated and is measured as the transmission of the y-polarized signal with an intensity of 30% in the up and down states. However, in the mid-state, there was no rotation of polarization observed. Thus, polarization of ±20° is realized at 3.05 THz and is proposed for efficient tunable THz polarizers.

The electrostatically-driven in-plane reconfiguration approach provides a direct means of dynamically altering the coupling distance in coupled mode resonator systems. Two asymmetric split-ring resonators (ASRRs) were used as the unit cell, where one of them was fixed to the substrate, while the other was patterned on a movable frame, as shown in Figure 13.4a [44]. By adjusting the distance between the two ASRRs as shown in Figure 13.4c–e, the strength of the dipole–dipole coupling was continuously tuned. For the electric field of the incident THz wave along the gap-bearing side of the ASRRs, the resonance was switched from electrical dipole (when the two ASRRs touch each other) to magnetic dipole (LC) resonance (when there is a gap introduced between them). However, for orthogonal polarization excitation, the resonance of the metamaterial remains electrical over various reconfiguration states. Even when a gap is introduced, the induced

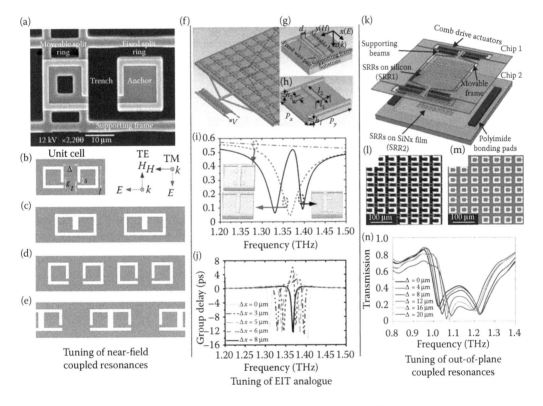

Figure 13.4 (a) Fabricated unit cell of in-plane MRM for active control of coupled mode resonance. (Adapted from Y. H. Fu et al., *Advanced Functional Materials*, 21, 3589–3594, 2011.) Schematic of the metamaterial—(b) The gaps of the ASRRS are shifted from the central line by s = 4.5 μm. (c–e) represent the unit cell in different configurations, (c) for the face-touch state (i.e., Δ = 0), (d) for the separated state (i.e., 0 < Δ < 20 μm), and (e) for the back-touch state (i.e., Δ = 20 μm). (f) Schematics of in-plane MRM for active control of EIT analogue. (Adapted from X. J. He et al., *Integrated Ferroelectrics*, 161, 85–91, 2015.) (i) Simulated transmission spectrum of the bright resonator (dash curve), dark resonator (dash-dot curve), and coupled resonator (solid curve), respectively. (j) Varying group delay of the EIT metamaterial with changing Δx from 0 μm to 8 μm. (k) Schematics of broadside-coupled SRR metamaterial, and the fabricated device in (l) misaligned state and (m) aligned state, respectively. (Adapted from X. Zhao et al., *Microsystems & Nanoengineering*, 2, 2016.) (n) Measured THz transmission response of broadside-coupled SRRs with varying alignment between the two SRR layers, Δ = 0–20 μm.

magnetic dipole in two ASRRs has a phase difference of π, thus cancelling the effective magnetic response. In-plane reconfiguration of coupled resonators using comb drives was also proposed for active control of electromagnetically induced transparency (EIT) analogue in the THz spectral region. The metamaterial unit cell consists of a cut wire resonator (CWR) acting as the bright mode and two shorter wires placed below the CWR as the dark mode, as shown in Figure 13.4f–h. The bright and dark mode resonators were coupled along the out-of-plane direction. EIT appears as a sharp transmission peak among the broader absorption spectra of the bright resonator by virtue of exciting the dark resonator through near-field coupling, as shown in Figure 13.4i [45]. The relative positioning of the bright resonator over the dark resonator is a key parameter for the active control of EIT. As the bright resonator is traversed along the dark resonator, the EIT peak strength increases with increasing asymmetry of excitation. Hence, by integrating the bright resonator on a movable frame that is connected to the comb drive actuators, the overlap distance between the bright and dark mode resonators was continuously varied from 0 µm (symmetry) to 6 µm (maximum asymmetry). Numerical calculation shows the complete modulation of the electromagnetically induced transparency analogue from 0.6 at 1.35 THz. Furthermore, the EIT analogue provides a means of achieving slow light behavior at the transparency peak, hence this metamaterial also provides an actively controllable slow light property. A maximum tunable group delay of 6 ps was achieved for the case of maximum asymmetry when the relative spacing was 6 µm, as shown in Figure 13.4j [45]. A similar approach was also reported for active control of out-of-plane coupled resonators operating in the THz spectral range. The metamaterial unit cell consists of two SRRs which are coupled in the out-of-plane (wave propagation) direction, and are called broadside coupled split ring resonators (BC-SRRs), as shown in Figure 13.4k [46]. One of the SRRs was fabricated on the fixed membrane, while the other SRR layer was fabricated on the movable frame driven by electrostatic comb drive actuators. The two SRR layers were bonded on top of each other with an air gap of 20 µm. When the SRRs are perfectly aligned, there is strong coupling between the layers, which causes the resonance to split into symmetric and antisymmetric modes. When one of the SRR layers is laterally shifted, the interactions between the SRRs weaken from the initial aligned state, as shown in Figure 13.4m, to the misaligned state, as shown in Figure 13.4l, respectively. This weakening of the coupling results in reduced splitting of the resonant modes as shown in Figure 13.4n. With the maximum lateral shift of ~20 µm, the symmetric mode blue shifts by ~60 GHz, and the antisymmetric mode red shifts by ~50 GHz, as shown in Figure 13.4n. More interestingly, a 180° phase shift is achieved at 1.08 THz between the aligned and misaligned states of the metamaterial and hence can be used as efficient THz phase shifters.

Hence, the electrostatically-driven comb drive-based in-plane MRMs have been reported for a wide range of tunable THz properties. The ease of integration of the comb drive actuators with metamaterial unit cell geometry, along with electrical control, makes it very attractive for miniaturized THz devices.

13.2.2 Electroactive polymer actuator-based in-plane MRMs

Electroactive polymer-based in-plane actuators are reported for the active reconfiguration of THz metamaterial response. The π-conjugated polymer actuator is based on an electrochemical reduction/oxidation (redox) reaction, in which doped ions move into or out of the π-conjugated polymer by applying and reversing voltages, producing mechanical deformation of the polymer in a reversible manner. Heavily doped polypyrrole (PPy) film was reported to work as a polymer actuator without an electrolyte and can be operated in ambient air. Such heavily doped PPy films are also known to behave as metals in the THz and lower frequency ranges. Two layers of complementary electrical SRRs (CESRRs) were placed in close vicinity, such that their near-fields interact through out-of-plane coupling. The resonance characteristics of the metamaterial can be tuned by altering their relative position and/or orientation with respect to each other, as shown in Figure 13.5 [40]. One of the CESRR layers is fixed on the quartz substrate, while the other CESRR layer is attached with

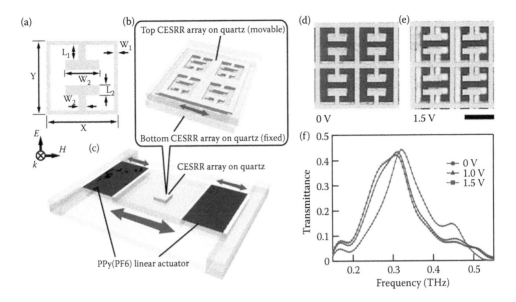

Figure 13.5 Schematics of (a) CESRR metamaterial unit cell geometry, (b) two-layer broadside-coupled resonator system, and (c) in-plane movable top CESRR layer integrated with electroactive linear actuators coupled with bottom fixed CESRR. OM image of metamaterial unit cell with (d) showing no shift at 0 V and (e) laterally shift layers at 1.5 V. (f) Measured THz transmission of the CESRR metamaterial at different applied voltage. (Adapted from X. Li et al., *Microsystem Technologies*, 19, 1145–1151, 2013.)

linear polymer actuators on either side to provide lateral reconfiguration, as shown in Figure 13.5c. By applying a few volts to one of the polymer actuators, the free metamaterial layer was laterally shifted, as shown in Figure 13.5e, from the initial aligned state shown in Figure 13.5d. The misalignment between the CSERRs caused the resonance frequency to red shift. With an applied voltage of 1.5 V, the resonance shift from 0.318 THz to 0.3 THz was experimentally observed, as shown in Figure 13.5f. By turning off the applied voltage, the system recovers its initial state and, along with the transmission response. The actuator fixed on the other side of the CESRRs can be electrically excited to achieve reconfiguration in the opposite direction. Hence, the π-conjugated polymer-based actuator enables large and repeatable linear deformation of the overlap region between the two CESRR regions. Furthermore, they require lower actuation voltages and manufacturing cost.

13.2.3 Thermally controlled in-plane MRMs

Thermal stimulus is one of the popular actuation mechanisms used for realizing microscale deformations. Thermal actuators are simpler to design and provide larger stroke and displacement, while consuming higher power. The most popular thermal actuators for achieving in-plane displacement are the V-beam actuators and U-beam actuators. They are both unimorph microstructures, and hence are made of a single material layer. Recently, THz MRMs based on V-beam and U-beam thermal actuators were proposed.

A V-beam actuator is primarily a fixed-fixed beam with a pre-displacement along the in-plane direction at the mid-point. When current is passed through the V-beam, Joule heating causes the material to expand. As the beam is fixed at both ends, the beam will displace in the lateral direction along the midpoint. When the current is removed, the V-beam will return to its original position. Active control of the THz magnetic response has been reported by employing a thermally actuated V-beam-based MRM [41,47]. The metamaterial unit cell consists of a two-cut SRR, with a metallic slab closely spaced over the gap regions. The metallic slab is attached to the mid-point of V-beam actuators. As current is passed through the V-beam actuator, the slab closes in towards the gap region of the SRR, and at the extreme case, can completely close the gap. The in-plane deformation increases as a square dependence with input voltage. Also, with increasing length of the V-beam actuators, the in-plane displacement of the V-beam actuator can be increased at a given voltage. Based on the displacement, the gap between the SRR and metallic slab can be controlled, which in turn determines the resonance frequency of the metamaterial. A large tunable frequency range from 1.374 to 1.574 THz has been theoretically achieved.

Alternatively, a U-beam actuator consists of two geometrically different arms connected in series and made of the same material. The geometrical variation causes a difference in electrical resistance between the two arms. When current is passed through the arms, the arm with the higher resistance heats up rapidly and hence deforms more, while the other beam resists this expansion, thereby causing the entire actuator to deform along the colder arm. With increasing current, the metallic slabs attached to the far end of two hot-arm actuators close the split gap in the SRR. This provides dynamic switching of the magnetic resonance [48]. Based on the input voltage, the displacement of the U-beam actuator can be increased. This displacement translates to a reduced distance between the metal slab and the SRR gap, and hence provides a tunable transmission response along with tunable magnetic permeability. Even though U-beam actuators provide a large displacement along the in-plane direction, they require high input power and have a slow response time.

13.2.4 Piezoelectrically actuated in-plane MRM

Piezoelectricity is the property of a material to produce strain when a potential difference is applied across it. Lead zirconium titanate (PZT) is one of the most popular piezoelectric materials used in a wide range of MEMS devices, owing to its large piezoelectric coefficient and relative ease of fabrication. PZT-based in-plane MRM has been proposed for active control of the THz response [49]. The pair of PZT actuators is mounted with a metal gripper and placed close to a metallic SRR. Similarly, on the other side, another pair of metal grippers supported by PZT actuators is placed closed to another SRR facing the first SRR. When a voltage is applied, the actuators latch the grippers onto the SRRs, which can now be pushed toward each other, thereby closing the air gap. With increasing voltage, the distance between the SRRs is reduced and a red shift of the SRR resonance frequency is observed. More interestingly, by reversing the applied voltage, the direction of deformation can be switched. Hence, based on the polarity of the applied voltage, the distance between the SRRs can be either increased or decreased, and this allows for both a blue shift and a red shift of the SRR resonance, accordingly. Hence, the tunable range can be doubled by employing a piezoelectric actuation mechanism. Numerical calculation shows that the resonance frequency of metamaterial can be tuned from 4.14 THz at 0 V to 4.04 THz for −200 V and 4.175 THz for +200 V, respectively [49]. Piezoelectric actuators provide continuously tunable characteristics with low operating power. However, they demand more complex processing because of the use of piezoelectric materials and the need for higher voltages, with relatively smaller displacements compared to electrostatic mechanisms.

13.2.5 Mechanically controlled in-plane MRMs

This class of in-plane MRMs takes advantage of the flexible nature of the substrate on which the metamaterial patterns are fabricated. The elasticity of the substrates allows for change in the metamaterial resonator geometry or coupling distance between them due to applied strain. This allows for a large tunable range for the THz response. In earlier reported work, a generic wrinkled layout with a honeycomb-shaped metamaterial was fabricated in a polydimethylsiloxane (PDMS) layer for mechanical tuning of the THz electrical resonance, as shown in Figure 13.6a–b [49,50]. The PDMS backing layer was prestressed and the metallic honeycomb structure was fabricated on top. When the PDMS is relaxed, the entire structure wrinkles, thereby bringing the resonators closer to each other as shown in Figure 13.6c, and its corresponding zoomed-in image is shown in Figure 13.6d.

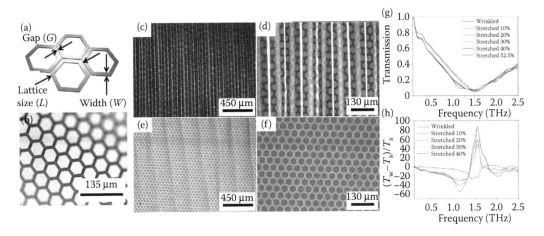

Figure 13.6 (a) Honeycomb metamaterial unit cell with geometrical parameter definition and (b) the fabricated metamaterial. OM images of reversibly wrinkled (c,d) and stretched (e,f) honeycomb metamaterials. OM images of (d) and (f) are magnified images of (c) and (e), respectively. (g) Measured THz transmission response of the metamaterial at varying stretched position and (h) relative changes in the transmission of wrinkled honeycomb metamaterials with varying strain ratios normalized to the transmission of fully stretched honeycomb metamaterial. (Adapted from S. Lee et al., *Advanced Materials*, 24, 3491–3497, 2012.)

As the PDMS layer is stretched, the distance between the honeycomb metamaterial unit cells greatly increases, as shown in Figure 13.6e, and its corresponding zoomed-in image is shown in Figure 13.6f. This increase in the interspacing distance between resonators reduces their coupling and causes the resonance to shift accordingly. The THz transmission response of stretchable metamaterial at varying strains is shown in Figure 13.6g. A PDMS stretchability of 52.5% was experimentally demonstrated, and the relative THz transmission was modulated by 90%, as shown in Figure 13.6h. The fabricated devices had fully reversible stretchability and compressibility. Following this work, a closely coupled "I" shaped metamaterial unit cell fabricated on a highly elastomeric substrate was reported for continuous tunability of the resonance frequency through a small applied strain, as shown in Figure 13.7a–d

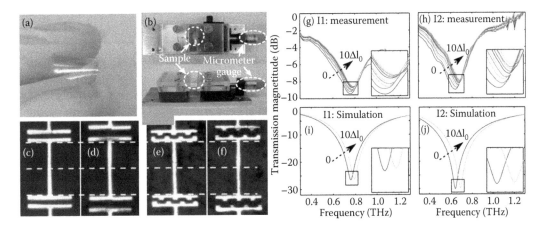

Figure 13.7 (a) Fabricated stretchable in-plane coupled metamaterial with inset showing two patterns—"I" shaped and "I" shaped with corrugated edges. (B0 shows the physical change in coupling distance between the resonators under 0 strain and increased strain. Measured transmission spectra at varying strain of the metamaterial formed with "I" shaped patterns under (c) no strain and (d) applied strain and "I" shape patterns with corrugated edges under (e) no strain and (f) applied strain, respectively. (Adapted from J. Li et al., *Applied Physics Letters*, 102, 121101, 2013.)

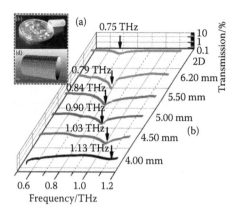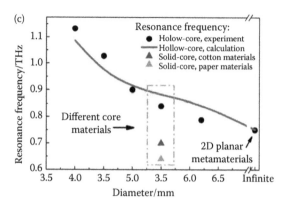

Figure 13.8 (a) Fabricated THz metamaterial tube, (b) measured THz transmission spectra of the tube with varying radius of curvature as schematically shown in (d) and (c) resonance frequency shift of THz waves traveling along the inside of the tube with respect to hollow tube diameter change and different filling materials. (Adapted from C. Zaichun et al., *Advanced Materials*, 24, OP143–OP147, 2012.)

[43]. By adding interdigitated gaps to the resonators, as shown in Figure 13.7e–f, a higher Q-factor and larger frequency shift were experimentally obtained, as shown in Figure 13.7g–j. The response of the metamaterials is fully recoverable in repeated stretching cycles after an initial priming cycle.

Alternatively, the metamaterials fabricated on a bendable substrate, such as polyethylene terephthalate (PET), were reported to show continuously tunable THz properties, as shown in Figure 13.8a [42]. As the diameter of the cylinder made of SRR on the PET substrate was reduced, a strong blue shift of the resonance frequency was observed, as shown in Figure 13.8b. The resonance frequency could also be changed by altering the materials that were filled in the cylinder, due to the refractive index changes as shown in Figure 13.8c.

Hence, in-plane reconfiguration can readily provide structural reshaping of the metamaterial geometry and varying periodicity and coupling distances. However, this class of MRMs usually possesses an external actuator to provide the reconfiguration, and has a continuously varying symmetry as the state of the metamaterial is dynamically reconfigured.

13.3 Out-of-plane reconfigurable metamaterials

The alternate class of MRMs is the out-of-plane reconfigurable metamaterials, in which the microactuators deform in the direction perpendicular to the plane containing the metamaterial unit cell. Hence, for normally incident THz waves, the reconfiguration of the unit cell geometry is along the direction of incident wave propagation. MEMS offers a wide range of actuation mechanisms to realize out-of-plane reconfiguration in microscale, such as thermal bimorphs, electrostatically-actuated cantilevers, plates and beams, pneumatically actuated plates, and many more. Various reports on out-of-plane MRMs for active control of THz properties will be elaborated based on the actuation methods in the following subsections.

13.3.1 Thermally actuated out-of-plane MRMs

Out-of-plane deformation with thermal input is achieved by using bimorph actuators. The bimorph structure consists of two materials with a large difference between their coefficients of thermal expansion (CTE). One end of the bimorph structure is fixed to the substrate, while the other end is free, thus they can form a cantilever structure. When the temperature to the bimorph cantilever is increased, one of the layers tends to expand more than the other due to the relative difference in CTEs between the two materials. This forces the bimorph cantilevers to move in the out-of-plane direction. The direction of actuation (up or down) is determined by the placement of the two layers forming the bimorph. If the material with the low CTE is placed below the material with the high CTE, then with increasing temperature, the bimorph will move in the downward direction. In the case of the low CTE material placed above the high CTE material, the bimorph will deform in the upward direction with increasing temperature. These simple structures provide a large displacement upon thermal excitation and were used for the initial demonstration of reconfigurable THz metamaterials. SRR was selected as the metamaterial unit cell and was supported by bimorph cantilevers made of gold (Au) and silicon nitride (SiN) layers, as shown in Figure 13.9a–b [51]. The Si substrate underneath the SRRs were completely etched away to allow the SRRs to deform in the out-of-plane direction. A rapid thermal annealing (RTA) technique was used to apply variable temperatures up to 550°C in steps of 50°C. The fabricated SRRs were relatively flat after the release step at room temperature, as shown in Figure 13.9c. The THz waves were normally incident on the SRR samples, with the electric field along the side arm of the SRR and the magnetic field along the gap-bearing side. Therefore, in the rest position at room temperature, the LC resonance was not excited. However, as the temperature was increased, the bimorph cantilevers lifted up the SRR unit cells, as shown in Figure 13.9c. This allowed for the direct excitation of LC resonance through the component of the magnetic field that oscillated through the SRR unit cells. The strength of the LC resonance increased with increasing out-of-plane deformation of the bimorph cantilever at elevated temperatures, as shown in Figure 13.9d. This is caused due to the increasing strength of the incident magnetic field component that directly excites the LC resonance in the SRR. However, when the electric field of the incident THz waves is along the gap-bearing side of the SRR, then an electrically induced LC resonance is observed even when the SRR is flat. The resonance strength drops as the SRR is lifted up, as shown in Figure 13.9e. This is caused by the reduced strength of the electric field component exciting the LC resonance at larger out-of-plane deformations of the SRR. Out-of-plane MRMs using thermal bimorph cantilevers were reported to achieve active control of both magnetic and electrical responses at THz frequencies.

The means of providing a thermal stimulus using RTA is not suitable for miniaturized THz systems. To overcome this limitation, electrothermal actuation was reported for active out-of-plane reconfiguration of metamaterial unit cells [52]. Electrothermal actuation takes advantage of the Joule heating that occurs when current flows though the conductive layers to provide the thermal stimulus. An omega ring with an inner disk resonator was used as the metamaterial unit cell, as shown in the inset of Figure 13.10a. The omega ring was released and was bent up due to residual stress in the bimorph layer made of 500 nm aluminum (Al) on top of 50 nm aluminum

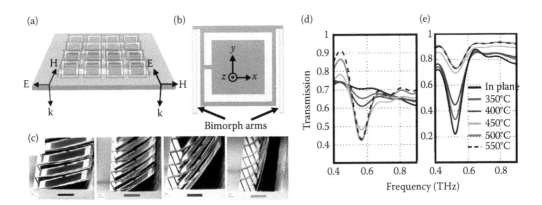

Figure 13.9 (a) Schematic drawing of thermally actuated out-of-plane deformable SRR metamaterial and (b) top view of its unit cell. (c) Out-of-plane reconfigurable metamaterial supported by bimorph beams at increasing temperature shows increased deformation. The measured THz transmission of MRM at different temperature for (d) *Ex* excitation and (e) *Ey* excitation, respectively, and (d) schematics of MRM unit cell. (Adapted from H. Tao et al., *Physical Review Letters*, 103, 147401, 2009.)

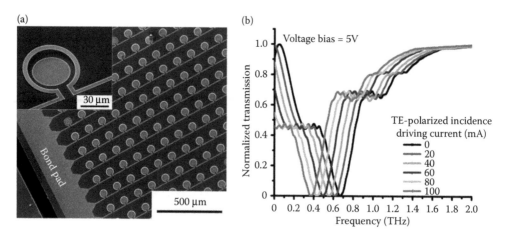

Figure 13.10 (a) Fabricated omega-shaped out-of-plane reconfigurable metamaterial with inset showing the unit cell and (b) shows the normalized transmission at varying input current. (Adapted from C. P. Ho et al., *Applied Physics Letters*, 104, 161104, 2014.)

oxide (Al_2O_3). All the unit cells are electrically connected to allow the current to pass through the Al layer. When current is passed through the Al lines, Joule heating occurs. Since the CTE of Al_2O_3 is much lower than that of Al, the released omega ring deforms toward the Si substrate upon heating. This change in gap between the Al layer and Si substrate causes the effective out-of-plane capacitance of the resonator to increase and the resonance frequency to red shift accordingly. The resonance frequency of the metamaterial red shifted from 0.7 to 0.35 THz with an input current of 100 mA, as shown in Figure 13.10b. The thermal actuation offers continuously tunable characteristics, but requires higher operating power and has a slower response time.

13.3.2 Electrostatically-actuated out-of-plane MRMs

Electrostatic actuation is one of the most popular MEMS approaches to achieve out-of-plane reconfiguration, due to its lower operational power, faster response time, and ease of fabrication. The generic electrostatic actuator for achieving out-of-plane deformation consists of a pair of electrodes with an air gap between them. One of the electrodes is kept fixed, while the other is made movable. When voltage is applied between these electrodes, the attractive electrostatic force pulls the movable electrode toward the fixed electrode. This mechanical deformation creates the restoring force, which acts in the opposite direction relative to the electrostatic force. Hence, at a given voltage, the position of the electrostatic actuator is determined as the point where the electrostatic force and restoring force balance each other. However, when the applied voltage increases, the electrostatic force increases much more rapidly than the restoring force. This causes the movable microstructure to snap down to the fixed electrode by completely closing the air gap between them and is known as the "pull-in" effect [53,54]. When the voltage is removed, the microstructures will return to their original position provided no stiction occurs. Geometrically, the movable microstructure can be a cantilever, fixed-fixed beam or diaphragm. The microcantilever is the simplest MEMS structure, and is fixed at one end and released at the other end. The fixed-fixed beam is fixed at both ends and released in the middle, while the diaphragm is fixed along the circumference and movable in the central part. The simplicity of electrostatic actuators along with the ease of integration with metamaterial unit cells has accelerated the research progress in the field of THz out-of-plane MRMs.

A simple microcantilever array was characterized as an out-of-plane MRM and was termed as a microcantilever metamaterial (MCMM) [55]. The periodicity of the MCMM was $P = 120\ \mu m$, cantilever length, $lc = 60\ \mu m$, and width, $wc = 5\ \mu m$, as shown in Figure 13.11a. Each cantilever was a

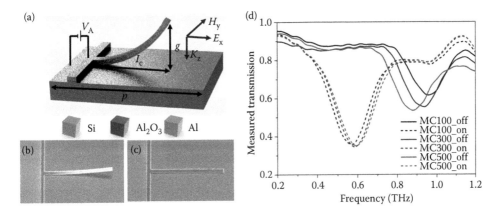

Figure 13.11 (a) Schematic representation of microcantilever metamaterial (MCMM) unit cell with air gap between the microcantilever and Si substrate. (b) SEM image of MCMM in OFF state and (c) SEM image of MCMM in ON state. (d) Measured THz transmission spectra for THz waves incident with electric field along the cantilever length for fabricated devices MC100, MC300, and MC500 devices in OFF (solid lines) and ON states (dashed lines), respectively. (Adapted from P. Pitchappa et al., *Journal of Microelectromechanical Systems*, 24, 525–527, 2015.)

bimorph structure formed by Al of t_{Al} thickness on top of an Al_2O_3 dielectric layer of thickness t_d. The microcantilever array was fabricated on an 8-inch Si wafer using a complementary metal oxide semiconductor (CMOS) compatible process with silicon dioxide (SiO_2) as the sacrificial layer. After the vapor hydrofluoric acid (VHF) release step, the bimorph cantilevers bend up due to the residual stress in the Al/Al_2O_3 layers, with a maximum air gap "g" between the tip of the released cantilevers and Si substrate, as shown in Figure 13.11a. This air gap between the Al layer and Si substrate contributes to the effective capacitance of the metamaterial. When voltage is applied across the Al layer of the microcantilever and Si substrate, the air gap is reduced. This causes the effective capacitance to increase, thereby leading to the red shift of the metamaterial resonance frequency. At the extreme case, when pull-in occurs, the air gap between the cantilever and Si substrate is completely closed. The thin layer of the Al_2O_3 dielectric layer prevents the electrical short between the Al layer and Si substrate at pull-in. This layer also had the added advantage of curving up the released cantilever upon release, and this significantly increases the tuning range of these MCMMs. Furthermore, the thickness of Al can also be reduced to enhance the initial tip displacement of the microcantilevers. This further increases the tunable range for the resonance frequency of the MCMMs. For the experiments, three thicknesses of Al layers were selected, $t_{Al} = 500$, 300, and 100 nm, and the MCMMs were termed as MC500, MC300, and MC100, respectively. Based on the position of the cantilevers, two states for MCMM are defined. The state when the cantilevers are released and no voltage is applied is termed the "OFF" state, and the state in which the cantilevers are in physical contact with the Si substrate after applying pull-in voltage (V_{PI}) is termed the "ON" state. The SEM images of the MCMM in the OFF and ON states are shown in Figure 13.11b and c, respectively. The initial gaps for MC500, MC300, and MC100 were measured as approximately 5.5, 10.2, and 20.2 μm, respectively. The OFF state resonance frequencies of MC500, MC300, and MC100 were observed at 0.88 THz, 0.935 THz, and 0.96 THz, respectively, as shown in Figure 13.11d. The red shift in resonance with the thicker Al layer in the OFF state is due to the decreased initial tip displacement of the released microcantilever. When the MCMM is switched to the ON state, the resonance frequency red shifts to around 0.6 THz for all three MCMMs, and a maximum modulation of 60% was achieved.

The simplicity of design and ease of integration of microcantilevers with various metamaterial unit cell geometries allowed for the realization of MCMMs for active control of various THz properties.

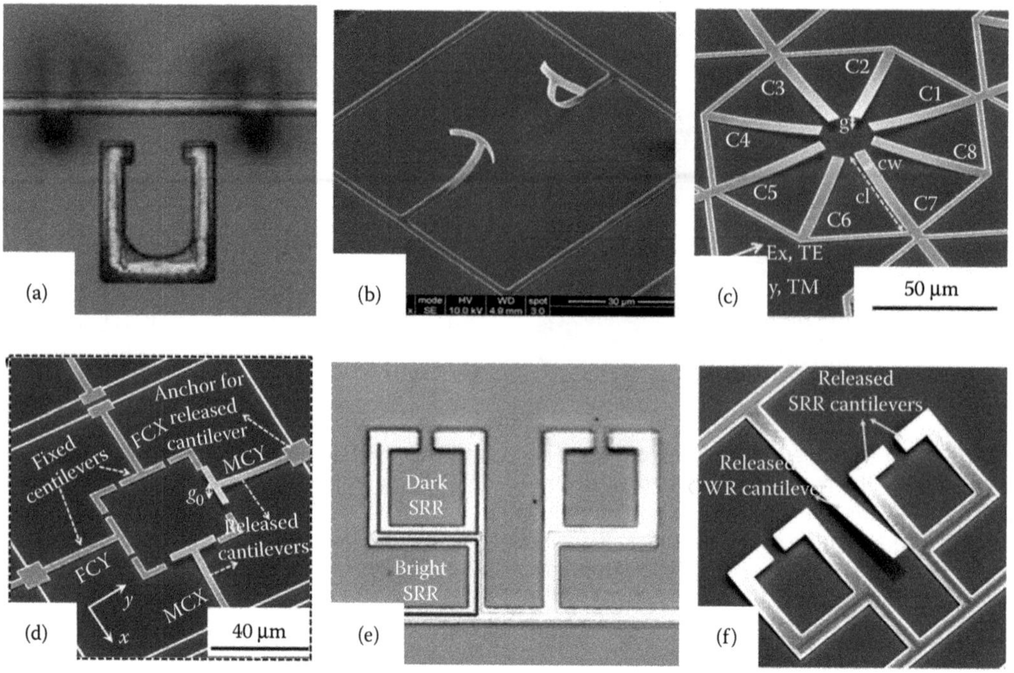

Figure 13.12 Electrostatically-actuated out-of-plane THz MRMs unit cells for active control of various THz properties—(a) SRR for magnetic response. (Adapted from Y.-S. Lin et al., *Applied Physics Letters*, 102, 111908, 2013.) (b) ESRR for electrical response. (Adapted from F. Ma et al., *Applied Physics Letters*, 102, 161912, 2013.) (c) Eight cantilever metamaterial for polarization independent resonance switching. (Adapted from P. Pitchappa et al., *Scientific Reports*, 5, 11678, 2015.) (d) Anisotropy switching. (Adapted from P. Pitchappa et al., *Advanced Optical Materials*, 19, 391–398, 2015.) (e) Conductively coupled resonator for active control of near-field coupling. (Adapted from P. Pitchappa et al., *Applied Physics Letters*, 108, 111102, 2016.) (f) Inductively coupled resonators for active control of EIT analogue. (Adapted from P. Pitchappa et al., *Advanced Optical Materials*, 4(4), 541–547, 2016.)

The unit cell geometries of various MCMMs are shown in Figure 13.12. The microcantilevers were integrated into the double SRR unit cells to achieve active control of the magnetic response, as shown in Figure 13.12a [56]. The side arms, along with the tip bearing side was released, while the inner SRR was kept fixed to the substrate. When voltage is applied across the cantilevers and Si substrate, the outer SRR cantilevers start to move toward the substrate and this increases the coupling between the outer and inner SRRs. This causes a strong red shift in the resonance frequency of the SRRs. Similarly, the microcantilevers integrated into the ESRR unit cell, as shown in Figure 13.12b, allowed for the dynamic control of electrical LC resonance at THz frequencies [57,62]. Furthermore, two "T" shaped released microcantilevers facing each other were demonstrated for active control of electrical resonance with improved tunable range [63]. The improvement in tunable range was achieved due to the additional change in coupling strength between the two cantilevers, along with the capacitance changes at varying reconfiguration states of the microcantilevers. More interestingly, the presence of an active element at the unit cell allowed for localization of the control signals. The enhancement in controllability was demonstrated by interpixelating SRR and ESRR unit cells to form a supercell with isolated electrical controls. This enabled the independent switching of magnetic and electrical responses using a single metamaterial and was termed as "multifunctional metamaterials" [64].

One of the key features desired for manipulation of EM waves is the polarization control. MCMMs have also been reported for advanced polarization control, such as polarization independent resonance switching, and linear and rotational polarization response switching. In order to achieve polarization independent resonance switching, the metamaterial unit cell should have rotational symmetry and the direction of reconfiguration should be along the wave propagation direction in order to the preserve the symmetry at all reconfiguration states. Hence, an MCMM with eight cantilevers placed at each corner of the octagon ring was reported to achieve active switching of dual band electrical resonances that were identical for both x-polarized and y-polarized THz incident waves, as shown in Figure 13.12c [58]. Furthermore, in order to achieve complete linear polarization response switching, the symmetry of the unit cells was made or broken on demand, along orthogonal directions. The possibility of isolating control at the subunit cell level in MCMM allowed for the demonstration of complete THz anisotropy switching [59]. The metamaterial unit cell for anisotropy switching consists of four cantilevers that are placed with $\pi/2$ rotation symmetry as shown in Figure 13.12d. Two of the adjacent cantilevers are fixed to the Si substrate at all times. The other two cantilevers are released and electrically isolated. This isolation in control allows for the making and breaking of symmetry along both the x-direction and y-direction, independently. Thus, the metamaterial anisotropy was switched between values below, equal to, and above unity by selectively reconfiguring two orthogonally placed microcantilevers. Furthermore, rotation polarization response control was also reported using electrostatically-controlled, out-of-plane reconfigurable spiral structures [65]. By displacing the right helical or left helical spirals from the initial flat position, the interaction of incident right circularly polarized light or left circularly polarized light was actively controlled. The amplitudes of the differential polarization azimuth rotation and the differential ellipticity angles reached 6° and 4°, respectively, at an applied voltage of 350 V. The proposed metamaterial thus exhibited tunable optical activity in the frequency range of 0.4 to 2.3 THz.

The integration of microcantilevers within the metamaterial unit cell greatly enabled the possibility of achieving active control of near-field coupling in multi-resonator systems. Near-field coupled resonators provide an access route for the realization of electromagnetically induced transparency (EIT) analogue and slow light effects in the THz spectral range as mentioned earlier. To actively control the near-field coupling using out-of-plane reconfiguration, two SRRs, rotated by 90° with respect to each other with conductive coupling, were proposed, as shown in Figure 13.12e. The electric field of the incident THz waves directly excited the fixed bright SRR, while the conductive coupling indirectly excited the LC resonance in the dark SRR. The side arms, along with the tip of the dark SRR, were released to achieve active tuning of near-field coupling. In the ON state, the resonance frequencies of the SRRs matched, and hence allowed for the excitation of the EIT peak. However, when the cantilevers are in the OFF state, the resonance frequency of the dark SRR blue shifts and this causes the EIT peak to be completely modulated [60]. More interestingly, an inductively coupled system was demonstrated as shown in Figure 13.12f, where a single cut wire resonator (CWR) acts as the bright mode and was released. Two SRRs placed on either side of the CWR act as a dark mode which is excited by virtue of the bright CWR. The side arms of the SRR were also released and electrical signals to the CWR and SRR cantilevers were isolated. It was observed that when the CWR and SRR are in the ON state, the EIT peak was observed due to strong coupling of the dark SRRs to the bright CWR. When only the CWR was in an OFF state, its resonance frequency blue shifts and the EIT peak was modulated. In the case where the CWR was in the ON state and the SRR was switched to the OFF state, the spectral shift in EIT was observed, and finally, when the CWR and SRR were in the OFF state, the system went to an uncoupled state. Hence, a single out-of-plane MRM was reported to achieve strong EIT excitation, EIT modulation, and EIT spectral tuning based on the various reconfiguration states of the metamaterial [61]. Even though the MCMM offers a rich variety of tunable THz properties, the presence of the Si substrate absorbs nearly half the incident THz waves. Hence, a quartz-based out-of-plane MRM was reported [66,67]. It consists of a pair of metallic plates—one fixed to the quartz substrate, while the other was made movable by applying voltage between them, as shown in Figure 13.13a–c. Due to the high transparency and low loss of quartz that was used as the substrate, the MRM exhibits a high contrast switching performance of 16.5 dB at 0.48 THz, as shown in Figure 13.13d.

Electrostatic actuation was also reported for active tuning of wavelength selective THz absorption. The metamaterial absorber is a trilayer structure, consisting of a continuous bottom metal and top metamaterial pattern with a dielectric layer. The bottom metal acts as the reflector and prevents any

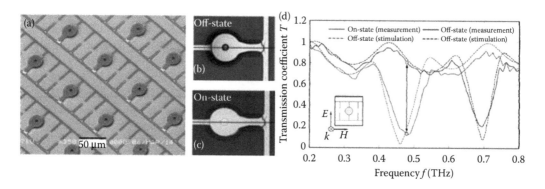

Figure 13.13 (a) OM image of the fabricated electrostatically-reconfigurable microcantilevers after release step, (b) unit cell in OFF state, (c) unit cell in ON state, and (d) measured THz transmission response in ON and OFF states. (Adapted from Z. Han et al., *IEEE Journal of Selected Topics in Quantum Electronics*, 21, 1–9, 2015.)

transmission through the device. The dielectric layer acts as the spacer between the two layers. Hence, by releasing the metamaterial patterns, an out-of-plane air gap can be achieved in between the bottom reflector and the metamaterial layers that act as the spacer layer. Through electrostatic actuation, the air gap, and hence the effective spacer thickness, along with angle of incidence can be actively altered. As the air gap was reduced with increasing voltage, the peak absorption frequency red shifted accordingly. [68].

Furthermore, modulation of THz transmission over a broad spectral range was reported using electrostatically-driven out-of-plane MRM. Broadband modulation is highly desired for high speed communication channels. This was experimentally reported by having two layers of metallic mesh structures, which could be actively brought into contact and non-contact modes, as shown in Figure 13.14a–c [69,70]. In the noncontact mode, capacitive (low pass filter) behavior is achieved, and in the contact mode, it becomes inductive (high pass filter). The top mesh filter consists of metallic anchors and moving membranes of an array of vertically oriented (along the *y*-axis) multi-contact MEMS switches in a fixed-fixed beam configuration. The bottom mesh filter consists of an array of vertical metallic slits on an Si substrate with arrays of horizontally oriented (along the *x*-axis) metallic patches extended outward to serve as the contact pads of the multi-contact MEMS switches. By applying voltage across the Si substrate and suspended top mesh switches, the response of the metamaterial was switched from low filtering to higher filtering behavior and vice versa, as shown in Figure 13.14d–e. The reported MRM provided a high modulation depth of more than 70% over a broad spectral range of 0.1–1.5 THz, with a switching voltage of 30 V and modulation speeds exceeding 20 kHz.

13.3.3 Pneumatically tunable metamaterials

Pneumatic actuators involve the use of pressure differences to deform the released microstructures. The pneumatic actuation mechanism has the advantage of enabling bidirectional actuation by reversing the gradient of pressure difference, which is highly critical for many applications. Furthermore, no metal interconnect lines are required, hence it does not cause any interference with the metamaterial resonances. They also provide much larger deformations compared to other actuation mechanisms. These critical advantages have been exploited to displace one of the near-field coupled SRRs along the out-of-plane direction [72]. This allowed for continuous tuning of the near-field coupling and the effective resonant frequency of the metamaterial. Recently, the bidirectional reconfiguration capability of pneumatic actuation was utilized for changing the handedness of the MEMS spiral structure to achieve dynamically tunable optical activity in the THz spectral region, as shown in Figure 13.15a–d [71]. The released spiral structures were in-plane and no optical

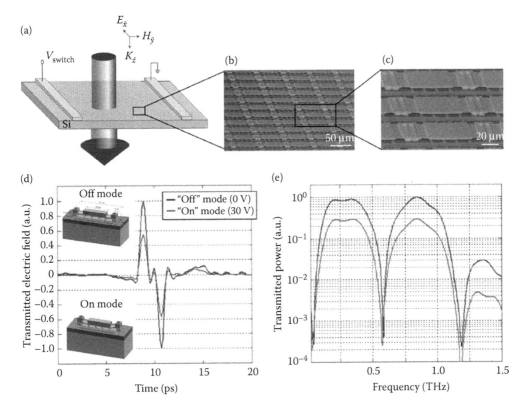

Figure 13.14 (a) Schematics of out-of-plane reconfigurable fixed-fixed beam-based metamaterial with (b) and (c) zoomed-in images of the fabricated metamaterial. (d) shows the dynamic change in THz pulse in OFF and ON state of the beams, and (e) measured broadband modulation characteristics of the MEMS metamaterial. (Adapted from M. Unlu and M. Jarrahi, *Optics Express*, 22, 32245–32260, 2014; M. Unlu et al., *Scientific Reports*, 4, 5708, 2014.)

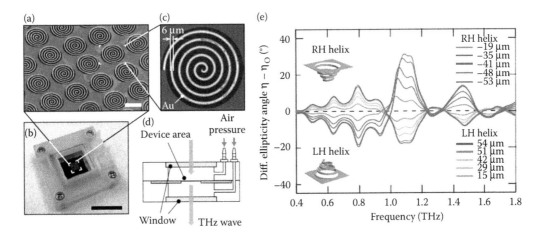

Figure 13.15 (a) Fabricated chiral MEMS metamaterial array, (b) corresponding unit cell, (c) assembled MEMS metamaterial with pneumatic actuation stage. (d) shows the working principle of the pneumatic actuation to achieve LH or RH helix and (e) shows the measured ellipticity of the chiral metamaterial with varying deformation of LH and RH helices. (Adapted from T. Kan et al., *Nature Communications*, 6, 8422, 2015.)

activity was observed. However, by changing the pressure difference across the top and bottom surfaces of the released structures, the upward spiral and downward spiral structures with continuously varying heights were experimentally demonstrated, as shown in Figure 13.15e. Hence, the response of the metamaterial can be switched between right and left circularly polarized THz wave incidence. This allowed for the 28° polarization rotation of incident THz waves.

The out-of-plane reconfigurable microactuator offers a wide range of performance characteristics with varied control stimulus that can be readily integrated with the metamaterial unit cell geometry. One of the most interesting features of the out-of-plane MRMs is that the actuators are integrated directly into each unit cell of the metamaterial, unlike their in-plane reconfigurable counterpart, where the actuators are placed externally. This allows for potentially isolated control of a unit cell, which is critical for the realization of dynamically programmable THz metamaterials. Furthermore, the sizes of out-of-plane reconfigurable actuators are much smaller and thus, offer a faster response time.

13.4 Conclusions

Structural reconfiguration of metamaterial unit cell geometry is the most direct and straightforward means of achieving tunable THz properties. MEMS offers a large variety of microactuators with varied performance characteristics to enable deformation at the microscale. The reconfiguration direction can be either in-plane or out-of-plane, and the displacement can be readily controlled through varied control stimuli, such as electrical, thermal, mechanical, magnetic or pneumatic. Electrical control is more suitable for miniaturized systems, where the metamaterial can be integrated with ASICs, while thermal and pneumatic offer large and continuous tunable ranges. Similarly, the in-plane reconfiguration is more suitable for dynamic reshaping of the unit cell geometry, while the out-of-plane reconfiguration aids in preserving the symmetry of the unit cell. More interestingly, the out-of-plane reconfigurable cantilever offers isolation of control at the unit cell level, which is highly critical for the realization of programmable THz metamaterials. MEMS-based structurally reconfigurable THz metamaterials are highly miniaturized, versatile, fabricated using scalable technology, easily integrated with ICs, and provide enhanced electro-optic performance. Thus, MEMS and THz metamaterials are perfectly complementing technologies that will allow for the realization of next generation high performance THz devices.

References

1. J. B. Pendry, A. J. Holden, D. Robbins, and W. Stewart, Magnetism from conductors and enhanced nonlinear phenomena, *Microwave Theory and Techniques, IEEE Transactions on*, 47, 2075–2084, 1999.
2. R. A. Shelby, D. R. Smith, and S. Schultz, Experimental verification of a negative index of refraction, *Science*, 292, 77–79, 2001.
3. N. Fang, H. Lee, C. Sun, and X. Zhang, Sub–diffraction-limited optical imaging with a silver superlens, *Science*, 308, 534–537, 2005.
4. N. Landy, S. Sajuyigbe, J. Mock, D. Smith, and W. Padilla, Perfect metamaterial absorber, *Physical Review Letters*, 100, 207402, 2008.
5. E. Plum, V. Fedotov, and N. Zheludev, Optical activity in extrinsically chiral metamaterial, *Applied Physics Letters*, 93, 191911, 2008.
6. X. Liu, T. Tyler, T. Starr, A. F. Starr, N. M. Jokerst, and W. J. Padilla, Taming the blackbody with infrared metamaterials as selective thermal emitters, *Physical Review Letters*, 107, 045901, 2011.
7. N. Papasimakis, V. A. Fedotov, N. Zheludev, and S. Prosvirnin, Metamaterial analog of electromagnetically induced transparency, *Physical Review Letters*, 101, 253903, 2008.

8. P. Tassin, L. Zhang, T. Koschny, E. Economou, and C. M. Soukoulis, Low-loss metamaterials based on classical electromagnetically induced transparency, *Physical Review Letters*, 102, 053901, 2009.

9. S.-Y. Chiam, R. Singh, C. Rockstuhl, F. Lederer, W. Zhang, and A. A. Bettiol, Analogue of electromagnetically induced transparency in a terahertz metamaterial, *Physical Review B*, 80, 153103, 2009.

10. P. Tassin, L. Zhang, T. Koschny, E. Economou, and C. M. Soukoulis, Planar designs for electromagnetically induced transparency in metamaterials, *Optics Express*, 17, 5595–5605, 2009.

11. N. Liu, L. Langguth, T. Weiss, J. Kästel, M. Fleischhauer, T. Pfau et al., Plasmonic analogue of electromagnetically induced transparency at the Drude damping limit, *Nature Materials*, 8, 758–762, 2009.

12. C. Sirtori, Applied physics: Bridge for the terahertz gap, *Nature*, 417, 132–133, 2002.

13. N. Katsarakis, T. Koschny, M. Kafesaki, E. Economou, and C. Soukoulis, Electric coupling to the magnetic resonance of split ring resonators, *Applied Physics Letters*, 84, 2943–2945, 2004.

14. W. Padilla, M. Aronsson, C. Highstrete, M. Lee, A. Taylor, and R. Averitt, Electrically resonant terahertz metamaterials: Theoretical and experimental investigations, *Physical Review B*, 75, 041102, 2007.

15. W. J. Padilla, A. J. Taylor, C. Highstrete, M. Lee, and R. D. Averitt, Dynamical electric and magnetic metamaterial response at terahertz frequencies, *Physical Review Letters*, 96, 107401, 2006.

16. H.-T. Chen, J. F. O'Hara, A. K. Azad, A. J. Taylor, R. D. Averitt, D. B. Shrekenhamer et al., Experimental demonstration of frequency-agile terahertz metamaterials, *Nature Photonics*, 2, 295–298, 2008.

17. J. Gu, R. Singh, X. Liu, X. Zhang, Y. Ma, S. Zhang et al., Active control of electromagnetically induced transparency analogue in terahertz metamaterials, *Nature Communications*, 3, 1151, 2012.

18. D. R. Chowdhury, R. Singh, A. J. Taylor, H.-T. Chen, and A. K. Azad, Ultrafast manipulation of near field coupling between bright and dark modes in terahertz metamaterial, *Applied Physics Letters*, 102, 011122, 2013.

19. W. Cao, R. Singh, C. Zhang, J. Han, M. Tonouchi, and W. Zhang, Plasmon-induced transparency in metamaterials: Active near field coupling between bright superconducting and dark metallic mode resonators, *Applied Physics Letters*, 103, 101106, 2013.

20. X. Su, C. Ouyang, N. Xu, S. Tan, J. Gu, Z. Tian et al., Dynamic mode coupling in terahertz metamaterials, *Scientific Reports*, 5, 2015.

21. Q.-Y. Wen, H.-W. Zhang, Q.-H. Yang, Y.-S. Xie, K. Chen, and Y.-L. Liu, Terahertz metamaterials with VO2 cut-wires for thermal tunability, *Applied Physics Letters*, 97, 1111, 2010.

22. B. Jin, C. Zhang, S. Engelbrecht, A. Pimenov, J. Wu, Q. Xu et al., Low loss and magnetic field-tunable superconducting terahertz metamaterial, *Optics Express*, 18, 17504–17509, 2010.

23. C. Kurter, P. Tassin, L. Zhang, T. Koschny, A. P. Zhuravel, A. V. Ustinov et al., Classical analogue of electromagnetically induced transparency with a metal-superconductor hybrid metamaterial, *Physical Review Letters*, 107, 043901, 2011.

24. J. Wu, B. Jin, J. Wan, L. Liang, Y. Zhang, T. Jia et al., Superconducting terahertz metamaterials mimicking electromagnetically induced transparency, *Applied Physics Letters*, 99, 161113, 2011.

25. H.-T. Chen, W. J. Padilla, J. M. Zide, A. C. Gossard, A. J. Taylor, and R. D. Averitt, Active terahertz metamaterial devices, *Nature*, 444, 597–600, 2006.

26. L. Ju, B. Geng, J. Horng, C. Girit, M. Martin, Z. Hao et al., Graphene plasmonics for tunable terahertz metamaterials, *Nature Nanotechnology*, 6, 630–634, 2011.

27. S. H. Lee, M. Choi, T.-T. Kim, S. Lee, M. Liu, X. Yin et al., Switching terahertz waves with gate-controlled active graphene metamaterials, *Nature Materials*, 11, 936–941, 2012.

28. Y. Cao, S. Gan, Z. Geng, J. Liu, Y. Yang, Q. Bao et al., Optically tuned terahertz modulator based on annealed multilayer MoS2, *Scientific Reports*, 6, 22899, 2016.
29. X. Yang, J. Yang, X. Hu, Y. Zhu, H. Yang, and Q. Gong, Multilayer-WS2: Ferroelectric composite for ultrafast tunable metamaterial-induced transparency applications, *Applied Physics Letters*, 107, 081110, 2015.
30. D. Shrekenhamer, J. Montoya, S. Krishna, and W. J. Padilla, Four-color metamaterial absorber THz spatial light modulator, *Advanced Optical Materials*, 1, 905–909, 2013.
31. D. Shrekenhamer, C. M. Watts, and W. J. Padilla, Terahertz single pixel imaging with an optically controlled dynamic spatial light modulator, *Optics Express*, 21, 12507–12518, 2013.
32. D. Shrekenhamer, W.-C. Chen, and W. J. Padilla, Liquid crystal tunable metamaterial absorber, *Physical Review Letters*, 110, 177403, 2013.
33. Z. Liu, C.-Y. Huang, H. Liu, X. Zhang, and C. Lee, Resonance enhancement of terahertz metamaterials by liquid crystals/indium tin oxide interfaces, *Optics Express*, 21, 6519–6525, 2013.
34. W. M. Zhu, A. Q. Liu, X. M. Zhang, D. P. Tsai, T. Bourouina, J. H. Teng et al., Switchable magnetic metamaterials using micromachining processes, *Advanced Materials*, 23, 1792–1796, 2011.
35. W. Zhu, A. Liu, W. Zhang, J. Tao, T. Bourouina, J. Teng et al., Polarization dependent state to polarization independent state change in THz metamaterials, *Applied Physics Letters*, 99, 221102, 2011.
36. W. Zhang, A. Liu, W. Zhu, E. Li, H. Tanoto, Q. Wu et al., Micromachined switchable metamaterial with dual resonance, *Applied Physics Letters*, 101, 151902, 2012.
37. W. Zhu, A. Liu, T. Bourouina, D. Tsai, J. Teng, X. Zhang et al., Microelectromechanical Maltese-cross metamaterial with tunable terahertz anisotropy, *Nature Communications*, 3, 274, 2012.
38. W. Zhang, W. M. Zhu, H. Cai, M.-L. J. Tsai, G.-Q. Lo, D. P. Tsai et al., Resonance switchable metamaterials using MEMS fabrications, *IEEE Journal of Selected Topics in Quantum Electronics*, 19, 4700306–4700306, 2013.
39. W. Zhang, W. Zhu, J. M. Tsai, G.-Q. Lo, D. L. Kwong, E. Li et al., Thz polarizer using tunable metamaterials, in *Micro Electro Mechanical Systems (MEMS), 2013 IEEE 26th International Conference on*, 2013, Taipei, Taiwan pp. 713–716.
40. T. Matsui, Y. Inose, D. A. Powell, and I. V. Shadrivov, Electroactive tuning of double-layered metamaterials based on π-conjugated polymer actuators, *Advanced Optical Materials*, 2015.
41. X. Li, T. Yang, W. Zhu, and X. Li, Continuously tunable terahertz metamaterial employing a thermal actuator, *Microsystem Technologies*, 19, 1145–1151, 2013.
42. C. Zaichun, M. Rahmani, G. Yandong, C. T. Chong, and H. Minghui, Realization of variable three-dimensional terahertz metamaterial tubes for passive resonance tunability, *Advanced Materials*, 24, OP143–OP147, 2012.
43. J. Li, C. M. Shah, W. Withayachumnankul, B. S.-Y. Ung, A. Mitchell, S. Sriram et al., Mechanically tunable terahertz metamaterials, *Applied Physics Letters*, 102, 121101, 2013.
44. Y. H. Fu, A. Q. Liu, W. M. Zhu, X. M. Zhang, D. P. Tsai, J. B. Zhang et al., A micromachined reconfigurable metamaterial via reconfiguration of asymmetric split-ring resonators, *Advanced Functional Materials*, 21, 3589–3594, 2011.
45. X. J. He, Q. X. Ma, P. Jia, L. Wang, T. Y. Li, F. M. Wu et al., Dynamic manipulation of electromagnetically induced transparency with MEMS metamaterials, *Integrated Ferroelectrics*, 161, 85–91, 2015.
46. X. Zhao, K. Fan, J. Zhang, G. R. Keiser, G. Duan, R. D. Averitt et al., Voltage-tunable dual-layer terahertz metamaterials, *Microsystems & Nanoengineering*, 2, 2016.
47. A. Lalas, N. Kantartzis, and T. Tsiboukis, Programmable terahertz metamaterials through V-beam electrothermal devices, *Applied Physics A*, 117, 433–438, 2014.
48. A. X. Lalas, N. V. Kantartzis, and T. D. Tsiboukis, Reconfigurable metamaterial components exploiting two-hot-arm electrothermal actuators, *Microsystem Technologies*, 21, 2097–2107, 2015.

49. A. Lalas, N. Kantartzis, and T. Tsiboukis, Tunable terahertz metamaterials by means of piezo-electric MEMS actuators, *EPL (Europhysics Letters)*, 107, 58004, 2014.
50. S. Lee, S. Kim, T. T. Kim, Y. Kim, M. Choi, S. H. Lee et al., Reversibly stretchable and tunable terahertz metamaterials with wrinkled layouts, *Advanced Materials*, 24, 3491–3497, 2012.
51. H. Tao, A. Strikwerda, K. Fan, W. Padilla, X. Zhang, and R. Averitt, Reconfigurable terahertz metamaterials, *Physical Review Letters*, 103, 147401, 2009.
52. C. P. Ho, P. Pitchappa, Y.-S. Lin, C.-Y. Huang, P. Kropelnicki, and C. Lee, Electrothermally actuated microelectromechanical systems based omega-ring terahertz metamaterial with polarization dependent characteristics, *Applied Physics Letters*, 104, 161104, 2014.
53. Y. Qian, L. Lou, M. J. Tsai, and C. Lee, A dual-silicon-nanowires based U-shape nanoelectro-mechanical switch with low pull-in voltage, *Applied Physics Letters*, 100, 113102, 2012.
54. P. Singh, C. G. Li, P. Pitchappa, and C. Lee, Tantalum-nitride antifuse electromechanical OTP for embedded memory applications, *Electron Device Letters, IEEE*, 34, 987–989, 2013.
55. P. Pitchappa, C. P. Ho, L. Dhakar, Y. Qian, N. Singh, and C. Lee, Periodic array of sub-wavelength MEMS cantilevers for dynamic manipulation of terahertz waves, *Journal of Microelectromechanical Systems*, 24, 525–527, 2015.
56. Y.-S. Lin, Y. Qian, F. Ma, Z. Liu, P. Kropelnicki, and C. Lee, Development of stress-induced curved actuators for a tunable THz filter based on double split-ring resonators, *Applied Physics Letters*, 102, 111908, 2013.
57. F. Ma, Y. Qian, Y.-S. Lin, H. Liu, X. Zhang, Z. Liu et al., Polarization-sensitive microelectrome-chanical systems based tunable terahertz metamaterials using three dimensional electric split-ring resonator arrays, *Applied Physics Letters*, 102, 161912, 2013.
58. P. Pitchappa, C. P. Ho, Y. Qian, L. Dhakar, N. Singh, and C. Lee, Microelectromechanically tunable multiband metamaterial with preserved isotropy, *Scientific Reports*, 5, 11678, 2015.
59. P. Pitchappa, C. P. Ho, L. Cong, R. Singh, N. Singh, and C. Lee, Reconfigurable digital metama-terial for dynamic switching of terahertz anisotropy, *Advanced Optical Materials*, 4, 391–398, 2016.
60. P. Pitchappa, M. Manjappa, C. P. Ho, Y. Qian, R. Singh, N. Singh et al., Active control of near-field coupling in conductively coupled microelectromechanical system metamaterial devices, *Applied Physics Letters*, 108, 111102, 2016.
61. P. Pitchappa, M. Manjappa, C. P. Ho, R. Singh, N. Singh, and C. Lee, Active control of elec-tromagnetically induced transparency analog in terahertz MEMS metamaterial, *Advanced Optical Materials*, 4, 541–547, 2016.
62. F. Ma, Y.-S. Lin, X. Zhang, and C. Lee, Tunable multiband terahertz metamaterials using a reconfigurable electric split-ring resonator array, *Light: Science & Applications*, 3, e171, 2014.
63. Y.-S. Lin, F. Ma, and C. Lee, Three-dimensional movable metamaterial using electric split-ring resonators, *Optics Letters*, 38, 3126–3128, 2013.
64. P. Pitchappa, C. P. Ho, L. Dhakar, and C. Lee, Microelectromechanically reconfigurable inter-pixelated metamaterial for independent tuning of multiple resonances at terahertz spec-tral region, *Optica*, 2, 571–578, 2015.
65. T. Kan, A. Isozaki, N. Kanda, N. Nemoto, K. Konishi, M. Kuwata-Gonokami et al., Spiral meta-material for active tuning of optical activity, *Applied Physics Letters*, 102, 221906, 2013.
66. Z. Han, K. Kohno, H. Fujita, K. Hirakawa, and H. Toshiyoshi, Tunable terahertz filter and mod-ulator based on electrostatic MEMS reconfigurable SRR array, *Selected Topics in Quantum Electronics, IEEE Journal of*, 21, 1–9, 2015.
67. Z. Han, K. Kohno, H. Fujita, K. Hirakawa, and H. Toshiyoshi, MEMS reconfigurable metamate-rial for terahertz switchable filter and modulator, *Optics Express*, 22, 21326–21339, 2014.
68. F. Hu, N. Xu, W. Wang, Y. e. Wang, W. Zhang, J. Han et al., A dynamically tunable terahertz metamaterial absorber based on an electrostatic MEMS actuator and electrical dipole reso-nator array, *Journal of Micromechanics and Microengineering*, 26, 025006, 2016.
69. M. Unlu and M. Jarrahi, Miniature multi-contact MEMS switch for broadband terahertz modulation, *Optics Express*, 22, 32245–32260, 2014.

70. M. Unlu, M. Hashemi, C. Berry, S. Li, S.-H. Yang, and M. Jarrahi, Switchable scattering meta-surfaces for broadband terahertz modulation, *Scientific Reports*, 4, 5708, 2014.
71. T. Kan, A. Isozaki, N. Kanda, N. Nemoto, K. Konishi, H. Takahashi et al., Enantiomeric switching of chiral metamaterial for terahertz polarization modulation employing vertically deformable MEMS spirals, *Nature Communications*, 6, 8422, 2015.
72. A. Isozaki, T. Kan, H. Takahashi, K. Matsumoto, and I. Shimoyama, Out-of-plane actuation with a sub-micron initial gap for reconfigurable terahertz micro-electro-mechanical systems metamaterials, *Opt Express*, 23, 26243–51, 2015.

SECTION III

Biomicro- and nanophotonics and optofluidics for health care applications

Optofluidic devices and their applications

Sung-Yong Park

Contents

14.1 Introduction

Optofluidics is an emerging research and technology area that combines the two disciplines of microfluidics and optics [1,2]. The integration of fluids and optics has a long history. The first effort to combine these two fields was in the late 1600s, when inventor Stephen Gray described what was called a water microscope [3]. A small hole punched through brass served as the aperture for a static water droplet. The image focusing was done by winding a thumbscrew to change the distance of an object. However, since there was no specific way to tune the lens's performance with a water droplet, the optical power of the microscope remained fixed. A tunable optical element using a liquid interface was demonstrated in 1872 [4]. To adjust the focal length, mercury contained in a small vessel was rotated, and the parabolic shape of the mercury's interface was controlled by the rotational speeds. In the early 1990s, the field of microfluidics emerged to show the potential for lab-on-a-chip applications [5] and, in the early 2000s, microfluidic elements began to be used for optical devices such as microfluidic tunable optical fibers and optical switching bubbles [6,7]. With further technology developments and advances in microfluidics for small scale liquid handling, the field of optofluidics formally began to emerge in the mid-2000s and was identified by researchers who were seeking synergies between these two areas [8,9].

By dealing with the fields of microfluidics and optics together, numerous advantages are provided for optofluidic devices [2,10,11]. One advantage is the smooth interface formed between two immiscible liquids as a result of a liquid's tendency to minimize its surface energy. Such optical-grade smoothness of the fluidic interface is very useful and cost effective as it eliminates the need for high precision fabrication or polishing processes typically required for solid type optics [12]. Another benefit is that the shapes and positions of liquid interfaces can be dynamically controlled using microfluidic technologies without the need for bulky and complex mechanical moving parts [13]. In addition, fluids offer many options to vary their optical and physical properties (e.g., refractive indices, density, transparency, diffusion, etc.). This feature makes optofluidic devices more versatile and reconfigurable to improve their optical performances [14], although optics itself provides more functionalities, flexibility, low cost, and high throughput for numerous biological and micro-/nanofluidic applications. Various optical sources have been used for manipulation of small scale objects such as single cells, micro-/nanoparticles, and liquid droplets. Furthermore, there are many optical technologies such as fluorescence, Raman scattering, polarization, and phosphorescence that offer the benefits of high sensitivity and fast response time for biological sensing and imaging.

Researchers have mainly developed two different types of optofluidic devices to take advantage of the synergies between these two fields of microfluidics and optics. The first type of optofluidic device utilizes existing microfluidic technologies and phenomena to adaptively control optical performances such as focal length, reflection/refraction, and waveguides. Technical advances in micro-/nanofluidics areas have allowed precise manipulation of light on small scales. Another type of optofluidic device is ones driven by light, such as lasers, to manipulate fluids and particles. By incorporating various spatial light modulators, optofluidic devices are expected to be adaptable, versatile, and reconfigurable. The following sections will describe the fundamentals of these two types of optofluidic devices and their optics, energy, and bio-related applications.

14.2 Light-driven microfluidic technologies

Due to the benefits of small volume requirement, fast analysis and diagnostic time, high sensitivity, and high throughput analysis [15–17], microfluidics has drawn much attention for a broad range of biomedical and chemical applications, including protein crystallization [18,19], polymerase chain reaction (PCR) [20], and enzyme kinetic assays [17,21]. Various microfluidic mechanisms have been investigated, such as electrowetting [22,23], dielectrophoresis (DEP) [24,25], thermocapillary force [26,27], surface acoustic waves [28], and magnetic force [29,30]. In recent years, these technologies have been integrated with light to provide more functionalities with greater flexibility, lower cost, and higher throughput [31,32]. Light is able to be patterned and reconfigured to achieve dynamic control of micro-/nanofluidic phenomena without the need for complex control circuitry on a chip. Using commercially available spatial light modulators, such as an LCD or DMD display, millions of optical pixels are readily generated to control millions of electrodes in parallel on low cost and disposable devices. In addition, since light can be propagated and focused in free space, optical methods allow spatial manipulation of small scale objects like single cells, particles, and liquid droplets.

There are three basic optical energy conversion mechanisms to induce micro-/nanofluidic forces: (1) direct optical, (2) opto-electro-fluidic, and (3) opto-thermo-fluidic. Figure 14.1 shows the optical energy transduced to achieve microfluidic manipulation. This section will describe the working principles of these light-driven microfluidic technologies and introduce their applications and limitations.

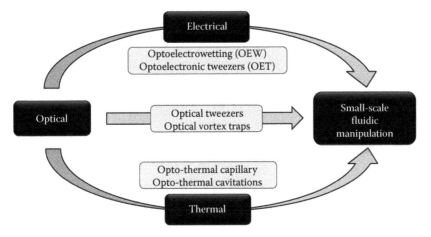

Figure 14.1 An illustration of three optical energy transduction pathways typically used for small scale fluidic manipulation; (1) direct optical, (2) opto-thermo-fluidic, and (3) opto-electro-fluidic manipulation.

14.2.1 Direct optical manipulation

14.2.1.1 Optical tweezers

Since Ashkin's first report on the concept of optical tweezers in 1970 [33], they have become a powerful tool widely used in the areas of physics, chemistry, and biology [34]. They generate direct optical forces typically in the range of pN to trap, transport, pattern, and sort microscopic objects such as dielectric particles, viruses, living cells, bacteria, and small metal particles without mechanical contact [34–41]. When light is focused into an aqueous medium through an objective lens, a suspended particle experiences two types of optical forces. A scattering force directs a particle in the direction of beam propagation [33], while a gradient force attracts it to the region of the highest intensity [34]. Figure 14.2 presents a schematic of optical tweezers to induce an optical trap dominantly using a gradient force. Applications of optical tweezers include trapping of cells and bacteria [37, 38], measuring of the forces exerted by molecular motors such as myosin or kinesin [42], and the study of the mechanical properties of single DNA strands [40, 43]. In the case of biological samples, light in the near infrared (700–1100 nm) is typically used to prevent radiation damage that would occur with shorter wavelengths. Biological applications of optical tweezers were overviewed in [44].

A theoretical description of optical trapping uses two different regimes (ray optics and Rayleigh), relying on the ratio of the incident light's wavelength (λ) to the diameter (D) of the irradiated particle. In the Rayleigh regime, the particle is very small compared to the wavelength ($D \ll \lambda$), thus the distinction between the components of reflection, refraction, and diffraction can be ignored. Since the perturbation of the incident wave front is minimal, the particle can be viewed as a point dipole behaving according to electromagnetic laws. A scattering force results from the radiation pressure of the incident light that is lightly focused. Incident radiation is absorbed and isotropically reemitted by atoms or molecules. During this process, two impulses are received by the molecule; one along the beam propagation of the incident light, and one opposite to the direction of the emitted photon. Since the photon emission has no preferred direction, a net force is directed along the propagation of light. When an intense laser beam is tightly focused through a high NA objective lens, a gradient force dominantly forms an optical trap which is able to attract a small particle, in the size from nanometers to micrometers, to the region of the highest intensity [45]. The gradient force's direction is toward the area of highest light intensity, that is, toward the beam axis in the case of a Gaussian beam profile, and toward the focus of the laser if the beam is focused.

In the ray optics regime, the size of the object is much larger than the wavelength of the light ($D \gg \lambda$), which includes typical biological situations handling single cells and dielectric beads

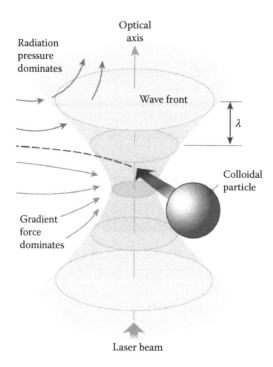

Figure 14.2 Optical tweezers use a strongly focused light beam to trap objects. Intensity gradients in the converging beam draw small objects, such as colloidal particles, toward the focus, whereas the radiation pressure of the beam tends to blow them down the optical axis. Under conditions where the gradient force dominates, a particle can be trapped, in three dimensions, near the focal point. (Adapted from D. G. Grier, *Nature Photonics*, vol. 424, pp. 810–816, 2003.)

(ranging from microns to tens of microns in size) surrounded by a biological medium. A ray emitted from the laser travels throughout the particle and the optical momentum changes in direction and magnitude. When the particle is not centered exactly on the optical axis of the Gaussian beam, the net force induced by the changes of all rays' momentums attracts the particle to the center for optical trapping. If the particle is located at the center of the beam, there is no net force in the lateral direction due to the symmetrical optical interaction through the particle. Along the axial direction, the net force cancels out the scattering force of the laser light and the bead is able to be stably trapped slightly downstream of the focal point.

14.2.1.2 Optical levitation

Optical levitation is a method to mainly utilize a scattering force to balance the gravitation force of an object by lightly focusing a light beam [46–48]. This technology has been typically used in measuring the object size and phase in thermodynamic equilibrium, which is controlled by temperature, composition of atmospheric aerosols, and relative humidity [49,50].

14.2.1.3 Optical vortex traps

Optical trapping uses a strongly focused laser beam with a Gaussian profile to trap high refractive index objects (n_o) in a low refractive index medium (n_m), such as beads or cells dispersed in an aqueous medium. However, optical trapping of aqueous droplets dispersed in immiscible oil, which is a common situation in digital microfluidics, cannot be achieved with this simple Gaussian beam. Since the refractive index of the droplet is lower than that of oil ($n_{water} < n_{oil}$), the optical gradient force repels an oil-immersed water droplet instead of trapping it. To overcome this issue, Gahagan and Swartzlander have demonstrated three-dimensional trapping of low index particles in water using a single dark optical vortex laser beam [51,52]. Instead of using Gaussian beam profile, optical vortex traps are based on the ring-shaped laser beam profile with a dark core in the focal point. An optical gradient force pushes a low index particle away from the region of high light intensity. As a result, the ring-shaped laser beam forms a potential barrier to trap a low index particle at the dark core where the net force of all gradient forces is zero [53]. Compared to optical tweezers, optical vortex trapping provides several advantages. First, it creates a stable optical trap of low index particles, such as water droplets dispersed in oil. Second, the potential barrier created by an optical vortex trap allows isolating a single droplet from potential contamination by other droplets. Utilizing the optical vortex trapping technique, Chiu's group has demonstrated optical manipulation of femtoliter volume droplets immersed in oil, such as droplet trap, fusion, and dynamic control of concentrations of dissolved species [54,55].

14.2.2 Opto-thermo-fluidic manipulation

Optics is often used as a heating source in many biomedical and manufacturing areas [56]. Recently, opto-thermal effects have been used for microfluidic manipulation. Compared to the direct optical methods discussed in the previous section, opto-thermal mechanisms are able to induce much higher forces, and hence faster microfluidic manipulation for high throughput performance.

14.2.2.1 Opto-thermal capillary

The Marangoni effect is a phenomenon of liquid movement induced by the surface tension gradient. Thermocapillary is the Marangoni effect associated with the temperature difference. This thermocapillary phenomenon was first investigated by Young et al. who observed air bubble motion in silicone oil induced by a temperature gradient [57]. In an unconfined fluid, thermocapillary migration speed, $\mathbf{U}_{\text{Th}}$, can be expressed by [58]

$$\mathbf{U}_{\text{Th}} = -\frac{2}{2\mu_o + 3\mu_i}\left(\frac{\partial\sigma}{\partial T}\right)\frac{R}{2 + \Lambda_i/\Lambda_o}\nabla T \qquad (14.1)$$

where R is the droplet radius, μ_o and μ_i are the shear viscosity, and Λ_o and Λ_i the thermal conductivities of the fluid inside and outside the droplet. The thermocapillary migration speed is proportional to droplet size (R), the ratio of change of surface tension with temperature ($\partial\sigma/\partial T$), and temperature gradient (ΔT). It is noted that when $\partial\sigma/\partial T < 0$ the droplet is attracted to the hot region heated by light illumination, while it is repelled if $\partial\sigma/\partial T > 0$.

Conventional thermocapillary devices typically use microfabricated electric resistors to provide the local temperature gradient that creates the thermocapillary effect for liquid transportation, trapping, and sorting on trajectories of prepatterned structures [59,60]. Instead of electric heating, opto-thermal capillary directly utilizes light to locally raise the temperature, thus the surface tension decreases at the heated site. The modification of such a light induced surface tension causes a droplet to move toward the colder region [61]. An opto-thermal capillary effect has been used for numerous microfluidic applications. For example, Baroud et al. demonstrated the interfacial flow at the water/oil interface ($\partial\sigma/\partial T > 0$) by focusing a laser beam at the front of the droplet to balance the light induced thermocapillary force, and the drag force from the hydrodynamic oil flow in a microchannel [62]. Figure 14.3 shows the use of a laser beam to sort the droplets moving at speeds up to 1.3 cm/s in microfluidic devices [63]. Recently, Ohta et al. have also reported droplet manipulation driven by the opto-thermal capillary effect on a light absorbing a-Si:H coated glass substrate using an optical projector [64,65].

14.2.2.2 Opto-thermal cavitation bubbles

Laser induced cavitation is a nonlinear optical phenomenon caused by tightly focusing an intense laser pulse in a liquid medium to generate a rapidly expanding vapor bubble [66]. Figure 14.4 presents schematic illustrations describing the working principle of laser induced plasma formation,

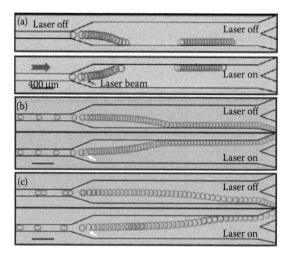

Figure 14.3 Superposition of successive frames illustrating drop switching by the light-induced thermo-capillary force at various ratios between oil and water flow rates (Qo/Qw), (a) 5/0.05, (b) 10/0.1 and (c) 15/0.2. The arrow indicates the location of the laser excitation observed by the fluorescence of the water-dye solution. (Reproduced with permission from M. R. d. S. Vincent et al. *Applied Physics Letters*, vol. 92, p. 154105, 2008.)

followed by the emission of a shock wave and the generation of a cavitation bubble [67]. When a laser pulse is intensively focused into a liquid medium such as water, the strong optical field breaks down water molecules and produces hot plasma at the focal point (Figure 14.4a). The plasma initiation, growth, and decay are completed within $25 \sim 30$ ns after the arrival of the laser pulse [68]. The emission of a shock wave is originated from the rapid plasma expansion and propagated to an outer medium. Its cooling and ion recombination lead to the bubble initiation within 25 ns after the laser pulse, followed by bubble expansion and collapse (Figure 14.4b and c) [68–72]. The expansion speed of a laser induced cavitation bubble is very rapid at up to hundreds of m/s within a 1 μs period. Such high speed cavitation bubbles have been used in many microfluidic applications, such as cell lysing [68], microfluidic mixing [73], pumping [74], and switching [75]. Figure 14.5 shows experimental demonstrations of the laser induced plasma formation followed by shock wave emission, cavitation bubble formation, and collapse in an open chamber [67].

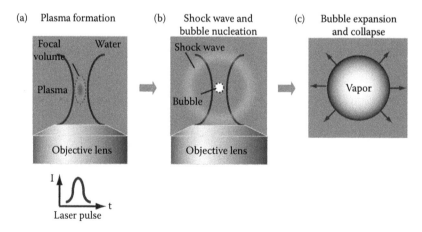

Figure 14.4 Schematic illustrations of (a) a laser pulse-induced plasma formation, (b) the following shock wave emission, and (c) generation of the cavitation bubble in liquid medium. (Reproduced with permission from P.-Y. Chiou et al. Pulse laser-driven ultrafast micro and nanofluidics system, *presented at the SPIE*, San Diego, 2010.)

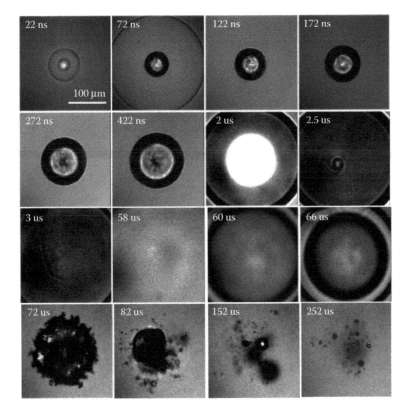

Figure 14.5 Experimental demonstrations of the laser induced plasma formation followed by shock wave emission, cavitation bubble formation, and collapse. A 400-μJ laser pulse was used to excite a water medium in an open chamber. (Reproduced with permission from Ref. P.-Y. Chiou et al. Pulse laser-driven ultrafast micro and nanofluidics system, *presented at the SPIE*, San Diego, 2010.)

Park et al. have used a pulse laser induced cavitation bubble for an ultrafast microfluidic actuation and demonstrated high speed, on-demand droplet generation at a speed of up to 10,000 droplets/sec with less than 1% volume variation [76]. This mechanism is called pulse laser-driven droplet generation (PLDG). Figure 14.6A shows a schematic of the PLDG device that simply consists of two microchannels, one for water and another for oil flows, connected by a nozzle-shaped opening in a single layer PDMS microfluidic chip. An intense laser pulse is focused in the middle of the water channel and induces a cavitation bubble to push nearby water into the oil channel for droplet formation. Figure 14.6B presents the time resolved images of the droplet generation processes. By adjusting the laser energy and pulsing location away from the sidewall of the microchannel, the volume of the generated droplets was tuned from 1 to 150 pL. Recently, Li et al. also utilized laser induced cavitation bubbles to trigger on-demand droplet fusion in microfluidic devices [77].

14.2.3 Opto-electro-fluidic manipulation

Electrowetting and dielectrophoresis (DEP) are the electrokinetic phenomena that are most commonly used for microfluidic manipulation. DEP typically uses the nonuniformity of the electric field for manipulating small scale objects like single cells, particles, and droplets [78], while electrowetting modifies the surface tension of liquid droplets on a hydrophobic-coated solid surface with an applied electric field [79,80]. Recently, researchers have used an optical method to control

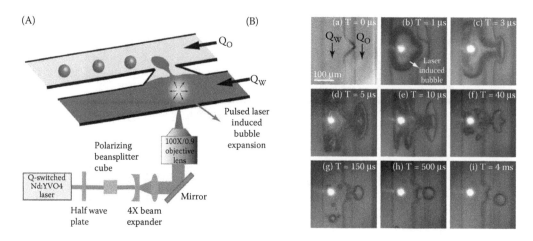

Figure 14.6 (A) A schematic of the PLDG device that consists of two microfluidic channels connected by a nozzle-shape opening. A tightly focused laser pulse induces a rapidly expanding bubble to push nearby water into the oil channel and form a droplet. (B) Time-resolved images of the on-demand droplet generation using a PLDG device. (Reproduced with permission from S.-Y. Park et al. *Lab on a Chip*, vol. 11, pp. 1010–1012, 2011.)

such electrokinetic phenomena. Light patterns are directly projected onto a photoconductive surface and generate virtual electrodes to locally modify the electric field distribution, which in turn, controls the DEP or electrowetting effects to optically achieve microfluidic manipulation.

14.2.3.1 Optoelectrowetting (OEW)

Electrowetting is a well-known microfluidic phenomenon such that externally applied electrostatic charges modify the surface tension at the solid-liquid interface. With the benefits of large forces in micro-/mesoscales, fast response time in the range of milliseconds, and easy implementation, the electrowetting technology has been used for numerous applications, including lab-on-a-chip [23,81], optical devices [82–84], thermal management [85–87], solar concentration [88,89], and surface science [90].

Figure 14.7 presents the working principle of electrowetting. A liquid droplet sits on a solid electrode plate coated with a hydrophobic and dielectric layer, forming its initial contact angle θ_0 in the absence of a bias voltage application (Figure 14.7a). When an electric potential is applied between a liquid and a solid electrode (Figure 14.7b), the charge redistribution modifies the surface tension from γ_{0_SL} to γ_{SL} at the solid-liquid interface where the like charges repulsion decreases the work by expanding the surface area. The resulting contact angle (θ) of a liquid droplet can

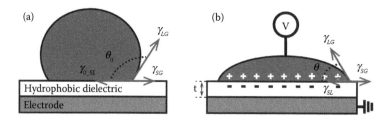

Figure 14.7 A working principle of electrowetting on dielectric (EWOD). (a) A liquid droplet sits on a solid surface. In the absence of an applied voltage, a droplet forms with an initial contact angle θ_0 along a three-phase contact line. (b) When a bias voltage is applied between a liquid and an electrode, the charge redistribution modifies the surface tension at the liquid-solid interface by decreasing the contact angle. γ denotes the surface tension at the interface between the solid-liquid (*SL*), solid-gas (*SG*), and liquid-gas (*LG*), respectively.

be mathematically estimated with the applied electric potential (V) by using the popular Young–Lippmann equation [91–93]

$$\cos\theta = \cos\theta_0 + \frac{1}{2\gamma_{LG}}\frac{\varepsilon_0\varepsilon_r}{t}V^2 \tag{14.2}$$

where γ_{LG} is the surface tension between two immiscible fluids, ε_0 is the permittivity of free space, ε_r is the dielectric constant, and t is the thickness of the capacitor formed between a liquid and a solid electrode. Droplet actuation in electrowetting devices is typically accomplished by electric activation on pixelated electrodes, while optoelectrowetting (OEW) utilizes optical activation either on patterned electrodes or a featureless photoconductive thin film to induce the electrowetting effect. As a result, OEW technology fully eliminates the fabrication issues of complex wiring and interconnection when a large number of electrodes or droplets need to be controlled in parallel without interference.

The concept of OEW was first reported by Chiou et al. [94,95]. Using light beams projected onto a photoconductive chip where tens of thousands of digitized electrodes were patterned with only two electrical bias wires, various droplet manipulation functions such as injection, transportation, merging, mixing, and splitting have been demonstrated. Figure 14.8 presents the configuration of an OEW device and its equivalent circuit model of one unit cell. A liquid droplet is sandwiched between two parallel plates composed of patterned indium-tin-oxide (ITO) electrodes. Each patterned electrode is connected by the photoconductive bridges of hydrogenated amorphous silicon (a-Si:H), on top of which silicon dioxide (SiO_2) and Teflon layers are deposited to provide dielectric and hydrophobic properties. An AC bias voltage of 500 Hz is applied between the two plates. In the absence of light illumination, the electrical impedance of R_{asi} is dominant and most of the voltage drop occurs across the photoconductive bridges. As a result, the ITO electrodes are not able to be activated and no movement of the droplet is induced. In contrast, the illumination of light beams causes the voltage switching between the oxide capacitors (C_{oxide}) and the photoconductors (R_{asi}). When light illuminates the photoconductive bridges near a droplet, the electric conductivity of the photoconductive bridges increases and ITO electrodes are activated to induce electrowetting for droplet actuation.

There are two main drawbacks in conventional OEW devices. First, the sandwiched configuration increases the difficulty in accessing droplets and integrating with other microfluidic components such as microchannels. Second, the minimum droplet size that can be manipulated is limited by the size of the pixelated electrode. To overcome these issues, several different OEW

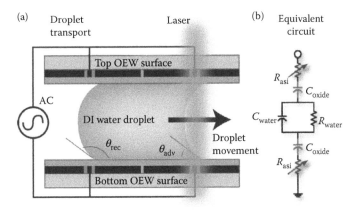

Figure 14.8 (a) A schematic of droplet actuation on an optoelectrowetting (OEW) device. (b) Its equivalent circuit model for one unit cell of the OEW device. (Reproduced with permission from P.-Y. Chiou et al. *Journal of Microelectromechanical Systems*, vol. 17, pp. 133–138, 2008.)

configurations have been proposed. Chuang et al. reported an open OEW configuration enabling droplet actuation on a single-sided photoconductive surface [96]. This open configuration provides a flexible interface to allow easy integration, but the patterned pixelated electrode still limits the minimum droplet size to be manipulated. Chiou et al. proposed a continuous optoelectrowetting (COEW) that allows continuous transportation of picoliter droplets sandwiched between two featureless photoconductive electrodes [97]. Nevertheless, it still has the issue of device interfacing with other microfluidic components such as detectors.

Park et al. demonstrated a new OEW mechanism, called single-sided continuous optoelectrowetting (SCOEW), to overcome these two main issues arising from conventional OEW devices [82]. Unlike conventional OEW, droplet actuation on SCOEW is based on optical modulation of lateral electric fields. Under uniform light illumination, no change in the contact angle of an oil-immersed water droplet occurs on top of the hydrophobic surface, hence no droplet movement occurs. As shown in Figure 14.9A, when the dark pattern illuminates near the droplet, the voltage drop across the dielectric layer locally increases and the contact angle correspondingly decreases. As a result, the droplet moves toward the dark region due to the pressure gradient inside a droplet. The SCOEW device provides several unique advantages over conventional EWOD and OEW devices. First, a single-sided open configuration allows easy integration with other microfluidic components, such as sample reservoirs and microchannels. Second, a continuous photoconductive

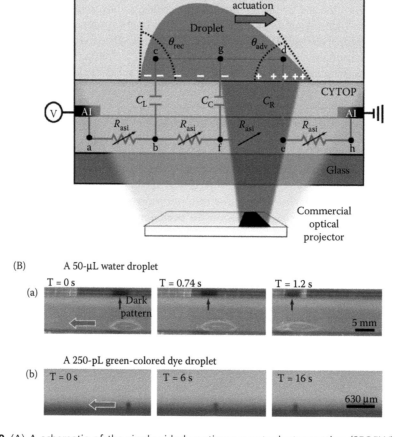

Figure 14.9 (A) A schematic of the single-sided continuous optoelectrowetting (SEOEW) device and its working principle with the equivalent circuit. (B) The continuous photoconductive surface (i.e., no pixelated patterns) in SCOEW devices allows transportation of droplets in large volume ranges from (a) 50 μL to (b) 250 pL simply by changing the optical pattern size. (Reproduced with permission from S.-Y. Park et al. *Lab on a Chip*, vol. 10, pp. 1655–1661, 2010.)

surface enables droplets to be continuously positioned at any location on a 2D surface. Third, the droplet size limitation determined by the size of physically pixilated electrodes is completely eliminated. Figure 9B shows the optical transportation of the droplets with a wide range of volume, from tens of microliters to hundreds of picoliters, through simply programming the projected light pattern. Lastly, compared to any previously demonstrated OEW devices, the lateral field-driven optoelectrowetting mechanism can be operated with extremely low light intensity, requiring three orders of magnitude lower light intensity ($\sim$400 µW/cm²) for droplet actuation than other OEW devices ($\sim$1 W/cm²). Such low light intensity requirement allows SCOEW to be operated by directly positioning a device on top of an LCD, an iPhone or an iPad screen without any extra optical sources or components such as lenses.

With technical advances in electrowetting over past years, recent attraction has been focused on three-dimensional (3D) devices capable of providing more flexibility and functionality with larger volumetric capacity than that of conventional 2D planar ones. Several 3D electrowetting devices have been reported to show a potential of the lab-on-a-wristband concept for portable point-of-care (POC) applications [98,99]. To meet such a demand for OEW, Park's group demonstrated light-driven 3D droplet manipulation on flexible SCOEW devices [11]. All prior optoelectrowetting (OEW) devices have used amorphous silicon (a-Si) as a photoconductive layer, which was typically fabricated through high temperature processes over 300°C, such as CVD or PECVD [100]. However, commercially available flexible substrates such as polyethylene terephthalate (PET) and polyethylene naphthalate (PEN) experience serious thermal deformation and melting under such high temperature processes [101,102]. Due to this compatibility issue encountered in previous OEW devices, light-driven 3D droplet manipulations have not been yet demonstrated on flexible substrates. A recent study conducted in Park's group overcame such a compatibility issue by using a polymer-based photoconductive material, titanium oxide phthalocyanine (TiOPc) [11]. The SCOEW device was simply fabricated on flexible substrates through a low cost, spin-coating method. Figure 14.10 presents light-driven 3D droplet manipulations demonstrated on various terrains such as inclined, vertical, upside-down, and curved surfaces. Our flexible SCOEW technology offers the benefits of device simplicity, flexibility, and functionality over conventional EWOD and OEW devices by enabling optical droplet manipulations on a 3D featureless surface.

14.2.3.2 Optoelectronic tweezers (OET)

Another electrokinetic principle popularly used for microfluidic manipulation is dielectrophoresis (DEP), which refers to the motion of a dielectric particle by the force subjected to a nonuniform

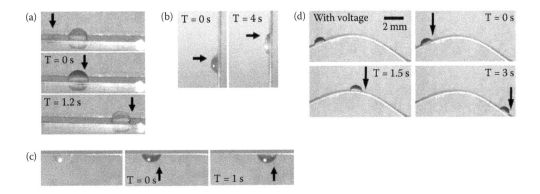

Figure 14.10 Experimental demonstrations of continuous light-driven droplet transportations on various 3D terrains such as (a) a flat ($\varphi = 0°$), (b) a vertical ($\varphi = 90°$), and (c) an upside-down ($\varphi = 180°$) surface. (d) The images present continuous, light-driven 3D droplet transportation on a flexible SCOEW device which was bent to provide uneven gravitational effects against droplet movement. Black arrows indicate the locations of the dark pattern. (Reproduced with permission from D. Jiang and S.-Y. Park, *Lab on a Chip*, vol. 16, pp. 1831–1839, 2016.)

electric field. Unlike electrophoresis, this force does not require the particle to be charged. Using the Taylor expansion of the electric field, ignoring the higher order terms, this Coulomic force on an electric dipole can be approximated [103] as

$$\mathbf{F} = (\mathbf{p} \cdot \nabla)\mathbf{E} \tag{14.3}$$

where $\mathbf{p}$ represents the dipole moment and $\mathbf{E}$ is the electric field. Assuming the effective dipole moment of a homogenous dielectric sphere, the well-known expression for DEP force is given as [104]

$$\langle \mathbf{F}_{dep} \rangle = 2\pi\varepsilon_m R^3 \, \text{Re}[K^*(\omega)]\nabla(\mathbf{E}_{rms}^2) \tag{14.4}$$

$$K^*(\omega) = \frac{\varepsilon_p^* - \varepsilon_m^*}{\varepsilon_p^* + 2\varepsilon_m^*}, \quad \varepsilon_m^* = \varepsilon_m + \frac{\sigma_m}{j\omega}, \quad \varepsilon_p^* = \varepsilon_p + \frac{\sigma_p}{j\omega} \tag{14.5}$$

where $\langle \mathbf{F}_{dep} \rangle$ represents the time average of $\mathbf{F}_{dep}$, $\mathbf{E}_{rms}$ is the root mean square magnitude of the electric field, R is the radius of the particle, ε_m and ε_p are the permittivity of the surrounding medium and the particle, respectively, σ_m and σ_p are the electric conductivities of the medium and the particle, respectively, ω is the angular frequency of the applied electric field, and $K^*(\omega)$ is known as the frequency dependent Clausius–Mossotti (CM). The real part of $K^*(\omega)$ has a value, $-0.5 \leq \text{Re}[K^*(\omega)] \leq 1.0$, depending on the polarization of the medium and the particle at a certain frequency. When $\text{Re}[K^*(\omega)] > 0$, the induced electric dipole is collinear with the imposed electric field vector and generates a positive DEP (pDEP) that pushes the particles towards the strong electric field region. On the other hand, when $\text{Re}[K^*(\omega)] < 0$, the induced electric dipole is antiparallel to the electric field. Thus, the particles move toward the weaker electric field region, known as a negative DEP (nDEP).

Equation 14.5 is a good approximation of the DEP force, which is strictly valid only when the particle size is much smaller than the characteristic length of the electric field gradient. To precisely calculate the DEP force, the Maxwell stress tensor method, which is regarded as the most rigorous approach, is typically used to estimate the field induced DEP forces by integrating the stress tensor $\mathbf{T}_{ij}$ over the spherical surface of an object

$$\mathbf{F} = \int_{surface} \mathbf{T} \cdot \mathbf{n} ds \tag{14.6}$$

$$\mathbf{T}_{ij} = \varepsilon_1 \left(E_i E_j - \frac{1}{2}\delta_{ij}E^2 \right) \tag{14.7}$$

where i and j refer to pairs of x, y, and z axes, δ_{ij} is the Kronecker delta, E is the electric field, and $\mathbf{n}$ is the unit vector normal to the surface [105–109].

Chiou et al. have first demonstrated a light-induced DEP mechanism, called optoelectronic tweezers (OET), which allows light images to pattern nonuniform electric fields to induce DEP forces for optical manipulation of micro- and nanoscale particles such as cells, carbon nanotubes, and nanotubes [110]. Figure 14.11A shows the OEW device structure consisting of the top and bottom transparent electrodes. The lower electrode is made of thin photoconductive layers (i.e., undoped and n+ doped) of hydrogenated amorphous silicon (a-Si:H) on a top of an ITO-coated glass substrate. These featureless layers are made without the photolithography step in fabrication, making the device inexpensive and attractive for disposable applications. Using an incoherent light source (a light emitting diode or a halogen lamp) and a digital micromirror spatial light modulator requiring 100,000 times less optical intensity than optical tweezers, they have demonstrated parallel manipulation of 15,000 particle traps on a 1.3 × 1.0 mm² area. When projected light illuminates the photoconductive layer,

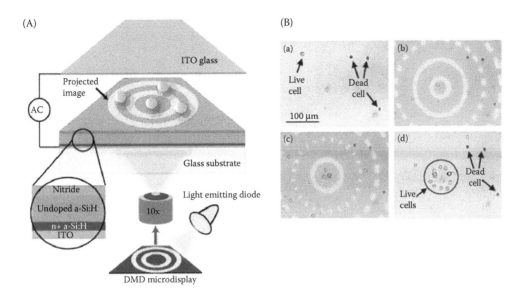

Figure 14.11 (A) A schematic of the optoelectronic tweezers (OET) device consisting of two transparent electrodes. The bottom photosensitive surface is composed of ITO-coated glass topped with 1 μm of undoped hydrogenated amorphous silicon (a-Si:H). The top and bottom surfaces are biased with an AC voltage. The optical source can be coherent or incoherent. (B) Selective collection of live cells from a mixture of live and dead cells. (a) Randomly positioned cells before OET.(b) and (c) Cell sorting. The live cells experience positive OET, trapping them in the bright areas, and pulling the live cells into the pattern's centre. The dead cells (stained with Trypan blue dye) leak out through the dark gaps and are not collected.(d) Sorted cells. (Reproduced with permission from P. Y. Chiou et al. *Nature*, vol. 463, pp. 370–372, 2005.)

it turns on the virtual electrodes, creating nonuniform electric fields and enabling particle manipulation via DEP forces. Figure 14.11(B) presents the OET capability for the selective concentration of live human B cells from a mixture of live and dead cells. An electric field is created in the device by applying an AC bias across the top and bottom electrodes, in which live and dead cells have different CM factors at an AC bias voltage at 120 kHz, and only the live cells respond to the positive DEP to be trapped. OET was further used for manipulating individual nanowires of 5 μm length and 50 nm radius suspended in an aqueous solution. Nanowires experience a torque in addition to the DEP force, which vertically aligns the long axis of the nanowire with the electric field [111].

OET technology has been a powerful tool for the parallel manipulation of small particles like single cells suspended in aqueous media. However, it is limited to operate OET in an air or oil environment due to the issue of electrical impedance matching. The large electrical impedance takes most of the voltage across these media, even without light illumination, which makes optical modulation of the electric field difficult. This means that OET is limited to be used for digital microfluidics where aqueous droplets are typically immersed in an electrically insulating air or oil medium. Park et al. have demonstrated an OET mechanism, aiming for manipulating aqueous droplets suspended in an electrically insulating oil medium, called floating electrode optoelectronic tweezers (FEOET) [112]. Figure 14.12 shows the device structure and the potential use for multifunctional, parallel processing such as one-to-one droplet dilution. The device structure of FEOET looks similar to SCOEW, except for the thickness of the dielectric layer between droplets and the photoconductive layer. In SCOEW, the dielectric layer has to be thinner than a few μm to dominate the electrowetting phenomenon, while the thick layer in the range from tens of μm to a few millimeters can be used in FEOET. Such a thick layer on top of the photoconductive surface allows versatile integration with other microfluidic components such as microwells and microchannels [32].

When a light beam illuminates the photoconductive layer, the nonuniformity of the electric field is formed and interacted with the droplet dipole induced by the application of a lateral field across

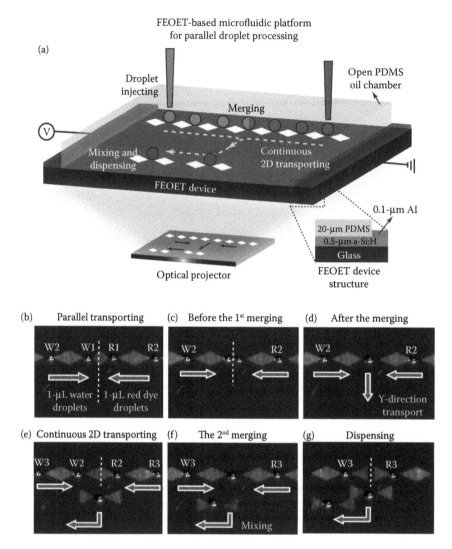

Figure 14.12 (a) A schematic of the FEOET platform for parallel droplet manipulation using direct optical images. Aqueous droplets containing different chemicals (denoted by different colors) are injected into the oil medium. 2D droplet manipulation functions including continuous droplet transportation, merging, mixing, and parallel droplet processing can be performed on FEOET. (b) through (g): Droplets carrying DI water and red dyes are injected at the two edges of the platform and transported to the middle, merged, mixed, and dispensed to target locations. This example demonstrates a one-to-one droplet dilution process. (Reproduced with permission from S.-Y. Park et al. *Lab on a Chip*, vol. 9, pp. 3228–3235, 2009.)

the entire device. Since the electric impedances of aqueous droplets are typically much smaller than that of an insulating oil medium in digital microfluidics, Re $[K^*(\omega)]$ in Equation 14.5 has a positive value close to 1, thus droplets experience positive DEP forces, moving toward the strong electric field regions. Figure 14.13 shows numerical simulation results of the electric field distributed around an aqueous droplet. Without the light illumination, the electric field is balanced, hence there is no movement of the droplet. In contrast, when a circular laser beam is projected near the droplet, such a balanced field distribution is broken, resulting in the droplet movement toward the strong electric field region (i.e., pDEP force).

Both FEOET and SCOEW devices require extremely low light intensity for droplet manipulation (~400 μW/cm²) owing to the fact that, in lateral electric field configurations, the virtual electrode is turned on as long as the light illumination can create a small photoconductivity difference

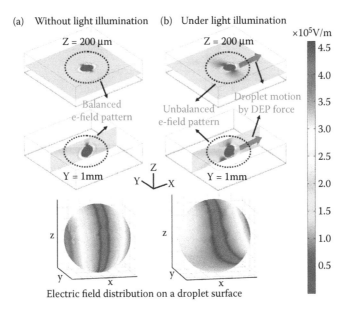

(a) Without light illumination   (b) Under light illumination

Electric field distribution on a droplet surface

Figure 14.13 Numerical simulation results of the electric field distribution around an aqueous droplet. (a) Without light illumination, the balanced electric field distribution is exhibited around an electric dipole of a droplet. (b) Optical illumination creates an unbalanced electric field distribution near a droplet. As a result, a droplet moves toward the strong field region (right side). (Reproduced with permission from S.-Y. Park et al. *Applied Physics Letters*, vol. 92, p. 151101, 2008.)

between the illuminated and dark regions. This is different from previous OET and OEW devices that require light to switch voltages between a photoconductor and a dielectric layer or a liquid layer. In other words, FEOET and SCOEW turn on virtual electrodes based on a relative photoconductivity difference, not the absolute photoconductivity as in other devices. Recently, Valley et al. also reported an integrated platform enabling both COEW and OET manipulation on the same chip [31]. This platform allows light beams to manipulate not only droplets, but also individual cells inside these droplets. They have demonstrated concentration enhancement of cells and single cell encapsulation in droplets.

14.3 Microfluidic-driven optical elements

In general, the optical performances of conventional solid type elements, such as lenses and prisms, are fixed once the material and geometry have been designed. Integration with microfluidic technologies allows these optical elements to have advantages such as optical grade smoothness at the fluidic interface, a high degree of optical tunability, and device reconfigurability. These features make optofluidic devices more versatile and reconfigurable for dynamically controlling their optical performances. Liquid lenses and prisms are typical examples of the optofluidic devices that have gained much interest over recent years. In this section, we will review technologies for liquid lenses and prisms as fundamental optical elements, which utilize microfluidic principles for tuning their optical performances.

14.3.1 Microfluidic tunable lens

A lens is no doubt the most widely used component in optics systems. The focusing power of the lens is determined by the lens's profile (i.e., convex or concave) and the refractive index relative

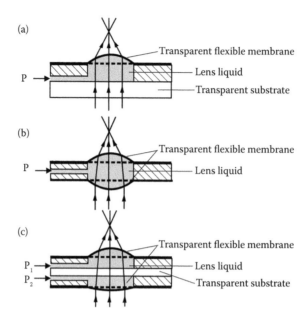

Figure 14.14 Out-of-plane optofluidic lens configured (a) in a planoconvex design and (b) a biconvex design, and (c) with two chambers where the lens profile covered by a flexible membrane is pneumatically tuned. (Reproduced with permission from N.-T. Nguyen, *Biomicrofluidics*, vol. 4, p. 031501, 2010.)

to the outside medium. Figure 14.14 shows several configurations of the out-of-plane optofluidic lenses that are designed as a circular chamber covered by a flexible membrane [113]. A highly refractive index liquid is made to fill the chamber. The liquid pressure provided by a pneumatic pump changes the curvature of the membrane to act as a tunable liquid lens and consequently tune the focal length. Many groups have reported out-of-plane type optofluidic lenses using PDMS as a deformable membrane and demonstrated focal length tuning from hundreds of microns to several millimeters [114,115].

Another type of the optofluidic lens is based on in-plane designs. Mao et al. used two miscible liquids with different refractive indices to act as a microlens [116]. These two liquids are co-injected to flow through a 90° curve in a microchannel, where centrifugal forces cause the fluidic interface to be distorted. Because the flow is laminar in microscale, the two liquids do not fully mix and allow for a contrast in refractive indices. The optical properties of the microlens are dependent on the interface between the two fluids, which can be controlled via their individual flow rates. Generally, higher flow rates cause a larger curvature in the interface and thus a shorter focal length. Shi et al. utilized the curved interface formed between two immiscible liquids (water and air) to achieve optofluidic tunability of the lens [117]. The flow rate of water adjusts the pressure of the trapped air in the chamber to tune the curvature of the liquid-air interface. They experimentally demonstrated the tunable lens by showing the fluorescence images and the results were further compared with the simulation data.

One critical issue of the optofluidic lens devices aforementioned is that the tuning of the lens's performance is based on the pressure or flow control, requiring extra microfluidic components such as pumps/valves and tubing. Therefore, they are bulky and consume large space and power, which defeats the true purpose of a lab-on-a-chip system. To achieve true-to-form scaled down systems, researchers have proposed to utilize active type microfluidic technologies that can eliminate the need for such auxiliary components. Electrical methods such as electromagnetic, electrowetting, dielectrophoresis, and electrochemical actuations are commonly used for tuning the lens performance [118,119]. Among them, the electrowetting mechanism has become a popular method for current optofluidic devices that can offer standalone systems with versatility in small scales, due to the benefits of large forces in the micro-/mesoscales, fast response time in the range of microseconds, and extremely low power operation. Numerous applications have been demonstrated using this technology, including lab-on-a-chip [23,82], electronic display [120], solar energy collection [89], and energy harvesting [121].

Gorman et al. first used the principle of electrowetting for tuning of the liquid lens [122]. By varying the voltage applied between the droplet and an underlying electrode, the solid-liquid interfacial tension and thus the curvature of the drop could be changed. The shape changes of the drop were used as a liquid lens for an optical switch which focuses/defocuses light. Several following studies showed the focal length modulation on planar substrates with liquid droplets in air [123] and oil [124]. However, it appeared difficult to center the laterally unstable drop on the optical axis. To overcome the centering problem arising from the previous designs, Kuiper and Hendriks demonstrated a liquid lens for which electrowetting is modulated at the sidewall (Figure 14.15) [125]. Two immiscible liquids are included in a cylindrical container for easy centering, compared to the previous planar devices. The performance of the variable lens improves upon miniaturization (Figure 14.15B). The concept of electrowetting on the sidewall was further developed by Krogmann et al. [126]. They used standard silicon-based MEMS processes for miniaturized device fabrication and showed the focal length tuning from 2.3 mm to infinity by applying a voltage of $0 \sim 45V$. Furthermore, Li and Jiang presented the tunable lens which was fabricated on a flexible polymer polydimethylsiloxane (PDMS) substrate, which improves the field of view size [127]. Heikenfeld's group reported the fabrication processes of an array of >12,000 microlenses [128]. Using various microfabrication techniques, they were able to fabricate cylindrical troughs 300 μm in diameter

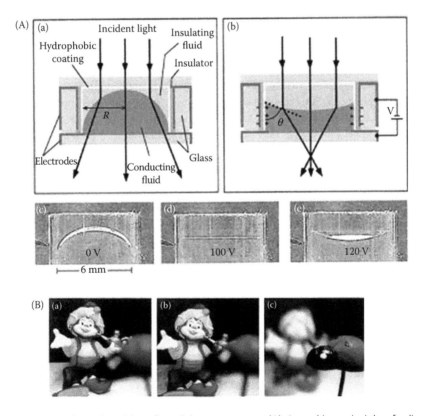

Figure 14.15 Variable-focus liquid lens for miniature cameras. (A) A working principle of a liquid-based variable lens in a cylindrical glass housing. (a) Before and (b) after a bias voltage application. The bottom images show the experimental demonstrations to modulate the liquid-liquid interface with various shapes with a voltage application of (c) 0V, (d) 100V, and (e) 120V. (B) Frames taken from a CMOS camera having (a) a fixed-focus lens and (b) a liquid lens focused at 50 cm and (c) a liquid lens focused at 2 cm. (Reproduced with permission from S. Kuiper and B. H. W. Hendriks, *Applied Physics Letters*, vol. 85, p. 1128, 2004.)

having inner walls lined with dielectric. A voltage was then applied throughout the whole array, allowing the focal length of each lens to be tuned, achieving flat, concave or convex profiles.

14.3.2 Microfluidic tunable prism

Another optical function for which optofluidic devices have been commonly used is beam steering based on a liquid prism. While a liquid lens requires the curved shape of the fluidic interface to focus incoming rays onto a single point, a liquid prism uses the flat interface between two liquids where an incident ray is refracted due to the refractive index difference of the two media. Xiong et al. present a tunable optofluidic prism using a laminar flow of two different liquids in microchannels [129]. Benzyl alcohol and ethylene glycol solutions in deionized water are made to flow into a triangular chamber, creating an angled prism from their interfaces. By separately controlling the flow rates of both liquids through three different channels, they were able to not only control the apex angle, but also change the prism shape from symmetric to asymmetric. With this method, the group reports being able to steer incoming light from $-13.5°$ to $22°$. Similar to the liquid lenses, the problem of such optofluidic prism devices requires extra microfluidic components, such as pumps/valves and tubing, to tune the prism's performance, which make the devices bulky and they consume large amounts of power and space.

To overcome this issue, the electrowetting principle has been popularly used for beam steering applications using a tunable liquid prism [130–133]. Figure 14.16a presents the device structure of the liquid prism and its working principle for beam steering [134]. Operation of the prism device is based on the electrowetting phenomenon, which was described in the previous section (see Figure 14.7). A liquid prism is fabricated by assembling four sidewalls on which a hydrophobic dielectric layer is coated on an electrode substrate (Figure 14.16b). Then, two immiscible liquids (e.g., water and oil or air) are filled in the assembled cuvette and covered by a top transparent plate. Initially, a curved meniscus is formed at the interface between two immiscible liquids with the contact angle of θ_0 at the sidewalls as a result of the Laplace pressure. To operate the prism, bias voltages are applied to the left and right sidewalls (V_L and V_R) of the prism and the contact angles are modified from θ_0 to θ_L and θ_R on the vertical sidewalls. Proper adjustment of θ_L and θ_R makes the interface

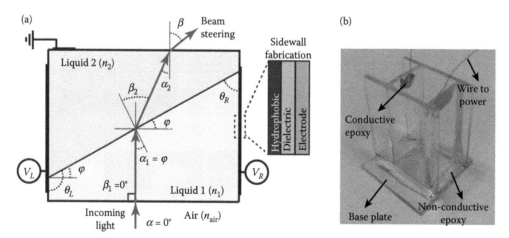

Figure 14.16 (a) A schematic of the single liquid prism filled with two immiscible liquids. Four prism sidewalls are assembled and dip-coated with a dielectric and a hydrophobic layer. An electrowetting effect adjusts the prism apex angle φ by applying bias voltages to the left and right sidewalls, separately. Due to the refractive index difference ($n_{air} \neq n_1 \neq n_2$) of each medium, incoming light is effectively steered without bulky mechanical moving components. (b) The assembled prism device with a 10×10 mm² aperture size before the two liquids are placed. (Reproduced with permission from C. E. Clement and S.-Y. Park, *Applied Physics Letters*, vol. 108, p. 191601, 2016.)

maintain a straight profile (i.e., $\theta_L + \theta_R = 180°$) with the apex angle ($\varphi$) of the prism. An applied electrical input varies to dynamically control the prism angle (φ). The Young–Lippmann equation in Equation 14.2 as previously described, can be rewritten for the apex angle (φ), which has the following relationships with bias voltage inputs (V_L and V_R) and the contact angles (θ_L and θ_R)

$$\theta_L = \cos^{-1}\left(\cos\theta_0 + \frac{1}{2\gamma_{LG}}\frac{\varepsilon_0\varepsilon_r}{t}V_L^2\right), \quad \theta_R = \cos^{-1}\left(\cos\theta_0 + \frac{1}{2\gamma_{LG}}\frac{\varepsilon_0\varepsilon_r}{t}V_R^2\right) \quad (14.8)$$

$$\varphi = \theta_L - 90° \quad \text{when} \quad \theta_L > 90°$$
$$\varphi = 90° - \theta_L \quad \text{when} \quad \theta_L \leq 90°. \quad (14.9)$$

Equations 14.8 and 14.9 allow us to predict how much voltage inputs (V_L or V_R) are required to obtain the desired prism angle (φ), while giving us insights in terms of material selection of the dielectric and liquid combinations. Figure 14.16a illustrates the optical pathway through a single prism, which consists of two immiscible liquids, 1 (bottom) and 2 (top), having refractive indices n_1 and n_2. Assuming that a ray comes in perpendicularly through the prism base, the beam steering performance was analytically studied for a single prism. By using Snell's law and the geometrical relations at the interfaces, the beam steering angle β can be expressed as

$$\beta = \sin^{-1}\left(n_1\sin\varphi\cos\varphi - n_2\sin\varphi\sqrt{1 - \left(\frac{n_1}{n_2}\sin\varphi\right)^2}\right) \quad (14.10)$$

which is a function of the refractive indices (n_1 and n_2) of the liquids and the prism apex angle (φ). Given two liquids, one can predict the prism angle φ required to steer the incoming ray by an output angle of β. It is indicated from Equation 14.10 that the beam steering performance varies with the liquid materials used and their arrangement. From the previous study [134], it is noted that the beam steering performance can be dominantly improved by increasing the refractive index of the liquid 1 (bottom), as well as using two immiscible liquids with a large refractive index difference, as indicated in the two graphs in Figure 14.17.

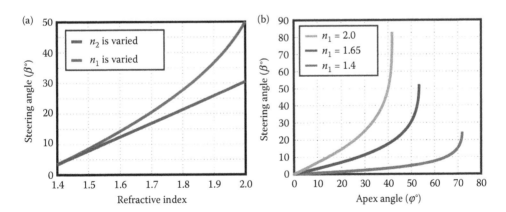

Figure 14.17 Beam steering capability for a single prism was studied to understand the effects of (a) the liquid arrangement and (b) the apex angle. (a) One liquid is assumed to be water ($n = 1.33$), while another is varied from $n = 1.4$ to 2. The prism apex angle is assumed to be $\varphi = 40°$. It is found that the refractive index of liquid 1 (i.e., a lower-side one) more dominantly affects beam steering. (b) The refractive index of liquid 1 varied, while that of liquid 2 is fixed as $n_2 = 1.33$. Beam steering performance is improved by increasing the apex angle φ and using the high-refractive-index materials for liquid 1. (Reproduced with permission from C. E. Clement and S.-Y. Park, *Applied Physics Letters*, vol. 108, p. 191601, 2016.)

Heikenfeld's group first reported a liquid prism device driven by electrowetting [130]. With various bias voltages applications, they experimentally demonstrated three different apex angles using a single liquid prism filled with water exposed to air. Due to the refractive index difference between air and water, incident light may be effectively steered without the need for bulky mechanical moving parts. Beam steering performance was estimated to be as large as $\pm 7°$ based on the refractive index ($n = 1.359$) of the aqueous glycerin solution [130]. However, evaporation became an issue as the water solution was exposed to air. To remedy this, a two liquid prism was developed using water and silicone oil [132]. This binary combination, for which electrowetting properties (e.g., surface tension, surfactant, contact angle hysteresis, etc.) are well known [135], offered more stability and larger modulation of the prism apex angle than that of the air-water system [132]. Nevertheless, silicone oils typically used for electrowetting studies do not have a high refractive index, ($n_{silicone} = 1.38$ ~ 1.45), compared to that of water ($n_{water} = 1.33$). As a result, the actual beam steering achieved by such a prism with water and silicone oil is very limited (less than 3.6°). To achieve a high degree beam steering performance, Park's group used water and the commercially available oil, 1-bromonaphthalene (1-BN), with a refractive index as high as $n = 1.65$, and experimentally demonstrated a significantly larger beam steering of up to 19.06° by using a double stacked prism configuration, which showed the highest beam steering performance ever demonstrated [134].

Electrowetting is the microfluidic phenomenon modifying the surface tension force which is dominant in micro-/mesoscales. Due to such a scaling issue, the aperture size of a single liquid prism is limited to a few centimeters at least. To open up the potentials of liquid prisms for more practical applications, optofluidic researchers have proposed an arrayed form of liquid prisms to offer an increased aperture area. Heikenfeld's group reported a full description of a scalable microfabrication process for the arrayed 1500 microprisms, and each element has $\sim$120 μm in size [136]. On the contrary, Park's group demonstrated an array form of liquid prisms in the mesoscale (each prism has an aperture area around 100 mm²) for an optofluidic tunable Fresnel lens enabling 3D focal control [84]. This feature provides a new degree of tunability that has not been yet demonstrated by any previous variable lenses.

Figure 14.18a shows a bulk lens where light refracts only at the surface of a bulk lens and the lens material in the middle (shown in the darker color) makes no contribution to the steering of incoming light. Thus, a bulk lens can be broken down into smaller subsections, where the curved

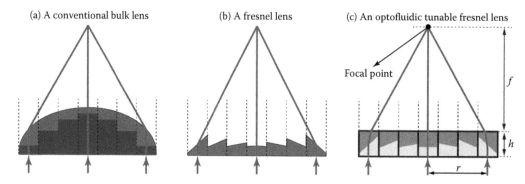

Figure 14.18 (a) A conventional bulk lens for light focusing. (b) A Fresnel lens approximates the performance of a bulk lens, while reducing the overall thickness by breaking it down into small subsections of the prism with different angles steeper at the edges. Such a thin configuration offers a low-cost and lightweight lens with a large aperture size. However, a solid Fresnel lens has fixed optical properties and is not capable of active focal control without any mechanical input. (c) To provide the tunability of spatial focal control for a Fresnel lens, we propose an optofluidic tunable Fresnel lens composed of linearly arrayed liquid prisms. By varying the prism apex angles at each individual prism using the EWOD technology, an array of the liquid prisms functions similarly to a solid Fresnel lens with the added versatility of adjustable light focusing. (Reproduced with permission from C. E. Clement and S.-Y. Park, *Applied Physics Letters*, vol. 108, p. 191601, 2016.)

surfaces in each section are replaced with flat surfaces of varying angles (Figure 14.18b). As a result, a Fresnel lens significantly reduces the lens material and cost, while offering a large aperture size. However, since conventional Fresnel lenses are made from solid materials (e.g., glass or plastic), their optical performance, such as a focal length, is fixed once the lens material and geometry has been designed. To provide a degree of tunability for a Fresnel lens, Park's group proposed an arrayed form of the liquid prisms where each prism is individually controlled by electrowetting to replicate the subsections of a solid Fresnel lens [84]. The set of discontinuous prisms in a Fresnel lens steers light through steep prism angles at the edges and incrementally smaller angles for inter-mediate prisms (see Figure 14.18c). By individually controlling each prism using electrowetting, they have achieved 3D focal control along the longitudinal ($263 \text{ mm} \leq f_{long} \leq \infty$) as well as the lateral ($0 \leq f_{lat} \leq 30 \text{ mm}$) directions. The optofluidic tunable Fresnel lens presented here would become a leading technology in compact and tunable optical systems, with applications ranging from imaging and motion tracking to multicollector solar systems.

The previous studies on liquid prisms have been typically conducted for various optics applications, such as free-space optical communication and laser detection and ranging. Recently, Park's group proposed to use arrayed liquid prisms for solar energy collection [89] and solar indoor lighting [137]. Figure 14.19 presents an arrayed microfluidic tunable prism panel that enables wide solar tracking and high solar concentration while minimizing energy loss. Each of the liquid prism modules is implemented by a microfluidic (i.e., nonmechanical) technology based on electrowetting for adaptive solar beam steering (Figure 14.19a). Therefore, the proposed platform offers a low cost, lightweight, and precise solar tracking system, while obviating the need for bulky and heavy mechanical moving parts essentially required for a conventional motor-driven solar tracker. By adjusting the contact angles at two and four sidewalls, various prism modulations are capable, such as single- as well as dual-axis solar tracking, as shown in Figure 14.19b. To further increase the aperture area, a basic prism module is arrayed, stacked up, and integrated with concentrating optics, as shown in Figure 14.19c. The arrayed microfluidic prism panel offers several advanta-geous features, including (1) reduced installation and maintenance cost by removing heavy and bulky mechanical moving parts such as dual-axis trackers, motors, and other related components, (2) reliable solar tracking and high concentration enabled by microfluidic-based precise sunlight modulation, (3) extremely low power consumption for operation in the range of $\sim$ mW, and (4) convenient installation and quiet rooftop operation due to the compact design of the system. This new tracking system, when used in conjunction with concentrating optics, can be revolutionary for rooftop solar applications and various other solar power technologies such as solar thermal heat-ing, solar indoor lighting, concentrated photovoltaic (CPV), concentrated solar power (CSP), and solar thermochemical reactions.

14.4 Summary

This chapter provides a brief review of the recent development of optofluidic technologies that combine the two disciplines of optics and microfluidics. Optofluidics has drawn great attention in broad areas of physical and biological sciences due to the advantages of optical grade smoothness at the interface, a high degree of optical tunability and adaptability, and device reconfigurability. There are two different types of optofluidic devices relying on these operational principles. The first type of optofluidic devices directly utilizes various light sources such as lasers, DMD, and LCD to achieve micro-/nanofluidic manipulation. Another type of optofluidic device utilizes exist-ing microfluidic principles for adaptive tuning of their optical performances, such as focal length, reflection/refraction, and waveguides. Based on this, we described the fundamentals and limita-tions of these two types of optofluidic devices and their applications in the areas of optics, energy, and biological sciences.

Light-driven microfluidic technologies can be basically categorized into three types based on how optical energy is converted to induce microfluidic forces: (1) direct optical, (2) opto-thermo-fluidic,

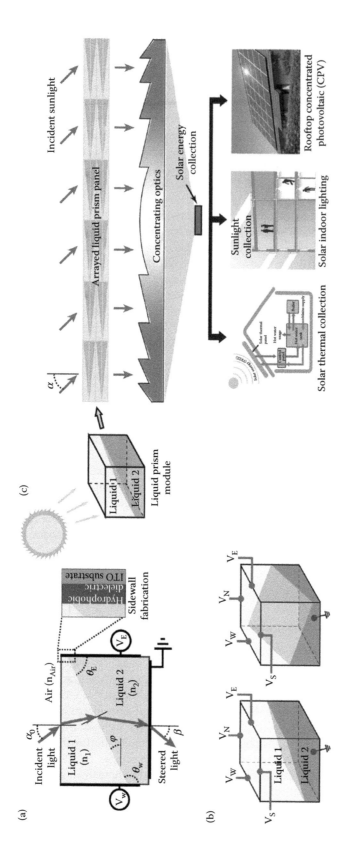

Figure 14.19 Schematic illustrations of (a) a basic liquid prism module, (b) various prism modulations and (c) an arrayed microfluidic solar energy collection system and its potential solar power applications. (a) Each of the liquid prism modules includes two immiscible liquids. The prism sidewall is fabricated by coating a hydrophobic dielectric layer on a conductive substrate. An electrowetting effect is induced by applying bias voltages and is able to adjust the apex angle (φ) of the prism. Due to the refractive index difference ($n_{air} \neq n_1 \neq n_2$) of two liquids, incident light is accordingly steered at each of the interfaces and thus the microfluidic device functions as an optical prism. (b) Modulation of the contact angles at two and four sidewalls allows for single- and dual-axis tracking, respectively. (c) A basic prism module is further stacked up and expanded to achieve an arrayed prism panel, which enables wide solar tracking and high concentration through concentrating optics. This microfluidic solar energy collector can be potentially useful for various solar power technologies such as solar thermal heating, solar indoor lighting, concentrated photovoltaic (CPV), concentrated solar power (CSP), and solar thermo-chemical reactions. (Reproduced with permission from V. Narasimhan et al. *Applied Energy*, vol. 162, pp. 450–459, 2016.)

and (3) optoelectrofluidic forces. Methods using direct optical forces are ideal tools for fundamental biological and physical science studies by precisely manipulating individual microscopic objects in 3D space. However, they require high power light sources, which limits their throughput and applications in other areas. Opto-thermal approaches utilize local optical heating to generate large forces for high throughput and high speed liquid manipulation. For example, the light-driven Marangoni effect generates the temperature gradient and modifies the surface tension force for manipulation of liquid droplets, while opto-thermal bubble actuation uses an intense laser pulse for high speed microfluidic actuation. Approaches using opto-electrical conversions could potentially provide a high throughput platform via massive parallel processing of a large number of droplets across a large area. The major difference of the opto-electrical method from opto-thermal and direct optical methods is that the optical energy is used to switch on an electrical driving force, which requires a much lower optical power and induces a minimum optical heating effect across a large area.

Another type of optofluidic device uses various microfluidic principles to overcome the optical performances that are typically fixed by conventional solid type optical elements such as lenses and prisms. Microfluidic integrated optical components allow for dynamic and adaptive control of the optical performances with the advantages of optical grade smoothness and device reconfigurability. We reviewed optofluidic technologies, particularly for lenses and prisms as fundamental optical elements. To provide the tunability of optical performances, an external pump is simply used either to deform the shape of the liquid lens or create the flat interface between two immiscible liquids for beam steering under the laminar flow condition. These pump-driven optofluidic devices are simple and straightforward, but they essentially require extra microfluidic components such as pumps/valves and tubes, which make them bulky and they consume a large amount of power. Electrowetting-driven optofluidic components have become popular to offer stand-alone and reconfigurable systems in small scales. By varying the applied bias voltages, the liquid interface is adaptively tuned either to form the curved shape for performing tunable lenses or to be flat for beam steering applications. An arrayed form of electrowetting-driven liquid prisms increases the aperture area to be used for solar energy collection and solar indoor lighting.

Despite many promising potential applications for optofluidic technologies, there are still several challenging issues. For the goal of portable integrated systems, device miniaturization is critical and the integrated system needs to reduce its footprint size, which is important for broadening future applications. With the rapid progresses in photonics and optoelectronics, low cost, compact, and high power light sources, switches, and other components can be realized. More fundamental studies and research efforts are required not only to find novel and useful optical actuation mechanisms that offer high optical energy conversion efficiency and low optical power requirement, but also to clearly understand small scale fluidic motions that can be integrated for more reconfigurable and versatile optical applications.

References

1. D. Erickson, D. Sinton, and D. Psaltis, Optofluidics for energy applications, *Nature Photonics*, vol. 5, pp. 583–590, 2011.
2. Y. Zhao, Z. S. Stratton, F. Guo, M. I. Lapsley, C. Y. Chan, S.-C. S. Lin et al., Optofluidic imaging: Now and beyond, *Lab on a Chip*, vol. 13, pp. 17–24, 2013.
3. S. Gray, A letter from Mr. Stephen Gray, giving a further account of his water microscope, *Philosophical Transactions*, vol. 19, pp. 353–356, 1695.
4. B. K. Gibson, Liquid mirror telescopes-history, *Journal of the Royal Astronomical Society of Canada*, vol. 85, p. 158, 1991.
5. D. J. Harrison, K. Fluri, K. Seiler, Z. Fan, C. S. Effenhauser, and A. Manz, Micromachining a miniaturized capillary electrophoresis-based chemical analysis system on a chip, *Science*, vol. 261, pp. 895–897, 1993.

6. P. Mach, M. Dolinski, K. W. Baldwin, J. A. Rogers, C. Kerbage, R. S. Windeler et al., Tunable microfluidic optical fiber, *Applied Physics Letters*, vol. 80, p. 4294, 2002.

7. J. J. Uebbing, S. Hengstler, D. Schroeder, S. Venkatesh, and R. Haven, Heat and fluid flow in an optical switch bubble, *Journal of Microelectromechanical Systems*, vol. 15, pp. 1528–1539, 2006.

8. D. Psaltis, S. R. Quake, and C. Yang, Developing optofluidic technology through the fusion of microfluidics and optics, *Nature*, vol. 442, pp. 381–386, 2006.

9. C. Monat, P. Domachuk, and B. J. Eggleton, Integrated optofluidics: A new river of light, *Nat Photon*, vol. 1, pp. 106–114, 2007.

10. Y. Fainman, L. Lee, D. Psaltis, and C. Yang, *Optofluidics - Fundamentals, Devices, and Applications*. USA: McGraw-Hill, 2009.

11. D. Jiang and S.-Y. Park, Light-driven 3D droplet manipulation on flexible optoelectrowetting devices fabricated by a simple spin-coating method, *Lab on a Chip*, vol. 16, pp. 1831–1839, 2016.

12. Y. C. Seow, S. P. Lim, and H. P. Lee, Optofluidic variable-focus lenses for light manipulation, *Lab on a Chip*, vol. 12, pp. 3810–3815, 2012.

13. W. Song and D. Psaltis, Electrically tunable optofluidic light switch for reconfigurable solar lighting, *Lab on a Chip*, vol. 13, pp. 2708–2713, 2013.

14. M. Pan, M. Kim, S. Kuiper, and S. K. Y. Tang, Actuating fluid-fluid interfaces for the reconfiguration of light, *IEEE Journal of Selected Topics in Quantum Electronics*, vol. 21, pp. 444–455, 2015.

15. H. Wang, K. Liu, K.-J. Chen, Y. Lu, S. Wang, W.-Y. Lin et al., A rapid pathway toward a superb gene delivery system: Programming structural and functional diversity into a supramolecular nanoparticle library, *ACS Nano*, vol. 4, pp. 6235–6243, 2010.

16. C.-C. Lee, G. Sui, A. Elizarov, C. J. Shu, Y.-S. Shin, A. N. Dooley et al., Multistep synthesis of a radiolabeled imaging probe using integrated microfluidics, *Science*, vol. 310, pp. 1793–1796, 2005.

17. L. S. Roach, H. Song, and R. F. Ismagilov, Controlling nonspecific protein adsorption in a plug-based microfluidic system by controlling interfacial chemistry using fluorous-phase surfactants, *Analytical Chemistry*, vol. 77, pp. 785–796, 2005.

18. B. T. C. Lau, C. A. Baitz, X. P. Dong, and C. L. Hansen, A complete microfluidic screening platform for rational protein crystallization, *Journal of the American Chemical Society*, vol. 129, p. 454, 2007.

19. D. L. Chen, C. J. Gerdts, and R. F. Ismagilov, Using microfluidics to observe the effect of mixing on nucleation of protein crystals, *Journal of the American Chemical Society*, vol. 127, pp. 9672–9673, 2005.

20. W. Li, H. H. Pham, Z. Nie, B. MacDonald, A. Genther, and E. Kumacheva, Multi-step microfluidic polymerization reactions conducted in droplets: The internal trigger approach, *Journal of the American Chemical Society*, vol. 130, pp. 9935–9941, 2008.

21. A. Huebner, M. Srisa-Art, D. Holt, C. Abell, F. Hollfelder, A. J. deMello et al., Quantitative detection of protein expression in single cells using droplet microfluidics, *Chemical Communications*, vol. 12, pp. 1218–1220, 2007.

22. J. Gong and C.-J. Kim, All-electronic droplet generation on-chip with real-time feedback control for EWOD digital microfluidics, *Lab on a Chip*, vol. 8, pp. 898–906, 2008.

23. A. R. Wheeler, Putting electrowetting to work, *Science*, vol. 322, pp. 539–540, 2008.

24. K. Ahn, C. Kerbage, T. P. Hunt, R. M. Westervelt, and D. R. Link, Dielectrophoretic manipulation of drops for high-speed microfluidic sorting devices, *Applied Physics Letters*, vol. 88, p. 024104, 2006.

25. J. R. Millman, K. H. Bhatt, B. G. Prevo, and O. D. Velev, Anisotropic particle synthesis in dielectrophoretically controlled microdroplet reactors, *Nature Materials*, vol. 4, pp. 98–102, 2005.

26. J. Z. Chen, S. M. Troian, A. A. Darhuber, and S. Wagner, Effect of contact angle hysteresis on thermocapillary droplet actuation, *Journal of Applied Physics*, vol. 97, p. 014906, 2005.

27. E. F. Greco and R. O. Grigoriev, Thermocapillary migration of interfacial droplets, *Physics of Fluidics*, vol. 21, p. 042105, 2009.
28. T. Franke, A. R. Abate, D. A. Weitz, and A. Wixforth, Surface acoustic wave (SAW) directed droplet flow in microfluidics for PDMS devices, *Lab on a Chip*, vol. 9, pp. 2625–2627, 2009.
29. M. Okochi, H. Tsuchiya, F. Kumazawa, M. Shikida, and H. Honda, Droplet-based gene expression analysis using a device with magnetic force-based-droplet-handling system, *Journal of Bioscience & Bioengineering*, vol. 109, pp. 193–197, 2009.
30. Y. Zhang, S. Park, K. Liu, J. Tsuan, S. Yang, and T.-H. Wang, A surface topography assisted droplet manipulation platform for biomarker detection and pathogen identification, *Lab on a Chip*, vol. 11, pp. 398–406, 2011.
31. J. K. Valley, S. NingPei, A. Jamshidi, H.-Y. Hsu, and M. C. Wu, A unified platform for optoelectrowetting and optoelectronic tweezers, *Lab on a Chip*, vol. 11, pp. 1292–1297, 2011.
32. S.-Y. Park, S. Kalim, C. Callahan, M. A. Teitell, and E. P. Y. Chiou, A light-induced dielectrophoretic droplet manipulation platform, *Lab on a Chip*, vol. 9, pp. 3228–3235, 2009.
33. A. Ashkin, Acceleration and trapping of particles by radiation pressure, *Physical Review Letters*, vol. 24, pp. 156–159, 1970.
34. A. Ashkin, J. M. Dziedzic, J. E. Bjorkholm, and S. Chu, Observation of a single-beam gradient force optical trap for dielectric particles, *Optics Letters*, vol. 11, pp. 288–290, 1986.
35. J. E. Curtis, B. A. Koss, and D. G. Grier, Dynamic holographic optical tweezers, *Optica Communication*, vol. 207, pp. 169–175, 2002.
36. K. Sasaki, M. Koshioka, H. Misawa, N. Kitamura, and H. Masuhara, Optical trapping of a metal particle and a water droplet by a scanning laser beam, *Applied Physics Letters*, vol. 807, p. 107427, 1992.
37. A. Ashkin and J. M. Dziedzic, Optical trapping and manipulation of viruses and bacteria, *Science*, vol. 235, pp. 1517–1520, 1987.
38. A. Ashkin, J. M. Dziedzic, and T. M. Yamane, Optical trapping and manipulation of single cells using infrared laser beams, *Nature*, vol. 330, pp. 769–771, 1987.
39. D. G. Grier, A revolution in optical manipulation, *Nature Photonics*, vol. 424, pp. 810–816, 2003.
40. U. Bockelmann, P. Thomen, B. Essevaz-Roulet, V. Viasnoff, and F. Heslot, Unzipping DNA with optical tweezers: High sequence sensitivity and force flips, *Biophysical Journal*, vol. 82, pp. 1537–1553, 2002.
41. J. E. Molloy and M. J. Padgett, Lights, action: Optical tweezers, *Contemporary Physics*, vol. 43, p. 241, 2002.
42. S. C. Kuo and M. P. Sheetz, Force of single kinesin molecules measured with optical tweezers, *Science*, vol. 260, pp. 232–234, 1993.
43. S. Chu, Laser manipulation of atoms and particles, *Science*, vol. 253, pp. 861–866, 1991.
44. K. Svoboda and S. M. Block, Biological applications of optical forces, *Annual Review of Biophysics and Biomolecular Structure*, vol. 23, pp. 247–285, 1994.
45. R. J. Hopkins, L. Mitchem, A. D. Ward, and J. P. Reid, Control and characterisation of a single aerosol droplet in a single-beam gradient-force optical trap, *Physical Chemistry Chemical Physics*, vol. 6, pp. 4924–4927, 2004.
46. A. Ashkin and J. M. Dziedzic, Optical levitation by radiation pressure, *Applied Physics Letters*, vol. 19, pp. 283–285, 1971.
47. A. Ashkin and J. M. Dziedzic, Optical levitation of liquid drops by radiation pressure, *Science*, vol. 187, pp. 1073–1075, 1975.
48. P. T. Nagy and G. P. Neitzel, Optical levitation and transport of microdroplets: Proof of concept, *Physics of Fluidics*, vol. 20, p. 101703, 2008.
49. M. I. Kohira, A. Isomura, N. Magome, S. Mukai, and K. Yoshikawa, Optical levitation of a droplet under a linear increase in gravitational acceleration, *Chemical Physics Letters*, vol. 414, pp. 389–392, 2005.

50. N. Jordanov and R. Zellner, Investigations of the hygroscopic properties of ammonium sulfate and mixed ammonium sulfate and glutaric acid micro droplets by means of optical levitation and Raman spectroscopy, *Physical Chemistry Chemical Physics*, vol. 8, pp. 2759–2764, 2006.

51. K. T. Gahagan and J. G. A. Swartzlander, Trapping of low-index microparticles in an optical vortex, *Journal of the Optical Society of America B*, vol. 15, pp. 524–534, 1998.

52. K. T. Gahagan and J. G. A. Swartzlander, Optical vortex trapping of particles, *Optics Letters*, vol. 21, pp. 827–829, 1996.

53. R. M. Lorenz, J. S. Edgar, G. D. M. Jeffries, Y. Zhao, D. McGloin, and D. T. Chiu, Vortex-trap induced fusion of femtoliter-volume aqueous droplets, *Analytical Chemistry*, vol. 79, pp. 224–228, 2007.

54. G. D. M. Jeffries, J. S. Kuo, and D. T. Chiu, Dynamic modulation of chemical concentration in an aqueous droplet, *Angewandte Chemie International Edition*, vol. 46, pp. 1326–1328, 2007.

55. D. T. Chiu and R. M. Lorenz, Chemistry and biology in femtoliter and picoliter volume droplets, *Accounts of Chemical Research*, vol. 42, pp. 649–658, 2009.

56. L. J. Radziemski and D. A. Cremers, *Laser-Induced Plasmas and Applications*: Marcel Dekker, New York, NY, 1989.

57. N. O. Young, J. S. Goldstein, and M. J. Block, The motion of bubbles in a vertical temperature gradient, *Journal of Fluid Mechanics*, vol. 6, pp. 350–356, 1959.

58. K. D. Barton and R. S. Subramanian, The migration of liquid drops in a vertical temperature gradient, *Journal of Colloid and Interface Science*, vol. 133, pp. 211–222, 1989.

59. B. Selva, V. Miralles, I. Cantat, and M. C. Jullien. Thermocapillary actuation by optimized resistor pattern: Bubbles and droplets displacing, switching and trapping, *Lab on a Chip*, vol. 10, pp. 1835–1840, 2010.

60. A. A. Darhuber, J. P. Valentino, J. M. Davis, S. M. Troian, and S. Wagner. Microfluidic actuation by modulation of surface stresses, *Applied Physics Letters*, vol. 82, pp. 657–659, 2003.

61. K. T. Kotz, K. A. Noble, and G. W. Faris. Optical microfluidics, *Applied Physics Letters*, vol. 85, pp. 2658–2660, 2004.

62. C. N. Baroud, J.-P. Delville, F. Gallaire, and R. Wunenburger, Thermocapillary valve for droplet production and sorting, *Physical Review E*, vol. 75, p. 046302, 2007.

63. M. R. d. S. Vincent, R. Wunenburger, and J.-P. Delville, Laser switching and sorting for high speed digital microfluidics, *Applied Physics Letters*, vol. 92, p. 154105, 2008.

64. A. T. Ohta, A. Jamshidi, J. K. Valley, H.-Y. Hsu, and M. C. Wu, Optically actuated thermocapillary movement of gas bubbles on an absorbing substrate, *Applied Physics Letters*, vol. 91, p. 074103, 2007.

65. W. Hu and A. T. Ohta, Aqueous droplet manipulation by optically induced Marangoni circulation, *Microfluidics and Nanofluidics*, vol. 11, pp. 307–316, 2011.

66. Y. R. Shen, *The Principles of Nonlinear Optics*: Wiley, New York, 1984.

67. P.-Y. Chiou, T.-H. Wu, S.-Y. Park, and Y. Chen, Pulse laser driven ultrafast micro and nanofluidics system, In *Proceedings of the SPIE*, San Diego, California, vol. 7759, p. 77590Z, 2010.

68. K. R. Rau, P. A. Quinto-Su, A. N. Hellman, and V. Venugopalan, Pulsed laser microbeam-induced cell lysis: Time-resolved imaging and analysis of hydrodynamic effects, *Biophysical Journal*, vol. 91, pp. 317–329, 2006.

69. E.-A. Brujan, K. Nahen, P. Schmidt, and A. Vogel, Dynamics of laser-induced cavitation bubbles near an elastic boundary, *Journal of Fluid Mechanics*, vol. 433, pp. 251–281, 2001.

70. R. K. Chang, J. H. Eickmans, W.-F. Hsieh, C. F. Wood, J.-Z. Zhang, and J.-b. Zheng, Laser-induced breakdown in large transparent water droplets, *Applied Optics*, vol. 27, pp. 2377–2385, 1988.

71. A. Vogel, S. Busch, and U. Parlitz, Shock wave emission and cavitation bubble generation by picosecond and nanosecond optical breakdown in water, *Journal of Acoustic Society of America*, vol. 100, pp. 148–165, 1996.

72. E. Zwaan, S. v. L. Gac, K. Tsuji, and C.-D. Ohl, Controlled cavitation in microfluidic systems, *Physics Review Letters*, vol. 98, p. 254501, 2007.

73. A. N. Hellman, K. R. Rau, H. H. Yoon, S. Bae, J. F. Palmer, K. S. Phillips et al., Laser-induced mixing in microfluidic channels, *Analytical Chemistry*, vol. 79, pp. 4484–4492, 2007.

74. R. Dijkink and C.-D. Ohl, Laser-induced cavitation based micropump, *Lab on a Chip*, vol. 8, pp. 1676–1681, 2008.

75. T.-H. Wu, L. Gao, Y. Chen, K. Wei, and P.-Y. Chiou, Pulsed laser triggered high speed microfluidic switch, *Applied Physics Letters*, vol. 93, p. 144102, 2008.

76. S.-Y. Park, T.-H. Wu, Y. Chen, M. A. Teitell, and P.-Y. Chiou, High-speed droplet generation on demand driven by pulse laser-induced cavitation, *Lab on a Chip*, vol. 11, pp. 1010–1012, 2011.

77. Z. G. Li, K. Ando, J. Q. Yu, A. Q. Liu, J. B. Zhang, and C. D. Ohl, Fast on-demand droplet fusion using transient cavitation bubbles, *Lab on a Chip*, vol. 11, pp. 1879–1885, 2011.

78. P. R. C. Gascoyne, J. V. Vykoukal, J. A. Schwartz, T. J. Anderson, D. M. Vykoukal, K. W. Current et al., Dielectrophoresis-based programmable fluidic processors, *Lab on a Chip*, vol. 4, pp. 299–309, 2004.

79. H. J. J. Verheijen and M. W. J. Prins, Reversible electrowetting and trapping of charge: Model and experiments, *Langmuir*, vol. 15, pp. 6616–6620, 1999.

80. H. Moon, S. K. Cho, R. L. Garrell, and C.-J. Kim, Low voltage electrowetting-on-dielectric, *Journal of Applied Physics*, vol. 92, pp. 4080–4087, 2002.

81. K. Ugsornrat, T. Maturus, A. Jomphoak, T. Pogfai, N. V. Afzulpurkar, A. Wisitsoraat et al., Simulation and experimental study of electrowetting on dielectric (EWOD) device for a droplet based polymerase chain reaction system, in *The 13th International Conference on Biomedical Engineering*, 2008, pp. 859–862.

82. S.-Y. Park, M. A. Teitell, and E. P. Y. Chiou, Single-sided continuous optoelectrowetting (SCOEW) for droplet manipulation with light patterns, *Lab on a Chip*, vol. 10, pp. 1655–1661, 2010.

83. D. Y. Kim and A. J. Steckl, Electrowetting on paper for electronic paper display, *ACS Applied Materials and Interfaces*, vol. 2, pp. 3318–3323, 2010.

84. C. Clement, S. K. Thio, and S.-Y. Park, An optofluidic tunable Fresnel lens for spatial focal control based on electrowetting-on-dielectric (EWOD), *Sensors and Actuators B: Chemical*, vol. 240, pp. 909–915, 2017.

85. K. Mohseni and E. Baird, Digitized heat transfer using electrowetting on dielectric, *Nanoscale and Microscale Thermophysical Engineering*, vol. 11, pp. 90–108, 2007.

86. S.-Y. Park, J. Cheng, and C.-L. Chen, Active hot spot cooling driven by single-side electrowetting-on-dielectric (SEWOD), in *Proceeding of the ASME 2012 Summer Heat Transfer Conference (ASME HT 2012)*, Puerto Rico, USA, 2012.

87. S.-Y. Park and Y. Nam, Single-sided digital microfluidic (SDMF) devices for effective coolant delivery and enhanced two-phase cooling, *Micromachines*, vol. 8, p. 3, 2017.

88. J. Cheng, S.-Y. Park, and C.-L. Chen, Optofluidic solar concentrators using electrowetting tracking: Concept, *Design, and Characterization, Solar Energy*, vol. 89, pp. 152–167, 2013.

89. V. Narasimhan, D. Jiang, and S.-Y. Park, Design and optical analyses of an arrayed microfluidic tunable prism panel for enhancing solar energy collection, *Applied Energy*, vol. 162, pp. 450–459, 2016.

90. M. Barberoglou, V. Zorba, A. Pagozidis, C. Fotakis, and E. Stratakis, Electrowetting properties of micro/nanostructured black silicon, *Langmuir*, vol. 26, pp. 13007–13014, 2010/08/03 2010.

91. M. G. Lippmann, Relations entre les phénomènes electriques et capillaires, *Annales de Chimie et de Physique*, vol. 5, pp. 494–549, 1875.

92. M. G. Pollack, F. Richard B, and A. D. Shenderov, Electrowetting-based actuation of liquid droplets for microfluidic applications, *Applied Physics Letters*, vol. 77, pp. 1725–1726, 2000.

93. J. Lee, H. Moon, J. Fowler, T. Schoellhammer, and C.-J. Kim, Electrowetting and electrowetting-on-dielectric for microscale liquid handling, *Sensors and Actuators A: Physical*, vol. 95, pp. 259–268, 2002.

94. P.-Y. Chiou, Z. Chang, and M. C. Wu, Droplet manipulation with light on optoelectrowetting device, *Journal of Microelectromechanical Systems*, vol. 17, pp. 133–138, 2008.

95. P. Y. Chiou, H. Moon, H. Toshiyoshi, C. J. Kim, and M. C. Wu, Light actuation of liquid by opto-electrowetting, *Sensors and Actuators, A: Physical*, vol. 104, pp. 22–228, 2003.
96. H.-S. Chuang, A. Kumar, and S. T. Wereley, Open optoelectrowetting droplet actuation, *Applied Physics Letters*, vol. 93, p. 064104, 2008.
97. P. Y. Chiou, S.-Y. Park, and M. Wu, Continuous optoelectrowetting for picoliter droplet manipulation, *Applied Physics Letters*, vol. 93, p. 221110, 2008.
98. M. Abdelgawad, S. L. S. Freire, H. Yang, and A. R. Wheeler, All-terrain droplet actuation, *Lab on a Chip*, vol. 8, pp. 672–677, 2008.
99. S.-K. Fan, H. Yang, and W. Hsu, Droplet-on-a-wristband: Chip-to-chip digital microfluidic interfaces between replaceable and flexible electrowetting modules, *Lab on a Chip*, vol. 11, pp. 343–347, 2011.
100. Z. Shen, T. Masuda, H. Takagishi, K. Ohdaira, and T. Shimoda, Fabrication of high-quality amorphous silicon film from cyclopentasilane by vapor deposition between two parallel substrates, *Chemical Communications*, vol. 51, pp. 4417–4420, 2015.
101. V. Zardetto, T. M. Brown, A. Reale, and A. d. Carlo, Substrates for flexible electronics: A practical investigation on the electrical, film flexibility, optical, temperature, and solvent resistance properties, *Journal of Polymer Science Part B: Polymer Physics*, vol. 49, pp. 638–648, 2011.
102. H.-j. Ni, J.-g. Liu, Z.-h. Wang, and S.-y. Yang, A review on colorless and optically transparent polyimide films: Chemistry, process and engineering applications, *Journal of Industrial and Engineering Chemistry*, vol. 28, pp. 16–27, 2015.
103. T. B. Jones, *Electromechanics of Particles*: Cambridge University Press, New York, 1995.
104. H. Pohl, *Dielectrophoresis*: Cambridge University Press, Cambridge, 1978.
105. M. Abdelgawad, P. Park, and A. R. Wheeler, Optimization of device geometry in single-plate digital microfluidics, *Journal of Applied Physics*, vol. 105, p. 094506, 2009.
106. A. Al-Jarro, J. Paul, D. W. P. Thomas, J. Crowe, N. Sawyer, F. R. A. Rose et al., Direct calculation of Maxwell stress tensor for accurate trajectory prediction during DEP for 2 D and 3 D structures, *Journal of Physics D: Applied Physics*, vol. 40, pp. 71–77, 2007.
107. T. B. Jones, Basic theory of dielectrophoresis and electrorotation, *IEEE Engineering in Medicine and Biology Magazine*, vol. 22, pp. 33–42, 2003.
108. C. Rosales and K. M. Lim, Numerical comparison between Maxwell stress method and equivalent multipole approach for calculation of the dielectrophoretic force in single-cell traps, *Electrophoresis*, vol. 26, pp. 2057–2065, 2005.
109. X. J. Wang, X. B. Wang, and P. R. C. Gascoyne, General expressions for dielectrophoretic force and electrorotational torque derived using the Maxwell stress tensor method, *Journal of Electrostatics*, vol. 39, pp. 277–295, 1997.
110. P. Y. Chiou, A. T. Ohta, and M. C. Wu, Massively parallel manipulation of single microparticles using optical images, *Nature*, vol. 463, pp. 370–372, 2005.
111. A. Jamshidi, P. J. Pauzauskie, P. J. Schuck, A. T. Ohta, P.-Y. Chiou, J. Chou et al., Dynamic manipulation and separation of individual semiconducting and metallic nanowires, *Nature Photonics*, vol. 2, pp. 86–89, 2008.
112. S.-Y. Park, C. Pan, T.-H. Wu, C. Kloss, S. Kalim, C. E. Callahan et al., Floating electrode opto-electronic tweezers: Light-driven dielectrophoretic droplet manipulation in electrically insulating oil medium, *Applied Physics Letters*, vol. 92, p. 151101, 2008.
113. N.-T. Nguyen, Micro-optofluidic lenses: A review, *Biomicrofluidics*, vol. 4, p. 031501, 2010.
114. N. Chronis, G. Liu, K.-H. Jeong, and L. Lee, Tunable liquid-filled microlens array integrated with microfluidic network, *Optics Express*, vol. 11, pp. 2370–2378, 2003.
115. C. Jackie, W. Weisong, F. Ji, and V. Kody, Variable-focusing microlens with microfluidic chip, *Journal of Micromechanics and Microengineering*, vol. 14, p. 675, 2004.
116. X. Mao, J. R. Waldeisen, B. K. Juluri, and T. J. Huang, Hydrodynamically tunable optofluidic cylindrical microlens, *Lab on a Chip*, vol. 7, pp. 1303–1308, 2007.

117. J. Shi, Z. Stratton, S.-C. S. Lin, H. Huang, and T. J. Huang, Tunable optofluidic microlens through active pressure control of an air–liquid interface, *Microfluidics and Nanofluidics*, vol. 9, pp. 313–318, 2010.
118. H. Ren and S.-T. Wu, Tunable-focus liquid microlens array using dielectrophoretic effect, *Optics Express*, vol. 16, pp. 2646–2652, 2008.
119. S. W. Lee and S. S. Lee, Focal tunable liquid lens integrated with an electromagnetic actuator, *Applied Physics Letters*, vol. 90, p. 121129, 2007.
120. R. A. Hayes and B. J. Feenstra, Video-speed electronic paper based on electrowetting, *Nature*, vol. 425, pp. 383–385, 2003.
121. T. Krupenkin and J. A. Taylor, Reverse electrowetting as a new approach to high-power energy harvesting, *Nature Communications*, vol. 2, p. 448, 2011.
122. C. B. Gorman, H. A. Biebuyck, and G. M. Whitesides, Control of the shape of liquid lenses on a modified gold surface using an applied electrical potential across a self-assembled monolayer, *Langmuir*, vol. 11, pp. 2242–2246, 1995.
123. T. Krupenkin, S. Yang, and P. Mach, Tunable liquid microlens, *Applied Physics Letters*, vol. 82, p. 316, 2003.
124. C.-X. Liu, J. Park, and J.-W. Choi, A planar lens based on the electrowetting of two immiscible liquids, *Journal of Micromechanics and Microengineering*, vol. 18, p. 035023, 2008.
125. S. Kuiper and B. H. W. Hendriks, Variable-focus liquid lens for miniature cameras, *Applied Physics Letters*, vol. 85, p. 1128, 2004.
126. F. Krogmann, W. Mönch, and H. Zappe, A MEMS-based variable micro-lens system, *Journal of Optics A: Pure and Applied Optics*, vol. 8, p. S330, 2006.
127. C. Li and H. Jiang, Electrowetting-driven variable-focus microlens on flexible surfaces, *Applied Physics Letters*, vol. 100, pp. 231105–2311054, 2012.
128. N. R. Smith, L. Hou, J. Zhang, and J. Heikenfeld, Fabrication and demonstration of electrowetting liquid lens arrays, *Journal of Display Technology*, vol. 5, pp. 411–413, 2009.
129. S. Xiong, A. Liu, L. Chin, and Y. Yang, An optofluidic prism tuned by two laminar flows, *Lab on a Chip*, vol. 11, pp. 1864–1869, 2011.
130. N. R. Smith, D. C. Abeysinghe, J. W. Haus, and J. Heikenfeld, Agile wide-angle beam steering with electrowetting microprisms, *Optics Express*, vol. 14, pp. 6557–6563, 2006.
131. L. Hou, N. R. Smith, and J. Heikenfeld, Electrowetting manipulation of any optical film, *Applied Physics Letters*, vol. 90, p. 251114, 2007.
132. J. Cheng and C.-L. Chen. Adaptive beam tracking and steering via electrowetting-controlled liquid prism, *Applied Physics Letters*, vol. 99, p. 191108, 2011.
133. D.-G. Lee, J. Park, J. Bae, and H.-Y. Kim, Dynamics of a microliquid prism actuated by electrowetting, *Lab on a Chip*, vol. 13, pp. 274–279, 2013.
134. C. E. Clement and S.-Y. Park, High-performance beam steering using electrowetting-driven liquid prism fabricated by a simple dip-coating method, *Applied Physics Letters*, vol. 108, p. 191601, 2016.
135. D. Brassardab, L. Malicac, F. Normandina, M. Tabrizianc, and T. Veres, Water-oil core-shell droplets for electrowetting-based digital microfluidic devices, *Lab on a Chip*, vol. 8, pp. 1342–1349, 2008.
136. L. Hou, J. Zhang, N. Smith, J. Yang, and J. Heikenfeld, A full description of a scalable microfabrication process for arrayed electrowetting microprisms, *Journal of Micromechanics and Microengineering*, vol. 20, p. 015044, 2010.
137. D. Jiang and S.-Y. Park, Microfluidic-guided solar indoor lighting system, in *Proceeding of the 6th International Multidisciplinary Conference on Optofluidics*, Beijing, China, 2016.

15

Implantable CMOS microphotonic devices

Jun Ohta and Takashi Tokuda

Contents

15.1 Introduction

Recently, genetic technologies combined with optics have emerged and become necessary tools for life science. For example, green fluorescence protein (GFP) [1] is widely used as an optical tag for a target protein, and channel rhodopsin 2 (ChR2) is an effective tool for optically stimulating specific cells, which is called "Optogenetics" [2]. This can be used in transgenic mice or introduced by a virus such as an adeno-associated virus (AAV). Since "Optogenetics" is an established word for optical stimulation with a light sensitive protein, we define the technology that combines genetic technology and photonic device technology as "Photogenetics" in this chapter. Photogenetics includes fluorescent proteins such as GFP, as well as light sensitive proteins. The most important

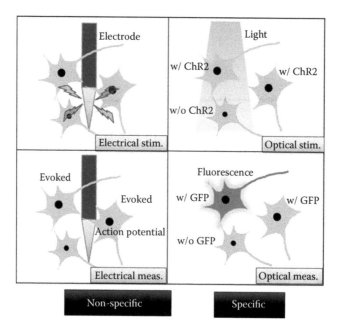

Figure 15.1 Comparison between electrical and optical methods for measuring and stimulating biological functions.

feature of photogenetics is its specific reaction with light: in GFP, a specific protein can be detected through fluorescence, and in ChR2, specific cells can be evoked through blue light. Electrical methods do not have such features, and hence are nonspecific. Figure 15.1 shows a comparison between the electrical and optical methods for measuring and stimulating biological functions.

To measure and control biological functions with photogenetics technology, photonic devices such as light emitting diodes (LEDs), image sensors, and optical fibers play an important role. The invention of blue LEDs is useful in photogenetics because a blue light is used to stimulate ChR2-introduced cells and to excite GFP-introduced cells for fluorescence. The advancement of CMOS image sensors has expanded their applications as biomedical devices. Biomedical devices that use image sensors can be classified into three configurations, as shown in Figure 15.2. The first type is a conventional configuration, such as in an optical microscope, where an image sensor is attached to the optics of the microscope (Figure 15.2a). Conventional sensors can also be used in this configuration. An example of this type is a fluorescent microscope equipped with a high speed CMOS image sensor for measuring neural activities using a voltage sensitive dye (VSD) [3]. Another example is an ultra-high-speed CMOS image sensor with a nanosecond response for fluorescent lifetime imaging microscopy (FLIM) [4]. The second type of device is a hermetic device, in

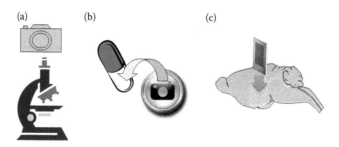

Figure 15.2 Classification of biomedical devices that use image sensors: (a) conventional type, (b) capsule type, and (c) direct implantation type.

which an image sensor is installed in an enclosure. A capsule endoscope [5] is a typical example of this type of device, as illustrated in Figure 15.2b. In this configuration, the image sensor is required to be small enough to fit into small enclosures, such as a capsule that can be swallowed. Recently, a third type of device has emerged, in which biomaterials or living tissues containing fluorophores are in direct contact with the surface of an image sensor. This allows for more compact measurement systems to be achieved [6]. Such a compact system can also be implanted into a living body, as shown in Figure 15.2c. The implantation of an imaging system also allows for new clinical device applications. A typical example of such a clinical application of this type is a retinal prosthesis [7], in which an imaging device of some type is implanted into the eye.

In the above examples, miniaturization technology is essential for implantable devices. Microphotonic devices such as microsized LEDs and miniaturized CMOS image sensors are suitable for implantation. In the following sections, we describe three examples of microphotonic devices for biomedical applications. First, we mention a brain-implantable imaging device for the third type of device in Figure 15.2. Second, we introduce an implantable glucose sensor as the second type of device in Figure 15.2. Third, microphotonic devices for optogenetics are described. Finally, this chapter is summarized.

15.2 Brain-implantable imaging device

15.2.1 Concept and fundamental structure of brain-implantable imaging devices

To measure brain activity in real time, several tools have been developed. These include electroencephalography (EEG), functional magnetic resonance imaging (fMRI), positron emission tomography (PET), and functional near infrared spectroscopy (fNIRS) or optical tomography. Although they are very powerful and noninvasive tools for investigating brain activity in humans, it is difficult to use these tools on small animals such as mice without tethering them. For small experimental animals, invasive measurements are acceptable, and electrical measurement by inserting electrodes into the brains of untethered animals is conventionally used. In addition to simple metal electrodes, several types of sophisticated Si probes have been developed, such as the Utah probe [8], the Michigan probe [9], and the Toyohashi probe [10]. These probes measure neuronal activity at multiple points in the brain. However, it is noted that electrical measurements are nonspecific, as mentioned in the previous section, and thus, in many experiments, optical measurement using fluorescence is required. In addition, an optical method using fluorescence can measure a large amount of data over a wide area. A fluorescence microscope is conventionally used for the measurements, and to some extent, it can be used to measure the neuronal activity of an awake and head-restrained mouse [11]. The mouse can walk on an air-supported styrofoam ball while its head is fixed. However, it is difficult to apply this to freely moving rodents.

To overcome this difficulty, three types of imaging devices have been developed: an optical fiber endoscope [12–14], a head-mountable device [15–17], and a brain-implantable device [18–24], as shown in Figure 15.3. In the optical fiber endoscope device and head-mountable device, an image sensor or a photodetector with a scanning mirror, optics, and excitation light sources are placed outside the body of an animal, as shown in Figure 15.3a and b. An optical fiber or an optical GRIN rod lens is inserted in the brain for an optical fiber endoscope or a head-mountable device, respectively. Although the optical fiber is flexible, it is still limited in the amount it can bend. In addition, the field of view is not very large. Compared with a fiber endoscope device, electrical wires are more flexible than optical fibers, so that an animal with a head-mountable device can move more smoothly than one with an endoscope system.

Recently, a micro imaging device was developed that can be implanted in a rodent brain and can measure brain activities in freely moving behavior [18–24]. The device has features that are different from those of any other head-mountable microscope devices [15–17] for several reasons.

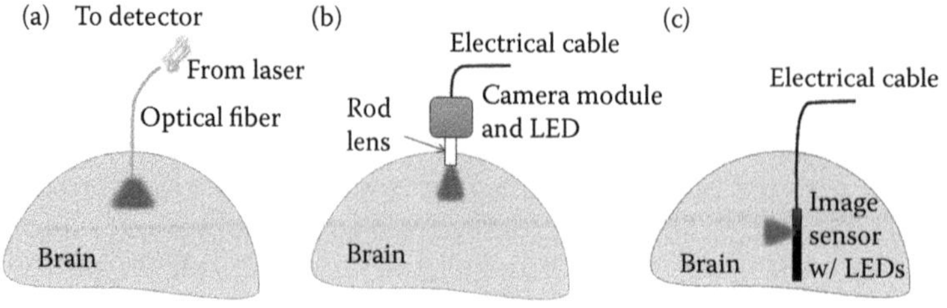

(a) To detector

From laser

Optical fiber

Brain

(b)

Electrical cable

Rod
lens

Camera module
and LED

Brain

(c)

Electrical cable

Image
sensor
w/ LEDs

Brain

Figure 15.3 Three types of imaging devices to measure brain activity in an untethered rodent: (a) optical fiber endoscope, (b) head-mountable device, and (c) brain-implantable device.

A head-mountable microscope device has high spatial resolution with a small area of view (like a conventional microscope), while an implantable device has a medium resolution of approximately a few tens of micrometers with a wide area of view, since it has no optics and does not make direct contact with tissues. In addition, the size of the device is so small that it is possible to implant multiple devices in different places of the brain. Since the device has no optics and only consists of a CMOS image sensor and LEDs for the excitation of the fluorescence, its weight is approximately 0.02 g. By contrast, the head-mountable microscope device is typically approximately 2 g. Such a light weight allows the rodent in which the device is implanted to move smoothly without any stress.

The brain-implantable CMOS imaging device is based on a dedicated CMOS image sensor fabricated using standard CMOS technology. The pixel structure is a three transistor active pixel sensor (3T-APS) with a parasitic photodiode composed of n-wells and p-substrate junctions. The number of inputs/outputs is limited to four in order to minimize the constraint on the implanted animals owing to the external wires [20]. The pixel size was designed to be $7.5 \times 7.5 \ \mu m^2$, which is sufficient for imaging neural cells because the spatial resolution is directly related to the pixel size. In this setup, only objects in the vicinity of the sensor can be clearly observed because the device has no imaging optics. The frame rate of the device can be varied by external control. The frame rate depends on the input light power: if the input light power is high, the frame rate can be higher. The device is thin and long enough to observe the corpus striatum in the basal ganglion with minimal invasion. In addition, the dimensions of the device are expected to allow implantation into deeper regions of the brain and to enable stable imaging in the deep brain regions of a freely moving mouse.

The device is compactly packaged in a polyimide substrate with LEDs and an excitation light filter for fluorescent imaging and improved biocompatibility. The polyimide substrate is fully flexible and has been shown to be biocompatible for the surgical insertion of implanted devices into the living body. The color filter blocks the excitation light, allowing only fluorescent emissions to reach the image sensor. LEDs are also implemented on the polyimide substrate. In order to protect the chip during the experiment, the device is coated with Parylene-C, which is optically transparent and waterproof.

Figure 15.4 shows two examples of the fabricated devices. One device is implanted on the surface of the brain (planar-type), whereas the other device is implanted in the deep brain (needle-type).

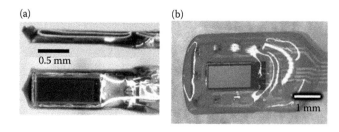

(a)

0.5 mm

(b)

1 mm

Figure 15.4 Two types of brain-implantable devices: (a) the side (upper) and top (lower) view of needle type and (b) the top view of planar type.

Table 15.1 Specifications of brain-implantable devices

Type	Needle type	Planer type
Die size	450 μm × 1500 μm	1000 μm × 3500 μm
Pixel number	40 × 120	120 × 268
Array size	300 μm × 900 μm	900 μm × 2010 μm
Pixel size	7.5 μm × 7.5 μm	
Fill factor	35%	
Pixel structure	3-transistor active pixel sensor	
Technology	AMS 0.35 μm CMOS 2P4M	

In Figure 15.4a, a close-up view of the needle-type device is shown. In the device, two LEDs are placed on both sides of the CMOS image sensor. The specifications for the two types are listed in Table 15.1. In the next section, typical experimental results for the two types of devices are mentioned.

15.2.2 Experimental results of the planar-type device

The planar-type device makes direct contact with the brain surface, as shown in Figure 15.5 [22]. Compared with the needle-type, the planar-type has some advantages. First, since the planar-type device can be tightly contacted to the tissue compared with the needle-type device, the distance between the surface of the device and the tissue can be smaller than that of the needle-type device. Consequently, the value of the spatial resolution is close to the pixel size (7.5 μm), which is much smaller than that of the needle-type device. Second, the surface area is larger than that of the needle-type device, and has enough room that a larger number of LEDs can be placed on the device. There are eight or nine LEDs in the planar-type device, while there are two LEDs in the needle-type device. Several LEDs can illuminate an object uniformly. Different colored LEDs can be placed on the planar-type device and such a device will be demonstrated later in this section.

The first example of the planar-type device is for measuring blood flow. In this measurement, green LEDs are used to obtain a clear image of the blood vessels on the brain surface. Figure 15.6 shows the experimental results of blood flow obtained with the implanted planar-type device in a rat brain [22]. The blood vessels as well as the blood flow are clearly observed because of the high spatial resolution of the device. The change in speed of the blood flow can be measured between

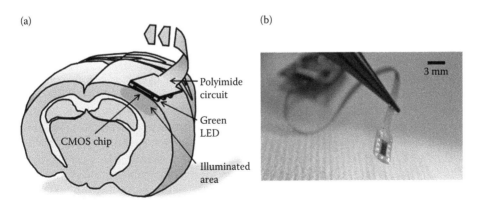

Figure 15.5 (a) Concept and (b) Photo of planar-type device. (Copyright 2014, The Japan Society of Applied Physics. Reprinted with permission from H. Haruta et al., *Jpn. J. Appl. Phys.*, 53, 04EL05, 2014.)

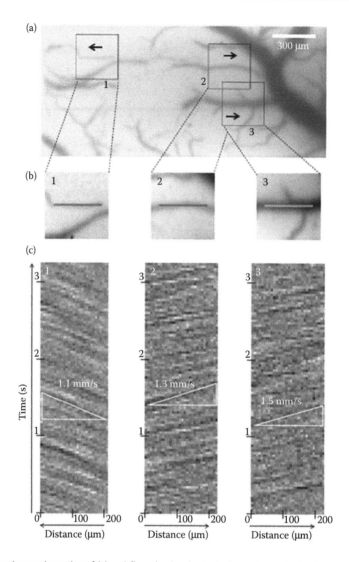

Figure 15.6 Experimental results of blood flow by implanted planar-type device in a rat brain. (a) Image captured by the device. (b) Magnification of the parts of (a). (c) Spatiotemporal plots for the selected blood vessels. The flow of blood is observed as parallel oblique features. (Copyright 2014, The Japan Society of Applied Physics. Reprinted with permission from H. Haruta et al., *Jpn. J. Appl. Phys.*, 53, 04EL05, 2014.)

resting and moving conditions, as shown in Figure 15.7 [22]. The device can be used to study cerebrovascular disease.

The second example of the planar device is for measuring the change in blood flow and change in hemoglobin conformation at the same point. In this experiment, two types of LEDs are placed on the device: green-colored LEDs to measure the change in blood flow, and orange-colored LEDs to measure the change in hemoglobin conformation, as shown in Figure 15.8 [23]. The absorption of green light in the local region changes with the changes in blood volume, while the absorption of orange light in the local region changes with the changes in oxidation and deoxidation conditions. Thus, the device can detect conformation changes in hemoglobin thorough the changes in absorbance as well as the changes in blood flow. A device employed with two colored LEDs can successfully measure the blood flow and the changes in hemoglobin conformation.

The third example measures neuronal activities through fluorescence by using voltage sensitive dye (VSD). In this experiment, RH795 was used as a VSD to measure fluorescent potentiometry.

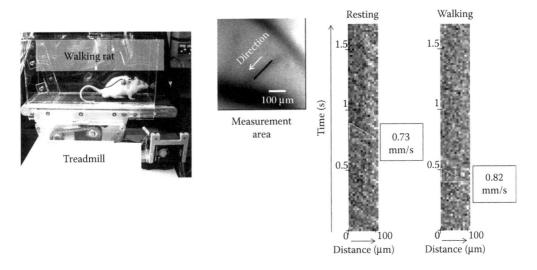

Resting   Walking

Figure 15.7 Experimental results of change in speed of blood flow can be measured between resting and moving conditions. (Copyright 2014, The Japan Society of Applied Physics. Reprinted with permission from H. Haruta et al., *Jpn. J. Appl. Phys.*, 53, 04EL05, 2014.)

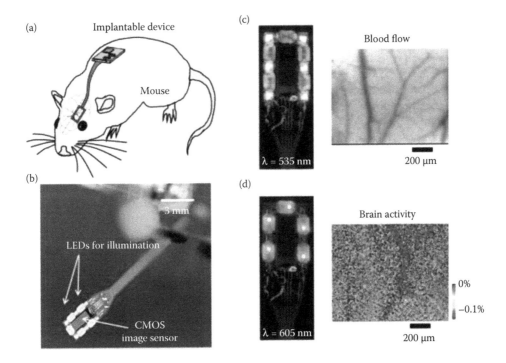

Figure 15.8 Photo of planar-type device implanted with green-colored LEDs and orange-colored LEDs for multifunction. (a) Total system. (b) Photo of the device. (c) The photo of the device when green LEDs are turned on and its captured image. (d) Photo of the device when orange LEDs are turned on and its captured image, which corresponds to the conformation change of hemoglobin. (Copyright 2014, The Japan Society of Applied Physics. Reprinted with permission from M. Haruta et al., *International Conference on Solid State Devices and Materials (SSDM)*, Sept. 11, 2014, Tsukuba, Japan.)

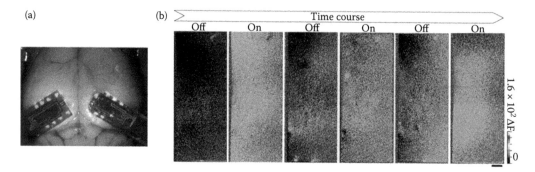

Figure 15.9 Experimental results of how turning light on and off affected the response of the visual cortex. (a) Photo of the devices that were placed on visual cortex. (b) Captured images. (Reprinted from 38(1), T. Kobayashi et al., *Biosens. Bioelectron.*, 321–330, Copyright 2012, with permission from Elsevier.)

A red filter, used as a high pass filter (>600 nm), was coated on the image sensor as a fluorescent filter, and nine blue LEDs were used as an excitation light source [21]. The two planar type devices were placed on the right and left sides of the visual cortex. The response in the visual cortex was measured through fluorescent changes according to the light input to one eye of the mouse. Figure 15.9 shows that turning the light on and off affected the response of the visual cortex [21]. This experiment also clearly demonstrates the multipoint measurement of neuronal activities in the brain of a freely moving animal.

15.2.3 Experimental results of the needle-type device

The advantage of the needle-type device is that it can be implanted into the deep brain of a mouse. A close-up view and schematic cross-sectional view of the implantation of the device are shown in Figure 15.10. Some experiments require an injection of chemical substances for molecular imaging, for which a specially designed cannula is attached to the device.

In the experiment, a device was implanted in the hippocampus of a mouse brain. The device was fixed to the skull with dental cement, as shown in Figure 15.10b, and was connected to a wiring harness through a slip-ring connector. As a fluorophore, a chemical substance was introduced into the mouse brain through a cannula. The substance changes from a nonfluorophore into a fluorophore when neural activity occurs in the hippocampus, which means that the sensor can detect a specified molecular activity. It was confirmed that the device implanted deep in the mouse brain could be successfully operated, and the mouse implanted with the device was alive and able to move freely one month after the implantation. Based on a previous experiment, it was confirmed that the mouse brain was intact (except for the region in which the device was inserted), and that the brain tissue was not severely damaged. Little foreign body reaction was observed. Figure 15.11 shows typical experimental measurement results for deep brain neuronal activity obtained using a device implanted in a mouse hippocampus [24]. In the experiment, artificial epilepsy was induced by injecting Kainic acid. After the injection, the mouse exhibited specific actions of epilepsy. In Figure 15.11a, some spots appear in the image when the mouse nods. In Figure 15.11b, a strong signal occurs in the image when the mouse exhibits clonic convulsions. After the convulsions cease, calm signals appear in the image. Spatiotemporal neuronal activity was successfully measured. It is noted that for large actions such as clonic convulsions, images can be taken without any noise, and after the action ceases, images can still be taken. This demonstrates that the device is tightly fixed to the skull because of its light weight.

The size of the needle-type device is so small that it can be implanted not only in the brain, but also in other organs. It also has potential applications in theranostics [25]. The image sensor implemented in the device is customized for implantation, that is, it has a small footprint and only

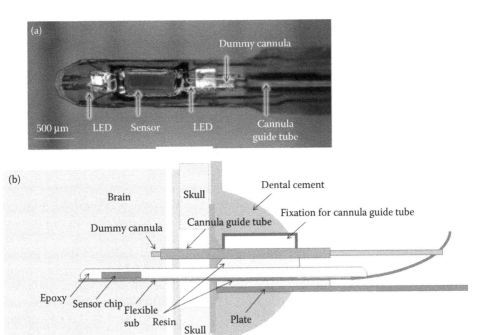

Figure 15.10 (a) Close-up view and (b) schematic cross-sectional view of implantation of the needle-type device.

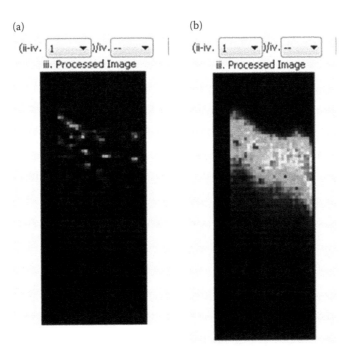

Figure 15.11 Experimental results from using the needle-type device. (a) Before and (b) after a medicine-induced convulsion. (K. Sasagawa et al., An implantable micro imaging device for molecular imaging in a brain of freely-moving mouse, *IEEE International Symposium on Bioelectronics and Bioinformatics* (*IEEE ISBB 2014*), S1.4, Copyright 2014 IEEE.)

four I/O pads. It is, however, optimized for the detection of fluorescence. Recently, a self-resetting CMOS image sensor that is suitable for detecting small changes in fluorescence was developed [26]. Because the background intensity of fluorescence is much higher than changes in its intensity, a self-resetting architecture that only measures changes in the value of the light intensity is effective.

15.3 Implantable glucose sensor

15.3.1 Concept of the implantable glucose sensor based on implantable image sensor technology

Implantable image sensor technology can be applied not only to observe living tissue in an animal, but also to conduct optical measurements on any other implanted materials. In this section, the technology of a fluorescence-based implantable glucose sensor is presented.

Glucose sensing is an important biosensing technology related to diabetes mellitus. In diabetes, owing to various reasons, a high blood glucose level is maintained. This causes various complications, such as kidney failure, high blood pressure, and blindness. In order to monitor a patient's status and suppress possible complications, it is very important to monitor blood glucose in daily life. Currently, self-monitoring of blood glucose (SMBG) is widely used to check blood glucose [27]. A small amount of blood is sampled from a fingertip using a needle device, and glucose is measured with a portable sensing device. Enzyme-based measurement schemes are employed for SMBG systems. Although SMBG is significantly effective for the daily healthcare of diabetes patients, there are some limitations. Since it requires the sampling of blood for every measurement, patients can check their blood glucose level only at certain intervals. Although it is known that the blood glucose sometimes shows drastic changes related to food ingestion (and insulin injection for some patients), the patients can only check their blood glucose level occasionally with SMBG technology. Because of the demand to keep an eye on the blood glucose, continuous glucose-monitoring (CGM) systems have been realized [27,28]. CGM systems employ partially implantable enzyme-based sensor tips that can be used for several weeks. However, because of some reasons (including risk management for infections), CGM systems are used only in limited situations.

Takeuchi et al. presented a glucose-monitoring scheme using implanted bead-shaped or fiber-shaped glucose responsive fluorescent hydrogel [29,30]. The fluorescent hydrogel is excited with UV light with a wavelength of approximately 400 nm, and emits blue fluorescence with a peak wavelength of 488 nm. The researchers demonstrated that a glucose concentration of interstitial solution in a rat's ear can be measured by observing the fluorescence intensity of the implanted hydrogel. In contrast to commonly used enzyme-based measurement technologies, the fluorescence-based measurement scheme does not include any electrochemical reactions; thus, no electrodes are required for measurement. Another attractive characteristic of fluorescent hydrogel is its long life span. Heo et al. reported that the fluorescent hydrogel retained its glucose-monitoring capability 140 days after implantation [30]. In addition to measuring from outside of the body, "measuring from inside the body" is a reasonable approach to perform glucose monitoring using the glucose-responsive fluorescent hydrogel. An implantable image sensor is suitable for intrabody fluorescence measurements of the glucose-responsive hydrogel.

15.3.2 Structure and building blocks of the implantable glucose sensor

Figure 15.12 schematically shows the structure of the implantable glucose monitoring device based on implantable CMOS image sensor technology [31]. A small-sized CMOS image sensor is integrated on a polyimide substrate with one or two GaInN UV LEDs for excitation. A color filter is integrated on the pixel array of the CMOS image sensor to distinguish the blue fluorescence from

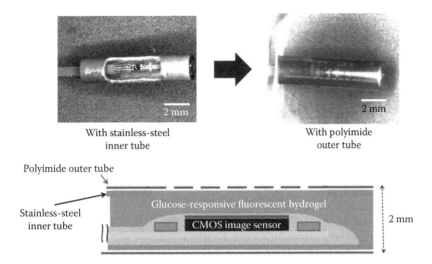

Figure 15.12 Structure of glucose monitoring device based on implantable CMOS image sensor technology. (T. Tokuda et al., CMOS image sensor-based implantable glucose sensor using glucose-responsive fluorescent hydrogel, *Biomed. Opt. Express*, 5(11), 3859–3870, 2014. With permission of Optical Society of America.)

the scattered excitation UV light. The typical width of the sensor core is 500 μm. The sensor core is mounted to an outer structure. (We have several variations of the outer structures, as shown in a later subsection.) After the outer structure was attached to the sensor core, the glucose-responsive fluorescent hydrogel was poured in the outer structure. The glucose-responsive fluorescent hydrogel was poured over both the CMOS sensor and the GaInN LEDs.

Figure 15.13 shows (a) the layout and (b) the block diagram of the CMOS image sensor integrated on the implantable glucose sensor [31]. Table 15.2 lists typical specifications of the CMOS image sensor [31]. The sensor was designed to minimize the I/O lines, and we can operate the sensor with only four wires: VDD (3.3 V) and GND for power supply, CLK for pixel address increment, and

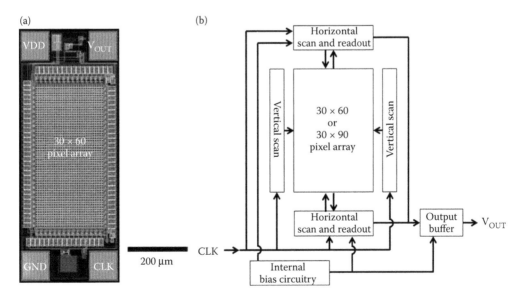

Figure 15.13 (a) Layout and (b) block diagram of CMOS image sensor for implantable glucose sensor. (T. Tokuda et al., CMOS image sensor-based implantable glucose sensor using glucose-responsive fluorescent hydrogel, *Biomed. Opt. Express*, 5(11), 3859–3870, 2014. With permission of Optical Society of America.)

Table 15.2 Specifications for implantable CMOS image sensor for glucose sensing

Process	AMS 0.35 μm CMOS 2P4M
Sensor size	320 μm × 790 μm
Pixel size	7.5 μm × 7.5 μm
Array size	30 × 60 or 30 × 90
Pixel type	3-transistor active pixel sensor
Fill factor	35.1%
Photodiode type	N-well/P-sub
Operation voltage	3.3 V
Number of I/Os	4 (VDD, GND, CLK, V_{out})
Output signal	Analogue voltage

Source: T. Tokuda et al., CMOS image sensor-based implantable glucose sensor using glucose-responsive fluorescent hydrogel, *Biomed. Opt. Express*, 5(11), 3859–3870, 2014. With permission of Optical Society of America.

V_{out} for readout. The CMOS image sensor was designed with a conventional three transistor active pixel sensor architecture. The pixel signal is relayed to the chip output (V_{out}) via serially connected source-follower circuits. The pixel to be accessed is incremented by pulses applied to the CLK input. We can adjust the operation condition by changing the pulse frequency and durations of specific pulses during the operation, which is described in Reference 32.

The GaInN LED is a commercially available bare chip formed on a sapphire substrate. The typical peak wavelength of the LED is 400 nm, and the dimensions of the LED were 300 μm × 300 μm. The rated emission power of the LED is approximately 10 mW with a driving current of 20 mA. During the measurement, we operated the LED with a driving current of 100 μA–4.5 mA. When we needed to adjust the sensitivity, we mainly changed the operation condition of the CMOS image sensor. In general, we can obtain a higher optical sensitivity with a low CLK frequency for the CMOS image sensor, because a lower frequency gives a longer accumulation time for the active pixel sensor circuit.

The current implantable sensor is designed for tethered operations. We designed two kinds of external control systems for the glucose sensor: a PC-based imaging system and an MSP430 (Texas Instruments) microcontroller-based recording system. Both systems can supply power to both the CMOS image sensor and the GaInN UV LED on the implantable glucose sensor. The system also supplies a CLK signal to the CMOS image sensor, and reads V_{out} from the sensor. The analogue voltage V_{out} is converted to digital values by an analogue-to-digital converter (ADC) chip or the ADC function of the microcontroller. For a PC-based imaging system, all pixel values are transferred to and recorded by the PC, and captured images are reconstructed. For a microcontroller-based recording system, the pixel values are processed in the microcontroller, and only calculated results are transferred to the recording PC. The microcontroller-based recording system can be wirelessly connected to the PC. Details of the sensor control systems are included in Reference 32.

15.3.3 *In vitro* performance evaluation and structural optimization

The basic measurement capability of the implantable glucose sensor was characterized in some *in vitro* (out-of-body) experiments. The glucose sensor was dipped in saline solution, and continuous measurement was performed. During the measurement, glucose solution was added to the saline solution to increase the glucose concentration. To decrease the glucose concentration, we added saline solution to dilute the measurement solution. At first, we performed an imaging experiment

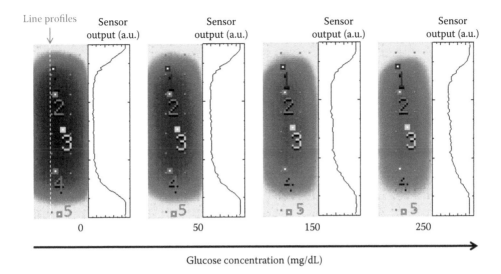

Glucose concentration (mg/dL)

Figure 15.14 Typical fluorescent images obtained with the sensor. (T. Tokuda et al., CMOS image sensor-based implantable glucose sensor using glucose-responsive fluores hydrogel, *Biomed. Opt. Express*, 5(11), 3859–3870, 2014. With permission of Optical Society of America.)

and observed the fluorescent distribution of the glucose responsive fluorescent hydrogel. After checking the fluorescence distribution, we evaluated the sensitivities of pixel values to the glucose concentration. A long term *in vitro* performance evaluation was also performed.

Figure 15.14 shows typical fluorescent images obtained with the sensor [31]. The fluorescent hydrogel was placed almost uniformly on the pixel array, and no structural features were observed. Since the GaInN LEDs were placed next to the top and bottom sides of the pixel array, the pixels around the top and bottom sides showed saturation, and no glucose dependent changes were observed. On the other hand, at the center of the pixel array, the pixel values increased as the glucose concentration increased.

Figure 15.15 shows pixel values for several pixels (as indicated in Figure 15.14) as functions of the glucose concentration. Each pixel shows an almost linear dependence on the glucose concentration, except for the cases of saturation (pixel 5). As shown in Figure 15.15, sensitivity to the glucose concentration depends on the pixel position. This result is quite reasonable because the intensity of the

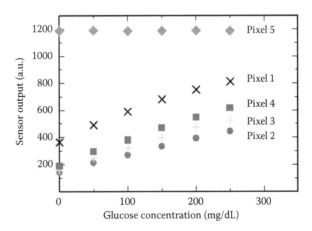

Figure 15.15 Pixel values for several pixels as functions of glucose concentration. (T. Tokuda et al., CMOS image sensor-based implantable glucose sensor using glucose-responsive fluorescent hydrogel, *Biomed. Opt. Express*, 5(11), 3859–3870, 2014. With permission of Optical Society of America.)

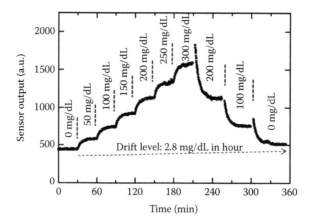

Figure 15.16 Typical result obtained in glucose tracking experiment. (T. Tokuda et al., CMOS image sensor-based implantable glucose sensor using glucose-responsive fluorescent hydrogel, *Biomed. Opt. Express*, 5(11), 3859–3870, 2014. With permission of Optical Society of America.)

UV excitation light has a positional distribution. We can take advantage of this position dependent sensitivity. Based on the nature of an image sensor pixel, no information is obtained when the pixel signal is saturated. However, using the image sensor-based measurement scheme, we can choose a pixel that shows an intermediate pixel value after a single shot image acquisition. Since images are continuously captured under certain conditions, even some of the pixels show saturation. Thus, we can keep measuring the glucose concentration. This feature is also advantageous from the viewpoint of the life span of the glucose responsive fluorescent hydrogel. Like most fluorophores, the glucose responsive fluorescent hydrogel shows photo breaching. Using a simultaneous multisensitivity measurement, we can reduce the frequency of the measurement and, in turn, the frequency of the UV illumination onto the glucose responsive hydrogel. The reduction of the UV illumination is effective to suppress the photo breaching of the glucose responsive fluorescent hydrogel.

Figure 15.16 shows typical results obtained in a glucose tracking experiment. In the *in vitro* setting, we increased and decreased the glucose concentration at certain intervals. As shown in Figure 15.16, the sensor can continuously measure the glucose concentration. Signal drift in the glucose tracking experiment was typically in the range of several mg/dL per hour, which is acceptable for the present development stage.

Figure 15.17 shows (a) sensitivity and (b) drift levels obtained in an *in vitro* long term performance evaluation [32]. The sensor was kept soaked in saline solution, and glucose tracking experiments

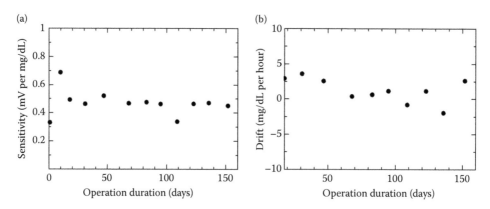

Figure 15.17 (a) Sensitivity and (b) drift levels obtained during *in vitro* long-term performance evaluation. (T. Tokuda et al., *In-vitro* long-term performance evaluation and improvement in the response time of CMOS-based implantable glucose sensors, *IEEE Des. Test*, Copyright 2016. IEEE.)

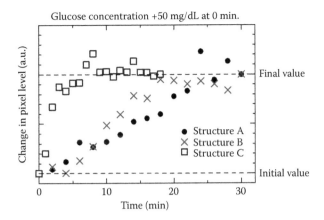

Figure 15.18 Response performance for various device structures. (T. Tokuda et al., *In-vitro* long-term performance evaluation and improvement in the response time of CMOS-based implantable glucose sensors, *IEEE Des. Test,* Copyright 2016. IEEE.)

were performed every two weeks. Both the sensitivity and drift level were acceptable for up to 150 days. The result suggests that the performance of the glucose responsive fluorescent hydrogel does not deteriorate with the integrated sensor structure.

In addition to sensitivity and long term performance, the response to changes in glucose concentration is an important characteristic for a glucose monitoring device. During the first development stage, as shown in Figure 15.12, we used a combination of metal (SUS) and polyimide tubes for the outer structure of the sensor. The diameter of the sensor was as wide as 2 mm [31]. We found that the response time for this sensor was typically approximately 30 min for a change in glucose concentration. We optimized the outer structure of the implantable glucose sensor to shorten the response time. To shorten the time to introduce glucose into the glucose responsive fluorescent hydrogel, we reduced the sensor diameter and widened the opening of the sensor. Figure 15.18 shows the response curves of implantable glucose sensors with different structures [32]. In Figure 15.18, structure A is the basic structure, structure B is a sensor with a smaller diameter, and C is a sensor with a smaller diameter and wider opening. The SUS tube was omitted for structures B and C. Details of the structures are presented in Reference 32. As can be seen in Figure 15.18, we successfully shortened the response time to less than 10 min. Widening of the sensor opening leads to a significant improvement in the response time of the sensor.

15.3.4 *In vivo* functional demonstration

In vivo functional demonstrations were performed using rats [33]. A sensor was implanted in the back of the rat's ear to measure the glucose concentration in an interstitial solution. The experiment was performed under anesthesia. To create glucose-changing events, glucose solution and insulin were injected during the experiment. Figure 15.19 shows a result obtained during an *in vivo* glucose tracking experiment. The timings of the glucose/insulin injection events are indicated in Figure 15.19. As control data, the blood glucose level measured with a conventional SMBG sensor is plotted in Figure 15.19. The result shows that the present implantable glucose sensor can monitor the glucose concentration of the interstitial solution that reflects the blood glucose. For the experimental trial shown in Figure 15.19, the measurement lag was almost negligible because the sensor structure was optimized [32].

In Figure 15.19, there is a discontinuity at 111 min, which coincides with the first insulin injection. We assume this is caused by an irregular movement of the mouse's body. This means

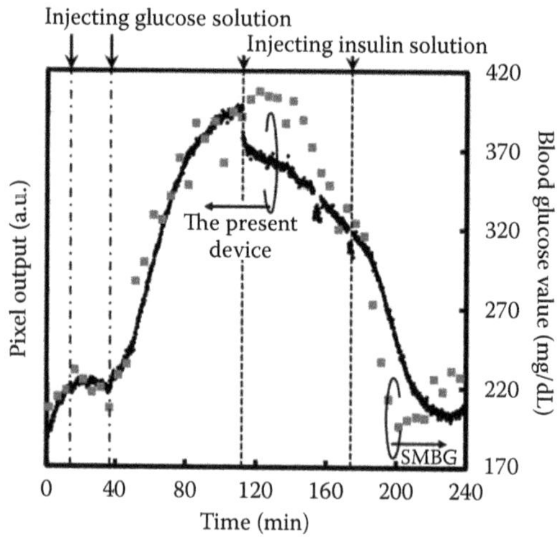

Figure 15.19 Result obtained for *in vivo* glucose tracking experiment. (Copyright 2015, The Institution of Engineering & Technology (IET). Reproduced from T. Kawamura et al., *Electron. Lett.*, 51(10), 738–740, 2015, with permission.)

that the sensor output was affected by a movement in the body of the animal, which in turn, disturbs the stability of the measurement. Although a background subtraction was implemented in the measurement system, it does not completely eliminate the ghost signal. We need to develop an improved background cancellation scheme, such as a modulated excitation.

We also require additional steps to apply this sensor technology to a real application. We need to evaluate the *in vivo* long time performance and biocompatibility of the sensor. Integrating a wireless operating platform is also essential to make the device fully implantable. We are developing a wireless power supply and data transfer system for the implantable glucose sensor.

15.4 Microphotonic devices for optogenetics

15.4.1 Concept of integrating light source for optogenetics

Optogenetics is a technology with which we can introduce light sensitivity to neural cells with the help of genetics. In optogenetics, light sensitive channel proteins such as channelrhodopsin2 (ChR2) are introduced into the cell membrane. The proteins are activated by light with a specific wavelength, and initiate or suppress the neural activity of the cell [34–36]. We can use either the transfection or transgenic inclusion of protein DNA for optogenetics.

Electrodes are required for electric stimulation and have been widely used in neurophysiology; however, they cause various drawbacks. The selectivity of the stimulation is mainly based on the structure and placement of the electrodes, which means that a sharp needle must be used for localized stimulations or measurements. On the other hand, optical stimulations can be performed using not only an optical probe structure, but also various remote illuminations. Nonoptical selectivity of the stimulation based on genetic methodology is also available.

In optogenetics, genetic methodologies and optical stimulation technologies play very important roles. Various light delivery methods for optogenetics have been proposed and realized. Optical fiber is one of the basic materials that deliver light to target tissue such as the brain [36–38]. We can use an external light source with a variety of wavelengths and intensities. Applicability to freely moving animals is another significant advantage of optical fiber-based optical stimulation systems. However, because of the structure of the optical fiber, it is difficult to realize a multisite localized stimulation with optical fiber-based systems.

The microscopy-based system is another promising platform for optogenetics [39–41]. The largest advantage of the microscopy-based system is its capability of localized stimulation. With the help of projection optics in a microscope, we can realize high resolution local optical stimulation. Desktop systems and miniaturized microscopes have been developed for optogenetics and are applicable to rodents and other animals.

In order to realize a small implantable optical stimulation device with a reasonable spatial resolution, the GaInN LED is one of the most promising light sources [42,43]. It can emit UV-visible light with high intensity, and its size can be as small as 50 μm [43]. Furthermore, arrayed LEDs are available within conventional manufacturing technologies. We proposed the integration of arrayed GaInN LEDs as an integrated optical stimulator for implantable CMOS image sensor technology. Currently, most GaInN LEDs are manufactured on a sapphire substrate, and both GaInN layers and sapphire substrate are transparent to visible light. This suggests that we can realize a device with the multiple functions of on-chip 2D localized light stimulation and on-chip imaging.

15.4.2 Optogenetic neural stimulators based on implantable CMOS image sensor

We proposed and realized implantable CMOS image sensors with on-chip GaInN LED arrays for optogenetics. Figure 15.20 schematically shows the structure of the device. We developed two kinds of optical neural stimulator with an imaging function [44,45]. One is a device with an integrated monolithic GaInN LED array (Figure 15.20a), and the other is an arrayed separated set of LEDs (Figure 15.20b). The former device was designed for optical stimulation and imaging, and the latter device was designed for optoelectric multimodal operation. To integrate an arrayed LED array, we used multifunctional image sensor technology. We implemented arrayed on-chip electrodes with addressing functions. Using the addressing function, we can choose one of the on-chip electrodes. We can use these electrodes to operate an LED for optical stimulation or electric stimulation/measurement, depending on the component bonded to the electrode. When an LED electrode is bonded to the on-chip electrode, we can inject current to operate an LED. When a neural electrode is formed on the on-chip electrode, we can perform electric stimulation or observation via the electric pathway established with the addressing function.

For optical-only devices (Figure 15.20a), electrodes of each LED in the array are bonded to the on-chip electrode. Since both GaInN layers and the sapphire substrate are almost transparent to visible light, we expect to capture images through the GaInN array chip. In this configuration, the bottom surface of the sapphire substrate is used as the contacting surface for biological targets, such as ChR2-expressed cells or the brain surface.

For an optoelectric multimodal neural stimulator, we prepared a chip with integrated electrodes and LEDs. Figure 15.20b includes a drawing that shows the structure of the LED-electrode mixed array chip [45]. We used a Si wafer with through silicon via (TSV) structures for the integrated chip. We formed rectangular cavities for blue LEDs. We placed discrete blue LED chips in the cavities and molded them with epoxy resin. The surfaces of the TSV and the electrodes of the LEDs were aligned flat at the bottom surface of the chip. Finally, this chip works as a mixed array of blue LEDs and neural electrodes.

15.4.3 Design of the multifunctional CMOS sensor chip and device packaging

We designed a multifunctional CMOS image sensor with on-chip current injection capability. We used a 0.35-μm 2-poly, 4-metal standard CMOS process. Figure 15.21 shows a block diagram of the sensor, and Table 15.1 lists the specifications for the sensor [45]. The multifunctional CMOS image sensor consists of two blocks that are almost independently designed. One block is a CMOS image

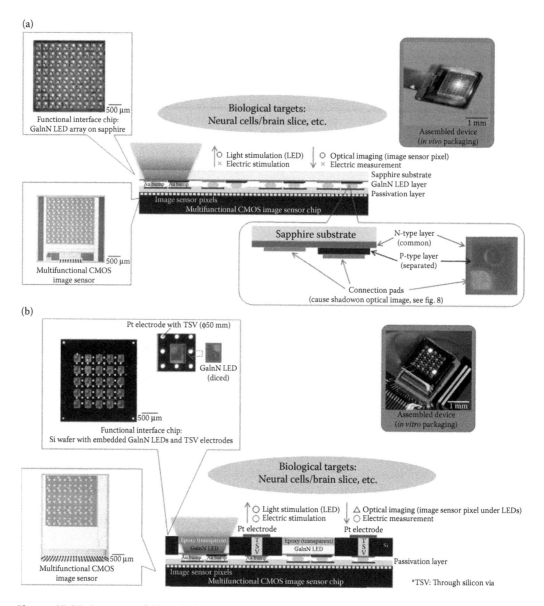

Figure 15.20 Structure of (a) optical-only and (b) optoelectronic neural interface device for optogenetics. (Copyright 2016, IEICE. Reproduced from T. Tokuda et al., *IEICE Trans. Electron.*, E99, 165–172, 2016, with permission.)

sensor designed with 3.3-V MOS transistors, and the other block consists of on-chip addressable electrodes designed with 5-V MOS transistors. The image sensor was designed based on a conventional three transistor active pixel sensor array. The pixel size is 7.5 μm × 7.5 μm, and the array size is 260 × 264. The pixel signal is read out as analogue voltage and is converted to a digital value by an ADC chip implemented on an external interface board between the sensor and a Windows PC. The addressable on-chip electrode array for LED operation is designed to match the electrode alignment of the LED array. The LED array was diced in 8 × 10 arrays from an LED wafer prepared for a commercial blue LED. Although each LED has both anode and cathode electrodes, only the anodes are bonded to the addressable electrode array, because the n-layers of the LEDs are monolithically connected and connected to the CMOS chip at one end of the LED array. The 8 × 10 on-chip electrodes are internally configured as one-dimensional exclusively selectable electrodes switched by a

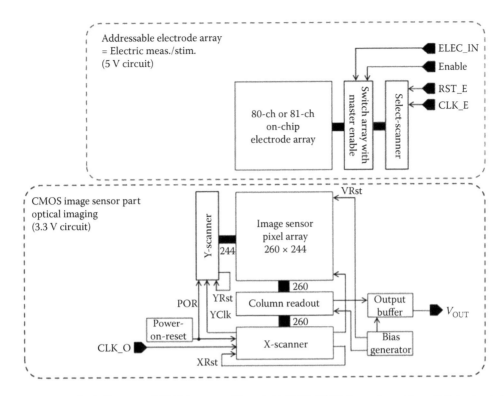

Figure 15.21 Block diagram of CMOS sensor. (Copyright 2016, IEICE. Reproduced from T. Tokuda et al., *IEICE Trans. Electron.*, E99, 165–172, 2016, with permission.)

scanner. Pulses applied to the CLK_E line increment the electrodes connected to the common line. This means that we can perform single-site stimulation/measurement with this device architecture. A quasi-two-dimensional patterned operation is also available by quickly switching the electrodes.

Figure 15.22 shows the bonding process for the multifunctional CMOS image sensor and the GaInN LED array chips, and a photograph of the bonded device. We used a flip-chip bonding

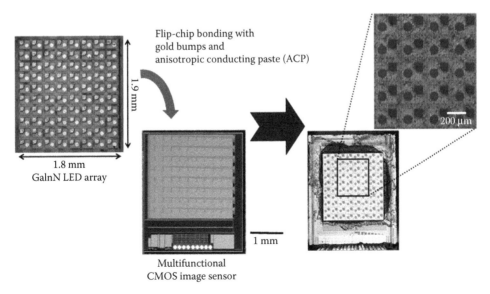

Figure 15.22 Bonding process of multifunctional CMOS image sensor and GaInN LED array chips, and a photograph of the bonded device.

process with anisotropic conducting paste (ACP). Gold bumps were formed on the CMOS chips prior to the bonding process. As shown in Figure 15.22, we can see the grid structure of the image sensor pixel array of the CMOS chip through the GaInN chip. This means that an optical on-chip imaging capability is available with this device structure. For the optoelectric multi-modal device, we use the same process to integrate the LED-electrode mixed array chip on the multifunctional CMOS image sensor chip. However, in contrast to the optical-only device, the LED-electrode mixed array chip was fabricated with Si, and the optical imaging function is not available.

15.4.4 Functional evaluation of the integrated device

Figure 15.23 shows the operation of the integrated device observed by (a) an external microscope and (b) the imaging function of the CMOS sensor. An LED in the array was successfully selected and operated with the current injection via the multifunctional CMOS image sensor. The output power of the LED is shown in Figure 15.24. An illumination density of approximately 50 mW/mm^2 was obtained. Considering that optical stimulation on the order of mW/mm^2 is typically used in the literature, the intensity range of Figure 15.24 suggests the applicability of the fabricated device for various optogenetic applications. As shown in Figure 15.23b, we can operate the imaging function simultaneously with the optical stimulation. We can obtain images to monitor the operation of the LEDs. This is an advantage of the present device, especially for *in vivo* optogenetic applications. We can monitor the LED operation even in situations where the device is completely implanted in the body of a target animal. In addition to observing LED operation, we can use the imaging function to observe the biological target on the device. However, it should be mentioned that shadows of the metal electrodes of GaInN LEDs were observed in the images, as shown in following subsections.

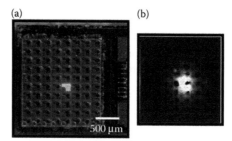

Figure 15.23 Operation of integrated device captured by (a) external microscope and (b) CMOS image sensor function.

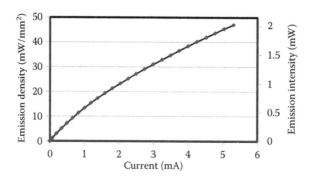

Figure 15.24 Output power of an LED.

15.4.5 *In vitro* functional demonstration

An *in vitro* experimental demonstration of the on-chip optical stimulation was performed using ChR2-expressed neuro 2A cells. We assembled the integrated device shown in Figure 15.20a into a device structure for *in vitro* applications. Figure 15.25 shows the structure of the *in vitro* optogenetic device. The device shown in Figure 15.20a was bonded on a rigid printed circuit board (PCB), and the sides of the integrated chips and bonding wires were molded with epoxy resin. A cell culturing dish with an open bottom was attached to the PCB to culture cells on the device surface, thus, the backside of the sapphire substrate. Prior to cell culturing, the surface of the chip was treated with poly-L-Lysine for 24 h to enhance the cell adhesion. Neuro-2A cells were cultured on the device, and transfection to introduce ChR2 was performed. The detailed process was presented in Reference 45.

The integrated LED optically stimulated the cultured neuro 2A cells. The response of the cell was observed with a conventional patch-clamp method in a voltage-clamp mode. Figure 15.26 shows

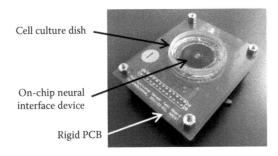

Figure 15.25 Device packaging for *in vitro* applications. (Copyright 2016, IEICE. Reproduced from T. Tokuda et al., *IEICE Trans. Electron.*, E99, 165–172, 2016, with permission.)

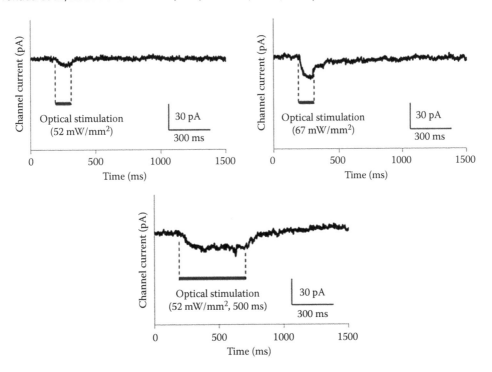

Figure 15.26 Optically evoked membrane current of ChR2-expressed neuro 2A cell observed in an *in vitro* experiment. (Copyright 2016, IEICE. Reproduced from T. Tokuda et al., *IEICE Trans. Electron.*, E99, 165–172, 2016, with permission.)

the changes in the membrane current evoked by the optical stimulation from the device under different conditions. A clear increase in the membrane current caused by the optical stimulation was observed. The duration and amplitude are dependent on the duration and intensity of the optical stimulation. The results shown in Figure 15.26 suggest that the present device is feasible for on-chip *in vitro* optogenetic applications.

15.4.6 *In vivo* functional demonstration

We also performed *in vivo* experimental demonstrations. Figure 15.27 shows a device packaging for *in vivo* applications. We put this device on an anesthetized mouse's brain to perform simultaneous on-chip optical stimulation and imaging. The mouse is a ChR2-venous mouse.

Figure 15.28 shows images obtained during simultaneous on-chip brain imaging. With the help of the external light source of a microscope system, blood vessels were successfully imaged. Although shadows caused by the electrodes on the GaInN chip are observed in the images, they are still informative for observing the blood flow and contrasting the blood vessel and brain tissue. As shown in Figure 15.28, we can observe the illumination from the on-chip blue LED for optical stimulation as a bright spot in the images.

Figure 15.29 shows a neural response evoked by optical stimulation. An epoxy insulated tungsten microelectrode with a 125-μm tip diameter was used to observe the brain response. Optical stimulation at 2 Hz with a pulse duration of 250 ms was applied on the primary sensory cortex of the ChR2-expressed mouse's brain. In the trace of the brain signal, we can see responses evoked by the optical stimulation. This result shows that our device can be applied to *in vivo* optogenetics. It was also confirmed that simultaneous on-chip brain imaging is also available. No deterioration caused by optical stimulation was observed. In addition, even during simultaneous optical stimulation and imaging, when the stimulation duration is short, we can obtain images without spots of LED illumination. We can check the position, intensity, timing, and duration of the LED illumination.

In summary, we proposed to integrate a GaInN LED array to implement an on-chip optical stimulation capability for optogenetic applications. A dedicated multifunctional CMOS image sensor was designed, and packaging structures were developed. The functionality of the on-chip optical stimulation/imaging device was demonstrated in both *in vitro* and *in vivo* experiments. Currently, improvements to chip performance and reliability are underway. We are also developing a device structure that covers a larger area, with the goal of wide-area coverage in the brains of animals.

 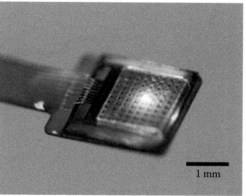

1 mm

Figure 15.27 Device packaging for *in vivo* applications with capability of imaging with green light.

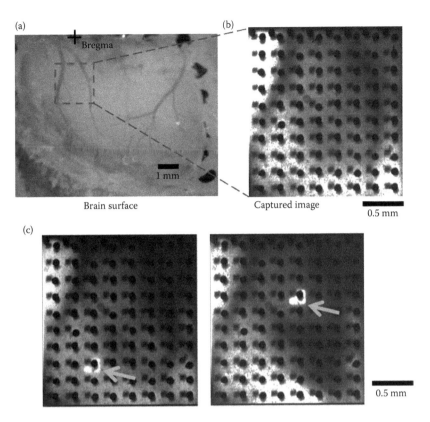

Figure 15.28 Images obtained from *in vivo* brain imaging experiment. (a) is a brain surface image captured with an external microscopy. (b) and (c) are images captured by the imaging function of the CMOS chip, (b) without and (c) with the integrated LED operation.

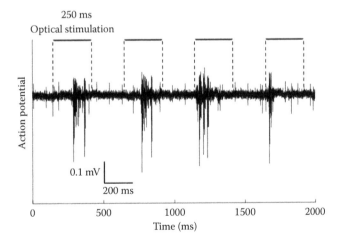

Figure 15.29 Neural response evoked by optical stimulation from the present device.

15.5 Summary

This chapter reviewed recent developments in microphotonic devices that can be implanted in a living body. We introduced three examples: a brain-implantable imaging device, implantable glucose sensor, and microphotonic device for optogenetics. In life science, optical technologies combined with genetic engineering are essential, thus microphotonic devices play an important role in designing implantable biomedical devices. In the near future, microphotonic devices will be applied to the medical field for optically measuring and controlling biological functions in a living body.

Acknowledgments

The glucose-responsive fluorescent hydrogel was provided by Prof. Shoji Takeuchi at the University of Tokyo, the Terumo Corporation, and BEANS laboratory. The ChR2 vector for *in vitro* experiments was provided by Prof. Karl Deisseroth at Stanford University. The transgenic mice for the *in vivo* experiments were provided by Prof. Hitoshi Hashimoto at Osaka University. The CMOS sensors were designed with the support of the VLSI Design and Education Centre (VDEC), University of Tokyo, in collaboration with the Cadence Corporation and the Mentor Graphics Corporation.

References

1. S.J. Remington, Green fluorescent protein: A perspective, *Protein Sci.*, 20, 1509–1519, 2011.
2. H. Yawo, H. Kandori, A. Koizumi (Eds.), *Optogenetics: Light-Sensing Proteins and Their Applications*, Japan: Springer, 2015.
3. http://www.brainvision.co.jp/en/
4. Z. Li, S. Kawahito, K. Yasutomi, K. Kagawa, J. Ukon, M. Hashimoto, H. Niioka, A time-resolved CMOS image sensor with draining-only modulation pixels for fluorescence lifetime imaging, *IEEE Trans. Electron Dev.*, 59, 2715–2722, 2012.
5. G. Iddan, G. Meron, A. Glukhovsky, P. Swain, Wireless capsule endoscopy, *Nature*, 405, 417, 2000.
6. J. Ohta, T. Tokuda, K. Sasagawa, T. Noda, Implantable CMOS biomedical devices, *Sensors*, 9, 9073–9093, 2009.
7. J. Ohta, T. Tokuda, T. Fujikado, CMOS technologies for retinal prosthesis, in *Handbook of Bioelectronics: Directly Interfacing Electronics and Biological Systems*, by S. Carrara, K. Iniewski (Eds.), Cambridge: Cambridge University Press, 2015.
8. P.K. Campbell, K.E. Jones, R.J. Huber, K.W. Horch, R.A. Normann, A silicon-based, three-dimensional neural interface: manufacturing processes for an intracortical electrode array, *IEEE Trans. Biomed. Eng.*, 38(8), 758–768, 1991.
9. K.D. Wise, K. Najafi, Microfabrication techniques for integrated sensors and microsystems, *Science*, 254, 1335–1342, 1991.
10. T. Kawano, Y. Kato, R. Tani, H. Takao, K. Sawada, M. Ishida, Selective vapor-liquid-solid epitaxial growth of micro-Si probe electrode arrays with on-chip MOSFETs on Si (111) substrates, *IEEE Trans. Electron Devices*, 51(3), 415–420, 2004.
11. D.A. Dombeck, A.N. Khabbaz, F. Collman, D.W. Tank, Imaging large-scale neural activity with cellular resolution in awake, mobile mice, *Neuron*, 56, 43–57, 2007.
12. J. Sawinski, D.J. Wallace, D.S. Greenberg, S. Grossmann, W. Denk, J.N.D. Kerr, Visually evoked activity in cortical cells imaged in freely moving animals, *Proc. Natl. Acad. Sci. USA*, 106(46), 19557–19562, 2009.
13. B. Flusberg, A. Nimmerjahn, E. Cocker, E. Mukamel, R. Barretto, T. Ko, L. Burns, J. Jung, M. Schnitzer, High-speed, miniaturized fluorescence microscopy in freely moving mice, *Nat. Methods*, 5(11), 935–938, 2008.
14. C.D. Saunter, S. Semprini, C. Buckley, J. Mullins, J.M. Girkin, Micro-endoscope for in vivo widefield high spatial resolution fluorescent imaging, *Biomed. Opt. Express*, 3(6), 1274–1278, 2012.
15. K.K. Ghosh, L.D. Burns, E.D. Cocker, A. Nimmerjahn, Y. Ziv, A.E. Gamal, M.J. Schnitzer, Miniaturized integration of a fluorescence microscope. *Nat. Methods*, 8(10), 871–878, 2011.

16. J.H. Park, J. Platisa, J.V. Verhagen, S.H. Gautam, A. Osman, D. Kim, V.A. Pieribone, E. Culurciello, Head-mountable high speed camera for optical neural recording, *J. Neurosci. Methods*, 201(2), 290–295, 2011.

17. K. Murari, R. Etienne-Cummings, G. Cauwenberghs, N. Thakor, An integrated imaging microscope for untethered cortical imaging in freely-moving animals, In *Proceedings of 2010 Annual International Conference of the IEEE Engineering in Medicine and Biology Society (EMBC)*, 5795–5798, 2010.

18. D.C. Ng, H. Tamura, T. Tokuda, A. Yamamoto, M. Matsuo, M. Nunoshita, Y. Ishikawa, S. Shiosaka, J. Ohta, Real time *in vivo* imaging and measurement of serine protease activity in the mouse hippocampus using a dedicated CMOS imaging device, *J. Neurosci. Methods*, 156(1–2), 23–30, 2006.

19. H. Tamura, D.C. Ng, T. Tokuda, H. Naoki, T. Nakagawa, T. Mizuno, Y. Hatanaka, Y. Ishikawa, J. Ohta, S. Shiosaka, One-chip sensing device (biomedical photonic LSI) enabled to assess hippocampal steep and gradual up-regulated proteolytic activities, *J. Neurosci. Methods*, 173(1), 114–120, 2008.

20. A. Tagawa, H. Minami, M. Mitani, T. Noda, K. Sasagawa, T. Tokuda, H. Tamura et al., Multimodal complementary metal-oxide-semiconductor sensor device for imaging of fluorescence and electrical potential in deep brain of mouse, *Jpn. J. Appl. Phys.*, 49, 01AG02, 2010.

21. T. Kobayashi, M. Motoyama, H. Masuda, Y. Ohta, M. Haruta, T. Noda, K. Sasagawa et al., Novel implantable imaging system for enabling simultaneous multiplanar and multipoint analysis for fluorescence potentiometry in the visual cortex, *Biosens. Bioelectron.*, 38(1), 321–330, 2012.

22. H. Haruta, C. Kitsumoto, Y. Sunaga, H. Takehara, T. Noda, K. Sasagawa, T. Tokuda, J. Ohta, An implantable CMOS device for blood-flow imaging during experiments on freely moving rats, *Jpn. J. Appl. Phys.*, 53, 04EL05, 2014.

23. M. Haruta, Y. Sunaga, T. Yamaguchi, H. Takehara, T. Noda, K. Sasagawa, T. Tokuda, J. Ohta, A multi-modal implantable CMOS imaging device with two-color light source for intrinsic signal detection in a brain, *International Conference on Solid State Devices and Materials (SSDM)*, Sept. 11, 2014, Tsukuba, Japan.

24. K. Sasagawa, Y. Ohta, M. Motoyama, T. Noda, T. Tokuda, J. Ohta, S. Shiosaka, An implantable micro imaging device for molecular imaging in a brain of freely-moving mouse, *IEEE International Symposium on Bioelectronics and Bioinformatics (IEEE ISBB 2014)*, S1.4, Apr. 12, 2014, Chung Yuan Christian University, Taiwan.

25. H. Takehara, Y. Katsuragi, Y. Ohta, M. Motoyama, H. Takehara, T. Noda, K. Sasagawa, T. Tokuda, J. Ohta, Implantable micro-optical semiconductor devices for optical theranostics in deep tissue, *Appl. Phys. Express*, 9, 047001, 2016.

26. K. Sasagawa, T. Yamaguchi, M. Haruta, Y. Sunaga, H. Takehara, H. Takehara, T. Noda, T. Tokuda, J. Ohta, An implantable CMOS image sensor with self-reset pixels for functional brain imaging, *IEEE Trans. Electron Dev.*, 63(1), 215–222, 2015.

27. J.D. Newman, A.P.F. Turner, Home blood glucose biosensors: A commercial perspective, *Biosens. Bioelectron.*, 20, 2435–2453, 2005.

28. J.A. Tamada, M. Lesho, M.J. Tierney, Keeping watch on glucose, *IEEE Spectr.*, 39, 52–57, 2002.

29. H. Shibata, Y.J. Heo, T. Okitsu, Y. Matsunaga, T. Kawanishi, S. Takeuchi, Injectable hydrogel microbeads for fluorescence-based *in vivo* continuous glucose monitoring, *Proc. Natl. Acad. Sci. U.S.A.*, 107, 17894–17898, 2010.

30. Y.J. Heo, H. Shibata, T. Okitsu, T. Kawanishi, S. Takeuchi, Long-term *in vivo* glucose monitoring using fluorescent hydrogel fibers, *Proc. Natl. Acad. Sci. U.S.A.*, 34, 13399–13403, 2011.

31. T. Tokuda, M. Takahashi, K. Uejima, K. Masuda, T. Kawamura, Y. Ohta, M. Motoyama et al., CMOS image sensor-based implantable glucose sensor using glucose-responsive fluorescent hydrogel, *Biomed. Opt. Express*, 5(11), 3859–3870, 2014.

32. T. Tokuda, T. Kawamura, K. Masuda, T. Hirai, H. Takehara, Y. Ohta, M. Motoyaa et al., *In-vitro* long-term performance evaluation and improvement in the response time of CMOS-based implantable glucose sensors, *IEEE Des. Test*, 33, 37–48, 2016.

33. T. Kawamura, K. Masuda, T. Hirai, Y. Ohta, M. Motoyama, H. Takehara, T. Noda et al., CMOS-based implantable glucose monitoring device with improved performance and reduced invasiveness, *Electron. Lett.*, 51(10), 738–740, 2015.

34. E.S. Boyden, F. Zhang, E. Bamberg, G. Nagel, K. Deisseroth, Millisecond-timescale, genetically targeted optical control of neural activity, *Nat. Neurosci.*, 8, 1263–1268, 2005.

35. T. Ishizuka, M. Kakuda, R. Araki, H. Yawo, Kinetic evaluation of photosensitivity in genetically engineered neurons expressing green algae light-gated channels, *Neurosci. Res.*, 54, 85–94, 2006.

36. K. Deisseroth, Optogenetics, *Nat. Methods*, 8, 26–29, 2011.

37. V. Gradinaru, K.R. Thompson, F. Zhang, M. Mogri, K. Kay, M.B. Schneider, K. Deisseroth, Targeting and readout strategies for fast optical neural control *in vitro* and *in vivo*, *J. Neurosci.*, 27, 14231–14238, 2007.

38. A.M. Aravanis, L.-P. Wang, F. Zhang, L. a Meltzer, M.Z. Mogri, M.B. Schneider, K. Deisseroth, An optical neural interface: *in vivo* control of rodent motor cortex with integrated fiberoptic and optogenetic technology, *J. Neural Eng.*, 4, S143–156, 2007.

39. O.G.S. Ayling, T.C. Harrison, J.D. Boyd, A. Goroshkov, T.H. Murphy, Automated light-based mapping of motor cortex by photoactivation of channelrhodopsin-2 transgenic mice, *Nat. Methods*, 6, 219–224, 2009.

40. A.M. Leifer, C. Fang-Yen, M. Gershow, M.J. Alkema, A.D.T. Samuel, Optogenetic manipulation of neural activity in freely moving Caenorhabditis elegans, *Nat. Methods*, 8, 147–152, 2011.

41. S. Sakai, K. Ueno, T. Ishizuka, H. Yawo, Parallel and patterned optogenetic manipulation of neurons in the brain slice using a DMD-based projector, *Neurosci. Res.*, 75, 59–64, 2013.

42. D. Huber, L. Petreanu, N. Ghitani, S. Ranade, T. Hromadka, Z. Mainen, K. Svoboda, Sparse optical microstimulation in barrel cortex drives learned behaviour in freely moving mice, *Nature*, 451, 61–64, 2008.

43. T. Kim, J.G. McCall, Y.H. Jung, X. Huang, E.R. Siuda, Y. Li, J. Song et al., Injectable, cellular-scale optoelectronics with applications for wireless optogenetics., *Science* 340, 211–216, 2013.

44. T. Tokuda, H. Kimura, T. Miyatani, Y. Maezawa, T. Kobayashi, T. Noda, K. Sasagawa, J. Ohta, CMOS on-chip bio-imaging sensor with integrated micro light source array for optogenetics, *Electron. Lett.*, 48, 312–314, 2012.

45. T. Tokuda, H. Takehara, T. Noda, K. Sasagawa, J. Ohta, CMOS-based optoelectronic on-chip neural interface device, *IEICE Trans. Electron.*, E99, 165–172, 2016.

16

Microfluidic photocatalysis

Ning Wang and Xuming Zhang

Contents

16.1 Introduction

16.1.1 Mechanisms and kinetics of photocatalysis

Heterogeneous photocatalysis is the process of photochemical reactions on the surface of solid photo-catalysts under the irradiation of light, thus light and a photocatalyst are the prerequisites to promote the photocatalytic redox reactions. Only when photons with sufficiently high energy are absorbed, semiconductor photocatalysts can further facilitate the oxidation and reduction processes [1,2]. At present, most photocatalysts are reported as n-type semiconductor oxides, which actually depend on their own optical characteristics. Some popular photocatalysts have been extensively studied, such as TiO_2, ZnO, CdS, WO_3, SnO_2, Fe_2O_3, In_2O_3, and so on. TiO_2 is the most popular photocatalyst for industrial applications because of its nontoxicity, natural abundance, high stability, and easy availability in various reaction systems, including liquid-solid, gas-solid, and liquid-gas-solid. Although different photocatalysts are matched with different catalytic activities, the photocatalytic principles of the compound semiconductors are similarly consistent, especially for two popular semiconductor oxides, ZnO and TiO_2, which are both n-type oxide semiconductors with roughly the same band gap. This section will mainly discuss the basic principles of direct photocatalysis [3,4].

Photocatalysis can be regarded as a series of oxidation and reduction reactions induced by the photo-excited electrons and holes. The basic principle of photocatalysis is shown in Figure 16.1.

Generally, the incoming photon with energy $h\nu \geq E_0$ is absorbed by semiconductor photocatalysts (SC, e.g., TiO_2, ZnO) to excite an electron to the conduction band [5–7], leaving a hole in the valence band. Here, h is Planck's constant, ν the frequency of light, and E_0 the bandgap of the semiconductor photocatalyst, for example, $E_0 = 3.2$ eV for anatase TiO_2, corresponding to the wavelength $\lambda = 387$ nm. The activation equation can be expressed by

$$SC + h\nu \rightarrow SC\left(e_{cb}^- + h_{vb}^+\right) \tag{16.1}$$

Here, the excited holes can migrate to the surface of the SC and then oxidize the adsorbed reactants via the reactions as expressed below

$$h_{vb}^+ + OH_{ads}^- \rightarrow {}^*OH_{ads} \tag{16.2}$$

$${}^*OH_{ads} + reactant \rightarrow oxidized\ products \tag{16.3}$$

This represents the hole-driven oxidation pathway. Here, "ads" means adsorbed onto the photocatalytic reaction sites, and the ${}^*OH_{ads}$ means hydroxyl radicals, which are generated by losing one electron from OH^- and have extremely strong oxidability in the aqueous phase.

Similarly, the excited electrons can also migrate to the SC surface, and thus can initiate a reduction reaction (e.g., $Hg^{2+} + 2e^- \rightarrow Hg^0$). Alternatively, the electrons can be captured by the dissolved O_2 molecules and can also contribute to the oxidation through a series of different pathways [8]

$$e_{cb}^- + O_{2,ads} \rightarrow O_2^{\bullet -} \tag{16.4}$$

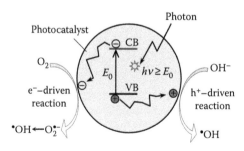

Figure 16.1 Mechanism of photocatalysis. The semiconductor photocatalyst absorbs a photon and excites an electron/hole pair. The electron and hole then migrate to the surface and initiate the reduction and/or oxidation to decompose the organic contaminants. (N. Wang et al., Microfluidic reactors for photocatalytic water purification, *Lab on a Chip*, vol. 14, pp. 1074–1082, 2014. Reproduced by permission of The Royal Society of Chemistry.)

$$O_2^{\cdot-} + H^+ \rightarrow HO_2^{\cdot} \tag{16.5}$$

$$HO_2^{\cdot} + HO_2^{\cdot} \rightarrow H_2O_2 + O_2 \tag{16.6}$$

$$H_2O_2 + e_{cb}^- \rightarrow {}^*OH + OH^- \tag{16.7}$$

$$\text{reactant} + O_2^{\cdot-} \text{(or }{}^*OH) \rightarrow \text{degradation products} \tag{16.8}$$

These form the electron-driven oxidation pathway. It is noted that the production of hydrogen peroxide (Equation 16.6) provides many more hydroxyl radicals (Equation 16.7), which have super oxidizability in the aqueous phase. Although both electrons and holes can lead to oxidation, some research studies have found that the electron-driven oxidation is apparently more efficient in degrading some organic contaminants [5,9].

During the last decades, thousands of researchers have been devoted to improving the photocatalytic efficiency for industrial applications, mainly including photocatalysts and the designs of reactors. Enhancement of the catalytic activity of the photocatalyst can be realized by increasing its specific surface area and by reducing its recombination centers of photogenerated electrons and holes. These have been mentioned by thousands of publications, but the direct evidence for recombination has rarely been presented [10–13]. Normally, the induced electron-hole pairs are mostly located at the surface of photocatalysts and the densities of electrons and holes go down rapidly from the surface into the interior. Electrons and holes located on the surface can recombine by both indirect and direct processes. Therefore, it is necessary to study the complex recombination processes.

The kinetics of electron-hole recombination may be affected by their recombination style. One recombination center is generated when one electron-hole pair is excited by the photon energy. The recombining rate should obey the first order law. However, if several electron-hole pairs are simultaneously exited on a photocatalyst particle, the recombination rate should obey the second order law [14,15]. Take a TiO_2 nanoparticle as an example, its absorption at 620 nm by the trapped electrons shows the second order decay with a baseline component when it is analyzed by the femtosecond pump-probe diffuse reflection spectroscopy, which can be expressed as [16]

$$Abs = \alpha \left\{ \frac{e_0}{1 + k_r e_0 t} + B \right\} \tag{16.9}$$

where α, e_0, k_r, t, and B, respectively, represent the constant, the initial concentration of trapped electrons, the second order rate constant, the time after the excitation pulse, and the baseline component. The last might be attributed to the electrons trapped in depth, but without participation in the reaction.

Beyond that, as reported by many research papers, photocatalytic reactions can be affected by many factors, such as the nature of the photocatalysts, the light source, the species of additive oxidants or reductants, the properties of waste water, the environment temperature, the fluidic dynamics and thermodynamics, the residence time in the reactor, and so on. These should be taken into comprehensive consideration for the overall design and operation of photocatalytic reactors. Therefore, it is very important to study the design of photocatalytic reactors. In this chapter, we will first give the details and limitations of current photocatalytic reactors. As a novel technology, microfluidics will be introduced to overcome the current limitations. By reviewing the current photocatalytic microreactors, we hope to reveal the hidden connection between microfluidics and photocatalysis and to find huge potential applications.

16.1.2 Major applications of photocatalytic reactors

The studies of heterogeneous photocatalysis can be roughly divided into two distinct areas. One focuses on the basic chemical transformations of semiconductor photocatalyst materials, including photocatalytic material science, surface interactions on photocatalysts, reaction mechanisms, and kinetics that impact on the processes at the molecular level. Many studies choose to improve the photon utilization efficiency by doping some elements (C, N, S, etc.) into TiO_2, causing a red shift of the light absorption band to absorb and utilize visible light [17]. The interface effects of some synthesized compound photocatalysts can cause the separation of electrons and holes. Some narrow bandgap semiconductor photocatalysts that can directly absorb visible light have also been used for visible light photocatalysis. However, due to their unsatisfactory performance and low efficiency, these photocatalysts are not ideal for large-scale industrial applications.

The second focuses on the reactor designs. In order to demonstrate the viability of semiconductor photocatalysis for a variety of industrial applications, reactor design is an equally critical factor. Research and development of effective reactor designs aim to scale up the laboratory-based processes to make them suitable for industrially feasible applications. Based on the above understanding, many photocatalytic reactors have been designed for their various applications. Some major applications are as follows.

16.1.2.1 Water treatment

As one of the main branches of photocatalytic applications, the treatment and recycling of industrial wastewater have been studied for a long time by thousands of institutes and researchers. From an economic point of view, chemical oxidation processes are of great interest due to their merits, such as operating at ambient temperature, being effective for a wide range of chemical pollutants, employing only dissolved oxygen in the wastewater as the oxidant, and utilizing solar energy to activate the photocatalyst [18,19].

Currently, photocatalysis seems to be the unique technique that meets all the referred prerequisites for industrial wastewater treatment. In this technique, photoredox reactions can be activated by the irradiation of a semiconductor photocatalyst with light energy over its bandgap energy. The absorption of these photons results in the formation of electron-hole pairs which subsequently migrate to the surface of the semiconductor photocatalyst. At the interface between the photocatalyst and the surrounding wastewater, these charge carriers can facilitate redox reactions, for example, degradation of the organic pollutants to the favored products in a series of coherent reaction steps. Thus, organic wastewater pollutants are decomposed and transformed to some small molecules such as CO_2, H_2O, and the respective mineral acids at ambient temperature.

In general, the efficiency of photocatalytic water treatment greatly depends on the design of an efficient photocatalytic reactor. Water phase contaminant removal by photocatalytic reactors is a surface reaction process which consists of three important steps: first, the transfer of pollutants to the catalyst surface; second, the adsorption/desorption of molecules; finally, the decomposition of pollutants into small molecules by the photocatalyst. Therefore, the main performance parameters of a photocatalytic reactor are the mass transfer efficiency, the kinetic reaction rate, and the reaction surface area. Normally, a photocatalytic reactor should contain two parts: the reactor structure and a light source. The reactor structure supports not only photocatalysts, but also the air/liquid flow channels. Therefore, an ideal photocatalytic reactor structure for wastewater treatment should have: (a) high specific surface area for a large reaction area; (b) appropriate light utilization efficiency; (c) high mass transfer efficiency; and (d) uniform residential time. Here, the residential time is defined as the reaction time of the reagent in the reaction chamber by the relationship: effective residential time = chamber volume/flow rate.

In view of these points, a wide variety of reactor configurations have been reported in the literature over the past 30 years, including the slurry reactors and the immobilized reactors. Some major designs are listed below:

- Annular photoreactor [20]
- Fluidized bed reactor [21]
- Membrane reactor (slurry and immobilized) [22]
- Swirl flow reactor [23]
- Packed bed photoreactor [24]
- Taylor vortex reactor [25]
- Rotating drum reactor [26]
- Coated optical fiber-based reactor [27]
- Hollow tubes reactor [28]
- Falling film reactor (slurry [29] and immobilized [30])
- Rotating disk reactor [31]
- Thin film fixed bed sloping plate reactor [32]
- Corrugated plate reactor [33]

16.1.2.2 Air purification

As air pollution becomes more and more serious, various methods have been developed to purify indoor waste air, including air filtration, ionizer purifier, activated carbon adsorption, ozone oxidation, ultraviolet germicidal irradiation, and many others. However, most of these technologies are not completely effective in degrading both biological contaminants and volatile organic compounds (VOVs). Recent studies show that photocatalytic oxidation is a promising technology for air purification. It is able to completely oxidize low concentrations of organic contaminants and is effective on different types of organics [34].

Air contaminant removal by photocatalytic reactors is similar to the wastewater removal process. What is different is that the gaseous phase organic molecules are adsorbed to the surface before decomposition. Except for the requirement of a photocatalytic reactor for water treatment, it is noted that an ideal photocatalytic reactor structure for indoor air purification should have a low pressure drop.

There are various photocatalytic reactors that have been reported for air purification purposes in the literature. However, most of the studies are only based on laboratory-scale tests instead of real applications. Most of these reactors employ immobilized photocatalysts on solid substrates. They have also been categorized based on their configurations. The reported reactor designs for research and some commercial applications are listed below:

- Flat plate reactor [35]
- Honeycomb monolith reactor [36]
- Fluidized bed reactor [37]
- Fiber/Membrane reactor [38]
- Optical fiber reactor [39]
- Filter reactor [40]
- Annular reactor [41]

16.1.2.3 Water splitting

The production of hydrogen by splitting water is a promising means to solving the future energy source crisis. However, it also faces some difficulties for practical applications. Photocatalytic water splitting utilizes sunlight and water as the sources for hydrogen production, which has been regarded as one of the important approaches and one of the cleanest energy sources. Although it is already an existing technology to produce solar fuel by using photovoltaic cells, direct synthesis of solar fuels by artificial photosynthesis has been found to be more scalable [42]. At least 1.23 V of energy is required to split water into hydrogen and oxygen, corresponding to a two-electron reaction. In this technique, one of the challenges is to develop an efficient photocatalyst that can absorb long wavelengths (particularly visible light), but still retain the ability to split water. There are many strategies to enhance the photocatalytic water splitting efficiency and the cut-off wavelength, such as doping with noble gas, plasmonic sensitization, impregnating a co-catalyst, loading a noble metal, and employing wire- or belt-shaped geometries.

Except for the difficulty of getting high performance photocatalysts, another challenge is to design an effective reactor for hydrogen production. However, fewer efforts have been made on novel and efficient designs of photocatalytic reactors for hydrogen production. Basically, the hydrogen produced by photocatalytic water splitting is produced by a series of complex photoelectrocatalytic reactions. Enhancing the photocatalytic reaction efficiency is still necessary to improve hydrogen production efficiency. Different from photocatalytic water treatment and air pollution purification, photocatalytic hydrogen production requires the reactor to be a three-phase reaction system on a gas-solid-liquid interface [43]. To improve the photon utilization efficiency and the efficient combination of the photocatalyst and a sacrificial agent, it is necessary to design and fabricate new photocatalytic reactors to enhance the three-phase (solid-liquid-gas) mass transfer efficiency. In addition, the efficiency of hydrogen production is strongly affected by other factors, such as the air tightness of the reactor, the testing ambient temperature, and the design of the photo-electrodes.

According to the design of the photo-electrodes, several photoreactors designed for water splitting are listed below.

- Non-biased monolithic photo-electrode [44]
- DSSC-biased monolithic photo-electrode [45]
- Photovoltaic-biased monolithic photo-electrode [46]
- Photoreactor with biphoto-electrodes [47]
- Dual-bed photoreactor [48]
- Photoreactors with electrolyte containment [49]

16.1.2.4 Other special applications

Except for the above mainstream applications, photocatalytic reactors have also been applied to many other more specific applications, such as organic photosynthesis, sterilization, deodorization, soil remediation, nitrogen fixation, and oil spill cleaning [5]. However, we will not discuss them in detail in order to avoid losing our focus.

16.1.3 Major designs and limitations of current photocatalytic reactors

In order to verify the photocatalytic characteristics of the semiconductor photocatalysts, the photocatalytic reactor should also be taken into consideration. The design and setup of the photocatalytic reaction system directly determines the overall photoreaction efficiency. Some configurations of photocatalytic reactors at the laboratory bench-scale have been demonstrated to work efficiently for pollutant treatment, which also paves the way to industrial-scale applications. Actually, the process of scaling up the photocatalysis reactors is extremely complicated and needs to consider not only the production technology, but also the economic cost.

Beyond that, many other aspects should be taken into comprehensive consideration for designing a photocatalytic reactor. These include, but are not limited to, the geometry of the reaction chamber, the distribution status of the photocatalysts, and the setting of the light source. For a scaled reactor, other aspects should also be considered, such as the concentration of pollutants, the throughput, and the irradiation density. Some detailed aspects have been distinctively highlighted for a photochemical reactor by Alfano et al. [50] when compared with conventional chemical reactors. First, the reactor geometry must be designed according to the irradiation range and the intensity of the light source. Second, the photocatalyst can be suspended or immobilized in the reactor, and this should be chosen to be the most beneficial for the reaction system. Finally, the light source directly influences the performance of photocatalytic reactors through elements such as the efficiency of output power, the utilization of the spectrum, the geometry, the cooling system, and some other auxiliary equipment for increasing the light utilization efficiency. For a configuration of a photoreactor for polluted water treatment, it is especially important to meet all these requirements.

Undoubtedly, reactor designs play a crucial role in tackling these limiting factors of photocatalysis, and have attracted numerous efforts in the last thirty years, leading to a number of innovative reactor designs, such as optical fiber-based photoreactors, fluidized bed reactors, thin film bed sloping plate reactors, and many others to be elaborated in the next section. Most of those reactors are referred to as bulk reactors due to the large dimensions of the reactor systems. Generally, the bulk reactors can be classified into two main types, depending on how photocatalysts are formed: (1) *slurry reactor*, in which the photocatalyst is in the form of nanoparticles and is suspended in water samples to generate a slurry (see Figure 16.2); and (2) *immobilized reactor*, in which the photocatalyst is immobilized on the substrates in the form of a film coating (see Figure 16.2). The former has a large surface area to volume ratio (SA:V) and enjoys fast mass transfer, but the absorption and scattering by the suspended photocatalyst nanoparticles cause a nonuniform distribution of light, and thus a low photon transfer. In addition, the suspended nanoparticles need to be filtered out after the purification, increasing the operation difficulty. In contrast, the latter type has good photon transfer and needs no post filtration, but the low SA:V causes a slow mass transfer.

Based on the above survey, photocatalytic reactors are currently facing several major limitations when they are applied in industrial applications, such as low mass transfer efficiency, low photon transfer efficiency, and deficiency of dissolved oxygen. The mass transfer efficiency affects how easily the contaminant particles are moved to the photocatalyst surface (mostly determined by the SA:V) and how fast the redox products are removed (affected by the desorption, diffusion, and stirring). The photon transfer refers to how to deliver the photons to the photocatalyst reaction sites. A uniform irradiation is often required for better utilization of the photons. For the dissolved oxygen, the electron-driven oxidation consumes oxygen, thus the limited concentration of naturally dissolved oxygen in water (typically 10 mg/L at room temperature and 1 atmosphere) would affect the photodegradation.

Various reactors have been attempted in order to break these limitations. For example, spin-disc reactors were designed to overcome the mass transfer limitation, optical fiber-based reactors were used to tackle the photon transfer limitation, and some reactors injected H_2O_2 or O_2 to solve the oxygen deficiency. However, most of them aimed at either one aspect or two, but none could get rid of all the limitations.

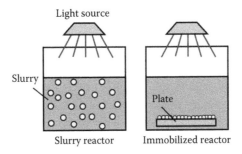

Figure 16.2 Typical bulky reactor designs—slurry reactor and immobilized reactor. (N. Wang et al., Microfluidic reactors for photocatalytic water purification, *Lab on a Chip*, vol. 14, pp. 1074–1082, 2014. Reproduced by permission of The Royal Society of Chemistry.)

16.2 Microfluidics for photocatalysis

16.2.1 Benefits of microfluidics for photocatalysis

Microfluidics technology employs microstructures to handle small volumes of fluids and exhibits remarkable capabilities in fine flow control, wide tunability, and parallel analysis [51]. Its great success in bioanalysis and drug discovery has triggered a boom of research to apply microfluidics to other areas, one of which is photocatalysis.

Microfluidics could bring many benefits to photocatalysis. The prominent ones are described below.

1. *Large surface area*: Microfluidic structures have an inherently large SA:V due to the small volume of fluid [52], typically in the range of 10,000–300,000 m^2/m^3, at least two orders of magnitude larger than that of the bulk reactors (typically <600 m^2/m^3) [53]. For this reason, a significant enhancement of the reaction rate has been observed in the microfluidic reactors (called *microreactors* hereafter) as compared to the bulk reactors. It is noted that SA:V here refers to the nominal surface area of the water sample over the water volume. For a microreactor with a rectangular bottom and a height h, it has SA:V $= 1/h$. In a real microreactor, SA:V could be much larger if the photocatalyst film is nanoporous.

2. *Short diffusion length*: The microfluidic layer is typically very thin (10–100 µm), making it easy for the organic pollutants to diffuse to the reaction surface. In a laminar flow, the velocity profile is parabolic and the mixing is limited by diffusion. In a macroscale reactor, the diffusion rates are often low because of the low concentration gradient between the bulk liquid and the catalyst surface. In a microreactor, however, the diffusion distance is much shorter and this is accompanied by much higher gradients. As a result, the diffusion rate enhances mass transfer efficiency. For example, microscale mixers can shorten the mixing times to be smaller than a second, whereas conventional stirrers often have mixing times of up to several tens of seconds.

3. *Uniform residence time*: The flow in microfluidic structures is typically laminar. This is not ideal for the mass transfer, but it ensures almost the same residence time (i.e., the time for the water sample to flow through the reactor, equivalent to the photocatalytic reaction time), and thus an equal level of degradation for different parts of the microflow. As the reaction rate (i.e., the percentage of pollutants converted per unit time) decreases with a longer residence time, an even distribution of residence time helps maximize the throughput (the processed water volume per unit time) at a targeted degradation percentage (i.e., what percent is being degraded) [8].

4. *Uniform illumination*: Microreactors usually have an immobilized photocatalyst film under the thin layer of the fluid, resulting in an almost uniform irradiation over the whole reaction surface, and thus a high photon efficiency. This is because the reaction rate constant of semiconducting photocatalysts is usually proportional to the square root of the power density [6].

5. *Short reaction time*: The combination of the above factors drastically improves the reaction speed and shortens the reaction time. In the microreactor, it takes only several to tens of seconds to obtain significant degradation (e.g., degraded by 90%), whereas bulk reactors usually needs several hours [54].

6. *Self-refreshing effect*: The running fluid acts naturally as a shear flow to refresh the reaction surface, which helps move away the reaction production and increase the stability of the photocatalysts. In the bulk reactors, the activity of photocatalysts typically fades noticeably after 10 runs of photocatalytic reactions, whereas in the microreactors, the photocatalysts can easily last for several hundred runs of reactions [5,55].

7. *Optimization of operation condition*: The fine control of fluids enables to optimize the operation condition of photocatalysis. For instance, by slowing down the flow rate and/or disturbing the laminar flows, the microreactors could clean most of the contaminants in one run, without resorting to the recirculation of flows, which is a common practice in the bulk reactors [7].

8. *More functionalities*: Microfluidics has the potential to add more functionalities to the photocatalysis, such as fast heat transfer, parallel process for rapid screening of photocatalysts, micromixing, on-chip monitoring of photocatalytic reactions, controllable delivery of light using optofluidic waveguides, selection of reaction pathways, and so on [5].

16.2.2 Photocatalytic microreactors classified by design

In view of the above benefits and the recent publications, photocatalytic microreactors can be mainly classified by their designs and applications.

Although the designs vary significantly, photocatalytic microreactors can be briefly classified into four configurations, as schematically shown in Figure 16.3. The major difference can be seen more clearly from the transverse cross-section (perpendicular to the flow direction) of the reactors. The *microcapillary reactor* coats a layer of photocatalyst on the inner wall of a capillary tube and runs the water sample inside the tube. The light can be irradiated from the outside. The *single microchannel reactor* makes use of a single straight microchannel to carry the water sample, whereas the *multi-microchannel reactor* exploits an array of the microchannels. The *planar microreactor* enlarges the microchannel in the lateral direction into a planar chamber. The first three configurations are based on microchannels, each of which has comparable dimensions of width and height (or diameter) in the range of 10–100 μm, whereas the last one has a much larger width (typically 1–100 mm) than height. For the photocatalysis, such a difference significantly affects the throughput, the photon utilization, the fabrication of photocatalysts, and the scalability to macroscale reactors.

16.2.2.1 Microcapillary reactors

An early work was reported by Li et al. [56], who fabricated a microreactor using capillaries with an inner diameter of 200 μm (see Figure 16.4). The inner wall was coated with TiO_2/SiO_2 film to degrade methylene blue (MB) solution. Microcapillary-based reactors have also been used to decompose various other organic dyes. It is the simplest design, but the coating on the inner wall is

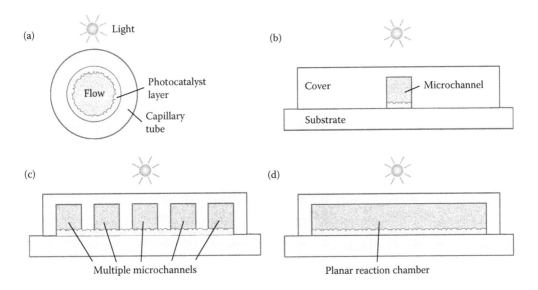

Figure 16.3 Typical designs of microfluidic reactors for photocatalysis water purification. (a) Transverse cross-section of micro-capillary reactor; (b) single-microchannel reactor; (c) multi-microchannel reactor; and (d) planar microreactor. (N. Wang et al., Microfluidic reactors for photocatalytic water purification, *Lab on a Chip*, vol. 14, pp. 1074–1082, 2014. Reproduced by permission of The Royal Society of Chemistry.)

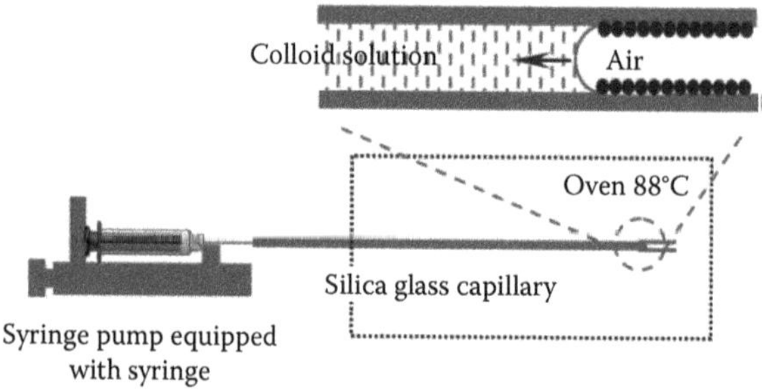

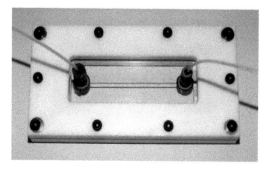

Figure 16.4 Micro-capillary reactor with the inner wall coated with self-assemble SiO$_2$/TiO$_2$ for methylene blue degradation, the dimensions of capillary: 5 cm (length) × 530 μm (outer diameter) and 200 μm (inner diameter). (X. Li et al., Modified micro-space using self-organized nanoparticles for reduction of methylene blue, *Chemical Communications*, pp. 964–965, 2003. Reproduced by permission of The Royal Society of Chemistry.)

cumbersome. And the external irradiation is not ideal for the utilization of light because the outer part of the photocatalyst layer absorbs light, but makes little contribution to the photodegradation.

16.2.2.2 Single microchannel reactors

With the rapid development of various etching techniques such as photolithography, micro-/nano-imprinting and dry/wet etching, some researchers have fabricated microchannels on glass, ceramics, and polymer substrates to examine photocatalytic reactions. Matsushita et al. [57] designed a single straight microchannel to degrade some organic models (see Figure 16.5). The bottom of the microchannel was immobilized with a sol-gel prepared TiO$_2$ thin film loaded with Pt particles. Similar reactors using single straight microchannels can be found in many other studies on the degradation of organic contaminants.

16.2.2.3 Multi-microchannel reactors

The microreactors based on microcapillary and single microchannels have small photon receiving areas and waste most of the external irradiation light, and the small cross-sectional area limits the throughput as well. To tackle these problems, multiple microchannels have been introduced. In 2004, Gorges et al. [59] designed a microreactor that branched out into 19 parallel microchannels. It immobilized a TiO$_2$ nanoporous film and fixed a UV-LED array above the area of the branched microchannels. The illuminated specific surface of the microreactor surpassed that of conventional bulk reactors by two orders of magnitude. In 2005, Takei et al. [60] fabricated a multi-microchannel reactor for photocatalytic redox-combined synthesis of L-pipecolinic acid using an anatase TiO$_2$ thin film (see Figure 16.6a). Pt nanoparticles were photodeposited from H$_2$PtCl$_6$ as the reduction

Figure 16.5 Single straight microchannel reactor with immobilized TiO$_2$-coated silica beads for degradation of 4-chlrophenol. (Reprinted from *Chemical Engineering Journal*, vol. 135, Y. Matsushita et al., Photocatalytic reactions in microreactors, pp. S303–S308, Copyright 2008, with permission from Elsevier.)

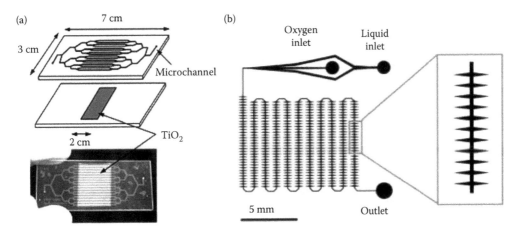

Figure 16.6 (a) Branched microchannel reactor for synthesis of L-pipecolinic acid; (b) serpentine microchannel reactor having 11 rows with 32 side lobes per row, coated with porous TiO_2 on the inner wall. (Reprinted from *Chemical Communications*, vol. 6, G. Takei, T. Kitamori, and H. B. Kim, Photocatalytic redox-combined synthesis of l-pipecolinic acid with a titania-modified microchannel chip, pp. 357–360, Copyright 2007, with permission from Elsevier and H. Lindstrom, R. Wootton, and A. Iles: High surface area titania photocatalytic microfluidic reactor. *AIChE Journal. 2007.* vol. 53. pp. 695–702. Copyright Wiley-VCH Verlag GmbH & Co. KGaA. Reproduced with permission.)

site. It was found that the conversion rate in the microreactor was 70 times larger than that in a cuvette using titania nanoparticles, with almost the same selectivity and enantiomeric excess. In another form of the multi-microchannel reactor, a single microchannel is folded up into a serpentine shape (see Figure 16.6b). This increases the photon receiving area and the residence time. However, the cross-section remains the same as that of a single microchannel, and thus limits the throughput.

16.2.2.4 Optofluidic planar microreactors

According to the previous designs of microfluidic reactors, most of them still suffer from reduced light absorption area, limited volume of reactants, and low throughput. Therefore, in order to enhance the specific area and light receiving area of a microreactor, a planar structure for a microreactor was developed. In 2010, as the pioneering work of optofluidic photocatalysis, Lei et al. [61] presented the first design of a photocatalytic planar reactor for water treatment by using solar light (see Figure 16.7a). Their study covered the device design, the material fabrication, and the experimental degradation of organic pollutions. This design demonstrated the great advantages of a microfluidic planar structure for photocatalysis, as it helps solve the fundamental problems of current photocatalysis technology, such as mass transfer limit and photon transfer limit. In 2013, Li et al. [62] developed an optofluidic microreactor with a catalyst-coated fiberglass based on the rectangular planar structure (see Figure 16.7b), which can dramatically improve the microreactor's performance as a result of the enhanced mass transport by shortening the diffusion transport length and inducing perturbation to the liquid flow.

16.2.3 Photocatalytic microreactors classified by application

Although the size of photocatalytic reactors in most of the previous studies has been miniaturized to microsize, their potential for industrial applications are still expected to be similar to that of the bulk reactors, because the studies using the microreactors can provide the guide for industrial designs. Recently, various microreactors have been explored for photocatalytic reactions, such as water purification, water splitting, photosynthesis, air purification, photocells, protein cleavage, and drug screening.

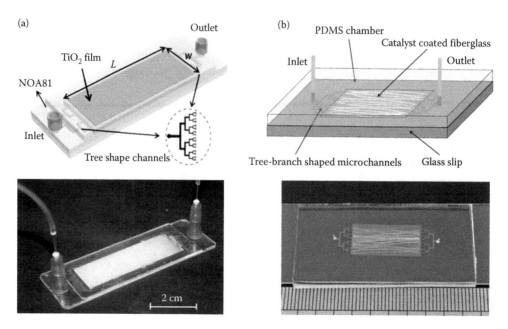

Figure 16.7 (a) The planar photocatalytic microfluidic reactor. The device is constructed by two TiO$_2$-coated glasses separated by a thin layer of microstructured UV-cured NOA81; (b) the optofluidic microreactor with the catalyst-coated fiberglass based on rectangular planar structure. (Reprinted with permission from L. Lei et al., Optofluidic planar reactors for photocatalytic water treatment using solar energy. *Biomicrofluidics*, vol. 4, p. 43004. Copyright 2010, American Institute of Physics and reprinted with permission from L. Li et al., Optofluidic microreactors with TiO2-coated fiberglass. *ACS Applied Materials & Interfaces*, vol. 5, no. 23, pp. 12548–12553. Copyright 2013 American Chemical Society.)

16.2.3.1 Water treatment

Recently, microreaction technology has evolved from a fringe area in catalysis research and reaction engineering into a valuable tool for the chemical industry and many other users who are looking for miniaturized and mobile applications for chemical systems. Water purification by using photocatalytic bulk reactors has been widely reported and seems more accessible. Due to the potential benefits brought by the microfluidics as summarized above, thousands of researchers have tried to exploit this field, though it is still in its infancy stage. Here, a brief survey of the reported microreactors for photocatalytic water treatment, divided by the referred designs, is listed in Table 16.1. Typical types of microreactors, the employed photocatalysts, and organic contaminant models are enumerated below.

16.2.3.2 Photosynthesis

Photosynthesis is a natural photochemical process and can be easily integrated into microfluidics. The research group of Matsushita fabricated straight single microchannel samples in different depths to conduct a photocatalytic process of N-alkylation of benzylamine in alcohol media [74]. After that, they designed serpentine microchannels to investigate the photoreactions, including photosensitized asymmetric reaction, photocatalytic oxidation, reduction, amine N-alkylation, and photocatalytic synthesis of L-pipecolinic acid from L-lysine. In fact, many photosynthesis processes have been investigated in microreactors, such as the natural photosynthesis and photochemical regeneration of the NADH cofactor, the reformation of ethanol, the photo-oxidation of CO, the photo-redox R-alkylation of aldehydes, and the measurement of the photosynthetic function of a single cyanobacterial cell. Takei et al. used a novel branched microchannel to implement an efficient photocatalytic synthesis of L-pipecolinic acid from L-lysine, as described in the last section (see Figure 16.6a) [60].

Table 16.1 Typical microfluidic reactors used for photocatalytic water treatment

Type of microreactor	Catalyst/light source	Model chemicals
Micro-capillary reactor	$TiO_2/SiO_2/UV$ light	Methylene blue [56]
	TiO_2/UV LED	Rhodamine 6G [63]
	TiO_2/UV lamp	Methylene orange [64]
	TiO_2/UV LED	Newcoccine, etc. [65]
	TiO_2/UV Nd-YAG laser	Salicylic acid [66]
Single straight microchannel reactor	TiO_2/UV led	Chelate (Cu-EDTA) [67]
	P25 TiO_2/UV light	4-chlorophenol [68]
	Pt-TiO_2/UV LED	Methylene blue [69]
Multi-microchannel reactor		
Branched microchannel	TiO_2/UV-A LED	4-chlorophenol [59]
Serpentine microchannel	Nanoporous TiO_2	Methylene blue [52]
	Nanofibrous TiO_2/UV	Methylene blue [70]
	$TiO_2/$Tungsten lamp	Methylene blue/phenol [71]
Planar microreactor	$TiO_2/$Solar light	Methylene blue [61]
	$TiO_2/BiVO_4/$Solar light	Methylene blue [72]
	$BiVO_4/$Blue LED	Methylene blue [73]

Source: N. Wang et al., Microfluidic reactors for photocatalytic water purification, *Lab on a Chip*, vol. 14, pp. 1074–1082, 2014. Reproduced by permission of The Royal Society of Chemistry.

16.2.3.3 Water splitting

For clean energy production, hydrogen generation by photocatalytic water splitting has attracted more and more attention. However, the productivity of hydrogen reported by many studies on bulk reactors has been shown to be barely satisfactory. Because the optofluidics combines fluids and their interaction with light, it is potentially the optimal platform for photocatalytic reactions and can increase the reaction rate by improving the mass and optical transfer efficiencies. In 2010, Matsushita et al. [75] successfully applied microstructured reaction vessels with an immobilized thin layer of photocatalyst to water splitting and CO_2 recycling. In 2013, the research group of David Ericson [76] demonstrated photocatalytic water splitting to produce hydrogen by designing a microfluidic reactor with a planar rectangular reaction chamber, in which Pt/TiO_2 thin film was employed and coated as the photocatalyst. In 2015, Chen et al. [77] fabricated a PDMS-based optofluidic microreactor for photocatalytic water splitting. In both works, redox reactions mediated by iodide/iodate (I^-/IO_3^-) pairs were employed to represent the water splitting reactions on the Pt/ TiO_2 catalysts.

16.2.3.4 Air purification

Although the specific surface area of microreactors is extremely high compared to the conventional bulk reactors, it is still low when compared with a photocatalyst with a nanoporous structure. Therefore, some researchers have mentioned that the inner surface area of microreactors can be increased to enhance the specific surface of the microreactor by chemical modification or physical deposition. Another technique is to directly fill microsized photocatalyst pellets into a microreactor. In 2005, Karim et al. [78] designed two different microreactors for methanol steam reforming, including a fixed-bed microreactor and a wall-coated microreactor. The authors also compared their performances. In the fixed-bed microreactor, obvious temperature gradients were observed and the reaction was limited by the slow heat transfer, while in the wall-coated reactor, neither the mass transfer nor the heat transfer was limited and the pressure drops were remarkably low.

In 2009, a microreactor was designed to evaluate a photocatalyst's compatibility with the current and proposed industrial techniques [79]. The microscale modeling was also developed to understand the interactions between the photocatalyst and gaseous contaminants.

There might be some reasons for the limited efforts on designing a microreactor for the decomposition of gaseous pollutants. These include the testing difficulty as compared to the water phase samples, and the requirement of high air-tightness for the reactor, which affects the processing and the pressure drops. Therefore, more researchers prefer the use of the degradation of aqueous phase contaminants to characterize the performances of microreactors or photocatalysts.

16.2.3.5 Photocells (PEC and PFC)

Many types of photo-electrochemical reactions have been studied using bulk reactors. However, as the size of the reactor shrinks from macroscopic to microcosmic, the mechanism of the reactions may be different. In 2012, Wang et al. [5] fabricated a photoelectrocatalytic (PEC) microreactor (see Figure 16.8a), that made use of the photocatalytic process, the electrocatalytic process, and their synergistic effect. This design introduced an external bias voltage and possessed new features. The first is the new mechanism of the PEC effect. The second is that the electric field helps with the separation of photo-excited electrons and holes. The third one is that the polarity of the external bias enables the selection of oxidation pathways, either electron-driven or hole-driven. The last one is that a bias greater than 1.23 V could electrolyze water to generate O_2, which supplies oxygen for the photo-oxidation and solves the oxygen deficiency problem in the previous microreactors. In 2014, Li et al. [81] developed a photocatalytic fuel cell (PFC) microreactor (see Figure 16.8b) with a membrane-free and air-breathing mode for efficient wastewater treatment and simultaneous electricity generation by using solar energy. In the photocatalytic fuel cell (PFC) system, the organic substance to be decomposited is formed as fuel and the chemical energy stored in organics can be directly converted into electrical energy accompanied by the organic compounds' degradation.

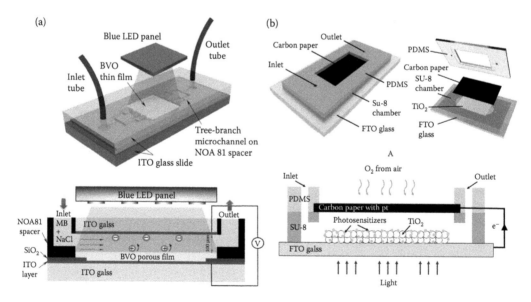

Figure 16.8 (a) Schematics of the PEC microreactor with 3D diagram and cross-sectional view of the reaction chamber; (b) schematic display of PFC microreactor with 3D structure and cross-sectional view. (N. Wang et al., Microfluidic photoelectrocatalytic reactors for water purification with an integrated visible-light source, *Lab on a Chip*, vol. 12, p. 3983, 2012 and L. Li et al., Optofluidics-based micro-photocatalytic fuel cell for efficient wastewater treatment and electricity generation, *Lab on a Chip*, vol. 14, pp. 3368–3375, 2014. Reproduced by permission of The Royal Society of Chemistry.)

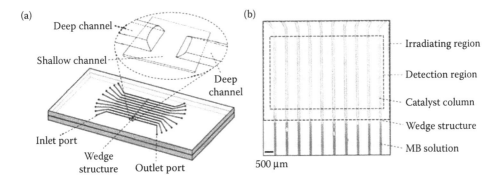

Figure 16.9 (a) Schematic diagram of the multi-channel array chip with a wedge structure in each channel; (b) image of the chip channels loaded with mesoporous TiO_2 and MB in each channel after a 15-min irradiation of UV light. (Reprinted from *Talanta*, vol. 116, H. Zhang et al., Microfluidic chip-based analytical system for rapid screening of photocatalysts, pp. 946–950, Copyright 2013, with permission from Elsevier.)

16.2.3.6 Peptide and protein cleavage

As relatively rare studies, Jones et al. [82] fabricated several types of photocatalytic microreactors for the cleavage of peptides and protein in 2005. These were based on the use of titania-coated glass or titania sol-gel-coated silica particles as the photocatalyst. The peptide angiotensin was fragmented using several different device configurations including: (1) small volume wells; (2) microfluidic channels; and (3) a microflow reactor.

16.2.3.7 Rapid screening of photocatalysts

Due to the limitations in the system scale and laborious operations, it is difficult for traditional bench-top chemistry to enhance the screening throughput. Multiple channels can be easily fabricated in microchip-based systems and thus enable parallel reactions to increase the screening throughput. In 2013, Qun Fang's group [83] first demonstrated the rapid screening of photocatalysts by a microfluidic chip-based system (see Figure 16.9). The microchip contains a multichannel array with a wedge structure with a shallow channel and two deep channels in each linear channel, which are designed for more convenient loading and keeping of the photocatalyst in the device. With this design, the photocatalyst could not only be immobilized in the microchip, but the channel could also remain unblocked. By using this kind of microchip, the photocatalysis performance of different photocatalysts can be rapidly investigated. In 2014, K. Katayama et al. [84] designed a photocatalytic microreactor based on a fused silica capillary to develop two analytical methods for evaluating and screening the oxidation and reduction abilities of photocatalytic materials, including product analysis and reaction rate analysis. It is expected that this novel analytical procedure could accelerate the development of new materials which are very useful for various purposes.

16.3 Discussion and outlook

Microfluidic reactors also have some limitations. Frequently quoted disadvantages of microreactors are the high fabrication cost, low throughput, incompatibility with solids, and the omission of cost reduction for the scaling-up effect. These would lead to poor industrial acceptance. In addition, distribution and fouling problems often prevent industrial applications from the complete miniaturization of reaction systems. Instead, it seems that multiscale approaches are increasingly successful in the applications for pilot or plant scales. This means that an appropriate scaling is selected for a respective purpose, and the size reduction is often limited according to the policy "as small as necessary."

First, the microreactor involves the fabrication of microstructures and the embedment of photocatalysts. Fortunately, the fabrication of microfluidic structures using soft materials (but they may have hard covers and hard substrates) is already a routine process, and a photocatalyst can be deposited onto the substrate before being bonded with the microstructure.

Another severe problem is the limited throughput, typically in the order of 1 mL/h. This is far from the threshold throughput of about 1000 L/h for industrial requirements. To boost the throughput, several possible approaches could be exploited. A simple way is to use a large array of microreactors to bring the throughput up to a meaningful level (like 1 L/h). A more practical way is to scale up the microreactors to the meter size, while maintaining the performance determining factors, such as surface-area-to-volume ratio and the residence time.

New designs could also be introduced, for example, coating the inner surface of capillary tubes with photocatalysts and then piling up the tubes into a large bundle (similar to the structure of a photonic crystal fiber [85], but having a much larger cross-section). This design is a combined use of an optofluidic waveguide and a photonic crystal fiber. The contaminated water could be run through the holes of tubes and the light may be irradiated directly onto the bundle cross-section. Each stream of flow inside the tubes acts as an optofluidic waveguide to carry the light, which is absorbed gradually by the photocatalyst layer coated on the tube's inner surface for photodegradation. This design ensures a long interaction length of the light with the photocatalyst layer and the water sample, and would lead to a full utilization of the light energy and a total cleaning of the contaminants. The large cross-sectional area and fast flow velocity (using long tubes to ensure sufficient residence time) join up to ensure a large throughput. It is feasible to achieve 1000 L/h. Many other innovative designs could be explored as well. With the techniques and physical understandings accumulated during the development of microfluidic reactors, the prospect for industrial water purification is very promising.

On the other hand, there are many scenarios that do not need a large throughput. In fact, the microreactors consume small amounts of liquid samples and photocatalysts. This could be a beneficial factor for some applications, for instance, rapid characterization of expensive photocatalysts, parallel performance comparison of different photocatalysts, and optimization of the operation condition. As the photocatalytic reaction is strongly dependent on many factors (e.g., type of photocatalysts, preparation details of photocatalysts, model chemicals, light sources, temperature, pH value, etc.), it has long been a headache to standardize photocatalytic efficiency tests and to make different tests comparable. The microreactor may provide a standard platform as it enables the convenient and fine control of operation conditions (such as flow rate, heat dissipation, etc.), rapid characterization, and parallel photoreactions (using an array of microchannels or reaction chambers). A similar case of the use of microfluidic chips for drug screening has achieved great success.

16.4 Concluding remarks

As has been discussed in this chapter, we highlight and elucidate the combination of microfluidics and photocatalysis into photocatalytic microreactors, covering the current understanding of the technology, the mechanism of photocatalysis, the benefits brought by microfluidics to photocatalysis, and various designs and applications of photocatalytic microreactors.

In this chapter, microreactors have been demonstrated to bring many benefits to photocatalysis, such as fast mass transfer, efficient photon transfer, high surface-area-to-volume ratio, low Reynolds number, short diffusion length, uniform irradiation and reaction time, self-refreshing effect, and many other functions. Although not reported as frequently as conventional photocatalytic devices, microreactors have shown high conversion rates in photodegradation experiments.

Beyond that, photocatalytic microreactors have great potential and have found uses in many other fields. As reviewed in this chapter, although some studies have already tried water splitting, protein cleavage, photosynthesis, and so on, most are in the infancy stage. There are still many more to explore. For example, bulk reaction systems have utilized photocatalysis for the destruction of bacteria and viruses, inactivation of cancer cells, nitrogen fixation, and remediation of oil spills. Their corresponding microfluidic designs have yet to come.

References

1. C. McCullagh, N. Skillen, M. Adams, and P. K. J. Robertson, Photocatalytic reactors for environmental remediation: A review, *Journal of Chemical Technology and Biotechnology*, vol. 86, pp. 1002–1017, 2011.
2. M. N. Chong, B. Jin, C. W. K. Chow, and C. Saint, Recent developments in photocatalytic water treatment technology: A review, *Water Research*, vol. 44, pp. 2997–3027, 2010.
3. B. H. Diya'Uddeen, W. M. A. W. Daud, and A. R. Abdul Aziz, Treatment technologies for petroleum refinery effluents: A review, *Process Safety and Environmental Protection*, vol. 89, pp. 95–105, 2011.
4. S. M. Rodríguez, C. Richter, J. B. Gálvez, and M. Vincent, Photocatalytic degradation of industrial residual waters, *Solar Energy*, vol. 56, no. 5, pp. 401–410, 1996.
5. N. Wang, X. Zhang, Y. Wang, W. Yu, and H. L. W. Chan, Microfluidic reactors for photocatalytic water purification., *Lab on a Chip*, vol. 14, pp. 1074–1082, 2014.
6. J. Herrmann, Heterogeneous photocatalysis: Fundamentals and applications to the removal of various types of aqueous pollutants, *Catalysis Today*, vol. 53, pp. 115–129, 1999.
7. Y. L. Chen, L.-C. Kuo, M. L. Tseng, H. M. Chen, C.-K. Chen, H. J. Huang, R.-S. Liu, and D. P. Tsai, ZnO nanorod optical disk photocatalytic reactor for photodegradation of methyl orange, *Optics Express*, vol. 21, pp. 7240–7249, 2013.
8. J. M. Herrmann, Fundamentals and misconceptions in photocatalysis, *Journal of Photochemistry and Photobiology A: Chemistry*, vol. 216, pp. 85–93, 2010.
9. A. L. Linsebigler, J. T. Yates Jr, G. Lu, and J. T. Yates, Photocatalysis on TiO_2 surfaces: Principles, mechanisms, and selected results, *Chemical Reviews*, vol. 95, pp. 735–758, 1995.
10. B. Ohtani, Preparing articles on photocatalysis—Beyond the illusions, misconceptions, and speculation, *Chemistry Letters*, vol. 37, pp. 216–229, 2008.
11. B. Jenny and P. Pichat, Determination of the actual photocatalytic rate of H_2O_2 decomposition over suspended TiO_2 - Fitting to the langmuir hinshelwood form, *Langmuir*, vol. 7, pp. 947–954, 1991.
12. D. Monllor-Satoca, R. Gómez, M. González-Hidalgo, and P. Salvador, The "Direct-Indirect" model: An alternative kinetic approach in heterogeneous photocatalysis based on the degree of interaction of dissolved pollutant species with the semiconductor surface, *Catalysis Today*, vol. 129, no. 1–2, pp. 247–255, 2007.
13. A. V. Emeline, V. Ryabchuk, and N. Serpone, Factors affecting the efficiency of a photocatalyzed process in aqueous metal-oxide dispersions, *Journal of Photochemistry and Photobiology A: Chemistry*, vol. 133, pp. 89–97, 2000.
14. M. Lazar, S. Varghese, and S. Nair, Photocatalytic water treatment by titanium dioxide: Recent updates, *Catalysts*, vol. 2, no. 4, pp. 572–601, 2012.
15. H. S. Fogler, *Elements of Chemical Reaction Engineering*, 4th edition, Prentice Hall of India PTR, 2006.
16. A. S. K. Sinha, N. Sahu, M. K. Arora, and S. N. Upadhyay, Preparation of egg-shell type Al_2O_3-supported CdS photocatalysts for reduction of H_2O to H_2, *Catalysis Today*, vol. 69, pp. 297–305, 2001.
17. R. L. Z. Hoye, K. P. Musselman, and J. L. Macmanus-Driscoll, Research update: Doping ZnO and TiO_2 for solar cells, *APL Materials*, vol. 1, no. 6, paper 060701, 2013.
18. M. Kositzi, I. Poulios, S. Malato, J. Caceres, and A. Campos, Solar photocatalytic treatment of synthetic municipal wastewater, *Water Research*, vol. 38, no. 5, pp. 1147–1154, 2004.

19. R. Dillert, S. Vollmer, M. Schober, J. Theurich, D. Bahnemann, H.-J. Arntz, K. Pahlmann, J. Wienefeld, T. Schmedding, and G. Sager, Photocatalytic treatment of an industrial waste-water in the double-skin sheet reactor, *Chemical Engineering & Technology*, vol. 22, no. 11, p. 931, Nov. 1999.

20. E. Sahle-Demessie, S. Bekele, and U. R. Pillai, Residence time distribution of fluids in stirred annular photoreactor, *Catalysis Today*, vol. 88, no. 1–2, pp. 61–72, 2003.

21. M. Zhang, T. An, J. Fu, G. Sheng, X. Wang, X. Hu, and X. Ding, Photocatalytic degradation of mixed gaseous carbonyl compounds at low level on adsorptive TiO_2/SiO_2 photocatalyst using a fluidized bed reactor, *Chemosphere*, vol. 64, no. 3, pp. 423–431, 2006.

22. S. Mozia, Photocatalytic membrane reactors (PMRs) in water and wastewater treatment. A review, *Separation and Purification Technology*, vol. 73, no. 2, pp. 71–91, Jun. 2010.

23. C. M. Schietekat, M. W. M. van Goethem, K. M. Van Geem, and G. B. Marin, Swirl flow tube reactor technology: An experimental and computational fluid dynamics study, *Chemical Engineering Journal*, vol. 238, pp. 56–65, 2014.

24. I. M. Arabatzis, N. Spyrellis, Z. Loizos, and P. Falaras, Design and theoretical study of a packed bed photoreactor, *Journal of Materials Processing Technology*, 2005, vol. 161, no. 1–2, SPEC. ISS., pp. 224–228.

25. J. G. Sczechowski, C. A. Koval, and R. D. Noble, A Taylor vortex reactor for heterogeneous photocatalysis, *Chemical Engineering Science*, vol. 50, pp. 3163–3173, 1995.

26. L. Zhang, T. Kanki, N. Sano, and A. Toyoda, Photocatalytic degradation of organic compounds in Aqueous solution by a TiO_2-Coated rotating-drum reactor using solar light, *Solar Energy*, vol. 70, no. 4, pp. 331–337, 2001.

27. R.-D. Sun, A. Nakajima, I. Watanabe, T. Watanabe, and K. Hashimoto, TiO_2-coated optical fiber bundles used as a photocatalytic filter for decomposition of gaseous organic compounds, *Journal of Photochemistry and Photobiology A: Chemistry*, vol. 136, pp. 111–116, 2000.

28. A. K. Ray, A new photocatalytic reactor for destruction of toxic water pollutants by advanced oxidation process, *Catalysis Today*, vol. 44, pp. 357–368, 1998.

29. G. Li Puma and P. L. Yue, Enhanced photocatalysis in a pilot laminar falling film slurry reactor, *Industrial & Engineering Chemistry Research*, vol. 38, pp. 3246–3254, 1999.

30. G. S. Shephard, S. Stockenström, D. De Villiers, W. J. Engelbrecht, and G. F. S. Wessels, Degradation of microcystin toxins in a falling film photocatalytic reactor with immobilized titanium dioxide catalyst, *Water Research*, vol. 36, pp. 140–146, 2002.

31. D. D. Dionysiou, G. Balasubramanian, M. T. Suidan, A. P. Khodadoust, I. Baudin, and J. M. Laîné, Rotating disk photocatalytic reactor: Development, characterization, and evaluation for the destruction of organic pollutants in water, *Water Research*, vol. 34, no. 11, pp. 2927–2940, 2000.

32. S. J. Khan, R. H. Reed, and M. G. Rasul, Thin-film fixed-bed reactor (TFFBR) for solar photocatalytic inactivation of aquaculture pathogen Aeromonas hydrophila, *BMC Microbiology*, vol. 12, pp. 1–11, 2012.

33. Z. Zhang, W. A. Anderson, and M. Moo-Young, Experimental analysis of a corrugated plate photocatalytic reactor, *Chemical Engineering Journal*, vol. 99, no. 2, pp. 145–152, 2004.

34. Y. Zhang, *Modeling and Design of Photocatalytic reactors for Air Purification*, PhD Thesis, January 2013.

35. R. J. Brandi, O. M. Alfano, and A. E. Cassano, Rigorous model and experimental verification of the radiation field in a flat-plate solar collector simulator employed for photocatalytic reactions, *Chemical Engineering Science*, vol. 54, no. 13–14, pp. 2817–2827, 1999.

36. J. Zhao and X. Yang, Photocatalytic oxidation for indoor air purification: A literature review, *Building and Environment*, vol. 38, no. 5, pp. 645–654, 2003.

37. J. Peral, X. Domènech, and D. F. Ollis, Heterogeneous photocatalysis for purification, decontamination and deodorization of air, *Journal of Chemical Technology and Biotechnology*, vol. 70, no. 2, pp. 117–140, 1997.

38. P. Pichat, J. Disdier, C. Hoang-Van, D. Mas, G. Goutailler, and C. Gaysse, Purification/deodorization of indoor air and gaseous effluents by TiO_2 photocatalysis, *Catalysis Today*, vol. 63, no. 2–4, pp. 363–369, 2000.
39. W. Wang and Y. Ku, Photocatalytic degradation of gaseous benzene in air streams by using an optical fiber photoreactor, *Journal of Photochemistry and Photobiology A: Chemistry*, vol. 159, no. 1, pp. 47–59, 2003.
40. Y. Paz, Application of TiO_2 photocatalysis for air treatment: Patents' overview, *Applied Catalysis B: Environmental*, vol. 99, no. 3–4, pp. 448–460, 2010.
41. A. A. Assadi, A. Bouzaza, C. Vallet, and D. Wolbert, Use of DBD plasma, photocatalysis, and combined DBD plasma/photocatalysis in a continuous annular reactor for isovaleraldehyde elimination—Synergetic effect and byproducts identification, *Chemical Engineering Journal*, vol. 254, pp. 124–132, 2014.
42. A. A. Ismail and D. W. Bahnemann, Photochemical splitting of water for hydrogen production by photocatalysis: A review, *Solar Energy Materials and Solar Cells*, vol. 128, pp. 85–101, 2014.
43. F. E. Osterloh and B. A. Parkinson, Recent developments in solar water-splitting photocatalysis, *Mrs Bullitins*, vol. 36, no. 1, pp. 17–22, 2011.
44. X. Xia, J. Luo, Z. Zeng, C. Guan, Y. Zhang, J. Tu, H. Zhang, and H. J. Fan, Integrated photoelectrochemical energy storage: Solar hydrogen generation and supercapacitor, *Scientific Reports*, vol. 2, paper 981, 2012.
45. A. H. Jahagirdar and N. G. Dhere, Photoelectrochemical water splitting using $CuIn_{1-x}Ga_xS_2$/CdS thin-film solar cells for hydrogen generation, *Solar Energy Materials and Solar Cells*, vol. 91, no. 15–16, pp. 1488–1491, 2007.
46. C. Longo and M. A. De Paoli, Dye-sensitized solar cells: A successful combination of materials, *Journal of the Brazilian Chemical Society*, vol. 14, no. 6, pp. 889–901, 2003.
47. V. M. Aroutiounian, V. M. Arakelyan, and G. E. Shahnazaryan, Metal oxide photoelectrodes for hydrogen generation using solar radiation-driven water splitting, *Solar Energy*, vol. 78, no. 5, pp. 581–590, 2005.
48. K. Lee, W. S. Nam, and G. Y. Han, Photocatalytic water-splitting in alkaline solution using redox mediator. 1: Parameter study, *International Journal of Hydrogen Energy*, vol. 29, no. 13, pp. 1343–1347, 2004.
49. W. E. Liss, Q. Fan, M. Onischak, Solar cell electrolysis of water to make hydrogen and oxygen, Patent US 7241950, 2007.
50. O. M. Alfano, D. Bahnemann, A. E. Cassano, R. Dillert, and R. Goslich, Photocatalysis in water environments using artificial and solar light, *Catalysis Today*, vol. 58, pp. 199–230, 2000.
51. T. M. Squires and S. R. Quake, Microfluidics: Fluid physics at the nanoliter scale, *Reviews of Modern Physics*, vol. 77, pp. 977–1026, 2005.
52. H. Lindstrom, R. Wootton, and A. Iles, High surface area titania photocatalytic microfluidic reactors, *AIChE Journal*, vol. 53, pp. 695–702, 2007.
53. J. C. S. Wu, T. H. Wu, T. Chu, H. Huang, and D. Tsai, Application of optical-fiber photoreactor for CO_2 photocatalytic reduction, *Topics in Catalysis*, vol. 47, no. 3–4, pp. 131–136, 2008.
54. M. Oelgemoeller, Highlights of photochemical reactions in microflow reactors, *Chemical Engineering and Technology*, vol. 35, pp. 1144–1152, 2012.
55. T. Van Gerven, G. Mul, J. Moulijn, and A. Stankiewicz, A review of intensification of photocatalytic processes, *Chemical Engineering and Processing: Process Intensification*, vol. 46, pp. 781–789, 2007.
56. X. Li, H. Wang, K. Inoue, M. Uehara, H. Nakamura, M. Miyazaki, E. Abea, and H. Maeda, Modified micro-space using self-organized nanoparticles for reduction of methylene blue, *Chemical Communications*, pp. 964–965, 2003.
57. Y. Matsushita, N. Ohba, T. Suzuki, and T. Ichimura, N-Alkylation of amines by photocatalytic reaction in a microreaction system, *Catalysis Today*, vol. 132, pp. 153–158, 2008.

58. Y. Matsushita, N. Ohba, S. Kumada, K. Sakeda, T. Suzuki, and T. Ichimura, Photocatalytic reactions in microreactors, *Chemical Engineering Journal*, vol. 135, pp. S303–S308, 2008.
59. R. Gorges, S. Meyer, and G. Kreisel, Photocatalysis in microreactors, *Journal of Photochemistry and Photobiology A: Chemistry*, vol. 167, pp. 95–99, 2004.
60. G. Takei, T. Kitamori, and H. B. Kim, Photocatalytic redox-combined synthesis of l-pipecolinic acid with a titania-modified microchannel chip, *Chemical Communications*, vol. 6, pp. 357–360, 2005.
61. L. Lei, N. Wang, X. M. Zhang, Q. Tai, D. P. Tsai, and H. L. W. Chan, Optofluidic planar reactors for photocatalytic water treatment using solar energy, *Biomicrofluidics*, vol. 4, p. 43004, 2010.
62. L. Li, R. Chen, X. Zhu, H. Wang, Y. Wang, Q. Liao, and D. Wang, Optofluidic microreactors with TiO_2-coated fiberglass, *ACS Applied Materials & Interfaces*, vol. 5, no. 23, pp. 12548–12553, 2013.
63. K. Oda, Y. Ishizaka, T. Sato, T. Eitoku, and K. Katayama, Analysis of photocatalytic reactions using a TiO(2) immobilized microreactor, *Analytical Sciences*, vol. 26, pp. 969–972, 2010.
64. Z. Zhang, H. Wu, Y. Yuan, Y. Fang, and L. Jin, Development of a novel capillary array photocatalytic reactor and application for degradation of azo dye, *Chemical Engineering Journal*, vol. 184, pp. 9–15, 2012.
65. N. Tsuchiya, K. Kuwabara, A. Hidaka, K. Oda, and K. Katayama, Reaction kinetics of dye decomposition processes monitored inside a photocatalytic microreactor, *Physical Chemistry Chemical Physics*, vol. 14, p. 4734, 2012.
66. G. Charles, T. Roques-Carmes, N. Becheikh, L. Falk, J. M. Commenge, and S. Corbel, Determination of kinetic constants of a photocatalytic reaction in micro-channel reactors in the presence of mass-transfer limitation and axial dispersion, *Journal of Photochemistry and Photobiology A: Chemistry*, vol. 223, pp. 202–211, 2011.
67. D. Daniel and I. G. R. Gutz, Microfluidic cell with a TiO_2-modified gold electrode irradiated by an UV-LED for in situ photocatalytic decomposition of organic matter and its potentiality for voltammetric analysis of metal ions, *Electrochemistry Communications*, vol. 9, pp. 522–528, 2007.
68. T. H. Yoon, L. Y. Hong, and D. P. Kim, Photocatalytic reaction using novel inorganic polymer derived packed bed microreactor with modified TiO_2 microbeads, *Chemical Engineering Journal*, vol. 167, pp. 666–670, 2011.
69. Y. Matsushita, N. Ohba, S. Kumada, K. Sakeda, T. Suzuki, and T. Ichimura, Photocatalytic reactions in microreactors, *Chemical Engineering Journal*, vol. 135, 2008.
70. Z. Meng, X. Zhang, and J. Qin, A high efficiency microfluidic-based photocatalytic microreactor using electrospun nanofibrous TiO_2 as a photocatalyst., *Nanoscale*, vol. 5, pp. 4687–4690, 2013.
71. H. C. Aran, D. Salamon, T. Rijnaarts, G. Mul, M. Wessling, and R. G. H. Lammertink, Porous photocatalytic membrane microreactor (P2M2): A new reactor concept for photochemistry, *Journal of photochemistry and photobiology. A: Chemistry*, vol. 225, pp. 36–41, 2011.
72. N. Wang, Z. K. Liu, N. Y. Chan, H. L. W. Chan, and X. M. Zhang, Microfluidic solar reactor for photocatalytic water treatment, in *MicroTAS*, Okinawa, Japan, paper W.9.195, 2012.
73. N. Wang, F. Tan, L. Wan, M. Wu, and X. Zhang, Microfluidic reactors for visible-light photocatalytic water purification assisted with thermolysis, *Biomicrofluidics*, vol. 8, no. 5, p. 054122, Sep. 2014.
74. Y. Matsushita, N. Ohba, S. Kumada, T. Suzuki, and T. Ichimura, Photocatalytic N-alkylation of benzylamine in microreactors, *Catalysis Communications*, vol. 8, pp. 2194–2197, 2007.
75. Y. Matsushita, H. M. A. Mohamed, and S. Ookawara, Micro-flow reaction systems for photocatalytic carbon dioxide recycling and hyrodogen generation, in *μTAS*, Okinawa, Japan, paper T.4.119, 2012.
76. S. S. Ahsan, A. Gumus, and D. Erickson, Redox mediated photocatalytic water-splitting in optofluidic microreactors., *Lab on a Chip*, vol. 13, pp. 409–414, 2013.

77. R. Chen, L. Li, X. Zhu, H. Wang, Q. Liao, and M. X. Zhang, Highly-durable optofluidic microreactor for photocatalytic water splitting, *Energy*, vol. 83, pp. 797–804, 2015.

78. A. M. Karim, J. A. Federici, and D. G. Vlachos, Portable power production from methanol in an integrated thermoeletric/microreactor system, *Journal of Power Sources*, vol. 179, no. 1, pp. 113–120, 2008.

79. J. Geng, D. Yang, J. Zhu, D. Chen, and Z. Jiang, Nitrogen-doped TiO_2 nanotubes with enhanced photocatalytic activity synthesized by a facile wet chemistry method, *Materials Research Bulletin*, vol. 44, no. 1, pp. 146–150, 2009.

80. N. Wang, X. Zhang, B. Chen, W. Song, N. Y. Chan, and H. L. W. Chan, Microfluidic photoelectrocatalytic reactors for water purification with an integrated visible-light source, *Lab on a Chip*, vol. 12, p. 3983, 2012.

81. L. Li, G. Wang, R. Chen, X. Zhu, H. Wang, Q. Liao, and Y. Yu, Optofluidics based micro-photocatalytic fuel cell for efficient wastewater treatment and electricity generation, *Lab on a Chip*, vol. 14, pp. 3368–3375, 2014.

82. B. J. Jones, L. E. Locascio, and M. A. Hayes, Radical activated cleavage of peptides and proteins: An alternative to proteolytic digestion, in *MicroTAS*, pp. 286–288, 2005.

83. H. Zhang, J. J. Wang, J. Fan, and Q. Fang, Microfluidic chip-based analytical system for rapid screening of photocatalysts, *Talanta*, vol. 116, pp. 946–950, 2013.

84. K. Katayama, Y. Takeda, K. Shimaoka, K. Yoshida, R. Shimizu, T. Ishiwata, A. Nakamura, S. Kuwahara, A. Mase, T. Sugita, and M. Mori, Novel method of screening the oxidation and reduction abilities of photocatalytic materials, *Analyst*, vol. 139, no. 8, p. 1953, 2014.

85. D. Erickson, D. Sinton, and D. Psaltis, Optofluidics for energy applications, *Nature Photonics*, vol. 5, pp. 583–590, 2011.

Index